Steel Connection Analysis

Steel Connection Analysis

Paolo Rugarli
Castalia S.r.l.
Milan, Italy

WILEY Blackwell

This edition first published 2018
© 2018 John Wiley & Sons Ltd

All rights reserved. No part of this publication may be reproduced, stored in a retrieval system, or transmitted, in any form or by any means, electronic, mechanical, photocopying, recording or otherwise, except as permitted by law. Advice on how to obtain permission to reuse material from this title is available at http://www.wiley.com/go/permissions.

The right of Paolo Rugarli to be identified as the author of this work has been asserted in accordance with law.

Registered Offices
John Wiley & Sons, Inc., 111 River Street, Hoboken, NJ 07030, USA
John Wiley & Sons Ltd, The Atrium, Southern Gate, Chichester, West Sussex, PO19 8SQ, UK

Editorial Office
9600 Garsington Road, Oxford, OX4 2DQ, UK

For details of our global editorial offices, customer services, and more information about Wiley products visit us at www.wiley.com.

Wiley also publishes its books in a variety of electronic formats and by print-on-demand. Some content that appears in standard print versions of this book may not be available in other formats.

Limit of Liability/Disclaimer of Warranty
While the publisher and authors have used their best efforts in preparing this work, they make no representations or warranties with respect to the accuracy or completeness of the contents of this work and specifically disclaim all warranties, including without limitation any implied warranties of merchantability or fitness for a particular purpose. No warranty may be created or extended by sales representatives, written sales materials or promotional statements for this work. The fact that an organization, website, or product is referred to in this work as a citation and/or potential source of further information does not mean that the publisher and authors endorse the information or services the organization, website, or product may provide or recommendations it may make. This work is sold with the understanding that the publisher is not engaged in rendering professional services. The advice and strategies contained herein may not be suitable for your situation. You should consult with a specialist where appropriate. Further, readers should be aware that websites listed in this work may have changed or disappeared between when this work was written and when it is read. Neither the publisher nor authors shall be liable for any loss of profit or any other commercial damages, including but not limited to special, incidental, consequential, or other damages.

Library of Congress Cataloging-in-Publication Data

Names: Rugarli, Paolo, 1963– author.
Title: Steel connection analysis / Paolo Rugarli.
Description: Hoboken, NJ : John Wiley & Sons, 2018. | Includes bibliographical references and index. |
Identifiers: LCCN 2017053858 (print) | LCCN 2018000726 (ebook) | ISBN 9781119303480 (pdf) | ISBN 9781119303534 (epub) | ISBN 9781119303466 (cloth)
Subjects: LCSH: Steel, Structural – Testing. | Finite element method. | Building, Iron and steel–Joints.
Classification: LCC TA475 (ebook) | LCC TA475 .R84 2018 (print) | DDC 624.1/821–dc23
LC record available at https://lccn.loc.gov/2017053858

Cover Design: Wiley
Cover Image: © tonymax/iStockphoto

Set in 10/12pt Warnock by SPi Global, Pondicherry, India
Printed and bound in Singapore by Markono Print Media Pte Ltd

10 9 8 7 6 5 4 3 2 1

To all my dears

Fiorenza dentro da la cerchia antica,
ond'ella toglie ancora e terza e nona,
si stava in pace, sobria e pudica.
Non avea catenella, non corona,
non gonne contigiate, non cintura
che fosse a veder più che la persona.

Dante, Paradiso, XV, 97-102

Contents

Preface *xv*

1 Introduction *1*
1.1 An Unsolved Problem *1*
1.2 Limits of Traditional Approaches *2*
1.2.1 Generality *2*
1.2.2 Member Stress State Oversimplification *3*
1.2.3 Single Constituent Internal Combined Effects Linearization *4*
1.2.4 Single Constituent External Combined-Effects Neglect *7*
1.2.5 Neglecting Eccentricities *8*
1.2.6 Use of Envelopes *9*
1.2.7 Oversimplification of Plastic Mechanisms Evaluation *11*
1.2.8 Evaluation of Buckling Phenomena *13*
1.3 Some Limits of the Codes of Practice *14*
1.3.1 Problem of Coded Standards *14*
1.3.2 T-Stub in Eurocode 3 *15*
1.3.3 Eurocode 3 Component Model *17*
1.3.4 Distribution of Internal Forces *20*
1.3.5 Prying Forces *20*
1.3.6 Block Tearing *21*
1.4 Scope of This Book *21*
1.5 Automatic Modeling and Analysis of 3D Connections *23*
1.6 Acknowledgments *24*
References *24*

2 Jnodes *27*
2.1 BFEM *27*
2.2 From the BFEM to the Member Model *29*
2.2.1 Physical Model and the Analytical Model *29*
2.2.2 Member Detection: Connection Codes *31*
2.2.3 An Automatic Algorithm for Straight Prismatic Member Detection *34*
2.2.4 Member Data Structure *36*
2.2.5 Member Classification at a Node *36*
2.2.6 Member Mutual Alignment Coding *37*

2.3 Jnodes *40*
2.3.1 Need for the Jnode Concept *40*
2.3.2 Jnode Definition *41*
2.4 Jnode Analytics *42*
2.4.1 Classification of Jnodes *42*
2.4.2 Simple Jnodes *42*
2.4.3 Hierarchical Jnodes *42*
2.4.4 Central Jnodes *43*
2.4.5 Cuspidal Jnodes *43*
2.4.6 Tangent Jnodes *44*
2.4.7 Constraints *45*
2.4.8 Summary of Jnode Classification *46*
2.4.9 Setting Connection Codes: Examples *46*
2.5 Equal Jnodes Detection *49*
2.5.1 Toponode *49*
2.5.2 Jnode Data Structure *49*
2.5.3 Superimposable Member Couples *50*
2.5.4 Criteria to Assess Jnodes Equality *51*
2.5.5 Algorithm to Find Equal Jnodes *52*
2.5.6 Examples *55*
2.6 Structural Connectivity Indices *56*
2.7 Particular Issues *59*
2.7.1 Symmetries *59*
2.7.2 Splitting of Jnodes *60*
2.7.3 Mutual Interaction of Different Jnodes, Jnode Clusters *61*
2.7.4 Tolerances *63*
2.8 Jclasses *63*
References *64*

3 A Model for Connection *65*
3.1 Terminology *65*
3.2 Graphs of Connections *66*
3.3 Subconstituents vs Layouts *69*
3.4 Classification of Connections *70*
Reference *72*

4 Renodes *73*
4.1 From Jnode to Renode Concept *73*
4.2 BREP Geometrical Description of 3D Objects *73*
4.3 The *Scene* *75*
4.3.1 Generality *75*
4.3.2 Members *77*
4.3.3 Typical Fittings *78*
4.3.4 Connectors *79*
4.4 Dual Geometry *83*
4.5 Automatic Connection Detection *85*

4.5.1 Faces in Contact *85*
4.5.2 Bolt Layouts *86*
4.5.3 Weld Layouts *89*
4.6 Elementary Operations *91*
4.7 Renode Logic and the Chains *93*
4.7.1 Minimum Compliance Criteria for Renode Good Design *93*
4.7.2 Chains *94*
4.7.3 Finding Chains *96*
4.8 Prenodes *102*
4.9 After Scene Creation *103*

5 Pillars of Connection Analysis *105*
5.1 Equilibrium *105*
5.1.1 Generality *105*
5.1.2 Statics of Free Rigid Bodies *108*
5.2 Action Reaction Principle *111*
5.3 Statics of Connections *115*
5.3.1 Equilibrium of Members in Renodes: Proper and Dual Models *115*
5.3.2 Force Packets for Compound Members *119*
5.3.3 Primary Unknowns: Iso-, Hypo-, and Hyperconnectivity *124*
5.4 Static Theorem of Limit Analysis *127*
5.5 The Unsaid of the Engineering Simplified Methods *130*
5.6 Missing Pillars of Connection Analysis *130*
5.6.1 Buckling *131*
5.6.2 Fracture *147*
5.6.3 Slip *150*
5.6.4 Fatigue *152*
5.7 Analysis of Connections: General Path *153*
References *154*

6 Connectors: Weld Layouts *155*
6.1 Introduction *155*
6.2 Considerations of Stiffness Matrix of Connectors *156*
6.3 Introduction to Weld Layouts *160*
6.4 Reference Systems and Stresses for Welds *162*
6.5 Geometrical Limitations *165*
6.5.1 Penetration Weld Layouts *165*
6.5.2 Fillet Weld Layouts *166*
6.6 Penetration-Weld Layouts (Groove Welds) *167*
6.6.1 Generality *167*
6.6.2 Simple Methods to Evaluate the Stresses *168*
6.6.3 Weld Layout Cross-Section Data *170*
6.6.4 Stiffness Matrix *172*
6.6.5 Special Models *185*
6.6.6 Example *188*
6.7 Fillet-Welds Weld Layouts *196*

6.7.1 The Behavior of Fillet Welds *196*
6.7.2 Numerical Tests of Fillet Welds in the Linear Range *207*
6.7.3 The Stiffness Matrix of a Single Fillet Weld *212*
6.7.4 Instantaneous Center of Rotation Method in 3D *214*
6.7.5 Computing the Stresses in Fillet Welds from the Forces Applied to the Layout *231*
6.7.6 Fillet Welds Using Contact and Friction *233*
6.8 Mixed Penetration and Fillet Weld Layouts *235*
References *235*

7 Connectors: Bolt Layouts and Contact *237*
7.1 Introduction to Bolt Layouts *237*
7.2 Bolt Sizes and Classes *238*
7.3 Reference System and Stresses for Bolt Layouts *240*
7.4 Geometrical Limitations *243*
7.4.1 Eurocode 3 *244*
7.4.2 AISC 360-10 *244*
7.5 Not Preloaded Bolt Layouts (Bearing Bolt Layouts) *244*
7.5.1 Shear and Torque *244*
7.5.2 Axial Force and Bending *249*
7.6 Preloaded Bolt Layouts (Slip Resistant Bolt Layouts) *266*
7.6.1 Preloading Effects *266*
7.6.2 Shear and Torque *274*
7.6.3 Axial Force and Bending *275*
7.7 Anchors *277*
7.8 Stiffness Matrix of Bolt Layouts and of Single Bolts *282*
7.8.1 Generality *282*
7.8.2 Not Preloaded Bolts *283*
7.8.3 Preloaded Bolts *292*
7.8.4 Non-Linear Analysis of Bolts *293*
7.9 Internal Force Distribution *296*
7.9.1 General Method *296*
7.9.2 Bearing Surface Method to Compute Forces in Bolts *302*
7.9.3 Instantaneous Center of Rotation Method *306*
7.9.4 Examples *307*
7.10 Contact *316*
References *317*

8 Failure Modes *319*
8.1 Introduction *319*
8.2 Utilization Factor Concept *320*
8.3 About the Specifications *326*
8.4 Weld Layouts *328*
8.4.1 Generality *328*
8.4.2 Penetration Weld Layouts *328*
8.4.3 Fillet Weld Layouts *332*

8.5 Bolt Layouts 337
8.5.1 Resistance of Bolt Shaft 337
8.5.2 Sliding and Resistance of No-Slip Connections 342
8.5.3 Pull-Out of Anchors, or Failure of the Anchor Block 345
8.6 Pins 346
8.6.1 Eurocode 3 346
8.6.2 AISC 360-10 347
8.7 Members and Force Transferrers 347
8.7.1 Generality 347
8.7.2 Local Failure Modes 350
8.7.3 Fracture Failure Modes 358
8.7.4 Global Failure Modes 373
References 382

9 **Analysis: Hybrid Approach** 385
9.1 Introduction 385
9.2 Some Basic Reminders About FEM Analysis of Plated-Structures 386
9.2.1 FEM Analysis as an Engineering Tool 386
9.2.2 Linear Models 387
9.2.3 Linear Buckling Analysis 388
9.2.4 Material Non-Linearity 390
9.2.5 Geometrical Non-Linearity 392
9.2.6 Contact Non-Linearity 394
9.2.7 Non-Linear Analysis Control 396
9.3 IRFEM 400
9.3.1 Goal 400
9.3.2 Hypotheses 401
9.3.3 Construction 402
9.3.4 Examples 408
9.3.5 Results 411
9.3.6 Remarks on the Use of IRFEM 413
9.4 Connector Checks 418
9.4.1 Weld Checks 418
9.4.2 Bolt Resistance Checks 419
9.4.3 Pull-Out Checks 419
9.4.4 Slip Checks 419
9.4.5 Prying Forces 419
9.5 Cleats and Members Non-FEM Checks 426
9.5.1 Action Reaction Principle 426
9.5.2 Bolt Bearing 428
9.5.3 Punching Shear 428
9.5.4 Block Tearing 428
9.5.5 Simplified Resistance Checks 429
9.6 Single Constituent Finite Element Models 430
9.6.1 Remarks on the Finite Element Models of Single Constituents (SCOFEM) 430

9.6.2 Stiffeners *432*
9.6.3 Meshing *433*
9.6.4 Constraints *437*
9.6.5 Loading *439*
9.6.6 Members: Deciding Member-Stump-Length *443*
9.6.7 Compatibility Issues *444*
9.7 Multiple Constituents Finite Element Models (MCOFEM) *445*
9.7.1 Goal and Use *445*
9.7.2 Mesh Compatibility Between Constituents and Connector Elements *446*
9.7.3 Saturated Internal Bolt Layouts and Contact Non-Linearity *447*
9.7.4 Constraints *448*
9.7.5 Stabilizing Springs and Buckling of Members *448*
9.7.6 Need for Rechecks *449*
9.8 A Path for Hybrid Approach *449*
References *450*

10 Analysis: Pure FEM Approach *451*
10.1 Losing the Subconnector Organization *451*
10.2 Finite Elements for Welds *455*
10.2.1 Introduction *455*
10.2.2 Penetration Welds *457*
10.2.3 Fillet Welds *460*
10.3 Finite Elements for Bolts *463*
10.3.1 Introduction *463*
10.3.2 Bolts in Bearing: No Explicit Bolt-Hole Modeling *464*
10.3.3 Bolts in Bearing: Explicit Bolt-Hole Modeling *465*
10.3.4 Preloaded Bolts: No Explicit Bolt-Hole Modeling *468*
10.3.5 Preloaded Bolts: Explicit Bolt-Hole Modeling *468*
10.3.6 Effect of the Bending Moments in Bolt Shafts *469*
10.3.7 Example: A Bolted Splice Joint Using PFEM *469*
10.4 Loads *478*
10.4.1 PFEM *478*
10.4.2 MCOFEM *479*
10.5 Constraints *480*
10.5.1 PFEM *480*
10.5.2 MCOFEM *480*
10.6 Checking of Welds and Bolts *480*
10.7 Checking of Components *481*
10.8 Stiffness Evaluation *482*
10.9 Analysis Strategies *484*
Reference *484*

11 Conclusions and Future Developments *485*
11.1 Conclusions *485*
11.2 Final Acknowledgments *486*
11.2.1 Reasons of This Project *486*

11.3 Future Developments *487*
References *488*

Appendix 1: Conventions and Recalls *489*
A1.1 Recalls of Matrix Algebra, Notation *489*
A1.2 Cross-Sections *490*
A1.3 Orientation Matrix *492*
A1.4 Change of Reference System *493*
A1.5 Pseudocode Symbol Meaning *493*

Appendix 2: Tangent Stiffness Matrix of Fillet-Welds *495*
A2.1 Tangent Stiffness Matrix of a Weld Segment *495*
A2.2 Modifications for Weld Segments Using Contact *499*
A2.3 Tangent Stiffness Matrix of a Weld Layout for the Instantaneous Center of Rotation Method *500*

Appendix 3: Tangent Stiffness Matrix of Bolts in Shear *503*
A3.1 Tangent Stiffness Matrix of a Bolt *503*
A3.2 Tangent Stiffness Matrix of a Bolt Layout for the Instantaneous Center of Rotation Method *505*

Symbols and Abbreviations *507*

Index *513*

Preface

Around 17 years ago, at the end of the 1990s, when I started my research on steel connections with the aim of developing some reliable and general software, able to tackle, hopefully, every connection, I often felt like giving up. The problem was tremendously complex, and the general rules of mechanics difficult to relate to the problem to be faced; there was a huge gap to be bridged.

Initially I thought that only a system able to learn from the analyst could deal with such a complex problem, learning ad hoc rules to be later applied, case by case. However, I was able to move some steps forward, finding what in the second chapter of this book is named the *jnode*, its analytics, and all the related concepts. My first useful result was detecting equal jnodes. Several years were then necessary to develop the tools needed to create the scene, that is, to place the constituents in their proper position, freely placing them interactively in 3D space, in the specific context of steel connection study. The mechanical problem of connection analysis, to be tackled with a general approach, was however still unsolved. I was prepared to develop an expert system able to learn from the user how to recognize specific subproblems to be faced, by simple ad hoc rules. This was the tentative generalization of the methods widely used by engineers, but was not the solution I was searching for.

Adopting the concept of the force packet, and recognizing that the connections could be classified as isoconnected or hyperconnected, I finally understood, in 2008, that a simplified finite element model that in this work is named IRFEM, could be used to compute the force packets flowing into the connectors for a generic set of connections. Then, by the action and reaction principle, a cornerstone for connection analysis, the forces loading the constituents could be known, and by finite element models of single constituents using plate–shells, coherent and well rooted Von Mises stress maps could be obtained. This is what I call the hybrid approach and is described in Chapter 9 of this book.

The door was then opened for the automatic creation of finite element models of constituents (2008), and from there, in 2012, to the complete automatic modeling of the whole node, using what I call here the pure fem approach (PFEM). This is seen as a special case of the hybrid approach and is discussed in Chapter 10.

What initially seemed an inextricable tangle could indeed be solved in strict observance of the main principles of mechanics and of plasticity theory.

Several issues are still to be better solved, but a general well rooted method is now available, that can be applied to every connection configuration, from the simplest to the most complex. Indeed, I think this is a useful result, because a part of the method can be implemented with relative ease.

I am well aware that several issues are pending and must still be better tackled. However, after many years of solitary work, I think the time has come to explain what I have researched and to propose my work for the attention of my colleagues.

Anything can be improved, but the structural analysis problem of analyzing steel connections having a generic geometrical configuration, regardless of the number of loading combinations, is now solvable with automatic tools.

Paolo Rugarli
Milan, 17 May 2017

1 Introduction

1.1 An Unsolved Problem

Steel connection analysis and checking is one of the most complex problems in structural engineering, and even though we use very powerful computing tools, it is still generally done using very simplistic approaches.

From the point of view of a typical structural engineer, the problem to solve is to design and check *nodes*,[1] not single connections, i.e. a number of connections between a number of different members – maybe tens or even hundreds of load combinations, inclined member axes, and generic stress states. In a typical 3D structure there may be several tens of such nodes (Figure 1.1), or maybe even hundreds, which may be similar, or may be different from one another; identifying nodes that are equal is one of the problems that the designer has to face in order to reduce the number of different possible solutions, and in order to get a rational design. However, this problem of detecting equal nodes has not been sufficiently researched, and there are currently no tools that are able to properly solve this issue.

If posed with the due generality, the problem of checking 3D nodes of real structures has not been solved by automatic computing tools. Also, because a general method of tackling all these problems is apparently still lacking, usually a few "cooking recipes" have been used to solve a limited number of typical, recurring (2D assimilated) nodes. Indeed, it often happens that true, real world nodes have to be analyzed by such recipes, despite the fact that the basic hypotheses needed to apply these recipes do not always hold true. This poses a serious problem because although these "cooking recipes" have been widely used, in the past few years they have been applied to 3D structures designed using computer tools, in the non-linear range, perhaps in seismic areas, and with the aim of reducing the weight of steel.

The effects of such oversimplification have already been seen in many structures where steel connections have failed, especially in seismic areas (e.g. Booth 2014), but even in non-seismic areas (e.g. White et al. 2013, Bruneau et al. 2011). Generally speaking, it is well known that connections are one of the most likely points of weakness of steel structures, one of the most cumbersome to design – indeed one of the least designed – and one of the least software-covered in structural engineering.

1 It will be seen that the term *node* is too generic for the aims of steel connection analysis. In this introductory chapter, however, it will be used due to its widespread diffusion.

Steel Connection Analysis, First Edition. Paolo Rugarli.
© 2018 John Wiley & Sons Ltd. Published 2018 by John Wiley & Sons Ltd.

Figure 1.1 A possible *node* of a 3D structure.

This book describes the research efforts made by the author since 1999 to tackle these issues, and it proposes a general set of methods to deal with these problems (see Section 1.6 for more details).

1.2 Limits of Traditional Approaches

1.2.1 Generality

Traditional approaches to connection design have been extensively used for many years, and are still widely used. Usually they imply several simplifying hypotheses, which are needed in order to apply them in by hand computation. The equivalent of by hand computation is today a "simple spreadsheet" often written very quickly for each given job. As with every other form of calculus, they are prone to serious errors (*slips* and *lapses* – see Reason 1990 for a general study of human error, and Rugarli 2014 for a discussion on validation of structural models; for spreadsheets programming errors, see the European Spreadsheet Risk Interest Group web site).

There are several possible design situations where the use of traditional approaches is completely justified. These approaches are rooted in the traditional 1D or 2D design. The use of 2D design needed intense by hand computation or the use of graphic tools up to the 1970s; at that time there was no need and no specific legal requirements for checking tens or may be hundreds load combinations, and safety factors were much higher than those used nowadays. When dealing with such situations, today – for example simple determined structures under elementary actions – the use of traditional approaches is still useful. So, it would not be sensible to exclude them completely. Indeed, they will never lose their utility, especially as one of several possible cross-checking tools that can be used to detect possibly unsound designs.

However, in current design practice, we almost always use 3D methods of analysis applied to highly redundant structures, sometimes in the plastic range, automatic computerized checks, with minimum weight often being a must, and safety factors have been reduced to their minimum. (Currently the material safety factor for limit state design is 1.0 in Eurocode 3. The load safety factor for dead loads is lower than that valid for live loads. The maximum loads are applied with a reduction factor ψ to take into account the reduced probability of contemporary occurrence. All these practices were not, as such, in traditional designs, which means that they used higher safety factors.)

In summary, while traditional design of structures was often simple, 2D, and was designed by making extensive use of safe-side envelopes both for loads and for resistance, today things are not so easy; indeed, they are much more complex. While virtually all design steps have been semi-automated (modeling, checking members, drawing them, and even cutting them into true 3D pieces by means of *computer numerical control,* CNC), the checking of connections has remained at the traditional level, more or less upgraded to the modern era by the use of spreadsheets and dedicated, ad hoc software.

As mentioned, several simplifications are widely used in traditional approaches. The following sections will briefly summarize them.

1.2.2 Member Stress State Oversimplification

Members in highly redundant 3D structures are often nonsymmetrical (such as in industrial plants or architects' innovative designs), and under the effect of combined load cases, they are always loaded in the most general way. If they are not: (a) fully hinged at both extremities, (b) straight, and (c) with no transverse load applied, they will in general exhibit all six internal forces components: an axial force, two shears, one torque and two bending moments, referred to the principal axes of the member cross-section.

Idealizing the connection in such a way that some member internal forces components are considered zero at the connection is still a widely used practice. While this is justified when the connections are specifically conceived with that aim, this is unjustified for connections that are not so designed. In a typical moment resisting frame (MRF) ideally designed to work in a plane, beam-to-column connections that must transfer bending and shear in one plane (and axial force) will always transfer the bending and the shear *also in the other plane* – and of course torque. So the internal forces to deal with are not three, but six. Sometimes it is said that the torque and out-of-plane bending are avoided by "the concrete slab", or by something equivalent, but often the concrete slab does not exist or cannot be considered a true restraint, or its true effect is questionable.

A simple beam hinged at an extremity (e.g. Figure 1.2), will transfer the shear, and will not transfer bending moment if the connection is light and does not use flanges, but it will also transfer the axial force and, if any, the shear in the other direction. However, textbooks usually refer to "shear connections" and only recently, under the flag of "robust design" (a replacement for *correct design*) has this axial force finally – sometimes, in some textbooks, – been considered (e.g. the Green Books by SCI).

This systematic neglect of some internal forces which have, however, been computed introduces a clear mismatch in the design process. Simply, load paths are interrupted (Figure 1.3) and the corresponding forces are thrown away: recalling the *variational crimes* of the finite element literature, this can be called a *connection-design crime,* more specifically an *equilibrium crime.* Usually no one cares, and no one mention it.

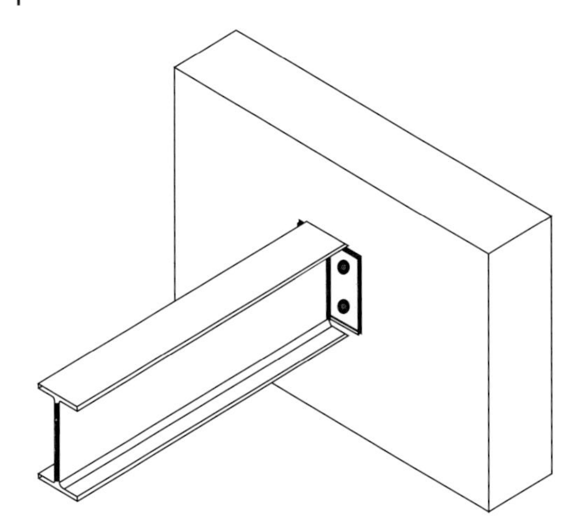

Figure 1.2 Flexible end plate connection ("shear" connection).

Figure 1.3 Traditional design applied to computerized analysis: no way for the load path.

1.2.3 Single Constituent Internal Combined Effects Linearization

Not only are some components of member internal stress states thrown away, but the remaining components are tackled one at a time, as if the connection were loaded only by a single member internal force component. The typical example is the axial force plus (one) bending moment loading condition, for beam-to-column connection or for a base plate. As already pointed out, this loading condition is itself usually the result of a connection-design crime. However, several possible combinations of N and M can be applied to the connection (two infinities), leading to an infinite number of possible stress states. This is usually tackled by computing the limit for the axial force, N_{lim}, and for the bending moment, M_{lim}, as if they were acting alone, and then the mutual interaction is computed by simply drawing a straight line in the (N, M) plane. So the design safety condition becomes

$$U = \frac{N}{N_{\text{lim}}} + \frac{M}{M_{\text{lim}}} \leq 1.0 \tag{1.1}$$

where the utilization ratio, U, can be considered as the reciprocal of the "limit" multiplier $\lambda = 1/U$. It must be underlined that this limit condition is not applied to the member cross-section, but to the member *connections*, implicitly considering all the possible failure modes: bolt bearing, block tear, generic resistance of constituents, buckling of plates, punching shear, weld-resistance, and so on.

As there are quite a number of plastic failures implicitly included in the typical design formulae (e.g. bolt bearing), this must not be considered as a superposition of effects, which would only be valid in linear range.

It must instead be considered a simplification of the limit domain, assuming that it can safely be considered convex, so that a straight line would be a safe simplification; as can be seen, the previous equation is the equation of a straight line in the (N, M) plane.

There are several issues to be discussed here.

The first is that this choice clearly lays aside every possible "realistic" computation of a safety measure. In particular, the *utilization ratio U*, a pure number and a much used safety index, which must be lower than 1 in the safety region, is usually computed as $U = \text{PO}/\text{AO}$ (Figure 1.4), that is, the ratio of the distance of the applied stress state P (N, M) from the origin, to the distance of point A from the origin, A being the point where the straight line joining P and O meets the limit (linearized) domain. If the true domain were convex the correct utilization ratio would have been $U = \text{PO}/\text{QO}$, which can be much lower. So this method is not very realistic, and can be too much on the safe side.

By posing

$$\lambda_N = N_{\text{lim}}/N$$
$$\lambda_M = M_{\text{lim}}/M$$

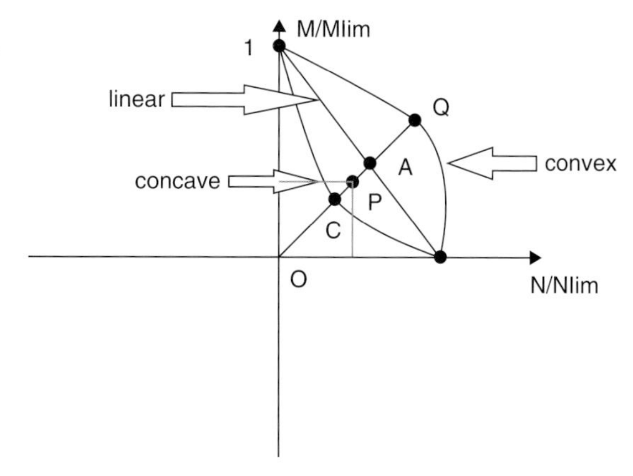

Figure 1.4 Limit domain for a connection; **P**(N, M) is the stress state for a single member assumed.

Equation 1.1. becomes

$$\frac{1}{\lambda_{linear}} = \frac{1}{\lambda_N} + \frac{1}{\lambda_M} \leq 1$$

A similar result can be found in Fraldi et al. 2010, a paper dealing with the problem of finding some bound of the limit multiplier under combined loadings, once the limit multipliers of single loadings are known; there it is formally proved, in the framework of classic plasticity, that the "true" combined-loading multiplier λ, is surely such as to satisfy the following inequality:

$$\lambda \geq \left(\frac{1}{\lambda_N} + \frac{1}{\lambda_M}\right)^{-1} = \lambda_{linear}$$

which, considering Figure 1.4, simply means

QO $\geq$ **AO**

The second issue is that in order to be confident that the limit domain, considering all the possible failure modes, is convex, no buckling effect must be possible at load levels lower than those leading to the first failure mode, that is, the failure mode which is met first, linearly increasing the stress state from (0, 0) to (N, M). If this is the case, then $U =$ **PO/CO** which is much higher than **PO/AO**. A good design of connection should always ensure that the first failure mode is plastic, i.e. ductile, and avoid brittle failure modes. However, this cannot be considered an implicit condition but must be assured by correct sizing and proper numerical checks, which presents a serious problem.

The third issue is related to signs. If the connection is not doubly symmetrical, then it can be expected that reversing signs can lead to different limit values, perhaps due to the buckling effects which must, however, always be kept in consideration, if only to be proven irrelevant. So, to be applicable when signs are reversed, two points in plane (N, M) are not enough and four must instead be evaluated, doubling the effort.

It can be concluded that the practice of drawing a linear domain, considering only the equilibrium-crime survivors, has several limits and can also be: (a) too much on the safe side because the "true" limit multiplier can be much higher than that obtained by the linearized limit domain, (b) *not on the safe side*, if some buckling mode (possibly associated with sign reversal) has not properly been accounted for and checked.

This is not an academic discussion. The point Q which would be obtained by increasing the couple (N, M) and searching for the first failure mode, in classic plasticity, may well be quite different from point A. This can be understood when considering for instance the *plastic* yield lines related to the possible failure modes: the plastic-limit resistance failure of parts of the member itself or of its connected constituents. These yield lines *are tightly related to the load configuration applied*. With only load N (load increasing along the horizontal axis of Figure 1.4), a yield lines set would be found, related to some mechanism. With only load M applied (load along the vertical axis of Figure 1.4), another yield lines set would be found. With a possible combination of N and M, a third, possibly completely different yield lines set would be found (load along the inclined straight line in Figure 1.4). So, linearizing means forgetting the true load state and doing a purely numerical simplification, with no physical meaning.

Finally, premature buckling of connections subconstituents is one of the most frequent failure modes, especially in seismic areas where internal forces sign reversal is a normal condition. So the problem of its correct evaluation is a real problem.

1.2.4 Single-Constituent External Combined-Effects Neglect

One typical simplification of traditional connection design is that when dealing with a node where *n* members are joined, the connections are evaluated one by one, considering two or at most three members at a time, and no more. So, if the node is like A-B-C, where B is the main member and A and C are secondary, connection design is often carried on by considering A-B, and so checking A, B, and their connectors, and then B-C, and so once more B, then C, and then their connectors. However, this working method is not correct, as the effects on B of A and C are contemporary, and not separate.

Consider for instance the node of Figure 1.5: all the five "slave" members act together on the column, which is the "master" member used as a reference for all other members. Considering the effects of each slave member on the column separately would only be possible if the superposition principle were to hold true (and it would not in plastic analysis), and only if the – very different – effects of each member on the column were correctly summed. But this is not what is usually done in designing connections. Usually, the effects will at most be grouped considering typical member configurations, like two opposite beams joined to a web, or two opposite beams joined to flanges, and traditional methods are definitely not able to take all these members into account.

What is clearly dangerous here is the possibility that combined effects could drive the common member to failure, or, more generally, the common constituents.

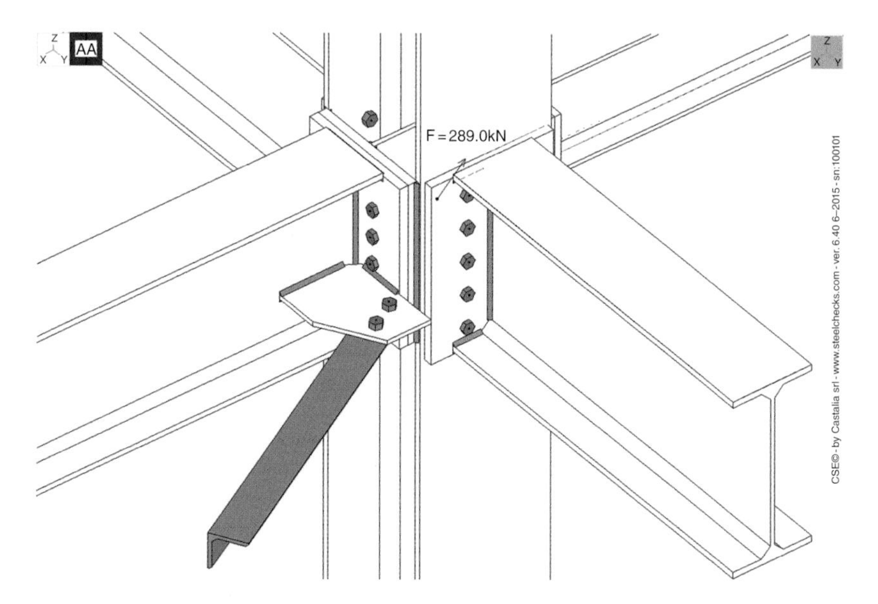

Figure 1.5 All the members connected to the column do act over it at the same time.

Without a clear and coherent computational method to consider the sum of the effects of all the connections, the evaluation of the combined effects is often left to improbable envelopes or the sum of physically meaningless quantities. And this is a very strong weakness.

Besides, it is important to note that the problem is not only related to *master members*, but in general to all constituents. As all the members in a node are directly or indirectly tied together, it is not unusual for the internal forces acting in a member can flow to the connections of another member, a possibility that is implicitly excluded by traditional design methods and that can instead be easily observed when a complete finite element model of a node is set up.

Connections are definitely more complex than a one-to-one, or a one-to-two relationship between members. Connections have inside themselves the same complexity as the whole structure.

1.2.5 Neglecting Eccentricities

True structures and true connections very often have relevant eccentricities that should be ideally considered, but that are very often neglected.

The first type of eccentricity is that of members' axis lines, which in the actual construction are often not as in the finite element model. This leads to possibly severe additional moments that can induce stresses comparable to, or even higher than, those computed considering members to be fully aligned with their computational schemata. As it is very lengthy to properly take into account all these eccentricities by hand, in a 3D context, it is still dangerously considered normal practice to neglect them. This choice is strengthened by the *equilibrium crime*, which, neglecting some internal forces components, also neglects the additional moments that they might drive in some constituents of the connections. However, if a force F is offset by e, the additional moment is Fe, which, assuming a resisting lever b, leads to an additional force equal to Fe/b. If e is much lower than b, then the additional force is negligible, but if it is not, the additional force may considerably increase the nominal one.

To compute the additional moment, in a 3D context, a more precise rule would be (**P** is the true point of application, **O** the point where the force is moved to, **F** is the force applied)

$$\mathbf{M} = (\mathbf{P} - \mathbf{O}) \wedge \mathbf{F}$$

and there are three moment components.

The second type of eccentricity is in the connection area where, due to constructional needs, the true layout of bolts, welds, stiffeners, and cleats may not be that effectively assumed in the connection simplified – often 2D – computational model, especially when considering forces and moments flowing into connectors. Indeed, this computation requires ideally simple, but in true practice quite boring and error-prone vector products. For instance, considering the angle connected to the node in Figure 1.6, while the eccentricity in an horizontal plane will probably be considered, the eccentricity in the vertical plane (the center of the diagonal is not at the elevation of plate mid-thickness) will probably be neglected, which is not correct.

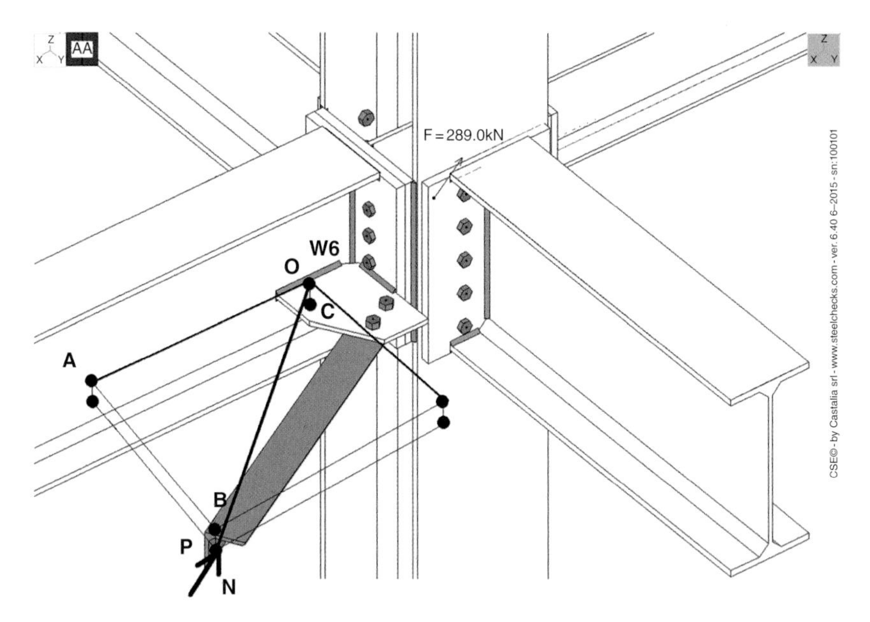

Figure 1.6 Eccentricity between the point of application of the force *N*, P, and the weld layout center, W6, O. The vector (P-O) has the three components (A-O), (B-A) and (P-B), but frequently (P-B) is neglected.

1.2.6 Use of Envelopes

One of the key features of traditional approaches is that they had to deal with a limited number of loading conditions, usually computed by considering envelopes of notional load cases that were themselves envelopes (maximum wind, plus maximum snow, and so on).

Nowadays the number of loading conditions, expressed in the form of load combinations, is quite high. Referring to Eurocodes and applying the combination rules there provided, it is not unusual to get hundreds or even thousands of combinations (Rugarli 2004). This means that the traditional way of computing connections, two or more loading conditions, can be obtained by only assuming a special kind of envelope, which considers maximum or minimum values of different internal forces components acting together. Owing to the equilibrium crime the number of different components of internal forces is often limited to two, so the notional combinations are usually quite a few.[2] Is that approach on the safe side? Well, provided that, when increasing the absolute value of the internal forces all the load effects do always increase, then this might be the case. Unfortunately, this is not always true, and it is not true when a decrease of some internal force, for some failure modes, is more dangerous than an increase, such as in slip checks under compression, or when a specific mix of internal forces leads to worst results.

2 The set of load combinations has to be really complete, otherwise the "maximum" is not really the worst value that the structure might experience, but this is another problem, well rooted in using "realistic" combination sets and not unrealistic, but enveloping.

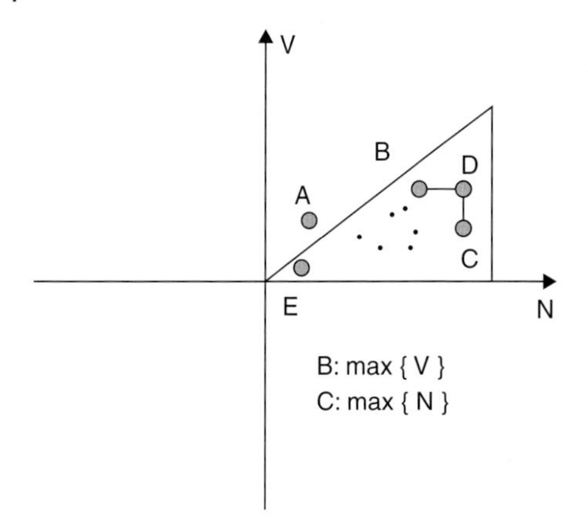

Figure 1.7 Friction connection. Point A is outside the limit domain, point D is inside.

Consider a very simplified example (Figure 1.7), a slip resistant simple support connection relying on the compression N acting over a friction plane having friction constant k. Let N_{lim} be the maximum compression allowable without failure (vertical line bound). A shear V is also applied, and it must be that $V < kN$ to avoid slip. Several couples (N, V) have been computed in a number of load combinations. To check the connection, the maximum N is taken with maximum V. Is that safe? Figure 1.7 clearly shows that it is not. The limit domain is here a triangle.

Point A, neglected by this "envelope" rule, is outside the limit domain. Point D is obtained by mixing maximum N (point C) with maximum V (point B). So considering the maximum absolute values as acting together is not always a safe approach, at least when slip can be a failure mode. It's interesting to note also that adding to checks point E beside point D, i.e. the point with minimum N, the problem would not have been solved, and only mixing the minimum N with the maximum V would have been on the safe side, but possibly, too much.

Indeed, mixing maxima and/or minima can be a quite over-safe approach. Assuming that the governing failure mode has a linear limit domain (Figure 1.8), this way of computing can sometimes really lead to an overestimate of utilization.

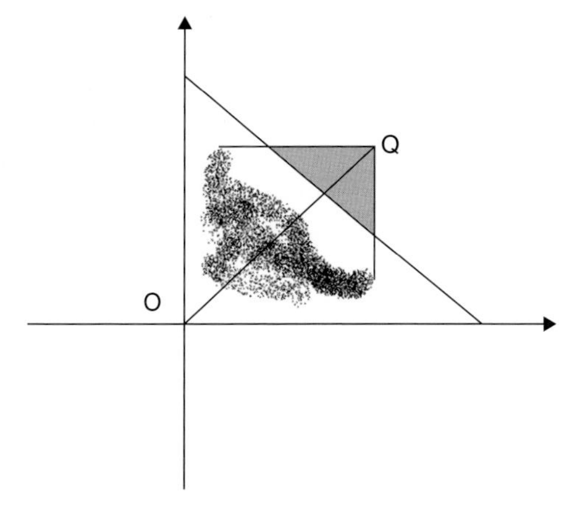

Figure 1.8 Overevaluation of utilization factor using contemporary maxima.

1.2.7 Oversimplification of Plastic Mechanisms Evaluation

Traditional by hand or by spreadsheet approaches all try to evaluate the plastic limit load of complex 3D assembly of steel plates. As this is in general a very complex task, it is not surprising that the number of such computations is reduced to the minimum possible (as it has been shown, by linearizing the limit domain), and that quite often the problem is tackled by using simplified and quite regular geometries.

One problem here is that real world connections do not always comply with such geometries, so the analyst is pushed to force his/her problem into one that is solvable. According to some computational tools, the problem to be solved is always a T-stub, pulled, compressed, or bent, with regular bolt "rows",. But this is not always realistic – for example see Figure 1.9.

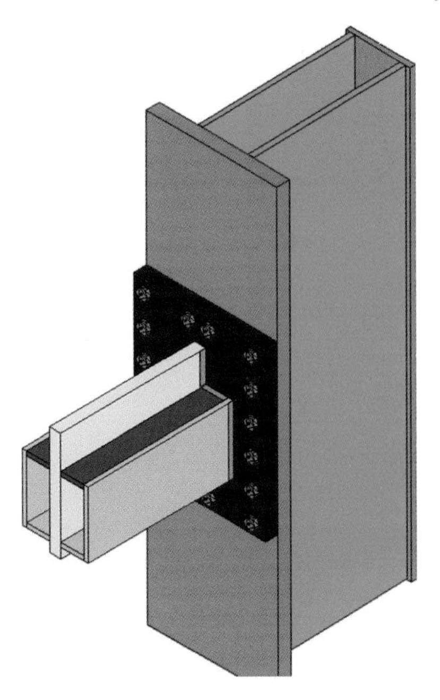

Figure 1.9 Real world moment connection (courtesy CE-N Civil Engineering Network, Bochum, Germany).

Not only can the geometry of the steel plates be quite different, but the bolt layout (not to mention the loading condition see Section 1.2.2), can be different. So, if for some good reason a bolt or a stiffener has to be shifted, or if a plate is not rectangular, or if the footprint of the cross-section and stiffeners is not as regular as in the textbooks, the computational model is simply not able to deal with the problem. Much serious effort has been spent in order to categorize the local failure modes related to typical connections, so that the limit multiplier under simple loads (axial force, bending moment) of typical assemblies could be evaluated. For instance, the *Green Book* referring to moment resisting connections (MRC), published by Steel Construction Institute (SCI 1995), is an excellent book which lists all the possible failure modes and partial yield lines related to typical connections, basically considering the T-stub idealization.

Evaluating the limit load is then a matter of summing up different contributions, analyzing different possible failure lines and modes (Figure 1.10), and finally getting the minimum value. Table 2.4 of the Green Book for moment connections lists 11 patterns for elementary yield lines,[3] used in order to evaluate a final effective length, L_{eff}, to be introduced in the formula for the problem at hand (see also Eurocode 3, Part 1.8, §6.2 and subsections):

$$M = \frac{L_{eff} t^2 f_y}{4}$$

3 They are: (i) circular yielding, (ii) side yielding, (iii) side yielding near beam flange or stiffener, (iv) side yielding between two stiffeners, (v) corner yielding, (vi) corner yielding near a stiffener, (vii) double curvature, (viii) group end yielding, (ix) corner yielding, (x) individual end yielding, (xi) circular yielding. For each of these patterns an L_{eff} formula is provided.

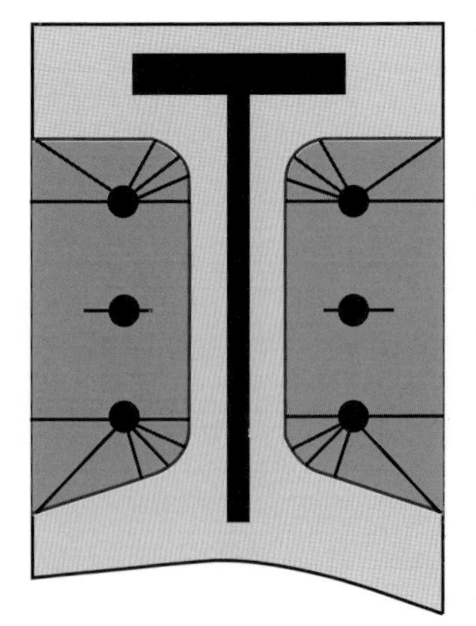

Figure 1.10 An example of evaluation of L_{eff} by summing effects. Red lines are the yield lines whose total length has to be evaluated.

where t is the thickness of the plate where yield lines will appear (e.g. a base plate, an end plate, or a cap plate of a column), f_y is its yield strength, and M is a limit bending moment. The formula is clearly notional and is exact when considering the plastic moment of a plate of length L_{eff} and of thickness t. The use of the typical patterns is not easy, nor particularly intuitive, and it may well lead to errors that are hardly detectable; it is up to the user of such tables to properly mix and compose the typical patterns in a reliable way (see Figure 1.10). Single effects are evaluated and then summed. For instance, to get the L_{eff} of "a bolt row below the beam flange of a flush end plate", we have to evaluate the final L_{eff} related to that bolt row as a function of the individual patterns $L_{eff,i}$ as follows:

$$Min\left\{Max\left\{\left(\frac{ii+iii}{2}\right), ii\right\}, i\right\}$$

but only if some specific geometrical limitations are met, otherwise we have to use

$$Min\{Max\{ii, iii\}, i\}$$

where "i" means "pattern number i". Of course this takes into account only that bolt row.

It is not necessary to get further into that here, but it has to be realized that the evaluation of the limit load multiplier is out of reach of these methods when applied to generic 3D models loading conditions. So, the problem should in general be tackled for what it is: there are six internal forces flowing at the end of each member, many load combinations, many failure modes, a geometry that may well not be forcible into a T-stub, bolt rows that could well be moved (or perhaps no *row* may be available), and so on.

Classical simplified resistance checks continued to use some kind of simplified geometry for the flow-lines of stresses, widely used for instance in the *strut and tie method* (STM), and these are simplified ways of computing plastic mechanisms. For instance, assuming a 30° line of stress flow (e.g. see AISC Steel Construction Manual, 14th ed. §9, "Whitmore section"), and computing some "effective" resistance cross-section to evaluate its limit or ultimate load, is one of these simplified and not always sufficient ways of computing a plastic mechanism. Often, approximations of this type are applied to gusset plates under complex membrane stress (Figure 1.11). Cutting a complex solid with an ideal plane and computing a *net-cross-section* to be checked in a beam-like way is another frequently used simplification. Using simple structural schemata, such as cantilevers, simply supported beams, or struts and ties, extracted from more complex 3D scenes, to be checked for their limit plastic loads is another way to try to assess plastic mechanisms of complex structural configurations. So, when considering what could be called "generic resistance checks", the traditional approaches try to evaluate a plastic mechanism by means of simplified tools. Basically, a generic resistance check is a check that the constituent – usually having an irregular shape – is able to carry the loads applied with no

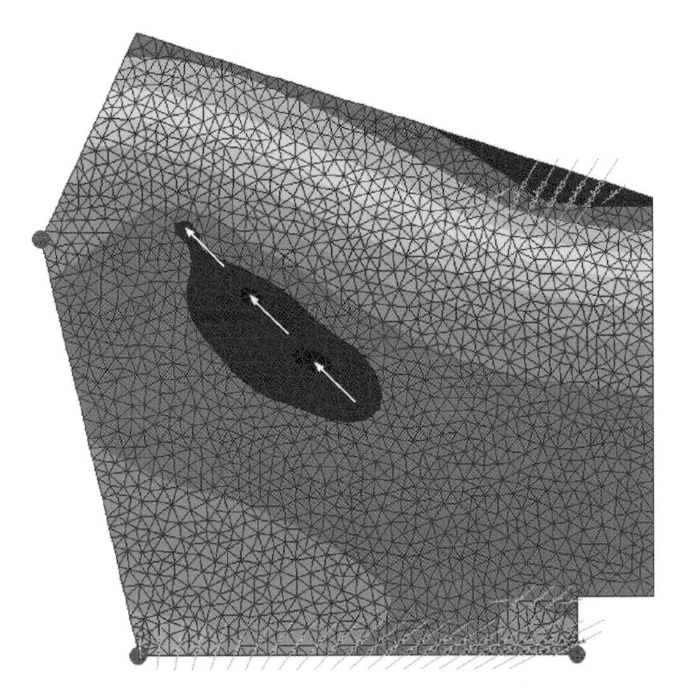

Figure 1.11 Gusset plate under membrane stresses: the three forces in the central area simulate the action transferred by the bolts; weld forces are balanced; fictitious constraints.

plastic mechanism. As local failure modes are tackled by specific checks which can neglect the global configurations (e.g. bolt bearing or punching shear), they are not mentioned here.

1.2.8 Evaluation of Buckling Phenomena

Buckling of some constituent is a dangerous failure mode when it happens before plastic mechanisms have fully developed. Traditional designs cope with this problem in two ways:

1) Size the constituents so that appropriate (low) width-to-thickness ratio of plates is used (e.g. the stiffeners have a thickness at least equal to that of the thickest constituents stiffened).
2) Use simple formulae that model the buckling of complex structural configurations (of geometry and loads) using simple schemata such as the one that is also used to evaluate plastic mechanisms (simple beams, usually "cut from" the existing constituents).

These methods of design work well in many situations, but there are configurations that need a more refined approach. As the load configuration is quite important, a buckling check should be done for every load combination. Often, in traditional approaches, the problem is simply neglected, and no formal proof that buckling modes may not occur is given.

Figure 1.12 Bridge gusset plate buckling failure (from Huckelbridge, A.A., Palmer, D.A. and Snyder, R.E., 1997, "Grand Gusset Failure," Civil Engineering, Vol. 67, September, pp. 50–52. Reprinted with permission of Arthur Huckelbridge).

Neglecting the importance of complex buckling modes related to complex geometries has caused severe problems in many real world structures.

This problem is serious and presently still often unmentioned.

For instance, one of the problems usually tackled by possibly oversimplified approaches is that of the gusset plates. They are under complex force and constraint patterns, and sometimes the methods traditionally used to deal with the problem have shown their limits (see the important photograph taken by Prof. Huckelbridge – Figure 1.12).

1.3 Some Limits of the Codes of Practice

1.3.1 Problem of Coded Standards

Coded standards have different use and meaning depending on the country where they are applied. Sometimes, coded standards are just "advice" with no enforcing value, but other are actual laws of the State and violating them is a legal infringement. In Italy, large books list all the coded rules, and the designer is forced by law to respect them. As well explained in the context of the Three Mile Island nuclear power plant accident, a "big book of rules" is the best way to guarantee a formal respect without actually accomplishing anything. In Kemeny's report on the Three Mile Island accident (Kemeny 1979), it says:

> We note a preoccupation with regulations. It is, of course, the responsibility of the Nuclear Regulatory Commission to issue regulations to assure the safety of

> nuclear power plants. However, we are convinced that regulations alone cannot assure safety. Indeed, once regulations become as voluminous and complex as those regulations now in place, they can serve as a negative factor in nuclear safety. The regulations are so complex that immense efforts are required by the utility, by its suppliers, and by the NRC to assure that regulations are complied with. The satisfaction of regulatory requirements is equated with safety. This Commission believes that it is an absorbing concern with safety that will bring about safety – not just the meeting of narrowly prescribed and complex regulations.

It is the author's opinion that in Italy and also possibly in Europe generally, the boundary line between reasonable and unreasonable regulations has been crossed, especially when considering that in some countries these regulations are enforced by law, and this is true not only for the most general principles, but also for the most detailed ad hoc formulae referring to very special cases. In any case, the hypothetical list of all the needed rules, on the one hand is surely incomplete and possibly even wrong (as these rules frequently assume regular schemata and loads and forget important parts of the problem), and on the other hand this so-specific list is the perfect alibi for people wishing to respect the letter of the rules while violating their deepest meaning. For instance, the Emilia Italian region was declared "not seismic" according to the law. Thousands of precast-concrete industrial buildings using portal frames with no true connection between the beam and the columns were built in the area, and so the moderate 2012 Emilia earthquake led to severe losses. A more discerning approach would perhaps have saved lives.

Referring to steel structure connections, it is not always clearly understood that the specific rules of the coded standards are – by definition, you might say – not complete or even applicable to all situations. For instance, not all connections' plastic mechanisms can be forced into those of a T-stub (described in part 6 of Eurocode 3, Part 1.8, "Structural joints connecting H or I sections), but it is not unknown to see local authorities denying the approval of a design, as the expected T-stubs are not even mentioned in the design reports. Coded standards cannot replace a serious professional judgment, and should never be considered a "by law" alternative to specific design considerations deeply rooted in the best "standards" we have: the laws of mechanics.

In summary, it is the author's opinion that the standards enforced by law should be short and should refer only to general principles. Specific guidance can be written for specific well-delimited problems, while the (humanly understandable) wish to cover all the matters by means of what are in fact specific tools, should be resisted. The way the coded standards are written should clearly push the readers (possessing different levels of knowledge on the matter) to understand that specific problems may well have specific solutions, and that only the general principles of mechanics should always hold true, not the ad hoc specific methods conceived to deal with well delimited problems.

1.3.2 T-Stub in Eurocode 3

One of the most complex failure modes to be investigated is that of "generic resistance", a generic set of constituents not falling into any of the available simplified theories (such as beam theory or rectangular or circular thin plate theory), that is loaded by a complex set of forces and must be checked against plastic limit and other failure modes.

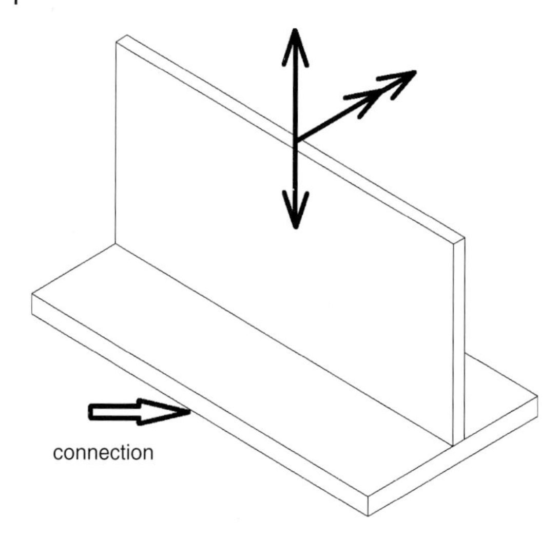

Figure 1.13 A T-stub.

The problem has no easy solution, in general. Some specific problems have been studied adding strong limitative hypotheses to geometry, and one of the most frequently used of these is a *T-stub* (Figure 1.13), which is a tee of short length connected, usually by a flange, to some other part, and loaded by a tensile or compressive axial force (force vector in the plane of the T cross-section, parallel to the web direction), or a bending moment (usually with the moment vector normal to the T-stub web). Under these simplified loads, and considering symmetrical and regular bolt "rows" connecting the flange, it is possible to assess the value of the limit plastic load, within the frame of yield line theory, also considering prying forces (see also Section 1.2.7).

In AISC 360-10, the T-stub is almost never explicitly mentioned, nor are given rules referring to its use in this specific context (but the T-stub model is explicitly used in prying force evaluation). Generally, the standard is less prescriptive and more flexible and open than the Eurocode.

In Eurocode 3, Part 1.8 the T-stub paradigm is introduced in Section 6, entitled "Structural joints connecting H or I sections", and specifically at subsections 6.2.4 "Equivalent T-stub in tension", and 6.2.5 "Equivalent T-stub in compression".

Eurocode 3, Part 1.8, Section 6.2 "Design resistance" uses the T-stub model in many of its subsections, and referring to problems that are not specifically those of a T-stub, for instance:

- §6.2.6.3(3) – "column web in transverse tension"
- §6.2.6.4(1) – "column flange in transverse bending"
- §6.2.6.5 (1), – "end-plate in bending"
- §6.2.6.6(1) – "flange cleat in bending"
- §6.2.6.8.(2) – "beam web in tension"
- §6.2.6.9(2) – "concrete in compression including grout"
- §6.2.6.10(1) – "base plate in bending under compression"
- §6.2.7.1.(8) – "extended end plate joint"
- §6.2.8.2(1) – "column bases only subjected to axial forces" (where three "non-overlapping" T-stubs are used – see Figure 1.14).

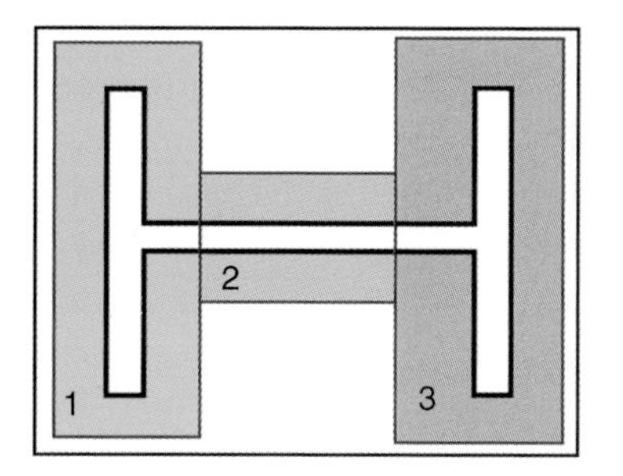

Figure 1.14 Eurocode 3, Part 1.8: column bases only subjected to axial forces.

Usually the typical sentence is that something like "...should be taken as equal to the ... of equivalent T-stub representing...".

Thus, in Eurocode 3, Part 1.8, the T-stub model is used as an all-rounder to cover almost all the real-world problems of "joints connecting H or I sections". Often, practitioners have extended the scope of Eurocode 3, Section 6, to joints using L, C, T, composite, CHS, RHS, and thin-walled cross-sections, and spreadsheets using the "T-stub" paradigm can be found almost everywhere.

The author is not convinced that this approach is necessarily always the best available in 2017. This is a paradoxical result: every sub-part of every joint appears to be necessarily modeled by a T-stub, and the use of by far more general and flexible approaches such as finite elements are not even explicitly mentioned by the standards (the words "finite elements" *do not appear* in Eurocode 3, Part 1.8: is that not so?!). This is probably due to the fact that the finite element technique is still considered somehow exotic, and that it is assumed that it would take too much time to prepare and run a model. In the author's view, this is technically just not true anymore because, from experience, for many problems detailed finite element models can be prepared automatically and run in a few tens of seconds.

1.3.3 Eurocode 3 Component Model

In order to compute the stiffness and the resistance of standard connections, Eurocode 3 uses a method called the "component method" (Jaspart 1991) which decomposes connections into a number of standard components, whose uniaxial elastic stiffnesses are then evaluated by means of simplified formulae (not necessarily simple). By composing the elementary stiffnesses it is ideally possible to model the behavior of sets of these elementary components. Much excellent work has been done by following this research path, and in-depth results have been obtained for specific typologies of connections, including in the non-linear range.

The following elementary components are listed in Table 6.1 of Eurocode 3, Part 1.8 (basic joint components):

- column web panel in shear (k_1)
- column web in transverse compression (k_2)
- column web in transverse tension (k_3)
- column flange in bending (k_4)

- end plate in bending (k_5)
- flange cleat in bending (an angle cleat is assumed as example) (k_6)
- beam or column flange and web in compression (k_7)
- beam web in tension (k_8)
- plate in tension or compression (k_9)
- bolts in tension (k_{10})
- bolts in shear (k_{11})
- bolts in bearing (k_{12})
- concrete in compression including grout (k_{13})
- base plate in bending under compression (k_{14})
- base plate in bending under tension (k_{15})
- anchor bolts in tension (k_{16})
- anchor bolts in shear (k_{17})
- anchor bolts in bearing (k_{18})
- welds (k_{19})
- haunched beam (k_{20}).

Note that the same physical part may appear more than once.

A uniaxial stiffness is then related to each of these components, evaluated as a function of simplified geometry, possibly of a T-stub model, and of the loads applied. Summing up the elementary uniaxial flexibilities related to each component would then allow us to model complex connections, which can thus be seen as an assembly of these simple "bricks" in series or parallel.

Some of the stiffness related to the elementary components is considered infinite, whatever the exact geometry. Specifically the following stiffnesses are considered infinite:

- column web panel in shear, if stiffened (k_1)
- column web in compression, if stiffened (k_2)
- column web in tension, if stiffened welded connection (k_3)
- beam flange and web in compression (k_7)
- beam web in tension (k_8)
- plate in tension or compression (k_9)
- bolts in shear, if preloaded (provided that design forces do not induce slip) (k_{11})
- bolts in bearing, if preloaded (provided that design forces do not induce slip) (k_{12})
- plate in bending under compression (k_{14})
- welds (k_{19})
- haunched beams (k_{20}).

The infinity of these stiffnesses must be understood in a relative sense: these are much higher than those of the other components, and so their reciprocal (flexibility) can be assumed null.

The working mode of beam elements (or columns) is apparently reduced to axial force plus strong axis bending. This bending plus axial force causes elementary forces with a lever arm z, also suggested by the code. A 2D model is implicitly assumed and these forces will find their path to the connected part (a column in beam-to-column connections, or a concrete slab in base joints, for instance).

The rotational stiffness S_j of the joint can be obtained by the following general formula (which is 6.27 of Eurocode 3, Part 1.8, E is Young's modulus):

$$S_j = \frac{Ez^2}{\sum_i \frac{1}{k_i}}$$

where the summation is extended to all applicable components having elementary stiffness k_i.

This model *seems* simple, and the idea to decompose complex systems down to elementary ones is indeed brilliant, but how general is it, really? Neglecting part of the member forces is a strong limitation. The lever arm is problem dependent and also load dependent. On the other hand, if the geometrical simplifications that led to the k_i stiffnesses evaluations for components using T-stubs are not fully applicable, the method cannot be used.

The method is difficult to apply even for simple systems, but it is simply not applicable for complex systems. However, it is applicable to a limited set of connections. The member forces and moments should be decomposed into simple forces, acting at well defined points, where some sort of load path must be drawn. In considering this load path, the previously enumerated standard components must necessarily be found. Every possible cause of non-compliance must be neglected.

The standard states at clause 6.1.1.(1) (emphasis added):

> This section contains design methods to determine the structural properties of joints in frames of *any* type. To apply these methods, a joint should be modeled as an assembly of basic components.
>
> ©CEN, reproduced with permission

And in a "Note" at clause 6.1.1(2) (emphasis added):

> The design methods for basic joint components given in this Standard are of general application and can also be applied to similar components in other joint configurations. *However the specific design methods given for determining the design moment resistance, rotational stiffness and rotation capacity of a joint are based on an assumed distribution of internal forces for joint configurations indicated in Figure 1.2.* For other joint configurations, design methods for determining the design moment resistance, rotational stiffness and rotation capacity *should be based on appropriate assumptions for the distribution of internal forces.*
>
> ©CEN, reproduced with permission

Figure 1.2 of Eurocode 3, Part 1.8 lists the following typical joints:

- major axis: single-sided beam-to-column joint configuration
- major axis: double-sided beam-to-column joint configuration
- major axis: beam splice
- major axis: column splice

- major axis: column base
- minor axis: double-sided beam-to-column joint configuration
- minor axis: double-sided beam-to-beam joint configuration.

So it is very clear that the rules coded by Eurocode 3, Part 1.8, Section 6, refer to, and are proposed for, a limited number of joint configurations, considering I or H cross-sections, that is, they are not a general tool to deal with the problem of connection checks.

In the present context, the member stress state is general and there is no practical distinction between "major axis" bending and "minor axis" bending as it is assumed that they do act at the same time, and the same applies to torque, shears and axial force.

In this book a different path will be tried, albeit with some broad concepts in common with that of "component model".

1.3.4 Distribution of Internal Forces

As we have seen in the previous section, the detailed formulations provided by Eurocode 3, Part 1.8 in order to check connections falling in the categories delimited by the code, hold true because "appropriate assumptions for the distribution of the forces" have been used. These *appropriate distributions* are behind all methods.

In turn, these appropriate distributions are tightly related to specific configurations for which it has been found that they are useful and "appropriate". Different node layouts will surely lead to different force distributions, and the lack of a general tool to compute such distributions is one important reason why general methods to check connections were not available.

Looking at the problem in a general way, we can say that, as will be shown in Chapter 5, finding an appropriate distribution of internal forces in a general context is the main problem of connection design.

Traditional approaches use ad hoc internal force distributions that are the result not of a calculation but of a free choice which implements some interpretation of connection experimental behavior. This can be done with the hope of being right if and only if the lower bound theorem of limit analysis is applicable. The use of somewhat arbitrary but balanced force configurations is behind all the traditional approaches as well as some of the new ones, and is also behind the success of many historic buildings. As Prof. Jacques Heyman has clearly shown in his marvelous book *The Stone Skeleton* (Heyman 1995) without the lower bound theorem no design could be carried out, as the "true" distribution of internal forces is basically impossible to determine. And this is also true for steel connections.

This is a key concept, and one of the main pillars to be considered in connection checks.

1.3.5 Prying Forces

Prying forces are the forces that arise at the contact between plates due to the flexure of the plates themselves induced by bolt tensile forces. They are statically undetermined, and strongly depend on the loads applied, on the geometrical configuration of the plates in contact, on their thicknesses, and on the position and diameter of the bolts.

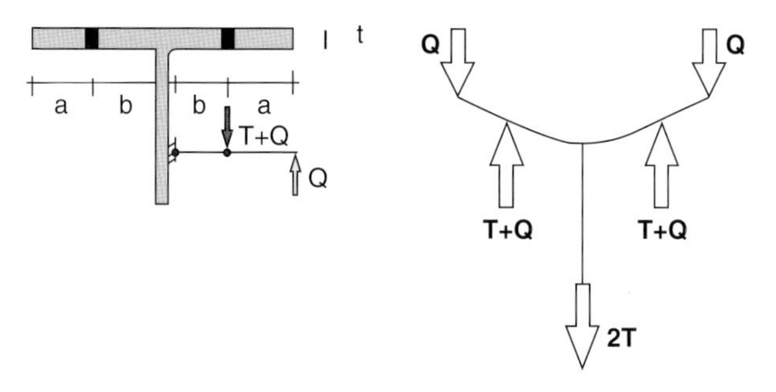

Figure 1.15 Prying forces Q increase tensile forces in bolts T. Simple T-stub model.

Prying forces imply an increase in the bolt tensile forces, and so must be taken into account in order to avoid the bolt tensile forces being underestimated. On the other hand, prying forces unload those plates that find, in the contact to another plate, a useful bearing of a free tip.

The widely used method of evaluating prying forces is due to Thornton (1985) and refers to a T-stub model (Figure 1.15). If the geometry is different, the method cannot be applied. Other simple models can be used instead (e.g. cantilevers or simply supported "beams"), but with questionable reliability. At the moment, the evaluation of prying forces is one of the most complex problems of generic steel connection analysis and can only be tackled for generic geometries (i.e. not T-stubs), by means of contact non-linearity and finite elements; the only possible way to better compute prying forces in a general context is to use plate–shell finite element models and consider contact non-linearity. This is a quite complex issue but, as we will see, it can be tackled by proper finite element analyses run automatically.

1.3.6 Block Tearing

Block tearing is the fracture and subsequent separation of a part of a steel plate, usually in a bolted connection, under the effect of the shear forces transferred to the plate by the bolt shafts.

In the available coded standards, block tearing is dealt with by considering shear stress paths and normal stress paths, mixed together so as to define cut lines for the plate at hand. Usually, also in the textbooks, it is apparently assumed that the rupture lines are aligned with plate sides (Figure 1.16), so that the geometry of the fracture lines is somehow forced to respect the external geometry of the constituent. However, this is not true in general. The forces transferred from the bolt shafts to the plate at hand are not only generally different from each other, but also have different inclinations. So, once again, it appears that the methods usually adopted to consider this failure mode are by far too simplified, and that a more general model should be used to properly evaluate this dangerous condition.

1.4 Scope of This Book

The aim of this book is to discuss the problems to be faced when trying to tackle steel connections analysis in a general way.

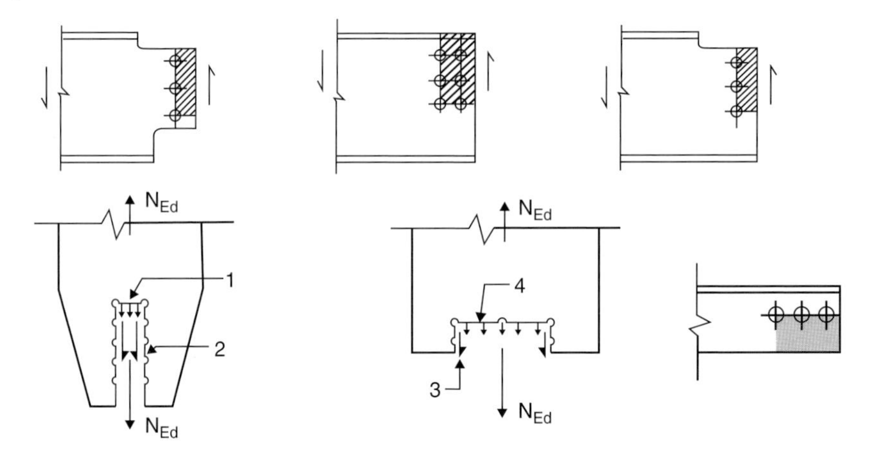

Figure 1.16 Block tearing (from Eurocode 3, Part 1.8): forces are aligned with plate sides, and so with the break lines. ©CEN, reproduced with permission.

The chapters will broadly follow the path of the author's research, starting from a generic 3D finite element model of a steel structure, like the thousands that are created every day by structural engineers all around the world. This is a model using may be hundreds of beam and truss elements, possibly inclined, and using generic cross-sections, tens of load cases, and many load combinations, and has probably been run in linear or non-linear range to check members according to some available standards (i.e. neglecting the problem of connections, and considering the members as wireframe sticks to be checked for member failure modes: resistance, stability, and deformability).

Now, the problem is to check connections according to the computed member forces, or according to the capacity design rules. The structure is 3D, members are inclined, so a first question is: how many different connections are there, and what stress levels are expected?

Then, once these different connections have been detected, it is necessary to describe how they are physically realized, one by one. So, it is necessary to set up what in this book is called the *scene* for each node. Finally, all the constituents, members, cleats, and connectors must be checked against all possible failure modes, and taking into account all load combinations applicable.

This is the task to be accomplished. The chapters of this book deal with this task.

Chapter 2 deals with *jnodes*, that is the connections as they can be described in a wireframe context. This chapter will explain how to detect equal jnodes automatically, and how to classify them (the so-called *jnode analytics*).

Chapter 3 is short and proposes a general model for connections. This is needed for the next steps. This chapter also introduces the concept of the connection graph.

Chapter 4 is related to *renodes*, the 3D counterpart of jnodes. Here every constituent is a 3D object in space, with a specific position, orientation, and shape. This chapter also deals with the problem of automatic detection of "connections", and automatically analyzing the coherence of the node. The *chain* concept will also be introduced here.

Chapter 5 is a review of the general principles that are always applicable and that must be complied with for a sound analysis of connections. They will be extensively used in the remaining chapters of this book. Also outlined here is the statics of connections, introducing iso-, hyper-, and hypo- connected nodes.

Chapters 6 and 7 are related to *connectors*, which are the devices used to physically implement a connection. Chapter 6 deals with welds, and Chapter 7 with bolts and contact. The aim is to review their mechanical behavior, the computational methods used to model them, and the hypotheses currently adopted by the methods in use.

Chapter 8 is a discussion of the most important failure modes that need to be checked in connection design, and of some of the models that can be used to check them, with particular reference to AISC 360-10 and Eurocode 3, Part 1.8. A more general method for block-shear is presented.

Chapter 9 deals with a first general method proposed to analyze connections: the *hybrid approach*, using several techniques to get quick results without merely relying on pure computational force. This basically explains how to check connections using a general method.

Chapter 10 refers to a pure fem approach, that is, the ideally unified, general, and single method that could be used to check all connections: it is the future for this subject, and is still being researched.

Chapter 11 draws some conclusions and discusses the open problems.

Finally, the appendices deal with specific problems such as tangent stiffness matrices, and with notation and symbols.

For the most part the proposed and discussed method has already been implemented, tested, and used by engineers, by running the software developed by the author and called CSE (Connection Study Environment).

1.5 Automatic Modeling and Analysis of 3D Connections

Times are changing: traditional approaches to steel connection design and checking will assume a secondary role, while the use of general methods, equivalent to those already used dealing with *structures*, will increase in importance and, gradually, will become the normal way to check connections. As has already happened with 3D framed structures, traditional approaches will still retain their value as cross-checking or initial sizing tools.

The computational power available nowadays makes it possible to create and run a finite element model in a matter of a few tens of seconds, and so, at least for some of the failure modes to be investigated, this way of checking is today the most efficient and promising. The limitations of traditional approaches to steel connection checks seem to be particularly evident when they are used together with modern computational tools, that is, the finite element software broadly used to model frames and structures by means of beam and truss elements. It is evident, then, that there is a clash between the need for using hundreds elements in a 3D context, loaded by tens or even hundreds combinations, and the oversimplified, much-throwing-apart, simple methods that forget to see the connections for what they are: complex structural subsystems that need general methods to be analyzed.

1.6 Acknowledgments

This book is the result of an individual research program carried on since 1999, with the aim of providing a general software tool able to check *generic* steel connections. This software tool has been written step by step, and has moved forward together with the results obtained by research. It is referred to as the *Connection Study Environment*, and has been on the market since 2008 (www.steelchecks.com/CSE). All the computational results presented in this book come from using CSE, and all the connection models in the figures have been modeled using CSE. The reader interested in a brief history of the project and the reasons why it was carried on, will find this information in the final Chapter 11.

The reason why this work was carried on by a single researcher is readily explained: none of the possible partners in Italy seemed interested, 10–15 years ago, in spending years of effort with the aim of developing a general tool dealing with steel connections. Some very important early results were presented at a National Conference on Steel Structures in Padova (Rugarli 2009), but at the time, nobody was interested – the project seemed too exotic.

The author wishes to thank, with deep feelings of gratitude, his unforgotten Professors at the Politecnico di Milano, and particularly Francesca Rolandi, Adelina Tarsi, and Laura Gotusso, who taught mathematical analysis and numerical analysis, Carlo Cercignani, who taught rational mechanics, Leo Finzi who taught elasticity, Giulio Ballio who taught steel structures, Leone Corradi dell'Acqua who taught instability, and Giulio Maier, who taught elasto-plasticity and finite elements during an unforgettable and life-changing year, some 33 years ago.

The author also wishes to thank all the colleagues around the world, of different cultures, languages, and origins, something marvelous in itself, who have trusted the work done during those years, and allowed it to be carried on, using the software procedures that, from year to year, have been developed and improved. This is what sometimes is reductively called *the market*. With some of them, the discussions were frequent and/or very useful, so thanks in particular to Emanuele Alborghetti, Giovanni Cannavò, Michele Capè, Marco Croci (who also helped greatly in developing tests and parameterizations and spent several years working on this project), Jason McCool, Andy Gleaves, Meda Raveendra Reddy, Sergio Saiz, Christos Saouridis and the colleagues at CCS, Harri Siebert, Edoardo Soncini.

While every effort has been made to deliver a text with no misprints or errors, it would be wise for the reader to check at: http://steelchecks.com/connections/SCA.asp to see whether any *errata corrige* is available, and also to be informed about the latest results of the ongoing research.

Finally, the author wishes to deeply thank Paul Beverley for his precious, careful, and skilled work in editing this book.

References

Booth, E. 2014, *Earthquake Design Practice for Buildings*, 3rd ed. ICE Publishing, London.

Bruneau, M., Uang, C.M., and Sabelli, R. 2011, *Ductile Design of Steel Structures*, 2nd ed., McGraw Hill.

European Spreadsheet Risk Interest Group, www.eusprig.org.

Fraldi, M., Nunziante, L., Gesualdo, A., and Guarracino, F. 2010, *On the Bounding of Limit Multipliers for Combined Loading, Proc. R. Soc., A* 2010, **466**, 493–514.

Heyman, J. 1995, *The Stone Skeleton*, Cambridge University Press, Cambridge.

Huckelbridge, A.A., Palmer, D.A., and Snyder, R.E. 1997, *Grand Gusset Failure*, Civil Engineering, Vol. 67, September, pp. 50–52.

Jaspart, J.P. 1991, *Etude de la semi rigidité des nœuds poutre-colonne et son influence sur la résistance et la stabilité des ossatures en acier*, PhD Thesis, Department MSM, University of Liège.

Kemeny, J. 1979, *The Need for Change: The Legacy of TMI*. Report of the President's Commission on the Accident at Three Mile Island, Government Printing Office, Washington DC.

Joints in Steel Construction – Moment Connections, P207/95, 1995, The Steel Construction Institute, ("Green Book").

Reason, J. 1990, *Human Error*, Cambridge University Press, 1990.

Rugarli, P. 2004, *Combinazioni di Verifica allo Stato Limite: il Non Detto delle Normative, Ingegneria Sismica*, **2**, 2004.

Rugarli, P. 2009, *Software Computation of Steel Joints Built Up Freely Placing Components* (Italian), Acta XXII CTA Congress, L'Acciaio per un Futuro Sostenibile, Padova September 28–30, 2009.

Rugarli, P. 2014, *Validazione Strutturale*, EPC Libri, Rome.

Thornton, W.A. 1985, *Prying Action – a General Treatment*, Engineering Journal, American Institute of Steel Construction, Second Quarter 1985.

White, D.W., Leon, R.T., Kim, Y.D., Mentes Y., and Bhuyan, M. 2013, *Finite Element Simulation and Assessment of the Strength Behavior of Riveted and Bolted Gusset-Plate Connections in Steel Truss Bridges – Final Report*, Georgia Tech, 2013.

2

Jnodes

2.1 BFEM

The most commonly used finite element models are those made up by beam and truss elements. These models are used to design, analyze, and check a wide range of structures, such as buildings, towers, pipe racks, industrial plants, roofs, and bridges.

Although not all designers seem aware of the fact, the wireframe models made up using beams and trusses are finite element models like those made up by more complex elements, like plate–shell elements, membrane elements, and solid elements.

Starting from the mid 1980s – not long ago from a historic perspective – engineers and architects modeled structures using a computer, and prepared 2D models and then, a few years later, 3D models. By the end of the 1980s and into the 1990s it was common to see 3D wireframe models of structures, having thousands degrees of freedom, solved in the elastic range for perhaps tens of load cases.

This was brand new: never before had so many engineers been able to create such complex models. Also, the requirements of the standards grew quickly. If at the end of the 1970s it was common to check structures for a few, unrealistic envelope combinations, by the mid 1990s it was common to study finite element models of 3D structures using tens of load cases and combinations. This opened a quite new scenario, which is still under research: how to use the computing power of personal computers to deal with complex analytical problems, with no loss of safety or comprehension by the analyst. Elsewhere (e.g. Rugarli 2014) it has been pointed out and discussed in depth that the validation of finite element models is a complex problem, involving several disciplines, and it is also an urgent problem as too often these models are unchecked by their creators and lead to completely wrong analyses.

These finite element models were usually based on the two most simple finite elements available: the *truss element*, modeling members carrying only an axial force, and the *beam element*, an element capable of absorbing one axial force, two shears, one torque and two bending moments, all variable along its axis. There are also curved elements, but by far the most frequently used elements were and are straight, using two *nodes*.

As I have tried to explain elsewhere in more detail (Rugarli 2010), a *finite element* is a portion of a continuum, governed by a set of usually complex differential equations, having a very simple shape. The shape of the finite element is fully described by a limited number of points that in the finite element method (fem) jargon are called *nodes*. The basic idea of fem is that what happens inside the finite element is entirely dependent

Steel Connection Analysis, First Edition. Paolo Rugarli.
© 2018 John Wiley & Sons Ltd. Published 2018 by John Wiley & Sons Ltd.

on the displacements of the nodes, which are the main problem unknowns (there are also fem approaches based on forces, but these are less common).

The structure as a whole is divided into finite elements, and the displacements of all the *nodes* form the unknown vector of the problem to be solved. From the nodal displacements, once known, we can derive the strains internal to the elements, and from the strains the stresses. This is done by means of simple interpolation formulae, usually polynomial: the polynomial functions used to interpolate are fixed a priori, and the displacements of the nodes are their weights. If the finite elements are small, when compared to stress gradient, then results are near the ones which can be obtained by solving the partial differential equations by other methods, such as finite differences, closed formulae, or series.

One important feature of beam and truss elements is that in elasticity problems their polynomial interpolation is exact if no internal loads are applied to the elements themselves: a linear variation of displacement is exact for a pulled bar, and a cubic displacement variation is exact for a Bernoulli beam in flexure without intermediate loads. This last can be seen by remembering the governing differential equation $EJw^{IV} = 0$, with E Young's modulus, J second area moment, and w^{IV} the fourth derivative of $w(x)$ to x. This means that beam and truss elements do not need to be very small, because their polynomial interpolation is exact, or very nearly exact (the additional effect of the internal loads can be re-added by a simple sum to the results acquired by interpolation).

Therefore, the subdivision of the structure in finite elements is easy: very often, a member is modeled with just one finite element. This explains why so many practitioners are not aware that they are indeed using finite elements, because they think they are using "bar" or "member" elements.

In non-linear analysis, each increment of displacement is related to an increment of strains, and that to an increment of stresses. Final stresses are obtained by summing up increments. However, in non-linear analysis, more refined meshes are needed for beam elements, and so in that context the need for small finite elements, typical of plate and membrane structural elements, or of solid elements, is also applicable to beam elements.

The starting model for designing a steel structure made by 1D members is thus a finite element model mainly made by truss and beam elements. In this book there will be several different types of finite element models, so it is necessary to distinguish them by a proper nomenclature. For this reason, this starting finite element model is named here the *BFEM*, where the "B" stands for beam or bar – or also for Bernoulli.

BFEMs are used very frequently in today's engineering practice. Sometimes they are not made *only* by beam and truss elements. For instance, in modeling a building, the bracing concrete core or shear walls may well be modeled by membrane or plate–shell elements, and also the reinforced concrete slabs. However, this model will always embed many beam and truss elements, which are currently used to model structural elements such as columns, beams, and bracings.

The existence of beam and truss elements is needed to properly model the skeleton of the structure, and after solving these elements are usually checked against pertinent standards, such as Eurocode 3 (EN 1993), AISC (AISC 360-10), or British Standard 5950, in order to test their proper design. This is the starting point for the most part of the steel structures, which are the object of this book.

Note that, when considering the problem of steel connection analysis, dealing with the familiar internal forces called "axial force", "bending moment", "shear", or "torque", the existence of an underlying model consisting of 1D structural elements is implicitly assumed. These follow the well known basic structural theories related to beam and truss elements.

Although it is not necessarily true that the original computational model is a finite element model, i.e. a BFEM, this is the most likely situation nowadays, and it is always possible to describe a structural layout, also perhaps computed by hand using virtual work principle, by means of a BFEM.

2.2 From the BFEM to the Member Model

2.2.1 Physical Model and the Analytical Model

In this book the term *member* refers to a component of a steel structure, usually straight and prismatic, but also possibly tapered or curved, which can be modeled by means of 1D structural theories (Euler–Bernoulli's beam theory, Timoshenko's beam theory, etc.). A member is fabricated as a unique piece and may possibly be connected to other members or to other structural elements such as plinths, concrete slabs, and walls, by means of fittings and connectors such as bolts and welds.

So, connection analysis is the analysis of the connection between members, and between members and other structural elements. It is easy to also conceive connections between structural elements which cannot be modeled by means of 1D structural theory, but these are not covered in this book: e.g. the connection between a plate and a wall or the connection of a tensile membrane to a ring.

The connections that are going to be considered here always involve at least one member, and refer to a limited part of its axis: usually the extremity, but also possibly, as will be seen, a limited internal part of its internal length (these will be named *passing members*).

When a steel structure is analyzed in order to compute internal forces and displacements under the applied loads, the analytical model is usually a finite element model. However, the physical model, which takes into account the members, is a different one, purely geometric.

Some software programs ask the users to prepare a physical model, which is later converted into an analytical model by adding the necessary finite elements. Some other programs do not use physical models, but only analytical models. If the need is only to analyze a structure in order to compute internal stresses, the physical model is not needed. Therefore, there are programs that deal with finite elements and do not have the *member* concept.

In order to hide the finite element model details, some software is designed so that the finite elements are kept in the background, possibly generated automatically according to internal rules. If this is the case, the user of the software has direct input *members*, and the finite element model has been generated by specific subsequent commands.

The automatic creation of a finite element model from a member model is not an easy matter and there are specific problems related to rigid offsets, releases, connectivity and so on. The danger of such automatic generation is that it can create finite element models that are hard to check and that the users have to trust with no direct control over them.

Sometimes it is then far better to generate the analytical model directly or, if a physical model is the starting point, to generate the analytical model with a step-by-step controlled procedure. For this reason, many design software programs are based on the analytical model, that is, they manage finite elements.

The advantage of this approach is that the expert analyst directly controls the analytical model, and is responsible for its meaningfulness. Usually this approach is preferred by structural engineers, expert in analysis, especially when the structures are complex; checking an analytical model created automatically from a drawing or from a physical model can be a nightmare, and the time necessary to clean up the fem may be much longer than the time needed to rebuild the analytical model from scratch.

Moreover, sometimes the rules used to generate the finite element model from the physical model or from the drawing cannot be shared by the expert analyst, as they introduce violations of basic concepts, such as keeping the axes line in the centers of the cross-sections, modeling properly the eccentricities, neglecting those unneeded, and so on. In many engineers' experience, it is rare that such automatic transforming of a physical model or of a drawing into a finite element model leads to sound and checked models. Very often, the models are trusted as such, and no real control is exerted, as it would be too expensive.

If the member model is already available, and a BFEM is not available, in order to properly consider the internal forces which try to detach a member from its neighbors, an underlying analytical model must be set up. Otherwise it is just not possible to properly assign forces to members, and to properly take into account the points where these forces are exchanged. This is not always well understood when a simple drawing of physical members is used to compute connections. To analyze connections an underlying analytical model is needed.

If a simple beam to column connection is considered (e.g. Figure 2.1), in order to check the connection several questions must be answered, and the pure geometric position of the parts (the plate, its thickness, the welds, and the bolts) is not enough to answer to such questions:

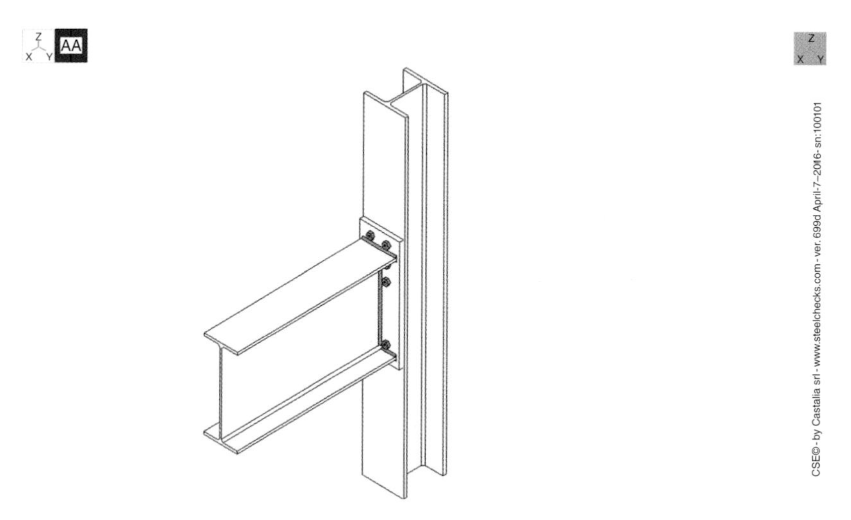

Figure 2.1 A beam to column connection.

1) What are the internal forces being exchanged? Is it just shear or also bending moment(s) and axial force and torque?
2) Where are these forces exactly exchanged, i.e. at which exact points? This is important because forces trigger moments, far from their line of action.
3) Have *the members* been checked for resistance and stability *coherently with these points of forces exchanged*?

So, willingly or not, a physical or geometrical model used to check connections is always supported by an underlying analytical model, which can be coherent or not with the analytical model used to check members. Ignoring the need for such coherence is a possible problem, as some basic principles are violated: the force distribution used to check connections is not the same force distribution used to check members, which violates the static theorem of limit analysis, as will be seen in Chapter 5. Sometimes this violation is done consciously, but other times not. Some people, starting from a drawing and being willing to check connections, are apparently unaware of the problem.

If, on the other hand, the analytical model is the only one available, a first necessary step in order to analyze connections between members is to convert it into a physical model by finding the members. This task could ideally be accomplished with direct selection of elements and then assigning them to members, but when dealing with a complex 3D structure this process may be long and cumbersome.

Therefore, if a BFEM has been prepared, the analyst needs to automatically convert a finite element model made by beam and truss elements into a member model.

2.2.2 Member Detection: Connection Codes

If the members are straight and prismatic, as usually happens, they can be detected by an automatic search.

The following will be assumed:

- A single truss element is always a member.

These elements are typically used for bracings.

If a set of n beam elements is such that:

- all the finite elements are connected one to one so that they can be ordered in such a way that element i is connected to element $(i + 1)$ in a node N_{i+1}
- all the finite elements are aligned between themselves (each element i is aligned to every element $j \neq i$) within a given tolerance
- all the finite elements have the same cross-section
- all the finite elements have the same material
- all the finite elements have the same orientation in space
- there is no end release applied to any element extremity connected to an internal node (that is a node N_k with $2 \leq k \leq n$)

then, it is assumed that this set of aligned beam elements is modeling a straight prismatic member, which is by far the most frequent type of member in steel structures.

Both the internal and the external nodes can be connected to other elements, which are part of other members; the nodes where this condition is met will be the place of connections between members.

When considering the alignment of two elements, say *Element i*, and *Element j*, the following considerations apply.

Let N_1 and N_2 be the nodes of *Element i*, and N_3 and N_4 be the nodes of *Element j*. If *Element j* is adjacent to *Element i*, then $N_2 = N_3$.

If no rigid offsets have been used to define beam elements, then the alignment condition will refer to the set of three 3D points (N_1, $N_2 = N_3$, N_4), in space.

On the other hand, if rigid offsets have been defined, we must distinguish between *element node* and *element extremity*.

The *element node* is the 3D point whose displacements are the primary unknowns in the displacement-based finite element method.

Element extremity is a 3D point obtained by adding to the node position the rigid offset. This point marks the end of the deformable part of the element axis. Extremity points can be named "true end points", as they are the true deformable element extremity points. If no rigid offset is applied, they collapse to nodes. Generally speaking *a node* of a beam element is not coincident with its *extremity*.

Let N_{T1}, N_{T2}, N_{T3}, and N_{T4} be the *true end points* of *Element i* and *Element j*. To consider aligned adjacent elements *Element i* and *Element j* the following conditions must be met (see Figure 2.2):

- N_{T2} must be coincident with N_{T3}.
- (N_{T1}, $N_{T2} = N_{T3}$, N_{T4}) must be three 3D points aligned in space.

Considering the three sets of elements in Figure 2.3, and assuming that all the elements have the same material and cross-section, the following considerations can be used (ordering of elements is from left to right):

- Elements of set A cannot be a member, as there is one end release applied to the second element from the left.
- Elements of set B cannot be a (straight) member as they are not aligned.

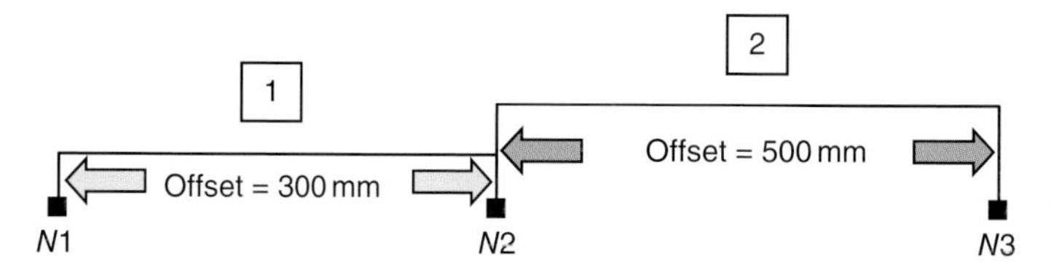

Figure 2.2 Example of two adjacent elements whose nodes are aligned but that, due to the existence of rigid offsets, cannot be considered aligned.

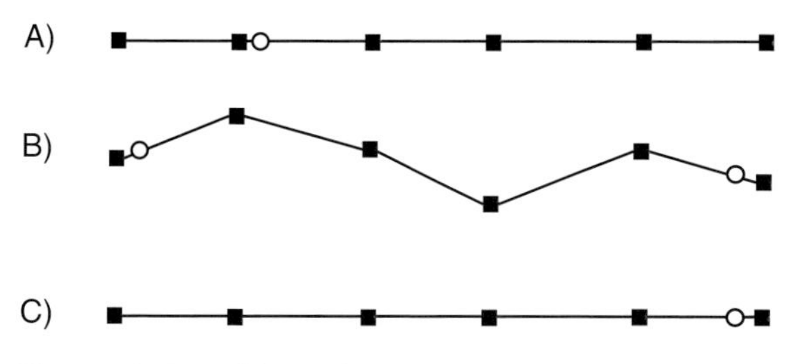

Figure 2.3 Sets of finite elements.

- Elements of set C can be a member, as the end release is applied to the end element second extremity. However, all the elements will have to be equally oriented.

By definition, *a member has no internal release.*

Given a set of aligned beam elements, possibly mapping a member, it is assumed that if an internal end release is found, no matter which component(s) are released, then this set of *elements* is divided into two straight *members* by the end release found. If the number of *internal* extremities with release applied is n, then there will be at least $(n + 1)$ members.

If a discontinuity of material, cross-section, or orientation is detected then the set is split, and it will give rise to more members.

This simple set of rules is able to automatically detect most of the members in an analytical model, but there are some specific cases, which must be dealt with.

1) Two aligned equal members may be connected in such a way that no release is to be applied to the underlying beam elements: a splice joint between two identical members is the typical example (Figure 2.4). Therefore we need the ability to split it into two members, equal beam elements aligned, with no end release inside and equally oriented, if needed.
2) A member may be tapered.
3) A member may be curved.

The first issue is solved considering *connection codes.* It will be assumed that a connection code can be optionally assigned to the extremities of beam elements.

If a connection code is assigned to a beam element extremity, then that extremity is also the end of the member to which the beam element belongs.

If the connection code is not assigned to the beam extremity, then the member to which the beam element belongs can be extended to more (aligned and member-compliant) beam elements, if any.

Besides, if an end release of any kind is applied to a beam element extremity this automatically implies a connection code. However, a connection code may be applied with no end release.

Connection codes are not usually dealt with in standard finite element software. However, they can be easily added or mapped to groups, as follows. If a software program deals with the concept of group or layer, three special groups or layers can be set up:

- group/layer “CCI”, whose elements are beam elements having a connection code at I node (first node)

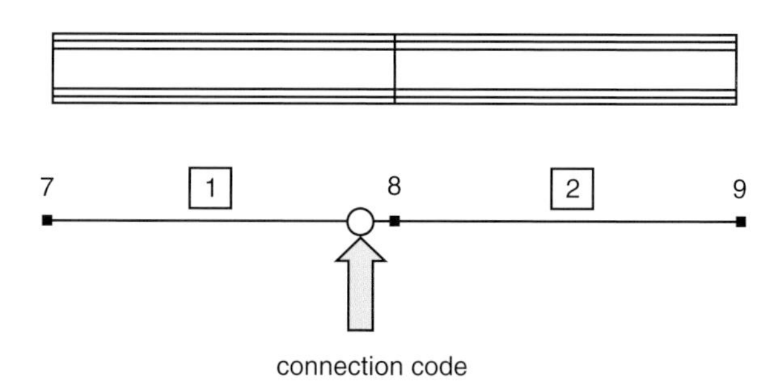

Figure 2.4 Aligned elements belonging to different members: no end release.

- group/layer "CCJ", whose elements are beam elements having a connection code at J node (second node)
- group/layer "CCICCJ", whose elements are beam elements having a connection code both at I and J extremities

Elements having end releases, by definition, have a connection code at the end released, so there is no need to add them to the groups/layers mentioned.

In standard finite element software, it is easy to add connection code management. The mask used to choose the end releases, may be of seven Boolean flags instead of six. The seventh code would be the connection code.

- If one end release is applied, then automatically the seventh flag (connection code) is switched on.
- However, if no end release is applied, the user might set the seventh flag on, without affecting the releases of the beam element.

This procedure has been applied in the code developed by the author (Sargon©), with excellent results. In view of special problems (Section 2.7.2) it is also possible to assign a color to a connection code.

Getting back to the specific cases listed, the second issue (tapered members) can be dealt with by appropriate checks on the element cross-sections, provided that single beam elements can be tapered. If this holds true, then the beam element will have two different cross-sections at the two extremities, and the conditions to check will also have to include that no abrupt change of cross-section may happen at member–element interfaces. For linearly tapered members, a more strict condition can be that the slope from one finite element to the adjacent one does not change.

The third issue (curved members) can be managed by releasing the alignment tolerance, that is, assuming that two straight adjacent elements may have a small difference in alignment, and still be considered part of a single member.

2.2.3 An Automatic Algorithm for Straight Prismatic Member Detection

A brief pseudocode description of the algorithm needed to automatically detect straight prismatic members will be now given.

```
For(each Truss Element i)
{
    AddMember(TrussElement i);
}
UnselectAllBeamElements();
For(each unselected Beam Element i)
{
    Select(Beam i);
    FindAllElementsAlignedContinuous(Set1, i, 1); // Set1 is a set
    of elements empty
    FindAllElementsAlignedContinuous(Set2, i, 2); // Set2 is a set
    of elements empty
    AddMember(Set1 + Beam i + Set2)
}
```

```
FindAllElementsAlignedContinuous(Elements, Beam n,
Extremity ext)
{
    If((Beam n).HasConnectionCode(ext) = = TRUE) return; //
    connection code reached, end

    Node = (Beam n).GetNode(ext);
    For(each unselected Beam i)
    {
        For(iext = 1, 2)
        {
            If((Beam i).GetNode(iext) = = Node) // connected I the
            same node
            {
              Point3D p1 = (Beam n).GetTrueEndPoint(ext); // beam n
              extremity
              Point3D p2 = (Beam i).GetTrueEndPoint(iext); // beam
              i extremity
              If(Distance(p1, p2) < sometolerance)  // true points
              coincide at connection
              {
                If(AreAligned(Beam n, Beam i)) // alignment
                condition
                {
                  If(GetCrossSection(Beam n) = = GetCrossSection
                  (Beam i))
                  {
                    If(GetMaterial(Beam n) = = GetMaterial
                    (Beam i)) // Equal material condition
                    {
                      If((Beam i).HasConnectionCode(iext) =
                      = FALSE)
                      {
                      Select(Beam i);
                      Elements.Add(Beam i);
                      If(iext = = 1) otherext = 2;
                      Else      otherext = 1;
                      FindAllElementsAlignedContinuous
                      (Elements, i, otherext); // recursive
                      return;
                      }
                    }
                  }
                }
              }
            }
        }
    }
}
```

This algorithm can be further improved by considering tapered members or curved members, but this is left to the reader.

At the end of the search, a set of members will be found. Each member will have:

- two end nodes, I and J (N_I and N_J)
- a single truss element, having a material, a cross-section, and an orientation. Or
- a set of underlying beam elements, ordered in such a way that the first element first node will be the first node of the member, and that the last element second node will be the end node of the member; moreover, the second node of *Element i* will be equal to first node of *Element* $(i + 1)$
- a material, equal for all the beam elements
- a cross-section, equal for all the beam elements
- an orientation, equal for all the beam elements (see Appendix 1).

2.2.4 Member Data Structure

The data structure for a member should include:

- the end-nodes identifiers
- the cross-section of the member
- the material of the member
- a flag indicating if the member is a truss or a beam
- the vector of the identifiers (numbers) of the finite elements of which it is composed
- the rigid offset relative to the nodes at the member ends
- the connection codes and the end releases of the elements at the ends
- the orientation of the member.

The orientation (Appendix 1) will be defined as the set of three unity vectors:

- The first vector, $\mathbf{v}_1$, is the member axis vector, from first to second extremity.
- The second vector, $\mathbf{v}_2$, is the strong principal axis of the cross-section.
- The third vector, $\mathbf{v}_3$, is the weak principal axis of the cross-section.

The definition of "strong" and "weak" axis is notional. For I, H, T, [,][, and similar cross-sections, it will be assumed that the "strong axis", axis 2, is normal to the web, and parallel to flanges. What is important, however, is just that a well defined rule to name principal axes is set up, for each cross-section type.

The set of the three unit (column) vectors $\mathbf{v_i}$ is able to define, by row as $\mathbf{v_i}^T$, a 3×3 square matrix $\mathbf{T}$, which is orthogonal, and is such that its transpose $\mathbf{T}^T$, multiplied by $\mathbf{T}$, leads to unit matrix $\mathbf{I}$: thus $\mathbf{T}^T\mathbf{T} = \mathbf{I}$.

Once the members have been searched, a global vector of members will be available.

2.2.5 Member Classification at a Node

Considering a node of a BFEM, and a member, the following classification can be set up.

If the node is not connected to any elements of those defining the member, the node is *unconnected* to the member, and the member is unconnected to the node.

If the node is one of the internal nodes of the member, then the node is *connected* to the member, and the member will be classified as *passing*, at the node considered

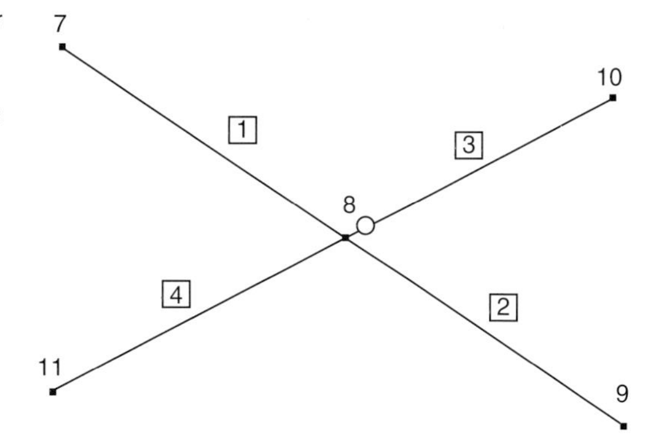

Figure 2.5 Member at a node. The member made up of beam elements 1 and 2 is passing at node 8. The member made of beam element 4 is cuspidal at node 8. The member made of element 3 is interrupted and unreleased. The full circle stands for "connection code". The small full squares mark the node positions.

(Figure 2.5). A passing member is by definition always connected, otherwise it will be considered unconnected to the node.

If a node is one of the two end nodes of the member, the member is connected to it, and connection codes must be considered. A member may or may not have, at each extremity, a connection code. The algorithm described at the previous section will end searching for more elements of a member, particularly if there is no connection code at the extremity. This will happen if:

- a discontinuity of alignment, material, cross-section or orientation is found
- no further beam element is connected to the end considered.

If the member *has a connection code* at one end, it is said to be *interrupted* at that end, or interrupted at a node (Figure 2.5). An interrupted member at a node is always connected to the node. Members interrupted at one end may or may not have there an end release. If they have one end release, they are said to be *released*, and cannot be, by definition, fully resistant members.

If a member *has no connection code* at one end, it is said to be *cuspidal* at that end, or cuspidal at a node. The word "cuspidal" is semantically related to the existence of a cusp, as the axis line comes to possibly meet other structural elements with no modification or smoothing ("connection"). As will soon be seen, such members are natural candidates for being the main member of a set of connected members.

Summing up, a member at a node can be:

- unconnected
- connected, divided into:
  - passing
  - cuspidal
  - interrupted, divided into:
    - released or
    - unreleased.

2.2.6 Member Mutual Alignment Coding

A useful classification can be set up when considering two members meeting at a node, depending on the mutual alignments of their axes (member axis and cross-section principal axes). This classification will be later used to help identifying nodes where member connections are similar.

Given an axis i of a member a, the following symbol will be used for its vector: $\mathbf{v_{ai}}$. Axis i can be axis 1, 2, or 3. The vector $\mathbf{v_{ai}}$ is a vector having unit norm.

Given an axis i of a member a, having unit vector $\mathbf{v_{ai}}$, and an axis j of a member b, having unit vector $\mathbf{v_{bj}}$, the two vectors are aligned if their dot product has modulus 1, allowing for some small tolerance. If this tolerance is *tol*, then $\mathbf{v_{ai}}$ and $\mathbf{v_{bi}}$ are aligned if and only if

$$\left|\left|\mathbf{v_{ai}} \cdot \mathbf{v_{bj}}\right| - 1\right| < tol \tag{2.1}$$

If the condition holds true, then members a and b are said to be *aligned ij*. In considering member alignment, member a will be considered the reference (*master*) and member b will be related to it (*slave*). So the alignment classification ij (j to i) may be named differently by ji (i to j).

We can set up a member-alignment classification according to the following list:

- Members a and b are aligned 11-22-33: *homogeneous splice alignment*
- Members a and b are aligned 11-23-32: *inverted splice alignment*
- Members a and b are aligned 11: *generic splice alignment*
- Members a and b are aligned 22-13-31: *normal homogeneous by 2 alignment*
- Members a and b are aligned 33-12-21: *normal homogeneous by 3 alignment*
- Members a and b are aligned 22: *inclined homogeneous by 2 alignment*
- Members a and b are aligned 33: *inclined homogeneous by 3 alignment*
- Members a and b are aligned 23-12-31: *normal inverted by 2 alignment*
- Members a and b are aligned 32-13-21: *normal inverted by 3 alignment*
- Members a and b are aligned 23: *inclined inverted by 2 alignment*
- Members a and b are aligned 32: *inclined inverted by 3 alignment*
- Members a and b are aligned 12: *to end by 2 alignment*
- Members a and b are aligned 13: *to end by 3 alignment*
- Members a and b are aligned 21: *to 2 by end alignment*
- Members a and b are aligned 31: *to 3 by end alignment*

Note that a simple alignment couple (*ij-kl*) cannot exist: the proof follows.

Let two members have alignments ij and kl. Let the third axis of member a be s: it must be perpendicular to i and k, which are also perpendicular to one another. On the other hand, if the third axis of member b is r, it must be perpendicular to j, which is aligned to i, and to l, which is aligned to k. So, it must also be as s, and then aligned with s, and the member a and b will be aligned *ij-kl-sr*.

There are six possible three couple alignments, and nine possible one couple alignments, no other alignment is possible: they are all listed.

Of course, two members can have no alignment at all.

When there is an alignment 11, members are *splice aligned*, that is, their axis lines are aligned.

Two not splice aligned members belong to a plane, and in that plane their axis lines form an angle. Trying to modify that angle, in the same plane, implies a rotation, which, depending on the alignment and on the connection type, may be a rigid rotation or a bending about a weak axis (usually axis 3); a rigid rotation or a bending about strong axis (usually axis 2); or a rigid rotation or bending about a different axis, not being principal for the cross-section.

When member axes 2 or 3 are aligned, the alignment is said to be "by strong [axis]" or "by weak [axis]" respectively. For aligned cross-sections of the type I, H, T, C, which are very frequent "by strong" may replace "by 2" and "by weak" may replace "by 3".

- Alignment **22** is *homogeneous* "by strong" or "by axis 2".
- Alignment **33** is *homogeneous* "by weak" or "by axis 3".
- Alignment **23** is *inverted* "by strong" or "by axis 2" (relative to reference).
- Alignment **32** is *inverted* "by weak" or "by axis 3" (relative to reference).

By repeatedly evaluating the condition expressed by Equation 2.1, for member axis 1, 2, 3, it is possible to assess the alignment codes of two members.

Figures 2.6 and 2.7 display two alignment examples.

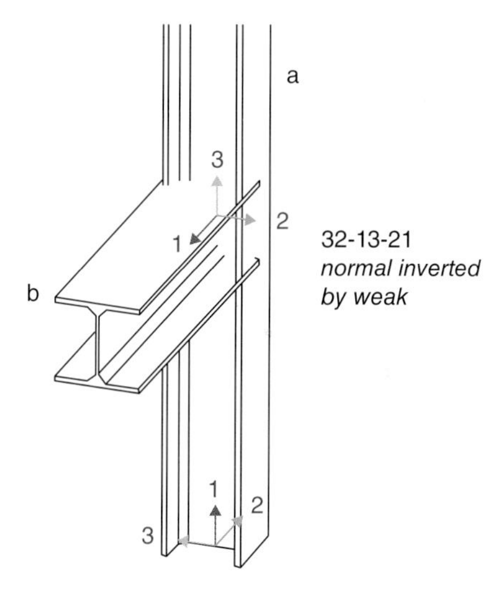

Figure 2.6 Two members alignment: 32-13-21.

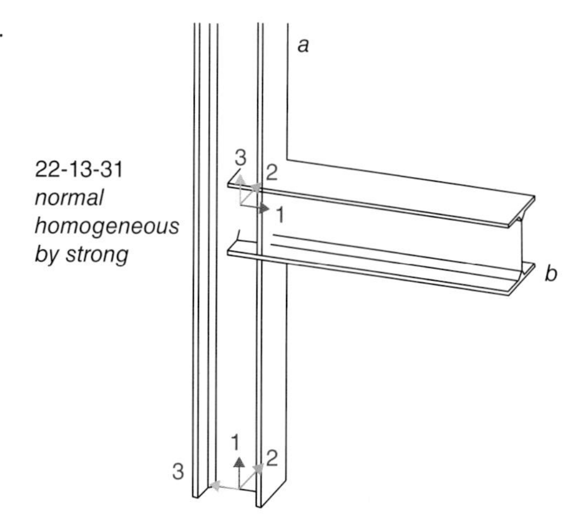

Figure 2.7 Two members alignment: 22-13-31.

2.3 Jnodes

2.3.1 Need for the Jnode Concept

Not all the nodes of a finite element model are places where members are connected. Most nodes of a finite element model are not related to any member connection. However, there are nodes in a finite element model that are also a place where:

- members connect to one another (e.g. a beam to column connection)
- members connect to other parts of the structure (e.g. a beam connected to a wall modeled by plate–shell elements)
- members connect to the reference system (e.g. a column base node, fully fixed).

So, it would be better to distinguish these nodes which are nodes where members are connected, between themselves or to *something*, from the simple normal nodes of a finite element model (BFEM). In order to do that, the term *jnode* (pronounced *dʒei-node*) is introduced here. The aim is to avoid misunderstandings and to quickly deliver a specific meaning.

The jnode concept is also introduced for further reasons.

Suppose we have a BFEM with hundreds of beam and truss elements, hundreds of members to be connected. From an engineering point of view, one of the first questions you would pose, as structural engineer, is how many different connections are there and how much are they loaded? To answer to such question you would ideally do the following:

1) Find all the members in your BFEM.
2) Find all the jnodes where members are connected.
3) Find all the identical jnodes, that is, the jnodes where the same members (as per cross-section and material) connect in the very same way (that is, with the identical mutual orientation and alignments in space). The identical jnodes will be considered as one jnode only, and will be distinguished as being different *instances* or *occurrences* of that jnode.
4) Consider the members that play the same structural role in the different occurrences of the jnode, and compute the relevant internal forces, when all the loading combinations are evaluated, as well as all the instances, to design the connections accordingly.

Step 1 was considered in the previous section: automatic member detection. Steps 2–4 can be fully automated, once the jnode concept is enriched by a wider set of information.

To this end, a jnode *instance* is defined as the set of the following information:

1) the *node number* in the BFEM related to the instance of the jnode
2) the *numbers of the members connected at the node*, identified in the global member vector by a number or some other kind of identifier
3) the local axes *orientation* of each member (inherited from the underlying beam or truss elements – see Appendix 1)
4) the *cross-section* of each member of the jnode (inherited from the underlying beam or truss elements)
5) the *material* of each member of the jnode (inherited from the underlying beam or truss elements)

6) the *connection code* (if any) and the *end releases* (if any) of the members at the node (inherited from the underlying beam elements)
7) a flag marking whether the node of the jnode is constrained or not, and how.

The previous information, together, is not enough to assess how the connections will be physically constructed. However, this information is completely sufficient to distinguish each jnode from all the others, and to identify the equal jnodes in a BFEM.

One jnode can be physically constructed in many different ways: welded, bolted, using stiffeners or not, using fin plates or angles, and so on. When considering all these details, a new term must be introduced, to avoid misunderstanding: the term is *renode*, which stands for *real node*.

So in this book these terms will all be used, with a well defined, different meaning:

- *node*: a node in a finite element model, usually the BFEM
- *jnode*: a node where at least one member must be connected to something else, and all the BFEM wireframe information related to it
- *renode*: the physical description of all the constituents, work processes, connectors, and computing choices that enable us to physically realize the connections and to check them

As *renodes* can also be parameterized, we will also be use the term *prenode*, which stands for *parameterized real node*.

2.3.2 Jnode Definition

Now that the concept has been explained it is possible to define the jnode more formally:

> *A jnode is a* node *of a finite element model where at least one member is connected to one or more different members or to a different structural part. It embeds the information related to the original* node *position and constraint, and to the connected members' unique identifiers, cross-sections, materials, orientations, connection code and end releases. If in a structure there are two identical jnodes, they will be considered as two different instances of the same jnode, and a unique mark will be assigned to them.*

The jnode is thus an intermediate step between the *node* in the finite element sense, and the *renode*, that is, the actual connections in the 3D world, with all the information about the constituents, connectors, bevels, cuts, and so on.

As each jnode instance is related to a single node, and its members are made by finite elements, it is clear that by knowing the jnode it is also possible to extract the internal forces flowing into members. This is of course of the utmost importance for connection analysis.

It is very important to be able to match equal jnodes, and to assign them to different instances of the same jnode. In this way, the renode will be created and checked only once, but considering all the relevant forces coming, instance by instance, from all the jnode instances.

Otherwise, if a jnode is repeated n times, it should have been constructed n times, and checked n times. It is implicitly assumed that equal jnodes will lead to equal renodes.

If this is not the case, it is always possible to split jnode occurrences into more jnodes, or also to distinguish between identical jnodes by means of the color of the connection code (see Section 2.7.2).

2.4 Jnode Analytics

2.4.1 Classification of Jnodes

It is very useful and very important to classify a jnode according to the existing relationships between the members it is made of, and the underlying node. This classification will be used when the renode is analyzed, so it is important to understand well how it works.

2.4.2 Simple Jnodes

The jnodes having just one member, and necessarily constrained to the reference system or to other non-member structural part, at the jnode-node, are said to be *simple* (Figure 2.8).

Simple jnodes are typically isolated column base nodes, or isolated beams with beam-to-wall connection. However, they can also be tendons, or struts, or ties.

2.4.3 Hierarchical Jnodes

The jnodes having just one member, cuspidal *or* passing, are named *hierarchical.* The cuspidal or passing member is named the *master member,* and all the other members are named *slave members.*

These jnodes are by far the most frequent jnodes in connection analysis. The master member can be seen as the member receiving the others, as in principal and secondary beams.

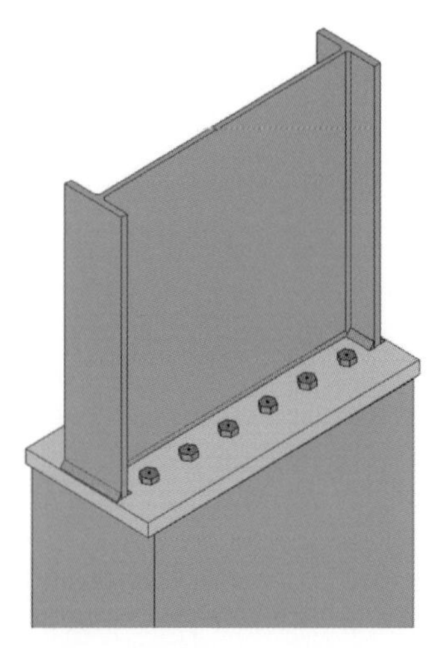

Figure 2.8 Example of a simple jnode in the real world: the column is constrained to a concrete block.

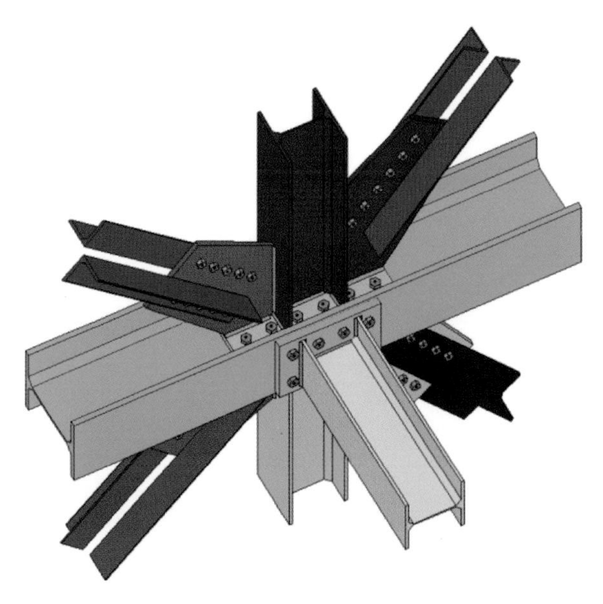

Figure 2.9 Example of a hierarchical jnode in the real world: the master is the horizontal (green) beam member, which is passing. All other members are interrupted (courtesy AMSIS Srl, Rovato, Italy).

In this book, the master member of hierarchical jnodes will always be the first member, i.e. member number 1, $m1$, in the jnode-instance member array.

It is very important to stress that no spatial feature is assigned to cuspidal or passing members: they can be inclined in any way in space, and the columns are not necessarily in "beam" to "column" connections. A vertical element may well be interrupted at a node and connected to a passing beam (Figure 2.9).

2.4.4 Central Jnodes

If all the members in a jnode are interrupted at the node, the jnode is named *central* (see Figure 2.10).

This type of jnode is determined when several members are connected to a central constituent, usually a gusset plate, but more generally every force transferrer. There is no special hierarchy between the members: in some sense, they all are equal, as they will be interrupted and connected to the very same constituent or to a set of constituents.

Once more, the spatial layout of the members is not important. What is relevant is just their connection coding.

2.4.5 Cuspidal Jnodes

If in a jnode more than one member is cuspidal at the jnode node, or if there is at least one cuspidal member and also a passing member, the jnode is named *cuspidal*.

These jnodes can always be avoided by properly assigning the connection codes to the finite elements. They introduce some ambiguity because more than one member should ideally arrive at the node with no modification (interruption), which is physically

Figure 2.10 Example of a central jnode in the real world: there is no master member, all members are interrupted and attached to the same constituent.

impossible. Also, they can be seen as missed hierarchical jnodes, because more than one member is a candidate to be master.

Generally speaking, there is no rule to decide which must be the unique cuspidal member. However, if some simple rule is set up, then an automatic search routine would be able to automatically apply the missing connection code to the slave members, in order to get an hierarchical jnode.

Possible rules can be:

- If two cuspidal members have identical cross-section, apply the connection code to the less important, considering:
  - vertical members more important than others
  - members aligned with global axis more important than members inclined
  - x (or y) members more important than y (or x); this may also be used considering constituents.
- If two cuspidal members have different cross-sections, apply the connection code to that having smaller area.

These rules are absolutely notional, and could possibly lead to improper situations. So, it would always be best to explicitly drive the automatic jnode classification, assigning the connection codes properly.

2.4.6 Tangent Jnodes

If in a jnode there is more than one passing member, the jnode is named *tangent*. Tangent jnodes are not frequent, as they imply two members that cross one another with no interruptions in their axis lines, but that are however connected. Usually this means weak connections and strong eccentricities.

The main example of a connection which seems like this one is the connection of a purlin of a roof with its transverse lattice chord. This connection is quite common, and it is usually designed with bolts put in place to avoid the displacement of the purlin, and to assure the load transfer by contact. Modeling this connection in a BFEM is not immediately obvious, because the rotation of the purlin axis line, being continuous, should not become a torsion in the chord. It is not possible to guarantee this kind of behavior (continuity of both members, but no continuity between rotations), with just one node. Usually a vertical connecting element will be added, leading to two different jnodes: the lower one, between the connector and the chord; the upper one, between the connector and the purlin. As the vertical connector will be released, it will become the slave in both jnodes, which will be hierarchical.

A tangent jnode can be transformed into a hierarchical one simply splitting one of the passing members into two parts, applying connection codes. The continuity can then be restored, assigning a dummy penetration complete welding excluded by the checks.

2.4.7 Constraints

A jnode is said to be a *constraint* if its underlying node is fixed to the reference system, or if other structural elements, which are not modeled by beam and truss finite elements, are connected to it.

If some of the degrees of freedom of the node are fixed, then the constraint is said to be *fixed*. This means that at least one degree of freedom of the node is null. It must be underlined that only degrees of freedom related to some structural stiffness need to be considered. Spurious constraints applied to avoid singularity need not be considered.

If at least one element of the following finite element types is connected to the jnode node, then the jnode is said to be an *elastic constraint*:

- springs (translational or rotational)
- membrane
- plate–shell (thin or thick)
- solid (bricks, tetrahedrons, etc.)
- links
- others, not being beam or truss elements

The term “elastic” does not mean that the surrounding finite elements have to have an elastic material or constitutive law. In this very specific context, the term will be used to mean that there will be some reaction exerted by these members, which will depend on the amount of displacement.

A jnode can be a constraint both fixed and elastic. This happens if some degree of freedom is fixed, while some other is connected to other finite elements stiffnesses.

If a jnode is a constraint, there must be *something* which will exert forces over the members connected to it. As it will be seen better later, this *something* can be modeled with no loss of generality by a *constraint block*, which mimics the surrounding part of the structure. Generally the constraint block is a rectangular box, which mimics a slab, a wall, a plinth, or something else; however, the constraint block can have any shape.

As the node must be in equilibrium, once the internal forces at the members-ends are known, the force exerted by the constraint is readily computed.

2.4.8 Summary of Jnode Classification

Table 2.1 lists the jnode classification that has been proposed in the previous sections.

2.4.9 Setting Connection Codes: Examples

In this section some example of jnodes will be provided (Figures 2.11 to 2.17), considering the connection codes of beam elements. For simplicity, it will be assumed that all members have the same cross-section and material, and that all splice-aligned members are equally oriented.

Table 2.1 Jnode classification.

Topological classification	Related constraint classification
Simple	Always constrained
Hierarchical	Constrained or not
Central	Constrained or not
Cuspidal	Constrained or not
Tangent	Constrained or not
Constraint classification	**Related topological classification**
No constraint	All but simple jnodes
Fixed constraint	Any
Elastic constraint	Any
Mixed Constraint	Any

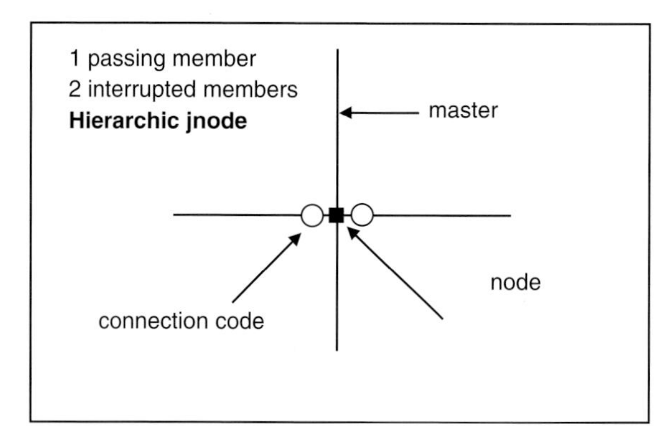

Figure 2.11 Hierarchic jnode. There is a vertical passing member, as no connection code has been specified.

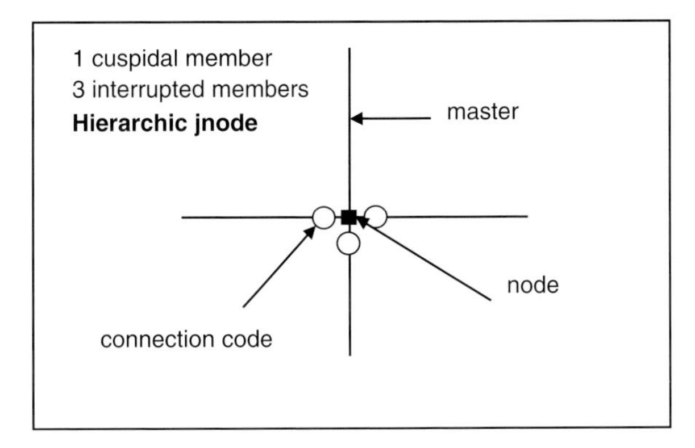

Figure 2.12 Hierarchic jnode. The upper vertical member is the master. All other members are interrupted, and so are slaves.

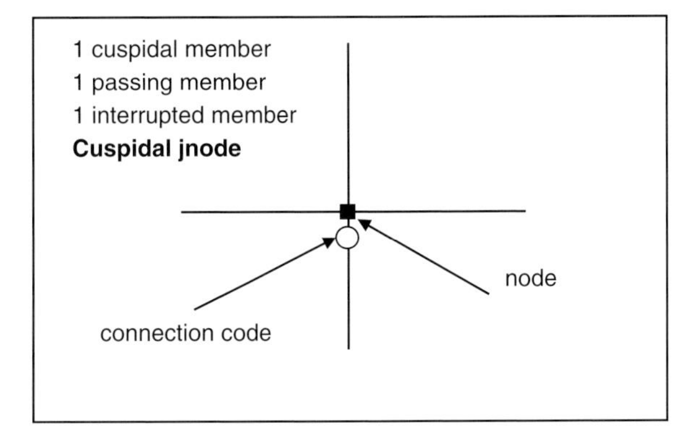

Figure 2.13 Cuspidal jnode. There are two possible candidates for being a master member: the horizontal passing member, and the vertical cuspidal member. So, there is no clear hierarchy and the jnode is cuspidal.

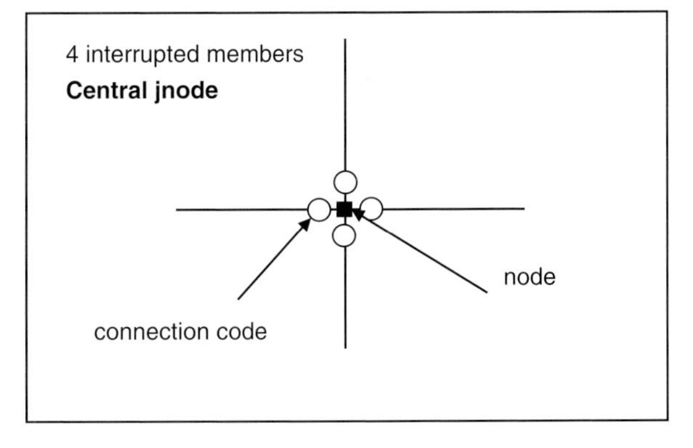

Figure 2.14 Central jnode. All the members are interrupted, so the jnode is central: it has no master.

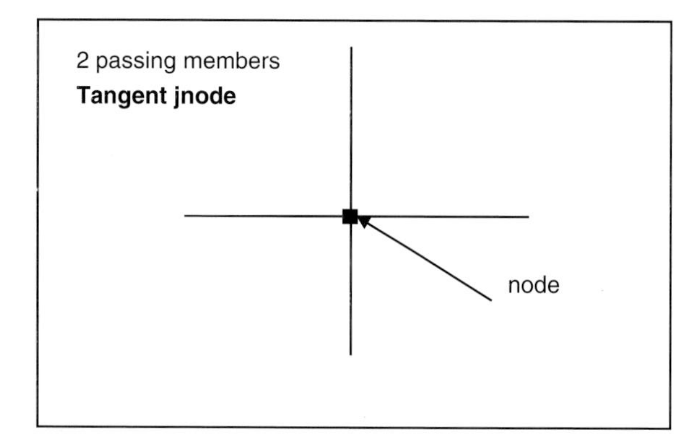

Figure 2.15 Tangent jnode. There are two passing members as no connection code has been specified, and the cross-section, material and orientation of splice-aligned members are the same. There is no clear master member.

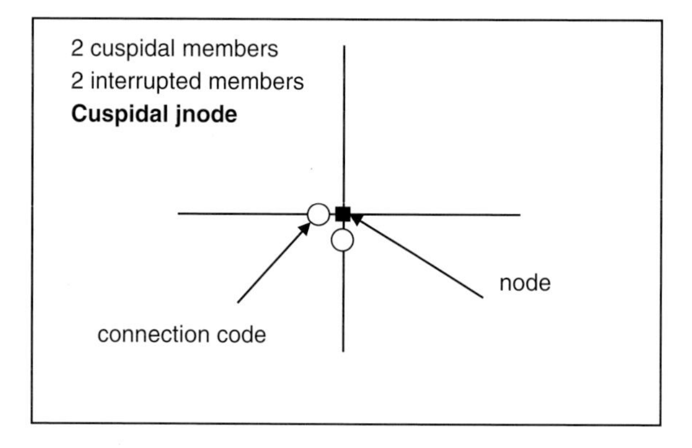

Figure 2.16 Cuspidal jnode. Here there are two cuspidal members and two interrupted members. No master can be clearly set.

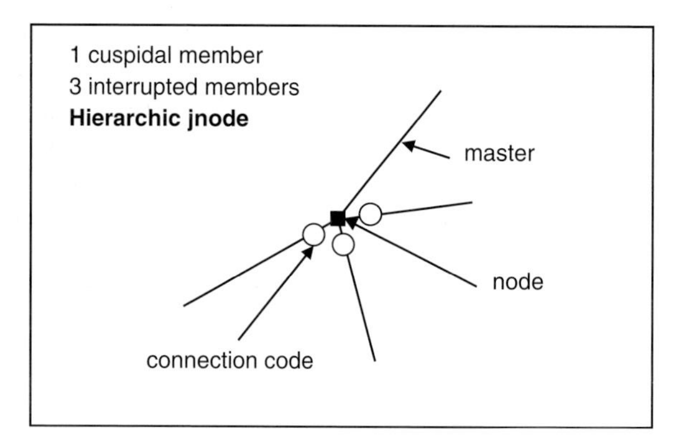

Figure 2.17 A hierarchical jnode with generic inclinations. This jnode is topologically equal to that of Figure 2.12.

To transform the jnode of Figure 2.13 into a hierarchical jnode, it would be enough, with no change of the static working mode of the finite elements, to assign a connection code to the cuspidal vertical member, or to assign two connection codes at the horizontal member elements ends.

When considering the scheme of Figure 2.16, to get a hierarchical jnode it would be enough to apply a connection code to one of the two cuspidal members, the horizontal or the vertical. The correct choice will depend on how the structure is built. In general, it is always possible to avoid a cuspidal jnode by properly assigning the connection codes.

Considering all the examples of the figures, it must be once more underlined that no specific orientation in the 3D world has been considered and that there may be any number of connecting members. Members are drawn vertical and horizontal, and at right angles, merely for the sake of drawing clearness. For instance, considering the scheme of Figure 2.12 we would get the same result of hierarchic jnode.

If also diagonal truss elements are added, the schemata jnode-classification can be upgraded easily, as by definition a truss element is always interrupted.

2.5 Equal Jnodes Detection

2.5.1 Toponode

In order to quickly classify jnodes, some specific data structures can be useful. The *toponode* is a piece of information related to jnode topology which embeds the following data:

- the number of members connected at the jnode (*nmembers*)
- the number of members made by beam elements (*nbeams*)
- the number of members made by truss elements (*ntrusses* = *nmembers* – *nbeams*)
- the number of passing members in the jnode (*npassing*)
- the number of cuspidal members in the jnode (*ncuspidal*)
- the number of interrupted members in the jnode (*ninterrupted*)
- a *constraint code*, which can be set to “no constraint”, “elastic constraint”, “fixed constraint”, or “both elastic and fixed constraint”
- a *constraint mask* (six Boolean codes)

Given two toponodes *ta* and *tb*, they will be said to be *equal* if and only if all the data of the toponode *ta* matches the related data of toponode *tb*. Checking the equality of two toponodes is a very quick filter to assess if two jnodes could possibly be equal.

The strict equality of the constraint mask or of the constraint code is in order to consider as equal jnodes having exactly the same constraint relationship with the surrounding structure, no matter what the layout of the members is. This condition might be relaxed to also consider as equal jnodes having different constraints (e.g. a clamped or hinged column will probably imply different renodes, but this assumption might be avoided).

2.5.2 Jnode Data Structure

The following data structure can be set up in order to store the information referring to a single jnode, with all its instances:

- the number of instances of the jnode (*nnodes*)
- the number of members of the jnode (*nmembers*)

- the vector of the node numbers related to the instances (one node number for each instance, array of *nnodes* element)
- the vector of all the member numbers of the jnode, for each instance (*nnodes* arrays of *nmembers* elements each; master members are by definition the first members in the arrays)
- the toponode of the jnode

2.5.3 Superimposable Member Couples

Given two jnodes A and B, let us consider two couples of members, one belonging and connected to jnode A, and one belonging and connected to jnode B. Let:

- Jnode A member couple be made by members i and j
- Jnode B member couple be made by members k and l.

If the related members, i and k, on the one hand and j and l on the other hand, have the same:

- cross-section
- material
- end release code (if any),

then it is perhaps possible that one couple can be obtained by properly applying a rigid rotation and translation to the other couple. To assess if this is true, the following steps should be made.

Given the matrix orientation of a member a, $\mathbf{T_a}$ (Appendix 1), and the matrix orientation of a member b, $\mathbf{T_b}$, it is always possible to find a matrix $\mathbf{Q}$ which can transform $\mathbf{T_a}$ into $\mathbf{T_b}$ by the matrix operation:

$$\mathbf{QT_a} = \mathbf{T_b}$$

The explicit form of matrix $\mathbf{Q}$ can be easily obtained by post multiplying by matrix $\mathbf{T_a}^T$

$$\mathbf{QT_aT_a^T} = \mathbf{Q} = \mathbf{T_bT_a^T}$$

The matrix $\mathbf{Q}$ is orthogonal, as can easily be proved:

$$\mathbf{Q^TQ} = \mathbf{T_aT_b^TT_bT_a^T} = \mathbf{T_aT_a^T} = \mathbf{I}$$

Each member of the couples has its own orientation matrix, so there will be:

- $\mathbf{T_{Ai}}$, the orientation matrix of member i of jnode A
- $\mathbf{T_{Aj}}$, the orientation matrix of member j of jnode A
- $\mathbf{T_{Bk}}$, the orientation matrix of member k of jnode B
- $\mathbf{T_{Bl}}$, the orientation matrix of member l of jnode B.

It is always possible to compute an orthogonal matrix $\mathbf{Q}$ which transforms $\mathbf{T_{Ai}}$ into $\mathbf{T_{Bk}}$, writing:

$$\mathbf{Q} = \mathbf{T_{Bk}T_{Ai}^T}$$

If the transformation of the orientation matrix $\mathbf{T_{Aj}}$ by the matrix $\mathbf{Q}$ used for members i and k gives the orientation matrix of member l, that is, if

$$\mathbf{QT_{Aj}} = \mathbf{T_{Bl}}$$

then the two couples *of orientations* are said to be *superimposable*. This means that it is possible to superimpose the two orientations of the two members of jnode B, to the two orientations of the members of jnode A, by simply applying a rigid rotation, which is equal for members i and j.

In order to prove the possibility of superimposing the two member couples, also the rigid offsets applied to members, if any, must be transformed and checked. These are vectors from the node to the element extremity.

If the rigid offset of member i of jnode A is $\mathbf{o}_{\mathrm{Ai}}$, two first necessary conditions for the two couples to be superimposable are

$$\|\mathbf{o}_{\mathrm{Ai}}\| = \|\mathbf{o}_{\mathrm{Bk}}\|$$
$$\|\mathbf{o}_{\mathrm{Aj}}\| = \|\mathbf{o}_{\mathrm{Bl}}\|$$

These conditions are quick to check if the length of the rigid offset is directly available.

Then, it must be checked that the same orthogonal transformation that was found with the orientation, that is $\mathbf{Q}$, can be applied also to rigid offsets:

$$\mathbf{Q}\mathbf{o}_{\mathrm{Ai}} = \mathbf{o}_{\mathrm{Bk}}$$
$$\mathbf{Q}\mathbf{o}_{\mathrm{Aj}} = \mathbf{o}_{\mathrm{Bl}}$$

The identity of vectors (as well as that of matrices) must be proved considering a proper tolerance.

2.5.4 Criteria to Assess Jnodes Equality

Let A and B be two jnodes in a structure. In this section the criteria needed to assess whether the two jnodes can be considered as two instances of the same jnode, will be outlined.

The first necessary condition for the equality of jnodes is that their toponodes are equal, but this condition is not sufficient. It must be stressed that if the toponodes are equal, then the topological classification of the two jnodes is also equal. So the first condition is:

1) The two toponodes of jnode A and B must be equal.

Considering jnode A and jnode B, members can be ordered in different ways. If the jnodes are hierarchical, the first member is fixed for both jnodes. If the jnodes are central, however, the first member is arbitrary. If there are n members in the jnode, these can be ordered in many ways, ideally $n!$. If the jnode is hierarchic, the orderings are $(n-1)!$.

Considering jnode A, one of all possible orderings of members can be arbitrarily chosen, saving the master member position 1 for hierarchical jnodes.

Now considering jnode B, let us assume that an ordering has been set. That ordering is correct and the two jnodes A and B are equal if all these necessary conditions are satisfied:

2) Given a member m_{Ai}, and the related member m_{Bi}, the cross-sections, material, connection codes, and end release, if any, of the two members must be equal, for $i = 1$, to n.
3) The member couples $m_{A1} - m_{Ai}$ and $m_{B1} - m_{Bi}$ must be *superimposable*, for each $i > 1$ (see Section 2.5.3).

In the next section we will outline an algorithm that is able to detect all equal jnodes in a BFEM, starting from scratch.

2.5.5 Algorithm to Find Equal Jnodes

A first loop over the BFEM nodes is aimed at finding and selecting all the nodes that are also jnodes. This task can be accomplished according to the following pseudocode:

```
UnselectAllNodes();
int count = 0;
For(each Member i)
{
        Node A = (Member i).GetNode(1));
        If(IsSelected(Node A) == FALSE)
        {
                If(IsConstrained(Node A) ||
                (GetNumMemberConnected(Node A) > 1))
                {
                        SelectNode(Node A);
                        count++;
                }
        }
        Node B = (Member i).GetNode(2));
        If(IsSelected(Node B) == FALSE)
        {
                If(IsConstrained(Node B) ||
                (GetNumMemberConnected(Node B) > 1))
                {
                        SelectNode(Node B);
                        count++;
                }
        }
}
njnodes = count;
```

where the function **IsConstrained** must assess if:

- a node has some not-spurious degree of freedom fixed
- a node is attached to some element not being truss or beam element.

At the end of the previous block, all the nodes being jnodes will be selected.

The next loop is to build the toponode data structure for each jnode:

```
ToponodeArray = AllocateToponodeArray(njnodes);
For(each selected Node i)
{
        ToponodeArray[i] = BuildToponode(Node i)
}
```

Finally, the procedure is going to find, for each jnode, all the other instances of that jnode, marking them to avoid future search.

```
JnodeArray = new EmptyJnodeArray();
For(each selected Node i)
{
```

```
        JnodeA = new Jnode();
        JnodeArray.Add(JnodeA);
        tA = ToponodeArray[i];
        memA = new IntArray(tA.nmembers); // array of member ids
        memA[0] = masternumA; // memA[0] = ma is the master-member
        number
        For(k=1; k < tA.nmembers; k++)
        {
                sa = GetMemberJoined(Node i, k+1); // kth member
                connected to Node i in Jnode A
                memA[k] = sa; // sa is the member number
        }
        JnodeA.AddInstance(Node i, memA); // adds an instance
        to Jnode
        For(each selected Node j > i)
        {
                tB = ToponodeArray[j];
                If(tA == tB)// Toponodes identity
                {
                        If(AreShapeMaterialEqual(Node i, Node j,
                        tA, tB))
                        {
                                memB = new IntArray(tB.nmembers); //
                                empty array of member ids
                                If(AreDefinitelyEqual(Node i, Node
                                j, tA, tB, memA, memB)
                                {
                                        JnodeA.AddInstance(Node
                                        j, memB);
                                        Unselect(Node j);
                                }
                        }
                }
        }
}
```

Two functions must now be explained: **AreShapeMaterialEqual()** and **AreDefinitelyEqual()**.

The first of the two functions is merely an accelerator. It is used to avoid unnecessary calls to the second function. **AreShapeMaterialEqual()** basically verifies that for each member in jnode A, there is another member in jnode B, having the same cross-section and material. If at least one member of jnode A has no related member in jnode B having the same cross-section and material, then of course the two jnodes cannot be equal.

If the function **AreDefinitelyEqual()** is called, then it is already probable that the two jnodes A and B are equal. However, this must be proved by a detailed search. The following pseudocode refers to a version of the function to be used for hierarchical jnodes. Its extension to central jnodes is left to the reader.

```
// version for hierarchical jnodes
BOOL AreDefinitelyEqual(Node A, Node B, tA, tB, memA, memB)
{
	int k=0;
	ma = masternumA;
	mb = masternumB;
	memA[k] = ma;// initializes the sequence of members with the master
	memB[k] = mb; // initializes the sequence of members with the master
	If(tA.nmembers == 1) return TRUE; // simple jnode
	If(GetMemberSection(ma) != GetMemberSection(mb)) return FALSE; //
	different masters
	If(GetMemberMaterial(ma) != GetMemberMaterial(mb)) return FALSE; //
	different masters
	For(i=1; i < tA.nmembers; i++) // first member of loop will be the second
	member
	{
		sa = GetMemberJoined(Node A, i+1);// 2nd member of A couple
		found = FALSE;
		For(j=1; j < tB.nmembers; j++)  // tA.nmembers =tB.nmembers
		{
			sb=GetMemberJoined(Node B, j+1);//2nd memb. of B couple
			If(IsMemberSelected(sb) == FALSE) // only untreated
			members
			{
				If(GetMemberSection(sa) == GetMemberSection(sb))
				{
					If(GetMemberMaterial(sa) ==
					GetMemberMaterial(sb))
					{
						If(AreCouplesSuperim posable(ma,
						sa,mb,sb,ta,tb))
						{
							k++; // one more member has
							found its place
							memA[k] = sa; // stores it
							memB[k] = sb; // stores it
							SelectMember(sb); //
							select it
							found = TRUE;
							j = tB.nmembers; // quick end
							of j loop
						}
					}
				}
			}
		}
		If(found == FALSE) return FALSE; // no match for this member
		→different
	}
	return TRUE;
}
```

The function **AreCouplesSuperimposable()** checks if the couple (*mb*, *sb*) can be superimposed to the couple (*ma*, *sa*).

2.5.6 Examples

The algorithm to detect equal jnodes has been extensively tested and used for years, and it guarantees quick and accurate results, even for complex BFEMs. In Figure 2.18 a simple example can be seen, where a 16 node BFEM has been considered, finding four different jnodes, AA, AB, AC, and AD (this last highlighted in Figure 2.18), each with four different instances.

This problem was a simple one, and also, even without an automatic search engine, it would have been solved.

At the other end of complexity scale is the problem shown in Figure 2.19, where a quite complex 3D model having thousands members has been considered.

The jnode search algorithm was applied successfully by the author acting as consultant, and it allowed us to get a clear idea of the number of different jnodes in the structure, and of the placement of all the instances. The number of jnodes was found to be equal to 569, for a number of nodes in the BFEM equal to 7450 and a number of members equal to 5464. The maximum number of instances in a single jnode was found equal to 123. The total number of instances of all jnodes was found to be 2246, so 30% of the BFEM nodes were actually also jnodes. The average number of members per jnode was 5.5, meaning that the structure was quite highly interconnected.

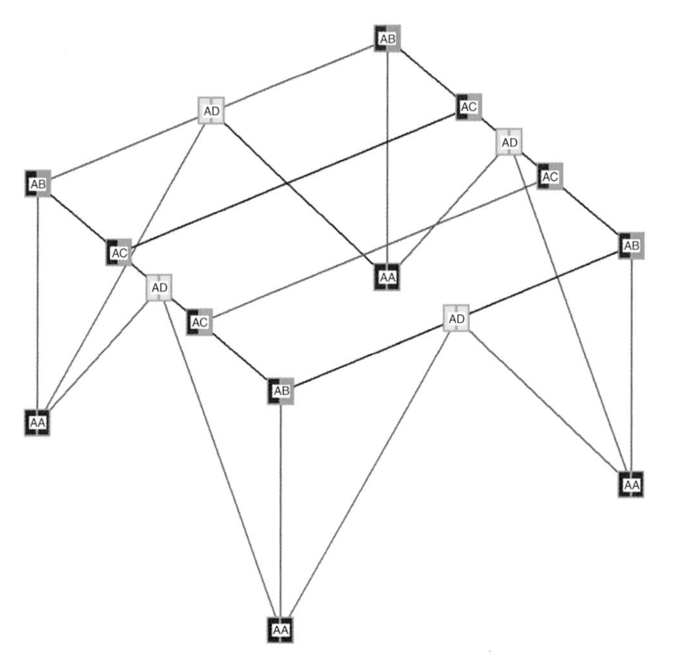

Figure 2.18 Simple example of equal jnode detection. Jnode AD (highlighted in yellow) has four instances. Also, jnodes AA, AB and AC have four instances.

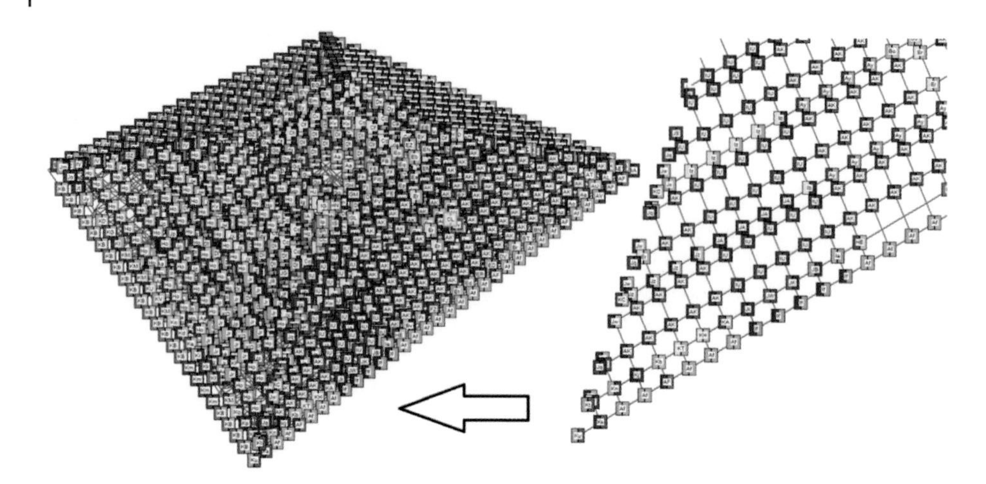

Figure 2.19 The new Mediateca in Colle Val d'Elsa, Siena (Galluzzi & Associati & Ateliers Jean Nouvel, Galluzzi Associati 2010). The steel structure was very complex and a large number of different jnodes were found (569). A detail of one of the four inclined planes is also sketched, considered as isolated from the surrounding part. Cuspidal and tangent jnodes have been discarded, so some jnodes do not have the jnode symbol. (Courtesy, Galluzzi Associati, Florence.)

The search took 130 seconds on a Intel Xeon CPU, 3.2 GHz, running under 32 bit subsystem of Windows© 7 Professional 64 bit.

The ability to deal with such a problem, believed to be one of the most complex that can be conceived in everyday structural engineering, can be considered final proof of the efficiency and effectiveness of the algorithm outlined here.

2.6 Structural Connectivity Indices

When dealing with a complex 3D structure, the availability of specific indices referring to connection analysis is very useful, with the aim of preliminarily evaluating the connection design effort.

Given a BFEM, it is assumed that the member search and the jnode search have already been done. The following numerical indices can be evaluated.

Let us first consider a single jnode.

The number of instances of the jnode is here named the *multiplicity* of the jnode. The jnode – which will be designed only once, leading to a renode, and to renode checks – will indeed refer to a number of different locations in the structural BFEM.

Then we define the *connectivity of the jnode*, the number of structural pairs to be connected. In turn this can be evaluated according to the following rules:

- hierarchic jnodes not being a constraint: $connectivity = nmembers - 1$
- hierarchic jnodes being a constraint: $connectivity = nmembers$
- central jnodes not being a constraint: $connectivity = nmembers$
- central jnodes being a constraint: $connectivity = nmembers + 1$
- simple jnodes: $connectivity = 1$

The connectivity is a quick index to evaluate the complexity of a jnode. Indeed, the number of structural pairs that must be connected is only an indirect measure of complexity, as a single connection can be so complex as to outweigh all the others. However, when considering a large number of jnodes, this metric is meaningful.

As a simple heuristic useful for formulating preliminary offers, in central or simple jnodes, the complexity of the jnode can be considered linearly varying with connectivity, while in hierarchic jnodes, the complexity must be related to some power of the connectivity, as the member-to-member connections are usually not independent of one another, as the master is the common link. So it can be assumed

- hierarchic jnodes: $complexity = k \times connectivity^m$
- central jnodes: $complexity = k \times connectivity$
- simple jnodes: $complexity = k$

The values of k and m, and the relationship between *complexity* and offer value, can be set by each consulting firm, depending on the typology of the connections, the number of combinations, the need for seismic design, the availability of software, and so on. As complexity is a notional measure, we can set $k = 1$. Besides, simple heuristic considerations applied to frequent nodes lead to the conclusion that m is usually between 1 and 2. Table 2.2 summarizes some complexity estimates for different choices of m, and for different *connectivity* values. The row referring to $m = 1.4$ has been highlighted as it seems a reasonable average choice according to the author's experience.

Assuming as reference the *connectivity* = 1 jnode, and setting $m = 1.4$, it is possible to give a rough preliminary average estimate of the relative complexity of hierarchic jnodes, as a function of connectivity (see Table 2.3). According to this heuristic, a hierarchic jnode having connectivity 5 (six members if it is not a constraint) is about 9–10 times more complex than a jnode having connectivity 1 (two members), and not just five times more complex, as linear variation would suggest.

It is now possible to consider the overall structure, and introduce more indices that might be useful to provide an estimate of the complexity of the connection analysis, and a useful measure of the structural design regularity.

The overall number of instances of all jnodes is named the *overall jnodes multiplicity*. This is also the number of nodes of the original BFEM that were found to be also jnodes.

The ratio of the multiplicity of a jnode to the overall jnodes multiplicity is the *jnode frequency*, by definition a number lower than or equal to 1. The jnode frequency is useful

Table 2.2 Heuristic complexity estimate for hierarchic jnodes of different connectivity.

Connectivity m	1	2	3	4	5	6
1.2	1	2.3	3.7	5.3	6.9	8.6
1.4	1	2.6	4.7	7.0	9.5	12.3
1.6	1	3.0	5.8	9.2	13.1	17.6
1.8	1	3.5	7.2	12.1	18.1	25.2
2.0	1	4	9	16	25	36

Table 2.3 Hierarchic jnodes relative complexity, assuming a heuristic evaluation formula with $m = 1.4$ and as reference complexity that of connectivity 2.

Connectivity	Relative complexity
1	1
2	2.6
3	4.7
4	7.0
5	9.5
6	12.3

for understanding how disseminated a jnode is in the structure, and being a relative measure can be compared between different models.

The sum of all jnodes frequencies must be 1. The average of all jnode frequencies, is the *average jnode frequency*, which is a measure of the diversity of connections in the model, and can be calculated as the reciprocal of the number of jnodes.

The *average connectivity* of all the jnodes is a measure of the average degree of connection of the jnodes in the structure. Simple, regular structures tend to have low average connectivity.

Summing up all the connectivities of the jnodes gives the *overall connectivity*, and summing up all the complexities finally gives an estimate of the *overall jnodes complexity* of a structural model. This measure can be preliminarily used to set offer values and to estimate the value of the work to be done.

It is now considered an example of application of the indices, to the structure of Figure 2.20.

As can be seen from Table 2.4, there are 12 jnodes in this BFEM. The maximum frequency is for jnode 6, AF, and it is 15.4%. The maximum connectivity and complexity is that of jnode number 2, AB: 5 and 9.5 respectively.

The average connectivity for the structure is 2.08, which is a low value. The average jnode complexity is 3, while the overall complexity is 36.

For comparison, the results from the new mediahouse in Colle Val d'Elsa "Fabbrichina" (see Figure 2.19), were the following:

- number of nodes of the BFEM: 7450
- number of members: 5464
- number of different jnodes: 569
- total number of instances of jnodes: 2246
- maximum frequency of a jnode: 5.48%
- maximum connectivity of a jnode: 13
- maximum complexity of a jnode: 36.27
- total connectivity: 2693
- total complexity: 5376
- average connectivity: 4.73
- average complexity: 9.44

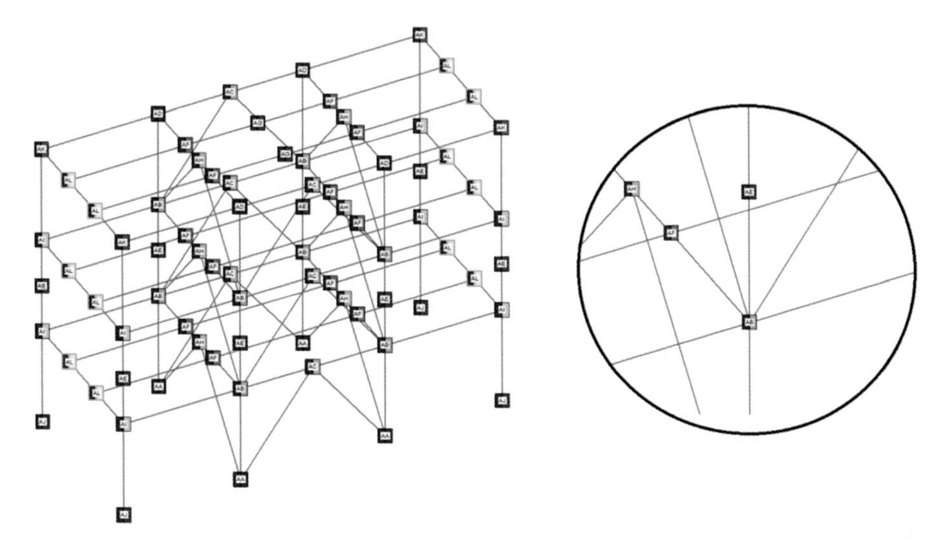

Figure 2.20 A framed structure with all the jnodes marked. Within the circle is jnode AB.

Table 2.4 Jnodes indices for the structure of Figure 2.20. N is the jnode number. Mark is the jnode mark. NO is the number of occurrences (instances) of each jnode. NM is the number of connected members. Typology is the topological classification of jnode. Constraint is the constraint status. Frequency is the frequency of each jnode. Connectivity is the jnode connectivity value. Complexity is the jnode complexity value, evaluated with $m = 1.4$.

N	Mark	NO	NM	Typology	Constraint	Frequency	Connectivity	Complexity
1	AA	4	3	hierarchic	rigid	5.13%	3	4.7
2	AB	8	6	hierarchic		10.26%	5	9.5
3	AC	6	3	hierarchic		7.69%	2	2.6
4	AD	4	4	hierarchic		5.13%	3	4.7
5	AE	8	2	hierarchic		10.26%	1	1.0
6	AF	12	3	hierarchic		15.38%	2	2.6
7	AG	2	2	hierarchic		2.56%	1	1.0
8	AH	6	3	hierarchic		7.69%	2	2.6
9	AI	8	3	hierarchic		10.26%	2	2.6
10	AJ	4	1	simple	rigid	5.13%	1	1.0
11	AK	4	3	hierarchic		5.13%	2	2.6
12	AL	12	2	hierarchic		15.38%	1	1.0

2.7 Particular Issues

2.7.1 Symmetries

When considering the identity of member orientation, no mention was made of the problem related to cross-section symmetry.

Table 2.5 Permutation of local axes saving physical layout, for cross-sections having one or two axes of symmetry.

Axis	Permutation 1	Permutation 2	Permutation 3
(1, **2**, 3)	(−1, **2**, −3)		
(1, 2, **3**)	(−1, −2, **3**)		
(1, **2**, **3**)	(−1, **2**, −**3**)	(−1, −**2**, **3**)	(1, −**2**, −**3**)

Two members having the same cross-section but different orientation matrix **T**, may indeed be the same physical member, as there is a degree of freedom related to the sign of the member axis and to the sign of the cross-section principal axis vectors.

If a cross-section has one symmetry axis, then that axis is certainly principal, and must be axis 2 or axis 3.

In this subsection, a principal cross-section axis which is also a symmetry axis will be marked in bold, e.g. **2** or **3**.

Considering the set of three axes (1, 2, 3), and assuming that one or two principal axes are axes of symmetry, a table of possible permutations can be obtained. This is Table 2.5.

Replacing the sets of axes in the first column, with one of the available sets in the second, third, or fourth column, does not imply a physical change in member orientation, but in fact implies a change in the orientation matrix **T**. So, when checking the *superimposable member couple* condition, allowance must be made for the possible existence of symmetries, and we must check the available permutations.

2.7.2 Splitting of Jnodes

During the analysis of a complex 3D structure it may happen that the same members, meeting at a node in the very same manner, and so belonging to the very same jnode, must be connected differently.

The main reason for this is that the level of internal forces may be quite different due to the different position (e.g. in elevation) of the jnode instances.

So, for this reason, it may be sometimes necessary to distinguish into different jnodes instances that would otherwise be decoded as belonging to the very same jnode.

An easy way to get this result is to assign to connection codes also a "color", and ask that in order to detect two jnodes as equal occurrences of the same jnode, also connection code colors must match. By assigning a different color to at least one member connection code of the jnode, that jnode and all the ones having the same color pattern will be assigned to a different jnode. In this way, different jnodes will possibly be defined into different real nodes (*renodes*), thus accomplishing the design requirement.

However, it may also be possible that once jnodes have been searched and detected, some instances need to be split into a different jnode. This operation is quite simple, as the jnode data structure embeds all the necessary information to generate, from a set of occurrences of a given jnode, a different jnode, marked in a different manner.

Of course, when the checks are done, each jnode will be checked according to the internal forces of the members of its current instances, not keeping track of the previous history.

2.7.3 Mutual Interaction of Different Jnodes, Jnode Clusters

Up to now it has been assumed that the analysis and checking of each real node associated to each jnode is independent of one the others. This is the by far the most probable condition, as members are long prisms having a length-to-width ratio higher than at least 8–10, and as local effects vanish quickly getting away from the connections area.

However, there are special cases in which two jnodes cannot be studied as being independent. This typically happens for hierarchical jnodes sharing the same master member, which receives slaves in different but very near points on its member axis. This condition is usually modeled in the BFEM by placing two nodes very close to each other along the member axis line. Two jnodes that interact with one another are said to be *interacting*.

A typical very important example is that of short links in eccentric bracing frames (EBFs). Here two diagonals are linked to a beam so that their axis lines are not meeting at a point over the beam axis line (as in CBF, concentric bracing frames).

This problem may be tackled in two different ways.

The first way is to use rigid offsets to take into account the difference in the arrival points of the slave members. By using this modeling trick, there will still be a single node, and a single jnode will be found.

However, when considering the EBF problem, this way of modeling has its drawbacks, as by definition the link, which should be highly deformable as it must get into plastic range, will not exist. This problem is specially related to the BFEM analysis results, not to the connection analysis. In fact, when considering the single jnode model, the forces acting over the master, being almost identical, will indeed be applied at their correct point of application, thus leading to an exact stress analysis (probably by fem) of the master, embedding the link and its stiffeners. However, the lateral displacement of the BFEM will be underestimated, and no study of the link plasticity would be accomplished, as no *link* appears in the BFEM.

The second way to tackle the problem is to keep track of the existence of interacting jnodes, so that they can be studied together. The problem of analyzing interacting jnodes will be detailed in Chapter 9. For the moment, it will be enough to know that the problem may exist, albeit that it must be considered a special not very frequent case.

It may be of some interest to outline an algorithm to detect interacting jnodes, once that the jnode vector is available. This is done in the following pseudocode.

At first all the couples of interacting jnodes along members are detected and stored:

```
int ndepths = 4; // number of depths of the cross-section to set
minimum interaction distance
For(each Member i)
{
    Mindistance = ndepths * (Member i).GetCrossSectionDepth();
    Startj = 1; // first node of the current member
    StartNode = (Member i).GetNode(Startj);
    while(IsItAJnode(StartNode) == FALSE)
    {
            Startj++;
            StartNode = (Member i).GetNode(Startj); // we will
            always find one
```

```
    }
    For(each Node j of Member i, with j > Startj, being also a Jnode)
    {
        EndNode = (Member i).GetNode(j)
        If(Distance(StartNode, EndNode) < Mindistance)
        {
            AddCouple(StartNode, EndNode); // possibly
            interacting jnodes
        }
        StartNode = EndNode;
    }
}
```

Then, the couples are scanned in order to possibly find clusters of interacting jnodes, which are the final result.

```
UnselectAllInteractingJnodesCouples();
For(i=0; i < NinteractingJnodesCouples; i++)
{
    If((Couple i).IsSelected() == FALSE)
    {
        Cluster = New JNodesCluster
        Node A = (Couple i).GetNode(1);
        Node B = (Couple i).GetNode(2);
        Cluster.Add(Node A);
        Cluster.Add(Node B);
        (Couple i).Select();
        While(Node A)
        {
            Node A= AddNodeToCluster(Node A, cluster);
        }
        While(Node B)
        {
            Node B= AddNodeToCluster(Node B, cluster);
        }
    }
}
```

Where the outline of the function **AddNodeToCluster**, follows:

```
Node AddNodeToCluster (Node A, Cluster) // Node A by val, Cluster by
reference
{
    For(i=0; i < NinteractingJnodesCouples; i++)
    {
        If((Couple i).IsSelected() == FALSE)
        {
            If((Couple i).GetNode(1) == Node A)
```

```
                        {
                                Cluster.Add((Couple i).GetNode(2));
                                // adds jnode to the cluster
                                (Couple i).Select();
                                return (Couple i).GetNode(2);
                        }
                        Else If((Couple i).GetNode(2) == Node A)
                        {
                                Cluster.Add((Couple i).GetNode(1));
                                (Couple i).Select();
                                return (Couple i).GetNode(1);
                        }
                }
        }
        return 0;
}
```

2.7.4 Tolerances

When considering the typical alignment condition between vectors, we must stress the importance of setting an appropriate tolerance value.

If the tolerance is too strict, this may avoid properly recognizing that two jnodes are equal. This, in turn, may happen when considering that the mesh nodes are often obtained by a CAD drawing, which may be quite imprecise, for a number of reasons, sush as:

- insufficient number of digits in ASCII transfer files
- improper rounding of angles or improper rounding of trigonometric functions
- improper snapping to near points during CAD operations.

If on the other hand the tolerance it too loose, this may lead to considering as equal jnodes some that are not, thus leading to possible problems in drawings and eventually in construction.

From practical experience of using these algorithms, the following settings have proved to be generally appropriate, when applied to unit modulus vectors:

$10^{-4} = 0.0001$ as the upper limit to consider a component null

$10^{-3} = 0.001$ as the upper limit of relative error to consider two components different

An unsuccessful initial experience was made, setting the first value to 10^{-6}.

2.8 Jclasses

So far, we have considered the problem of detecting equal jnodes, that is, jnode occurrences that appear in different parts of the structure, differently loaded, but that are physically superimposable on one another.

However, when considering the jnode problem, it is in fact found that jnodes that are different can actually share a lot of common features, so that their construction and checking can follow exactly the same route.

Consider, for instance, a jnode A made by two members, a master beam IPE300-S235 and a secondary beam IPE180-S235. Let the alignment be normal and homogeneous by 3.

Now consider a different jnode B, made by two members, a master beam IPE200-S355 and a secondary beam IPE140-S355. Let the alignment be normal and homogeneous by 3.

Jnodes A and B are different, but they are quite similar: the only difference is the size of the members (not their type) and the material of the members.

It is clear from this example that a new concept might be useful, the concept of *jclass*. A jclass is the set of all jnodes that differ only by one or more of these features:

- size of the member cross-sections, at equal type
- material
- inclination angle between the member axis, if this angle is not 90°

The third condition may well be better defined by assuming that the alignment code between any two related members is equal among all jnodes belonging to the same jclass. On the other hand in two jnodes belonging to the same jclass:

- The toponode structure is the same (also the number of passing, cuspidal, and interrupted members).
- The cross-section *type* and topology is the same.
- The mutual alignment codes are the same (non-normal angles may vary).
- The connection codes and end release codes of related members are the same.

The jclass concept is very useful for parameterizing the creation of renodes from jnodes. If the rules needed to construct a renode for a jnode A are known, those of the renode of jnode B, if jnode B belongs to the same jclass of A, are also known.

References

Galluzzi Associati, 2010, Fabbrichina – Mediateca Colle Val d'Elsa, Structural Design, http://www.galluzziassociati.it.

Rugarli, P. 2010, *Structural Analysis with Finite Elements*, Thomas Telford, London.

Rugarli, P. 2014, *Validazione Strutturale*, EPC Libri, Rome.

3

A Model for Connection

3.1 Terminology

In this book the problem of connection analysis will be studied with particular reference to the problem of connecting members, as defined in the previous chapter.

Considering the connection as a whole, there will be different sets of constituents, having different rules. The term "constituent" will be used to refer to a generic element of any of the sets.

The reason why the term *constituent* is used instead of the friendlier *component*, is that the term "component" has already been used by researchers in the field of connection analysis with a narrower meaning (in the *component method*). So, to avoid confusion, this book will use the term "constituent". In this work a constituent is a body, or set of bodies for connectors, which plays a mechanical role in a renode. A constituent is not a subpart or a sub-piece, so not a *web*, not a *flange*, and not a single bolt or single weld seam. In the author's opinion, the word *component* has not been used in a way fully coherent with its literal meaning. According to the Oxford Dictionary, a *component* (the word comes from the Latin *componere*, meaning "put together") is "any of the separate part of a motor vehicle, machine, etc." *Separate part* means that a component may appear only once, not more. But in the "component method", physical parts may appear twice or more, with different hats, depending on the loading: for instance "column web in compression", "column web in tension", "column web panel in shear". So, it seems that a fitter name for the *component method* would be for instance *fictitious springs method*, because that's what the method is, without extending the meaning of the word *component* far from its original meaning. This is certainly not a criticism of the method, which is brilliant, but merely to the use of the word *component* for it.

The constituents that have to be connected will be *members* or *constraint blocks*. Globally they will be named *connected main elements*. Their connections are the goal of connection design. The other sets are useful to physically realize the connections. A constraint block is only needed when the jnode is a constraint, that is, if the members are attached to another structural element, acting as restraint. It could be a wall, a concrete slab, a plinth, or more.

The constituents used to transfer forces and moments from and to the main elements will be called *force transferrers*. Sometimes, in the technical language they are also referred to as *fittings* or *cleats*. However, from the connection analysis point of view, the term force transferrer alludes to the ability to transfer forces and moments from

Steel Connection Analysis, First Edition. Paolo Rugarli.
© 2018 John Wiley & Sons Ltd. Published 2018 by John Wiley & Sons Ltd.

one constituent to another, and so it's preferable. A new synthetic term, would possibly be *fortran* (FORce TRANsferrer, a term already used in information technology for the homonymous programming language).

The new constituents, which are inserted into an existing constituent and which are only connected to that very single constituent, will be named *stiffeners*. In engineering practice, some of the constituents that will be here called force transferrers are named stiffeners, because they transfer forces and moments, but also stiffen. However, in this book, the term "stiffener" has a more specific meaning. It refers to constituents that merely stiffen other constituents without transferring forces or moments from one constituent to another. Stiffeners may indeed transfer forces from parts of a constituent to other parts of the same constituent, but they do not transfer forces from one constituent to another. Clearly, sub-parts of force transferrers, being connected to a constituent may also stiffen it, but their main goal is to transfer forces between two or more different constituents.

The constituents used to physically connect two or more other constituents are named *connectors*. These are usually made by bolts, welds, rivets, nails or other devices (fasteners) which connect two or more constituents. The number of different constituents connected by a connector will be named the *multiplicity* of the connector. Usually the multiplicity is 2, but bolts may easily have multiplicity 3 or even more. A connector connecting two constituents is said to be *simple*, a connector connecting three or more constituents is said to be *multiple*. Summing up the multiplicities of all the connectors, gives the total multiplicity of connectors. This is a very important index in connection analysis.

A connector must not be connected to other connectors. Besides, in the model of connections presented here, if the connector has sub-elements, *they all must connect the very same constituents*, and will be of the same type (all bolts, all welds, and so on).

So a connector can only connect force transferrers, stiffeners, members or constraint blocks. The connectors connecting stiffeners are named *internal connectors*. The other connectors are named *external connectors*.

Connectors can also be non-material. For instance, a body over a friction plane is somehow connected to it (for a subset of its acts of motion), so there is a connector but does not have a physical body or bodies.

The (ideal) 3D points at which a connector exchanges all the forces and moments with the constituents that it connects are named the *extremities of the connector*. The number of extremities of a connector is equal to its multiplicity.

3.2 Graphs of Connections

Sometimes it is useful to represent the logic of connections by using graphs, where connectors, connected main elements, stiffeners and force transferrers are represented with a specific symbol.

- The symbol used for connected main elements will be a rectangle or a regular polygon.
- The symbol used for force transferrers will be a circle.
- The symbol used for stiffeners will be a triangle.
- The symbol used for connectors will be a multi-arrow having as many heads as the multiplicity of the connector, and no tail. To distinguish between the bolts and the welds, if needed, a circle or a square will be applied to the arrow body.

For example, Figure 3.1 shows the graph of a simple column base plate to a plinth. The single force transferrer is the base plate. The upper simple connector is a weld layout connecting the column to the base plate, the lower simple connector is a bolt layout connecting the base plate to the constraint block.

As a second example a simple bolted splice joint will be considered (Figure 3.2).

The graph is shown in Figure 3.3.

When preparing the graphs it is important to understand that it is not important to know the point where the connector arrow-heads touch the connected parts, and that the aspect ratio of the polygons referring to the connected main elements is arbitrary, as is their orientation in the plane.

In Figures 3.4–3.7 two other examples of well known connections are shown.

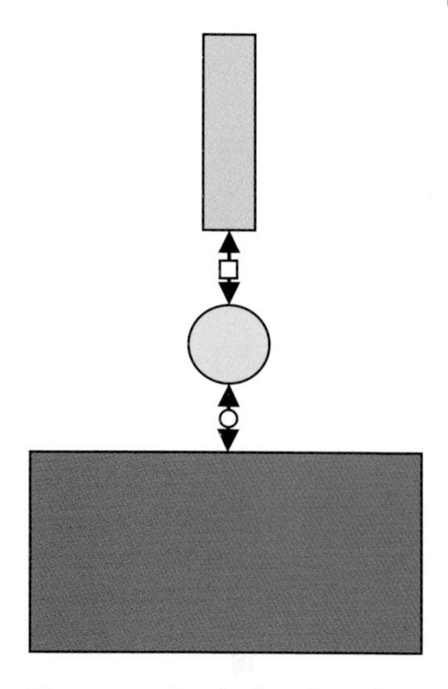

Figure 3.1 Graph of a column base plate connection.

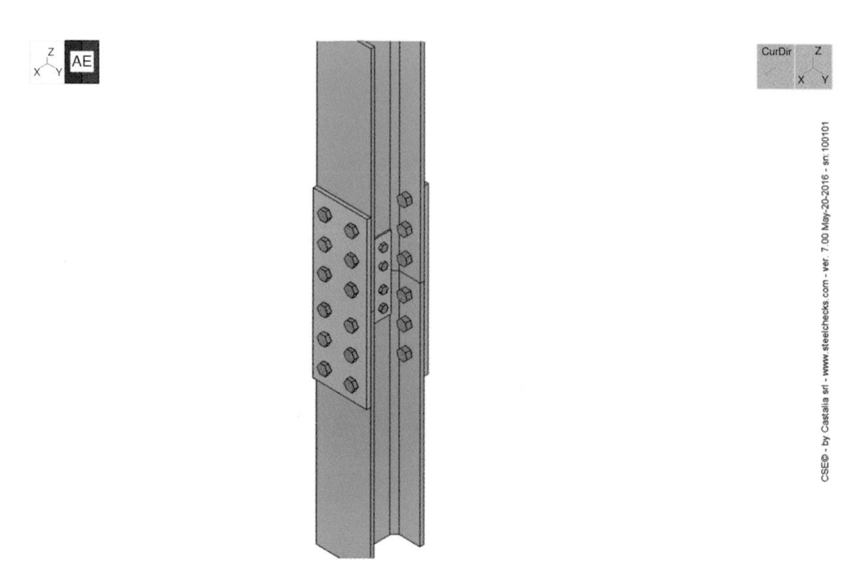

Figure 3.2 A simple bolted splice joint.

Figure 3.3 Graph of the splice connection of Figure 3.2. There are two bolt layouts connectors having multiplicity 3 and four bolt layouts connectors having multiplicity 2. The total multiplicity of connectors is 14.

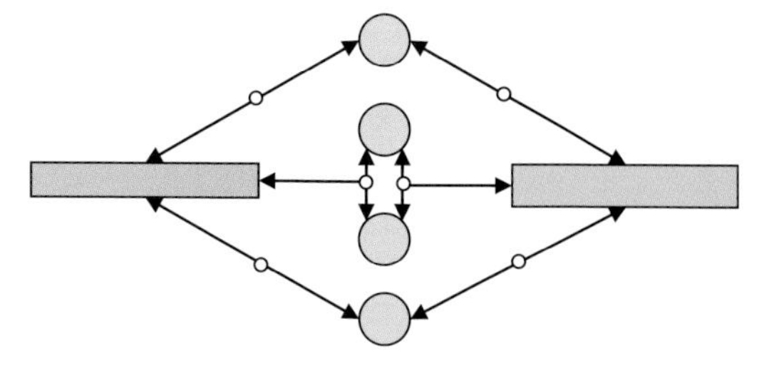

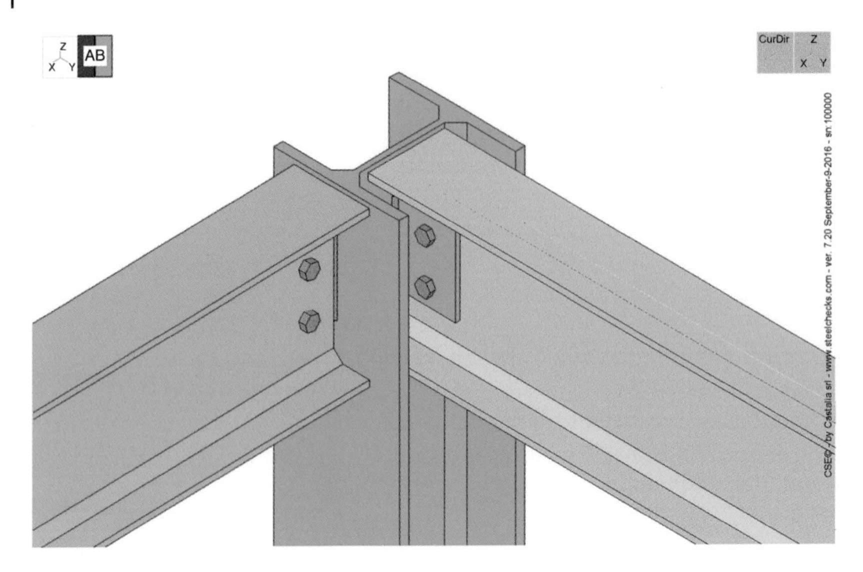

Figure 3.4 A two beam to column real node. Connection is by fin plates.

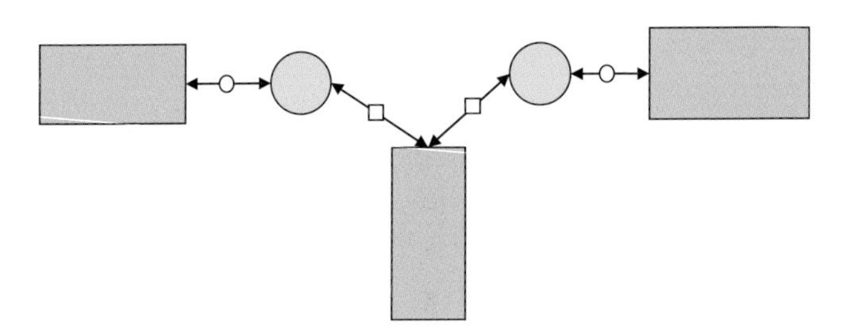

Figure 3.5 Graph of the real node of Figure 3.4.

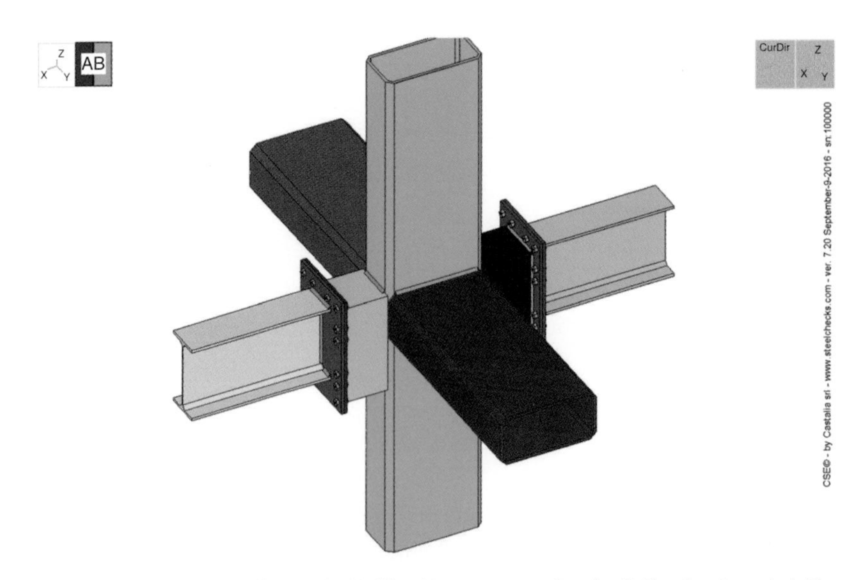

Figure 3.6 A more complex node (*Fabbrichina*, courtesy Studio Galluzzi e Associati, Florence).

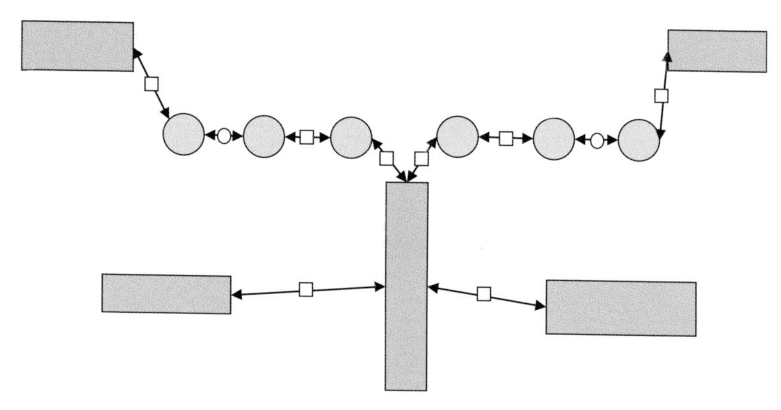

Figure 3.7 Graph of the real node of Figure 3.6.

3.3 Subconstituents vs Layouts

When considering connectors, they are commonly composed of a set of subconstituents, which all act together in an organized manner. Each subconstituent works basically like the others and by definition shares with the other subconstituents some basic features.

Connectors may well be realized by single constituents, such as pins, but they are usually obtained by adding more elementary constituents, with the aim of absorbing some set of forces and moments. These forces and moments arise when the connected constituents try to detach one another. Also, although the connectors transfer the forces from one ideal point of application to another, the geometrical extension of the connectors is usually limited, and their primary goal is to act as a link between more extended constituents.

"Connecting" means obliging two constituents to displace in the same manner, which in turn is accomplished by exchanging the necessary forces and moments needed to ensure equilibrium. The forces are often high, and single elementary connector-constituents are not able to absorb them, both due to the intensity and to the different nature of that forces (e.g. moments), therefore connectors are usually made of a number of subconstituents acting together. So, in this book the term *layout* will be used, assuming that the connector is made up of a set of organized subconstituents acting together. Most frequently the term *weld layout* and *bolt layout* will be used.

When all the subconstituents are considered as acting together, they will have to connect to the very same constituents. So if a layout is connecting constituents A and B, then all its subconstituents will connect A and B. Moreover, some rule will have to be available in order to assess the single contribution of each subconstituent to the layout, when compared to the overall structural action of the connector.

This last condition is accomplished by introducing *rules* which do not always have an exact counterpart in the experimental behavior of the connection. The introduction of rules able to predict the internal forces flowing into a single connector's subconstituents, starting from the overall connector forces is a very important step and is quite comparable to the analogous steps introduced in all the structural theories: beam, plates, plane stress, plane strain, and so on. Usually a kinematic model is introduced first.

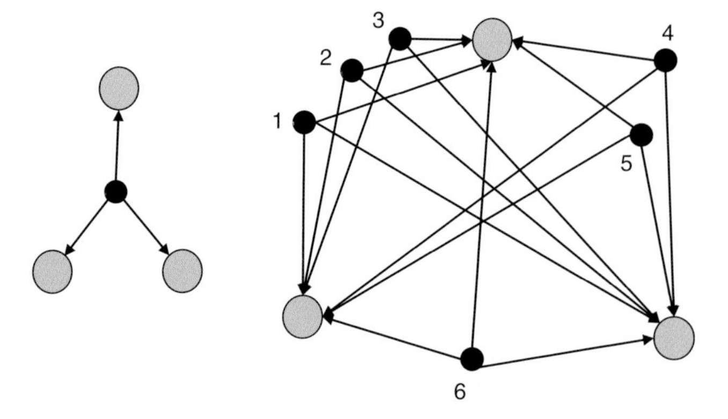

Figure 3.8 A single six-bolt layout connector with multiplicity 3 can be transformed into six elementary connectors having multiplicity 3.

This kinematic model may be suitable or not, depending on the structural problem at hand. Once the key hypothesis is introduced, a huge number of subconstituents are marshaled to behave in an organized manner and become elements of a set. This clearly applies, for instance, to the (infinite) fibers of a cross-section under axial force and biaxial bending in normal beam theory. Therefore, assuming that the subconstituents follow some rule, the approach is similar to others already existing in structural analysis.

If, in the real world, each connector subconstituent is isolated from its siblings, and is loaded according to the complex interaction between the constituents that it connects, in the analytical simplified models of a layout, the subconstituents will behave in a manner that is predictable once the overall forces and moments that the connector is required to exert are known. So the *organization* of the subconstituents finds a counterpart in the availability of analytical rules that will be able to compute the forces flowing into each subconstituent, once the overall forces are known.

Analytical models of the connectors may also consider the subconstituents (single weld seams and single bolts) as acting totally alone. In this case, there is no need to introduce the *layout* concept. As will be seen later, this kind of analytical model is necessarily more complex, as it needs a more advanced modeling of the local interaction between elementary connectors and the constituents connected. In turn, this requires modeling of local plasticization, plays, friction, imperfections, local buckling, contact, and so on.

However, the model of a connector having multiplicity n, and the related possible graph, can still be used if each single subconstituent is kept as a standalone connector. For instance, a single bolt layout made up by $2 \times 3 = 6$ bolts, having multiplicity 3, can be converted into six connectors each having multiplicity 3 (see Figure 3.8).

3.4 Classification of Connections

Connections can be classified in a number of different ways. Two of the most important connection classifications are based on the stiffness of the connection, and on its resistance.

When the problem of connection analysis is seen from a 3D perspective, these classifications partly lose their effectiveness. This is due to the presence of six generalized forces at the extremity of the members, and not just one or two, as expected when considering 2D models.

For instance, a beam to column connection can be considered "rigid", i.e. sufficiently stiff to be assumed rigid (or FR, *fully restrained*), if the strong axis bending moment is applied to the beam, and a moment-rotation curve is plotted, leading to a stiffness of the connection S_j sufficiently high. According to Eurocode 3, this condition is met when the rotational stiffness of the connection S_j (moment per unit rotation) is higher than $8EJ/L$ for braced frames, or $25EJ/L$ for unbraced frames, and a connection is classified "nominally pinned" if S_j is lower than 0.5 EJ/L, where J is the second area moment of the beam, E is the Young's modulus of steel, and L is the beam span (the classification is a bit more structured, but this is the core; also, the connection of columns to the base are classified using different bounds). Between the "nominally pinned" and the "rigid" connections are the *semi-rigid connections* (or PR, *partially restrained*), which in the past decades have been studied in depth. The problem is to determine the stiffness of the connection so as to properly tune the lateral stiffness of the frames, and so predict properly the P-Δ effects and the frequencies (to enter properly in a response spectrum analysis). Unfortunately not only is the problem very complex, but it also depends on the loads applied, so different stiffnesses can be obtained in different loading conditions, which makes the analysis even more complex. Even today, not all structural engineers are convinced that semi-rigid connection design is the best possible approach, although it must be recognized that it is apparently more economical (earthquake-design evaluations might be different, if the economic considerations have to also include the possibility of an earthquake, in seismic areas of the world). In the approach proposed here, the stiffness can be evaluated by means of finite element models created automatically. This issue will be covered in Chapter 10.

However, the displacement and ductility of the connection may well be affected by the presence of *other* internal member forces that may be able to modify the results obtained considering the strong axis bending moment only. For instance, compressive axial forces may help reach buckling of stiffeners, and weak axis bending combined to strong axis and torsion may imply load concentrations in limited regions of the connection. In seismic areas – a good part of the world – the seismic loading may drive the structure to be loaded in a very generic way, so a very complex and 3D state of stress can be expected. Usually the presence of a concrete slab is considered able to eliminate torsion and weak axis bending, but there are indeed moment connections not bearing a concrete slab, as not all the structures are buildings. Moreover, the real coupling effect between the concrete slab and the underlying steel structure is often unknown or barely predictable (for instance, after the Northridge earthquake it was questioned whether the asymmetric effect of a concrete slab could have had the some role in the typically bottom flange beam fractures observed, e.g. see Bruneau et al., 2011).

The classification according to *one* member force only is a useful indicator, but it should not and may not be considered exhaustive.

This is also true for classifications based on resistance; usually FS and PS acronyms are used for *full strength* and *partial strength*. But to which loading condition is this a strength referring? In theory, the connection is FS if it is able to resist to the full plasticization of the member cross-section, and this is usually measured (also in view of capacity design) using a single bending moment and not even axial force. This would be unsatisfactory in seismic areas, especially when it is remembered that (neglecting torsion and shears) there are three member internal forces defining the plastic domain – axial force, strong axis bending and weak axis bending – and not just one. If the interaction with shears and torque is also considered, there are six member forces, interacting one another and five infinities of plastic states, which should ideally be resisted by the connection. In 3D analysis, to mark a connection as FS, it must be assured that no stress state will ever load the connection different from what has been checked, something which is not easy to prove. A connection of a member to a renode is FS, if and only if it has been proved that for each stress state of the member belonging to the plastic domain, $(N, V_2, V_3, M_1, M_2, M_3)_{pl}$, the connection will resist. The disappearance of some of the internal forces, for instance keeping only one of them, M_{2pl}, may sometimes be justified by the existence of true restraints (e.g. a well connected concrete slab). Other times it may be justified just by our wish – the wish that the member will be loaded as desired – but cannot be justified by the connection, which will also resist to all the other five member forces, or by the true loading of the structure, which will never be so clean, straight, organized, and aligned to the axis of symmetry as we would like.

For all these reasons, it is the author's opinion that the classifications currently used are only a partial view of the problem to be faced, and that they can even be misleading. This happens because the code and the current practice refer the words "full strength" or "rigid" to well specified internal forces acting alone. Instead the analyst, the structural engineer and what is worse the owner, frequently understand "full strength" or "rigid" in their extended, ungrounded general meaning.

Reference

Bruneau, M., Uang, C.M. and Sabelli, S. 2011, *Ductile Design of Steel Structures*, McGraw Hill.

4

Renodes

4.1 From Jnode to Renode Concept

The jnode concept is useful for collecting generic information about the connections between the members, their orientation and layout, cross-section and materials, and so on, but in no way does it define how the connections are physically made.

To do that, a new concept is needed, which will be identified by a specific word, *renode*, which stands for *real node*. So, a renode is the set of all the constituents, connectors, work processes, and all the choices made in order to check all these constituents, under the effect of the applied loads.

Each jnode can be transformed into a renode in an infinite number of ways, using different force transferrers and different connectors.

However, each renode has one, and only one jnode related to it (Figure 4.1), and it can be determined by examining the underlying BFEM, as explained in Chapter 2.

The renode embeds all the needed information to actually check the connections, and so it is necessarily edited and modified in a true 3D environment, where all the constituents are described with the maximum possible accuracy, in order to define all their geometrical features: dimensions, thicknesses, cuts, bevels, holes, work processes in general, position, and orientation. Moving a single constituent in space, or changing its dimensions will necessarily lead to a different renode, and to different checks. And indeed, it does.

Also the connectors must be correctly described and positioned in space. The shift of a bolt layout or the addition of a single weld seam must imply some change to the static behavior of the renode, and so it must be correctly detected and recorded into a general format, able to comply with all this information.

4.2 BREP Geometrical Description of 3D Objects

A 3D object in space can be efficiently described by its boundary. If the boundary is a closed surface, the object body is by definition the part of space contained inside the boundary surface. A representation of an object by means of its boundary is called a BREP (Boundary REPresentation).

Steel Connection Analysis, First Edition. Paolo Rugarli.
© 2018 John Wiley & Sons Ltd. Published 2018 by John Wiley & Sons Ltd.

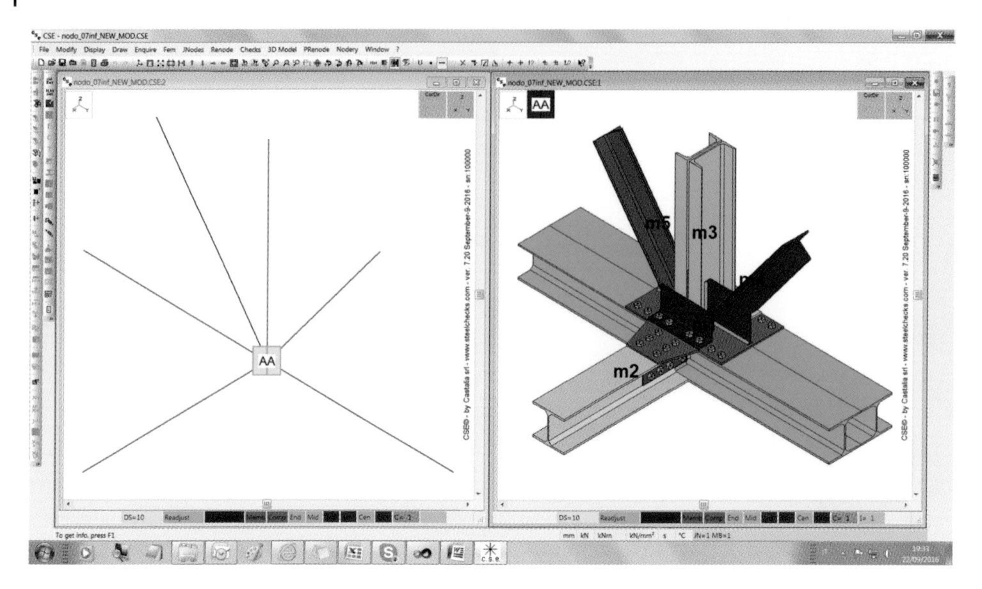

Figure 4.1 A jnode and its renode completed.

There are also other possible ways to describe and store the information related to a 3D object. For instance, extrusions are solids formed by translating a planar surface in a direction which must not be perpendicular to the planar surface normal. Another important example is constructive solid geometry (CSG) representation, which does not store the object in its final configuration, but stores a primitive object and the description of all the constructive operations done in order to modify it.

All these representations have pros and cons. All are possibly considered by 3D CAD CAE software, which is nowadays able to allow an interactive manipulation of objects in space. By means of CAD CAE software, it is possible to modify the objects, applying all the needed changes: of size, translations, rotations, cuts and bevels, and so on.

The BREP way to describe objects has a first very important feature, which is very useful in the context of steel connection analysis: it is general. Not all objects are extrusions, but all objects have a BREP.

CSG is also general, and has the important advantage of requiring much less memory space than BREP, to store an object on a hard disk. However, the object must be reconstructed at run time, before rendering it.

The boundary surface of a BREP can be divided into more elementary surfaces, which must have borders in common, and leave no gap. The easiest way to do that is using (part of) planes as subsurfaces. If this is done, each object can be seen as a collection of planar faces.

In turn, a planar face can be described by a set of straight or curved edges, by means of a number of points. Once the information referring to the edges is stored, a simple vector of points in 3D space, $\mathbf{P_i}$, will be able to describe a face (see Figure 4.2).

Faces must have sides in common, in order to leave no gap in the external surface of the body.

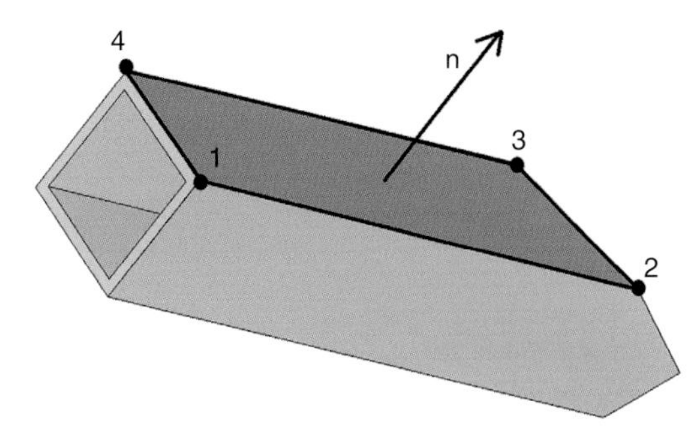

Figure 4.2 A face with its numbered points and outward normal.

By definition, the vector of points defining a face can be ordered in such a way that the unitary-modulus normal vector to the face, **n**, computed by means of vectorial products, points *outward* to the 3D object. This is done by the following rule:

$$\mathbf{v} = \sum_{i=1}^{n-2} (\mathbf{P}_{i+2} - \mathbf{P}_1) \wedge (\mathbf{P}_{i+1} - \mathbf{P}_1)$$

$$\mathbf{n} = \frac{\mathbf{v}}{\|\mathbf{v}\|}$$

So, all planar faces, by definition, have their normal vector pointing outward.

A 3D object is defined in its local reference system, using three orthogonal axes (1, 2, 3) and an origin, which is usually one of the points of the object. Local axes are chosen so as to make it easy to define the coordinates of the points of the faces, locally. These points are stored in a unique vector for each object, so as to avoid repetition. Each face, in turn, is defined by a vector of integers: the numbers of defining points in the 3D object point array.

The placement of the object in space is defined by an insertion point **Q**, and by an orientation orthogonal matrix **T** (having three rows and three columns) which defines how the object will have to be rotated so as to assume the right layout.

So, the final global position $\mathbf{P}_G$ of each point $\mathbf{P}_L$ of the array of points of the 3D object in the local system will be

$$\mathbf{P}_G = \mathbf{Q} + \mathbf{T}^T \mathbf{P}_L$$

Clearly it is the immediately opposite operation which allows us to find the local coordinate of a point once the global ones are known:

$$\mathbf{P}_L = \mathbf{T}(\mathbf{P}_G - \mathbf{Q})$$

4.3 The *Scene*

4.3.1 Generality

Once all the constituents have been placed in the space, the set of all the geometries of all the constituents, including their sizes, shapes, work processes, positions, and orientations will be called the *scene*. The scene is the geometrical part of a renode.

The constituents are, in general:

- members
- force transferrers (optional)
- stiffeners (optional)
- connectors
- constraint block (optional).

Almost always there will be force transferrers, but it is possible to conceive renodes where members are directly connected by connectors with no use of force transferrers.

A constraint block will be placed in the scene only if the jnode is a constraint (fixed or elastic) – see Figure 4.3. Otherwise it will not and must not be placed in the scene.

So, the renode is made by the scene and by all the choices made in order to check the constituents, that is, the set of all the choices (among the several possible ones) that entirely define the checks that will be done, and the mechanical behavior of connections (this will be discussed in the next chapters).

A renode can be loaded in vary many different ways, as the internal forces in the members can be very different. However, the loading is not part of the definition of the renode.

In a structure there can be many jnodes, having several instances. Each jnode is going to generate a single renode, that will be used for all the instances of the jnode. In this way, members that are connected in the very same manner, and that exhibit a similar structural behavior, will be designed identically and checked all together, with a great saving of time. If a jnode has been marked by a label like "AA" or "AB", its renode will have the same mark to identify it (see Figure 4.1).

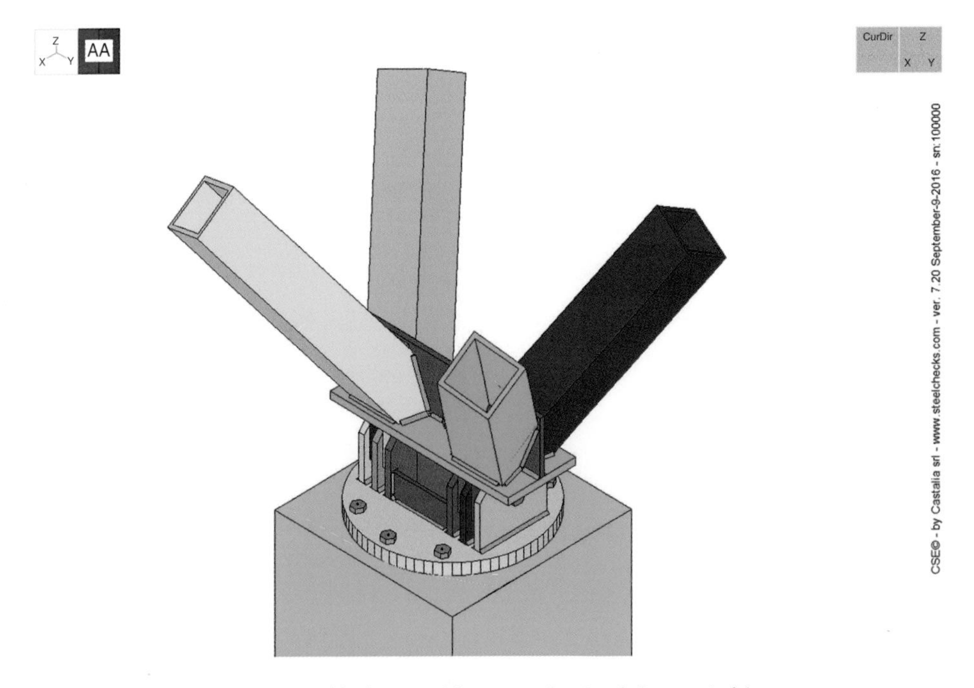

Figure 4.3 A *scene*: a constraint block is used (courtesy Amsis srl, Rovato, Italy).

In order to identify the constituents of the renode, a unique method must be set up. A good way to do that is to associate a single letter to the constituent (so as to have short names), followed by a number to distinguish it. Letters can be different, depending on the constituent type or on its geometry. For instance,

- members can be identified as m1, m2, ...mn_m (Figure 4.1)
- bolt layouts can be identified as B1, B2,....Bn_b
- weld layouts can be identified as W1, W2, ... Wn_w
- plates as P1, P2, ...Pn_p
- generic constituents as C1, C2, ...Cn_c
- the unique constraint block by a special identifier, such as *cb*, or "|---|".

So, for instance, member number 2 of renode AB is just "AB.m2".

The creation of the scene is an unavoidable, necessary, and preliminary step in order to check connections. Without a scene it is just not possible to properly check connections.

4.3.2 Members

Initially, members are considered to end at the ideal member extremities, as defined in the underlying BFEM. Unless proper rigid ends have been defined, members will overlap (see Figure 4.4). This is a normal situation at the start of the process, and the definition of the scene should always also imply the removal of every overlap, including those of the members.

The initial length of the member stumps can be set according to a general rule, and the stump length can be modified during the scene creation, if needed, by properly shifting the far end **F** of the members.

Initially, member are extrusions, but later they can be modified in very many different ways in order to adjust connections.

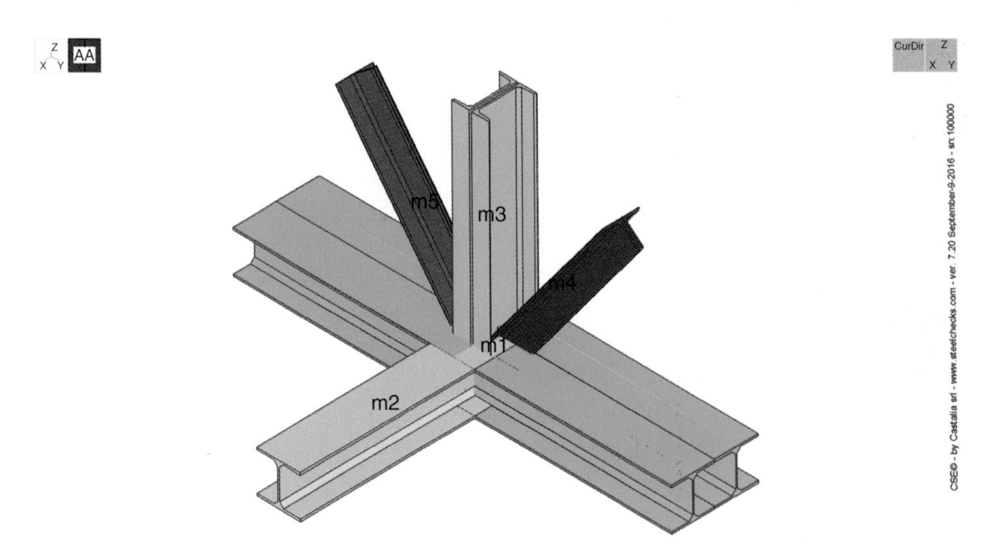

Figure 4.4 Initially, members overlap.

If members are shifted – that is if their axis line is moved in a direction normal to the axis itself – an important violation of equilibrium is used if the point of application of the member forces is shifted as well (nodal equilibrium is lost). However, if the member is shifted but the point of application of the member forces is left unchanged, the member will undergo internal forces different from those actually computed (and checked) in the BFEM. Besides, the stiffness of the system will also change, so in theory such a shift should not be allowed.

However, there are situations where the members must be shifted for unavoidable reasons, which could not be foreseen at the BFEM stage. If so, the best would probably be to leave the point of application unchanged, but redo the checks of the member for the additional moments due to the shift. This would save equilibrium, but imply a step backward in design, as the member will have to be rechecked.

So, generally speaking, the best would be to define properly the eccentricities in the BFEM, as this would guarantee the best coherence with the connection design.

4.3.3 Typical Fittings

The so-called “fittings”, in this book, are force transferrers, stiffeners, and a possible constraint block. They are defined as starting from a wide set of possible primitives, whose sizes are adapted to the connections at hand, and that may themselves be subjected to work processes such as cuts, bevels, and so on.

A very special kind of fitting is the constraint block (Figure 4.3), which must be added to the scene if and only if the jnode is a constraint. With no loss of generality it can usually be considered a stiff rectangular box, properly oriented in space. This mimics structural elements such as floors, walls, and plinths.

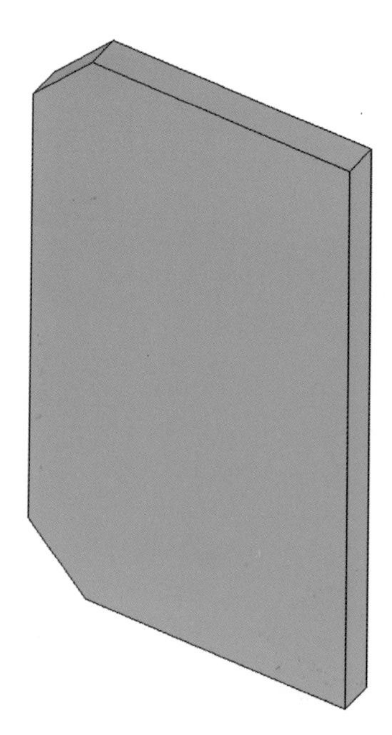

Figure 4.5 A stiffener for a rolled I or H cross-section.

A list of the possible primitives to be used for generating fittings would include:

- rectangular plates (possibly with rectangular or circular holes)
- circular plates (possibly with circular or rectangular holes)
- stiffeners for rolled H, I and C cross-sections (Figure 4.5)
- haunches
- cross-section trunks (L, I, H, T, U, C,...)
- planar plates with special shapes: triangular, polygonal, trapezoidal etc.
- generic planar plates
- ....

These constituents are placed in the scene by selecting in their body an insertion point, and then by choosing a point in the scene that will match (Figure 4.6). Constituents are rotated properly before insertion, using as reference the global axes or the local axes of some other constituent.

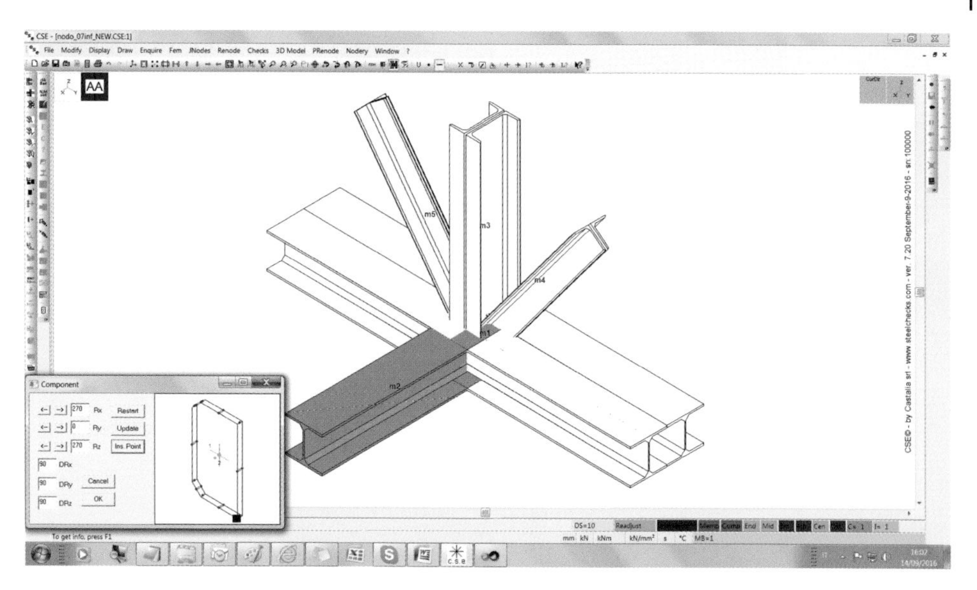

Figure 4.6 A stiffener is put in place by selecting its insertion point.

Constituents are usually placed in such a way that one of their faces will be coplanar with the face of some other constituent. As it will be seen, this is also a precondition to apply connectors.

4.3.4 Connectors

4.3.4.1 Bolt Layouts

Bolt layouts are inserted in the scene, choosing the face of the object that the inner faces of the bolt heads will be in contact with (see Figure 4.15).

It is not necessary to represent the bolt shaft, as it is normally hidden by the bolted objects. Each bolt can be conveniently represented by means of two hexagonal prisms, one of the size of the head and one of the size of the nut.

More structured geometrical representations, including the washers, are also possible, in order to have a more realistic geometrical rendering, and in order to better detect clashes. However, the speed of representation can also be an issue, and so the basic representation may have good reason to still be used.

So, a bolt layout having n bolts will be represented by $2n$ hexagon prisms, n for the heads and n for the nuts. These hexagon prisms will be rendered by a BREP, each having eight faces (see Figure 4.7). Once the faces are inserted in the scene, they are marked with a unique identifier that writes the bolt layout, and bolt number they belong to. The inner face of the hexagon receives a unique marker, as well as the outer face and the lateral faces. So, these faces can be easily recognized.

All the inner faces of the bolt heads will have to share the same face of the same other constituent (or constituent's subconstituent), or will have to belong to more coplanar faces of the same constituent.

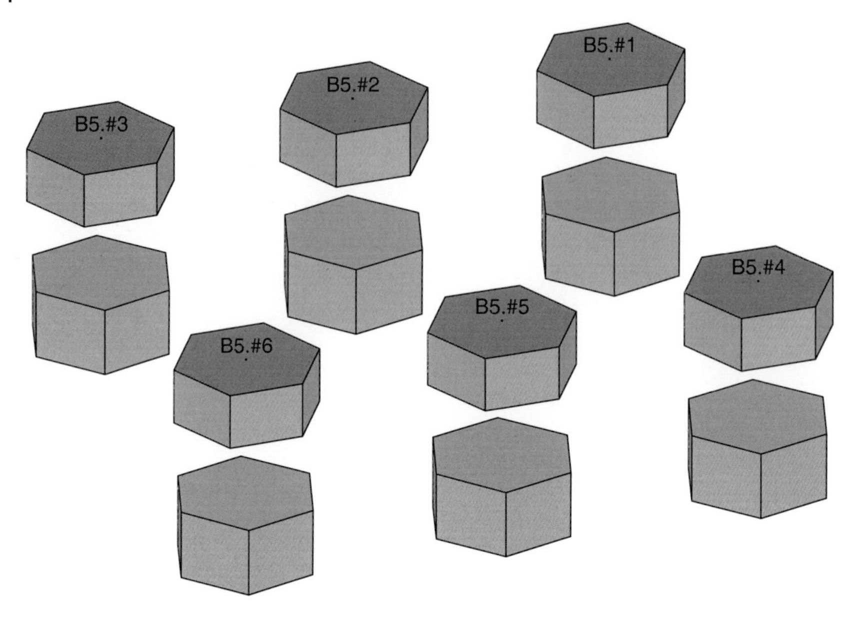

Figure 4.7 Synthetic representation of a bolt layout having six bolts. Connected parts are not displayed.

All the inner faces of the bolt nuts, will also have to share the same face of the same constituent (or constituent's subconstituent, different from the previous one), or will have to belong to more coplanar faces of the same other constituent.

These previous conditions are necessary preconditions, implying that all bolts drill the same first and last object. More refined checks will also ensure that the inner drilled constituents are the same for each bolt.

The minimum geometrical information needed to store a bolt layout, assuming that all bolts have the same diameter and class, is the following:

- A 3D insertion point lying over the first drilled face of the first constituent bolted.
- Axis 3 of the bolt layout will be equal to the normal of the aforementioned face. Axes 1 and 2 will lie over that face, possibly chosen so as to be parallel to one of the chosen face sides.
- For each bolt, a point $\mathbf{P_i}$ defined in the plane 1-2 – as an alternative, simple regular grids can be defined by the number of rows, the number of columns, their distance, the coordinate of the grid center, over the face, and a rotation angle. A single bolt position can be defined with complete generality and does not need to belong a regular grid. Each bolt can possibly be freely moved from here to there, and provided that the constituents drilled are the same for all the other bolts, this must be allowed.
- For each bolt, its class and diameter.

4.3.4.2 Weld Layouts

Weld layouts are placed in the scene by choosing a face, coplanar and at least partially in contact with another. The two faces in contact belong to different constituents, O_1 and

O_2, that the weld layout will connect. The two faces have opposite normals (see Figure 4.18). The plane of these opposite faces will be named the *faying plane*.

In fillet weld layouts, each weld seam is represented by a triangular prism, whose length is the length of the seam, and whose triangular cross-section's side lengths in contact with the welded objects are the side lengths of the seams (the so-called "legs"). With no loss of generality from the computational point of view, it will be assumed that the two sides have equal length. The two faces welded by each seam s will form different angles γ_s, from seam to seam, usually between 60° and 120°.

Each weld seam thus has five faces. The two triangular terminal faces are inactive. The two rectangular lateral faces in contact with the welded objects O_1 and O_2 will be the *active faces* or *fusion faces* of the weld seam.

Weld layouts connect two constituents, not more. All the weld seams must connect the same two constituents.

All the weld seams will have a plane in common, the weld plane or faying plane, which is the plane of the welded face of the first constituent welded, O_1. The other active faces will all weld constituent number two, O_2, generally using different planes (Figure 4.19).

So, fillet weld layouts can be represented by a number of triangular prisms, correctly positioned in space.

The minimum information to store weld data is the following:

- An insertion point laying over the chosen face, which is a face of the second constituent welded, O_2.
- Axis 3 of weld layout will be opposite to the normal of the aforementioned face. Axis 1 and 2 will lie over that face, possibly chosen so as to be parallel to one of the chosen face sides.
- For each weld seam, two points, $\mathbf{P_1}$ and $\mathbf{P_2}$, defined in the plane (1-2) and indicating the start and end of each weld seam. These two points belong to one of the oriented sides of the chosen face (Figure 4.8). Sides are oriented, like vertices (from i to $i+1$), to

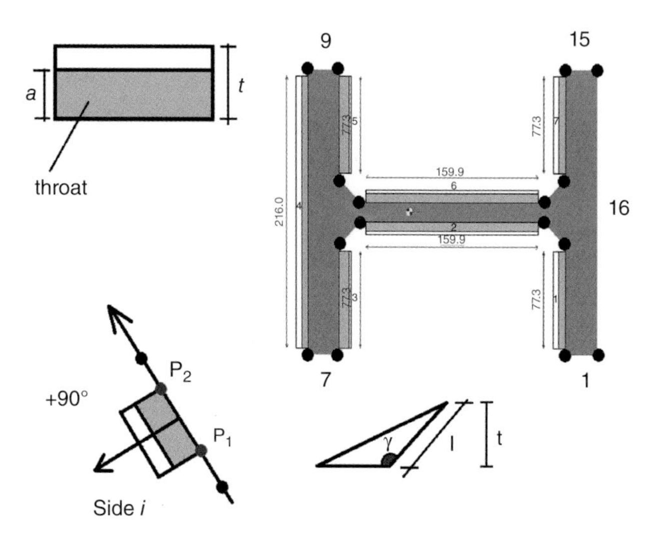

Figure 4.8 Fillet weld seams definition. The sides of the chosen face are numbered clockwise as the normal of the chosen (blue) face enters into the page.

properly define the face normal. The non-dimensional abscissas of their points are inside the interval (0–1), so that 0 means a point equal to the start of the side, while 1 means a point equal to the end of the side.

- For each weld seam s, a "thickness" t_s must be given, related to the size of the legs, l_s. If γ_s is the angle between the active faces, t_s is the thickness, and l_s the leg size, $l_s = t_s/\sin(\gamma_s)$.

The representation of the penetration weld layouts poses some problem, as they would appear inside the thickness of the objects welded. Besides, to correctly receive the weld seams, at least one of the two objects would have to be modified by specific work processes. From the strictly computational point of view, at the currently reasonable scale of the analysis, we don't really need to know how the constituents have been prepared in order to receive the welds. This is, of course, quite important from the constructional point of view, and in order to avoid defects in welding, stress raisers, and so on. What is really important from the analytical point of view is if the penetration welds are full penetration, or partial penetration, where they are, how long and how thick. So, in 3D, it would also be possible to display the weld seams as if they were fillets, but using a different graphical method in order to distinguish them. For instance, instead of using an isosceles triangle, the triangle might have a 1:2 ratio for the "legs". This would avoid the need to define bevels for the receiving part.

Care must be taken in order to avoid the addition to the scene of penetration welds having a thickness higher than the thickness available, also considering that a thickness might receive more than one weld.

Referring to Figure 4.9 the following considerations apply:

a) single weld seam modeling a full penetration weld
b) two weld seams modeling a full penetration weld
c) two weld seams modeling a full penetration weld, but unacceptable as overlapping
d) single weld seam modeling a partial penetration weld
e) single weld seam unacceptable as it is thicker than the available thickness

It must be emphasized that the welds represented in Figure 4.9 could possibly trigger lamellar tearing for high thicknesses and high strength parts (in the unrepresented thickness). However, from the strictly analytical point of view the definition of this detail can be avoided, and the detail of the work process applied to the other part is not needed.

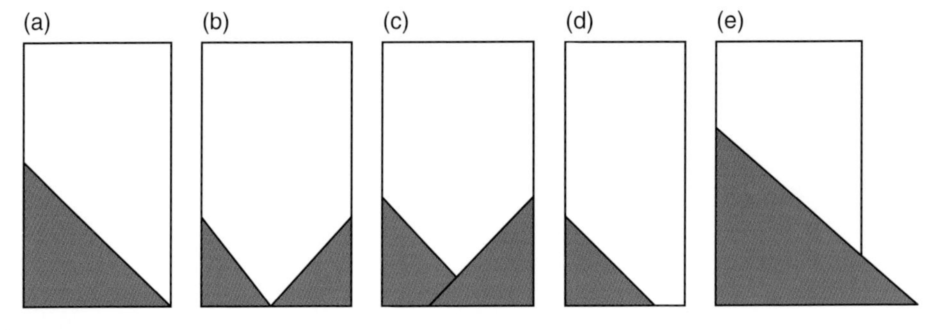

Figure 4.9 Several possible penetration weld seams applied to a thickness.

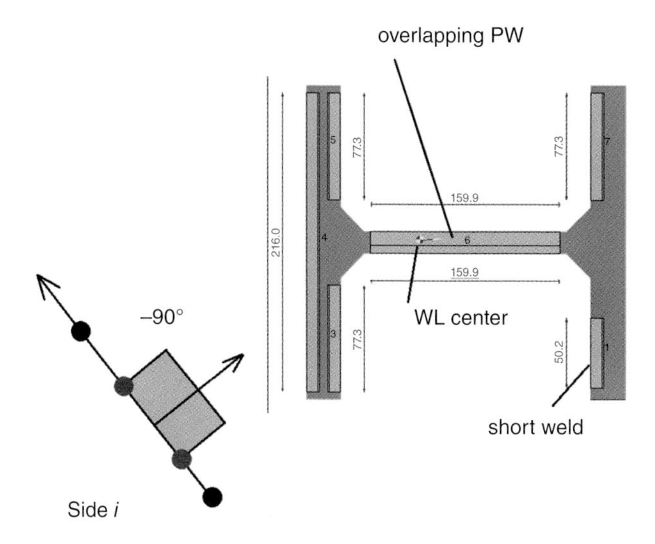

Figure 4.10 Penetration weld layout definition.

When penetration welds are defined, no throat size is needed, and the weld seams are applied using an angle −90° degrees relative to the angle of the side involved (Figure 4.10). This is opposite to what is done with fillet welds, which in planar view are applied using a +90° angle.

Fillet or penetration welds can be shorter than the side they are placed over (but this is usually forbidden for penetration welds), and more than one seam can be added to a side. Careful checks must be done in order to avoid situations as the one in Figure 4.9(c) or (e).

4.4 Dual Geometry

Each object placed in the scene, with the exception of the connectors, may be modified by work processes that modify its body (Figure 4.11).

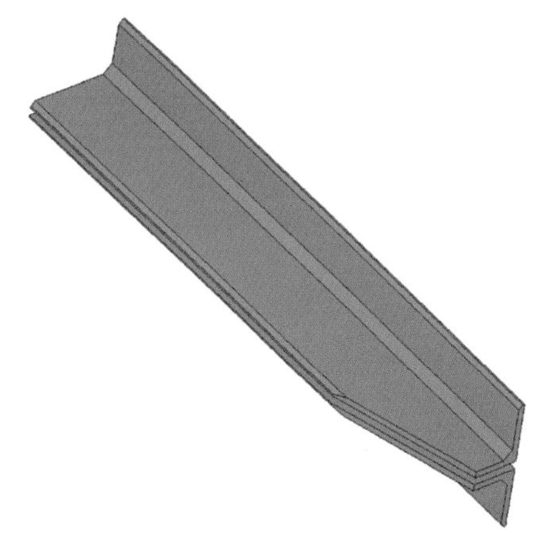

Figure 4.11 A compound member with a part removed.

Summing several work processes will transform the initial object into another one, perhaps completely different. However, this completely different object must be analyzed and its stress state must be determined. This is a very much important problem, probably one of the most important problems of all steel connection analysis. Usually, no simplified theory is available to quickly analyze the object as it is, once the work processes have been applied.

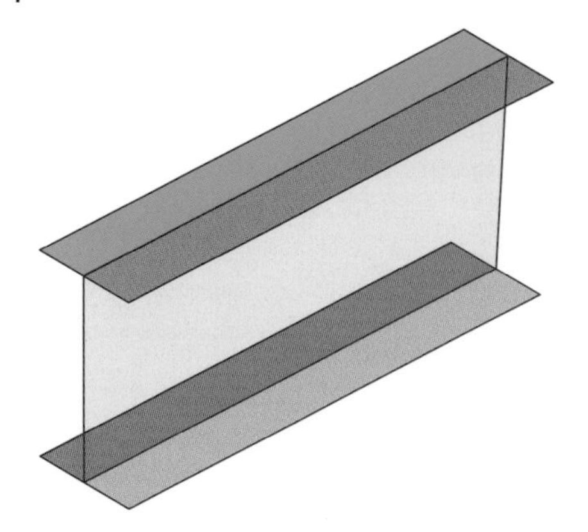

Figure 4.12 Subplates of a member stump having an I cross-section.

Initially, almost all the primitive constituents (members, force transferrers, stiffeners) can be modeled by the finite element method as a set of subplates (Figure 4.12), which fall within the theory of plates and shells. The planar plates may have a generic number of sides, and sides of these subconstituents must match one another.

So, initially, each constituent can be modeled by plate–shell finite elements meshing the subparts it is composed of, and ensuring proper node-match at the common sides. This kind of modeling is very useful as it allows a refined stress analysis of the constituent, once the forces that load it are known.

Generally speaking, the fem models of the constituents will also depend on the connectors applied to the constituent itself. "Hard" nodes will be placed along the weld lines, or at bolt centers, and meshing will have to respect such hard points (see Chapter 9).

Rigid translations and rotations applied to the constituent will not imply any change in its finite element modeling.

However, each work process like the application of bevels, cuts, and stretching, will imply a change in geometry that will have to be correctly taken into account by the mesh.

If the typical object is purely stored as a BREP, once at least one work process has been applied, it loses its original typical nature, so there is no general rule anymore to correctly model it. Generally speaking, if a 3D object is only known by its BREP, it is not trivial to *automatically* get its finite element model, if the much faster plate–shell elements have to be used. Of course, the tetrahedrization of a volume can be automated, and tetrahedrons can be merged into wedges or bricks. However, modeling constituents by solid elements is not presently the best choice. Solid meshes imply a larger number of degrees of freedom, and the analysis of the internal part of the bodies can be long and cumbersome, despite the automatic detection of maxima (often not very meaningful).

Given a solid described by a BREP, and assuming that the plate–shell theory can be used to model it (which should itself be proved), the correct *automatic* definition of an "average surface" to be then meshed is not trivial (e.g. see Figure 4.11). At the moment, also in very powerful 3D software like Rhinoceros©, this operation is not available. If the BREP uses planar faces, the problem could perhaps be tackled by searching parallel faces at a distance much lower than their longest side, and assigning *scores* to each set of parallel faces, e.g. related to surface extension. By removing the face-set one after another, using the score to set precedence, and replacing each pair of faces with their average surface, an algorithm could perhaps be set up. However, there is a better way to tackle the issue.

As the final aim is to get the plate–shell model of an object which has undergone a set of work processes, and as this constituent, initially typical, has well defined plate subconstituents, the solution of the problem is to record the history of the work processes that the object has received (CSG) and to keep a dual geometry available: the solid representation by BREP, when needed, and the plated representation when a fem model must be created.

The fem model of the final BREP will be obtained by applying the same work processes applied to the true solid also to its initial plated representation.

Each plate subconstituent will possibly be cut, or its corners smoothed (Figure 4.13). Later, it can be meshed in the usual way.

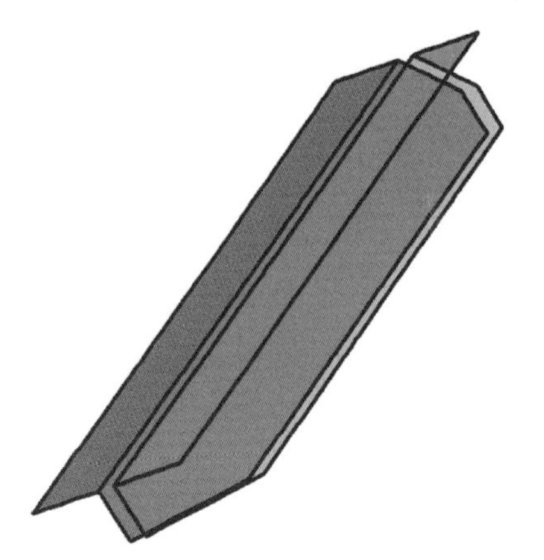

Figure 4.13 Plated subconstituents of a member after two work processes have been applied.

4.5 Automatic Connection Detection

4.5.1 Faces in Contact

Considering a face A belonging to a body “a”, and a face B, belonging to a body “b”, the two faces are in contact if and only if they are coplanar, they have opposite normal, and at least one point of A is inside B, or one point of B is inside A. If the two faces have parallel not opposite normal and are coplanar and contained one into the other, the two objects overlap, so this is an error condition.

The existence of two faces in contact is a preliminary necessary condition to set a connection between body a and body b.

If the two faces are merely coplanar, they can be external one another (Figure 4.14). Also, if one point of a face is inside the other, and this point belongs to its plane, but the two planes of A and B are not identical, there will not be enough surface to set a connection.

The connection between a body a and a body b will be assured by placing in the scene a connector able to connect a and b. The connector will have faces coplanar and in contact with one or more faces of the body a, and will have faces coplanar and in contact with one or more faces of the body b.

Placing a connector in the scene is a geometrical issue.

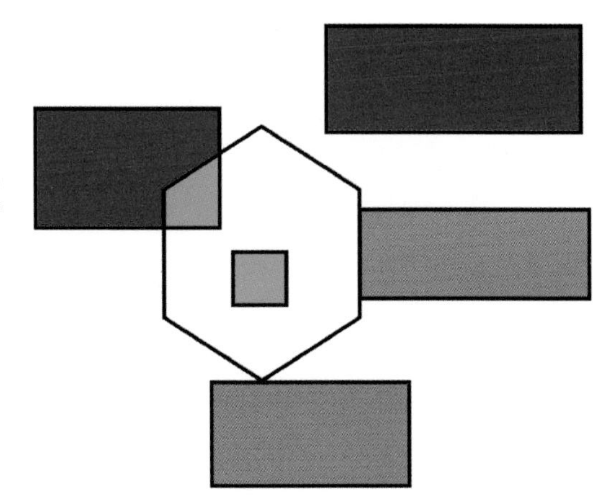

Figure 4.14 Coplanar faces. The hexagon face has a positive normal, all other faces have a negative normal. Green surfaces are contact areas. The square face is entirely contained in the hexagon face and is in contact. Leftmost rectangular face has only a part in contact. All other faces are external or tangent.

The existence of a connection between two bodies can be detected by examining all the connectors, and possibly finding a connector connecting a and b. The rules needed to automatically detect a connection between two objects, just by analyzing geometry, will be briefly summarized in the next two sections.

4.5.2 Bolt Layouts

To apply bolts to a number of objects in contact, a stack of parallel faces is set up.

This starts with a specific face C_1, that is chosen in the scene and whose normal will be the direction of the axis 3 of the bolt layout to be added (Figure 4.15).

The next faces in the stack are collected by repeatedly apply this *drilling algorithm* (Figure 4.16):

1) Find the nearest face of the same object, its "next face" N_i, having the opposite normal, and parallel to the current one, C_i.
2) Find a face that is in contact with N_i, and belonging to a different object.
3) This is the new current face, C_{i+1}, go back to step 1.
4) If no face in contact is found, end with face N_i.

More complex situations may arise if a constituent has more than one face coplanar and in contact with the face N_i currently under examination (think about the inner faces of the top flange of a member having an H cross-section). Generally, it may be possible that for a given drilled elevation, more than two faces are met. The total number of elevations is $(n + 1)$, if n is the number of thicknesses drilled. For each level, more than one face should sometimes be checked.

At the beginning, the current face C_i is the one chosen in the scene, C_1.

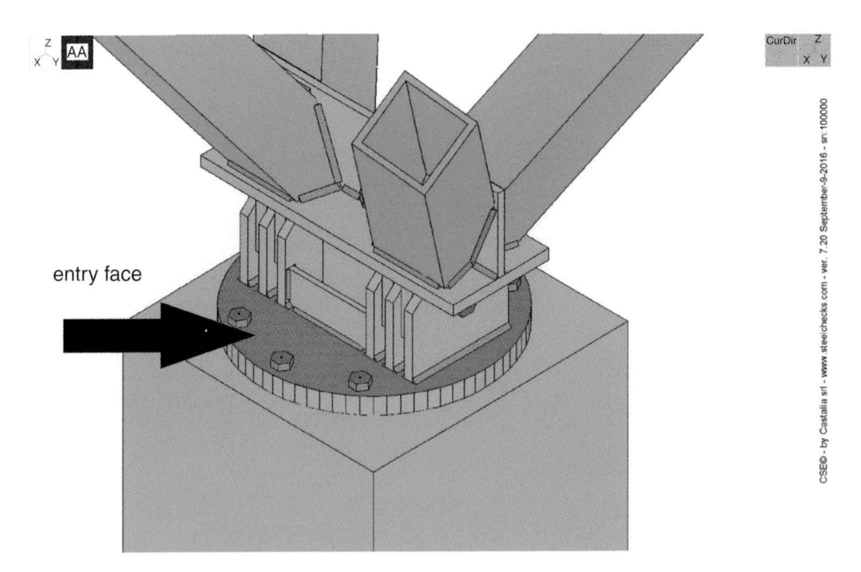

Figure 4.15 Entry face C_1 to add a bolt layout (also displayed). The model, developed using software CSE is by AMIS srl, Rovato, Italy.

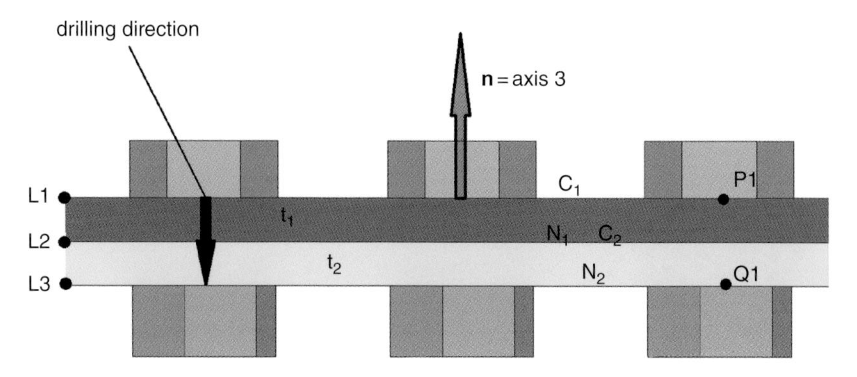

Figure 4.16 Drilling to air from face C_1.

Once the stack of faces is found, the thicknesses potentially drilled by the bolts are known. It is assumed that the faces are parallel, which means constant thickness from bolt to bolt: this does not include tapered flanges or plates having variable thickness. This is a quite general model which is able to cover the most part of the problems; however, it can be generalized also to tapered flanges considering faces N_i having normal sufficiently near to the normal of the face C_i.

In order to allow a reasonable placement of the bolts, other faces will also be collected. They are all the faces of other objects, in contact with the initial face C_1. These faces can be considered as "obstacles" to the placement of bolts, that will be positioned in such a way to avoid overlapping.

Bolts are then placed in the subscene (Figure 4.17), considering the stack of faces collected, and viewing this stack of faces from their normal. Bolts are added, and moved in the plane of C_1. For each point $\mathbf{P_b}$ of each bolt b, it is then possible to verify the following necessary conditions:

1) Point $\mathbf{P_b}$ is inside at least one face for each level in the stack from L_1 to L_{n+1}, where n is the number of thicknesses drilled.
2) Point $\mathbf{P_b}$ is sufficiently far from the boundary of each face in the stack.
3) Point $\mathbf{P_b}$ is sufficiently far from any other point $\mathbf{P_j} \neq \mathbf{P_b}$.
4) The sequence of the objects drilled by a bolt b is equal to the sequence of the objects drilled by each other bolt j.

If and only if all the points $\mathbf{P_b}$ and bolts b satisfy all the conditions above, the bolt layout can be accepted and added to the scene.

Each bolt b will have an entry point $\mathbf{P_b}$ and an exit point $\mathbf{Q_b}$ (Figure 4.16). It can be assumed, albeit not strictly necessary, that all bolts have the same diameter and material class. If so, the center of the set of points $\{\mathbf{P_b}\}$ will be the center $\mathbf{G_1}$ of the bolt layout at level 1. For each elevation level, a point in space can be determined which is the center of the bolt layout at level i. Clearly, if t_j is the j-th thickness drilled:

$$\mathbf{G_{i+1}} = \mathbf{G_1} + \left(\sum_{j=1}^{i} -t_j \right) \mathbf{n}$$

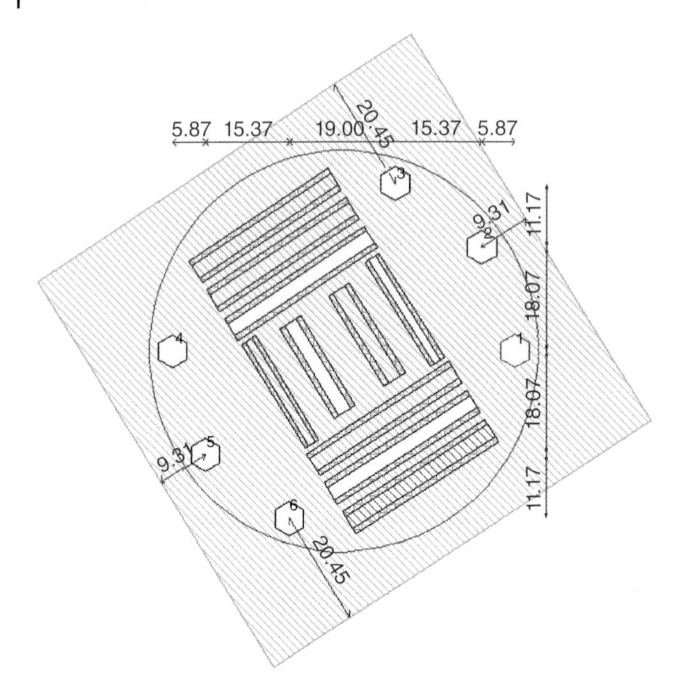

Figure 4.17 The two circular faces of a circular base plate and the two faces of a concrete plinth are the drilled faces stack. Other faces are the footprint of plates and welds and are used to avoid overlaps of the bolts (the bolt heads are drawn). See Figure 4.15.

and

$$\|\mathbf{P_b} - \mathbf{Q_b}\| = \sum_{k=1}^{n} t_k \quad \forall b$$

The multiplicity of the bolt layout is the number of thicknesses drilled, n. The extremities $\mathbf{E_e}$ ($e = 1$ to n) of the bolt layout can be computed as midpoints between the centers at adjacent levels:

$$\mathbf{E_e} = 0.5(\mathbf{G_e} + \mathbf{G_{e+1}})$$

that is, the bolt layout extremities lie over mid-thickness planes.

Having the points $\mathbf{P_b}$ and $\mathbf{Q_b}$, and the normal $\mathbf{n}$, the head and the nut of each bolt b can be described and added to the scene.

The bolt layout will be considered as a single object, which will be possible to modify by adding more bolts, removing them, displacing them, or changing the diameter of the bolt shafts or the bolt class. As a constituent, the bolt layout will have a set of three local axes, named 1, 2, and 3, which are useful to manipulate it. The insertion point of the bolt layout will, for example, be its center belonging to the first level, over the face C_1. Axis 3 is given by the normal $\mathbf{n}$ of the face C_1 (Figure 4.16). The axes 1 and 2 lie over the face C_1 and can be fixed considering one remarkable side of the face C_1 as direction for the axis 1. Axis 2 is then obtained by a vector product between axis 3 and axis 1. As will explained in more detail in Chapter 7, axes 1 and 2 are not important for checking the bolt layout: what is relevant is the position of the principal axes of the bolt layout, (**u**, **v**). However, the position in the plane of C_1 of each bolt, is more easily defined by using axis 1 and axis 2.

Finally, in order to detect which are the objects connected by a bolt layout, using only the geometrical position of its heads and nuts, the following steps are needed:

1) Let **n** be the normal of the heads' external faces (direction of bolt layout axis 3).
2) Find the plane α common to all head internal faces.
3) Find an object with a face A having normal equal to **n**.
4) Verify that all the centers of the inner faces of bolt heads belong to the face A. If not, find another face having normal **n**.
5) Using the *drilling algorithm*, build the stack of the faces and check that every bolt is inside at least two opposite faces, belonging to two different objects, for each level except the first and last. Only one face has to be searched for the first and last level.
6) Check that the sequence of the objects drilled is the same for all bolts.

4.5.3 Weld Layouts

As explained previously, both fillet weld layouts and penetration weld layouts are such that all their weld seams share a common plane.

Fillet welds are placed in such a way that each triangular prism has two adjacent faces, the *active faces* (the *legs* in the triangular cross-section of a weld seam), in contact with two different constituents, O_1 and O_2, that are connected by the welds. All the active faces connecting the first constituent share the same plane (the *faying plane*); the first extremity of the connector will belong to this plane and will be computed as the center of the projected throats of the weld seams. The other active faces of the weld seams are all in contact with some faces of the other connected constituent O_2, in general different faces (see the red faces in Figure 4.19). The second extremity of the connector is notionally fixed at an average distance from the first extremity, in a direction that will be that of axis 3 of the weld layout (Figure 4.19). That direction is opposite to the direction of the normal of the chosen face, and it's the same direction of the normal of the other face, in contact with the one chosen, C.

Once the face C is chosen, all the faces coplanar to it can be found, forming a stack of faces, possibly in contact. This is useful in order to detect possible clashes with bolts or with other constituents.

It is then possible to define a planar view, where all the coplanar faces in contact are displayed. The weld seams will be applied to the sides of the chosen face, C.

The angle between active faces, for each weld seam, is detected as follows.

The chosen face C in the scene has a number of sides *nside*. Each side of C is shared by two faces of the 3D object to which it belongs (the second object welded O_2): one is the chosen face C, the other is one of the lateral faces of the object, L (Figure 4.18).

A weld seam will belong to one of the sides of the chosen face. The angle between its active faces is the angle between the normal of the chosen face C and the normal of the lateral face L sharing the same side (Figure 4.19).

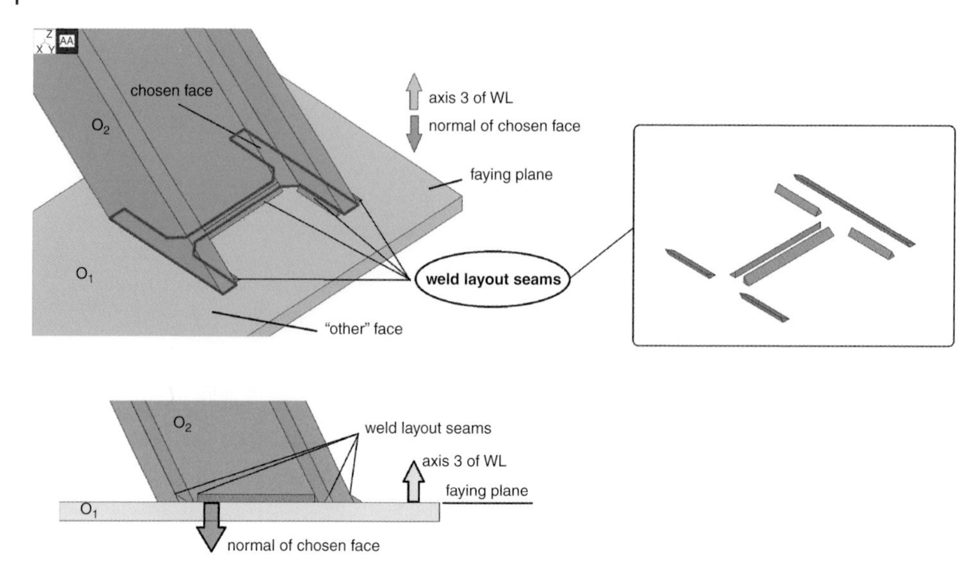

Figure 4.18 An example of addition of a fillet weld layout. The chosen face is usually contained in the other contact face. The plane of the chosen face is common to all weld seams.

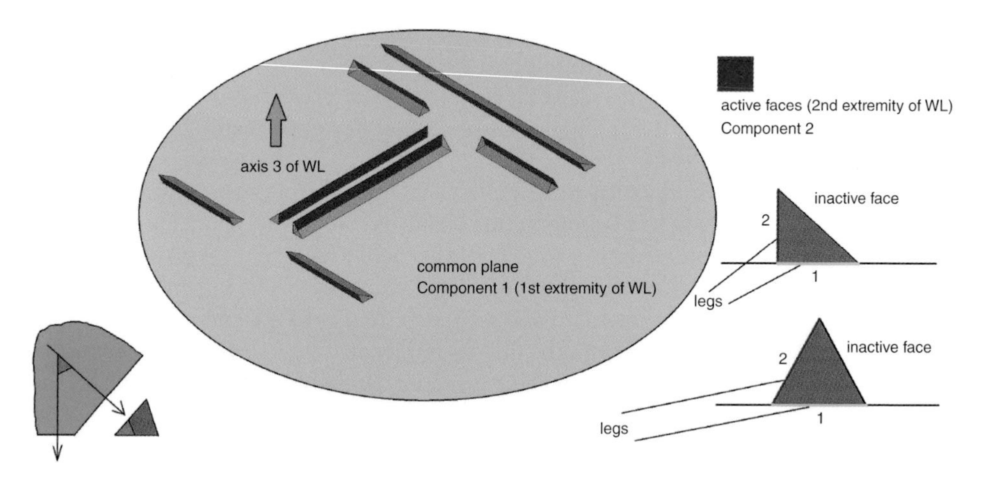

Figure 4.19 Active faces of weld seams (fillet weld layout).

To automatically detect if a weld layout is properly connected, and which objects it connects, the following steps can be executed:

- Search a face A of an object O_2 in the scene, such that the insertion point of the weld layout lies over this face plane (usually the face center is used).
- Check that the normal **n** of the face A found is opposite to axis 3 of the weld layout (their dot product must be equal to −1). If so, this is the face that had originally been chosen, C.

- Check that each weld seam *s* has its extremities $\mathbf{P}_{1s}$ and $\mathbf{P}_{2s}$ (see Figures 4.9 and 4.10) strictly over and inside one of the sides of the face, C.
- Find another object O_1 having a face B with normal – **n** (i.e. directed as axis 3 of weld layout), and in contact with the chosen face C. Check that for each weld seam *s*, the two extremities of the weld seam $\mathbf{P}_{1s}$ and $\mathbf{P}_{2s}$ are strictly inside the face B.
- Active faces have four points. Ensure that all the four points of each active face are strictly inside the face they weld. This ensures that possible shifts of the constituents have not led to a loss of connection.
- The weld layout connects O_1 and O_2.

4.6 Elementary Operations

Managing the scene properly requires the ability to perform a number of elementary geometrical operations:

Insertion is the ability to correctly place a constituent in space, usually in contact with others. As already explained, positioning of connectors usually needs the choice of a face, and specific commands to input bolt and weld numbers, sizes, and material, and checking choices.

Shift is needed to displace constituents in order to better comply to specific design needs. Shifting should be done along lines, or over specific planes.

Copy is very useful in order to quickly generate similar constituents. As the detection of the connections can be purely based on the geometrical position of the constituents, a copy of objects and related connectors will not need the addition of any information.

Overlapping check is a fundamental operation as it avoids scenes physically impossible to construct. An interesting solution is to make the overlap check automatically after a scene change, and to associate a metallic clash sound if an overlap is detected: this immediately indicates that the modification implies clashes. The no-clash sound can be a "click", perceived as correct indentation (this has been implemented by the author in CSE since 2004). Limited overlapping is often forgivable in real-world design. For instance, those referring to the corners of the welds are often too strict. So, it is possible to have different tolerances, depending on the pair of clashing objects.

3D object clash-check, if the objects are defined as BREP, is not easy. The problem is typical of CAD-CAE or animation software, and its description is beyond the scope of this book. However, analytic geometry is all that is needed, once the coordinates of the face vertex are known, as well as face normals.

In place editing can be very useful in order to apply work processes such as bevels or cuts, or in order to change the dimensions of an already existing constituent. Probably one of the most frequent edit operations is a change of thickness. The change of thickness of a force transferrer placed in an already well defined scene poses a tough problem, as there is not, in general, only one way to modify geometry in order to fit the new configuration. If the thickness is changed by *dt*, one of the two parallel faces can be displaced by *dt*, or the other, or both by 0.5*dt*. Each choice implies different changes in the vicinity of the connected objects. So, the automation of this process is not easy.

The algorithm that executes the readjust is quite complex, and not all the configurations can be repaired automatically.

Consider a constituent G (a force transferrer), which is part of one or more series of constituents in contact with each other, by tangent faces (not necessarily connected by connectors, but having faces parallel and in contact). In general, G may be in these situations:

1) G-C-C-C-C, or ....C-C-C-G: the constituent is at the end of a chain; it can be modified with no change of the other constituents. So, no other constituent is "offended".
2) M-C-C-G-C-C-C-M: the constituent is inserted in a chain of constituents which are "offended" by its modification. At the ends there are two members (M).
3) M-C-C-G-C-C: the constituent is inserted in a chain of constituents which are "offended" by its modification. At one end there is a member, but not at the other.
4) C-C-C-C-C-G-C-C: the constituent is inserted in a chain of constituents which are "offended" by its modification. There are no members at the ends.

The following rules apply:

a) The constituent modifies its shape. Some of its faces, shifting due to dimension change, may "offend" other constituents. If this happens, a readjust is necessary.
b) If the constituent is rigidly moved, after its shape has been modified, the faces that change position can be chosen, perhaps leading to an easier problem to solve.
c) The force transferrers and the connectors that are offended can be shifted.
d) The offended members cannot be shifted and will not be shifted: this would imply a *static crime* that cannot in general be allowed. If the offended member face is an end face of the member, the member can be trimmed or extended. If the offended face is the result of a previous cut (work process) it is possible to shift it, removing the offense automatically by shifting the offended face or faces.
e) If a constituent is not offended it has a null offending score.
f) The shift of a force transferrer or of a connector implies a low offending score.
g) Modification according to the rules explained of a member, implies a medium offending score.
h) Other modifications of a member (shift or application of work processes to faces not previously worked) are considered inadmissible and have a very high score.
i) The final score of the trial solution is the sum of the scores assigned to the offended objects.
j) The adjustment will be done so as to minimize the "offending score".
k) The weld layouts whose weld seams are involved in the shift are not resized, that is, the thickness of the weld seams is not modified.

Most of the typical cases are correctly managed by the readjust. Among them are (Figure 4.20):

- modification of the thickness of a plate T-welded to another plate or to a member (also for angles different from 90°)
- modification of the thickness of plates in base nodes or end plates
- modification of the thickness of plates where two or more thicknesses are bolted
- modifications of the work processes of hollow cross-sections which receive welded plate.

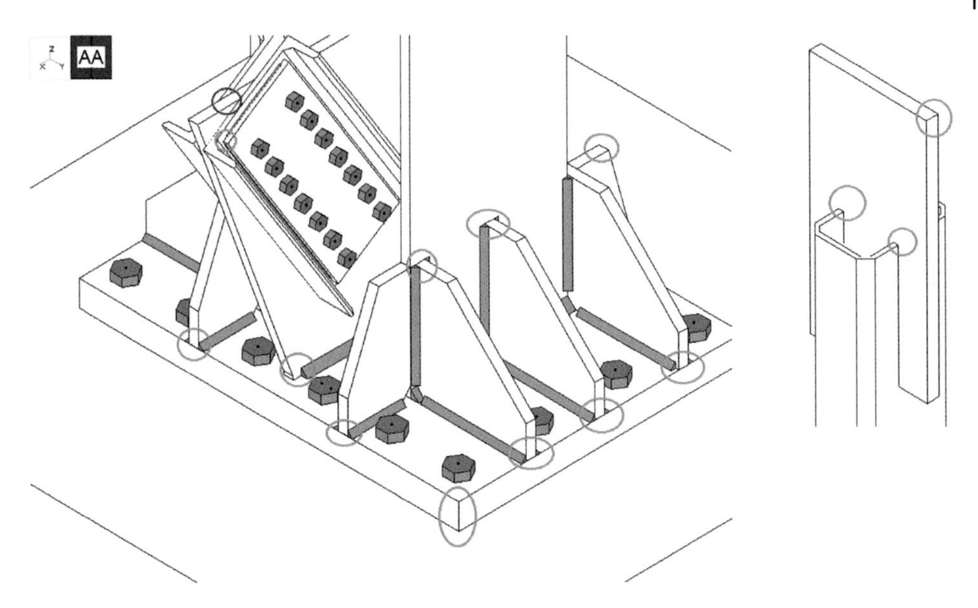

Figure 4.20 Thickness changes. Green ones can be adjusted automatically. The red ones cannot, because it would imply a member cross-section change and a lack of coherence with the underlying BFEM.

4.7 Renode Logic and the Chains

4.7.1 Minimum Compliance Criteria for Renode Good Design

Setting up a scene is a necessary condition for analyzing a renode; however, it is not sufficient. In order to allow the analysis and check of the connections, a scene must satisfy some necessary preconditions, which can all be automatically checked and this implies that a minimum good design has been set up.

These conditions, or *scene compliance criteria*, can be summarized thus:

1) Every constituent must be connected to at least one other constituent.
2) All the subconnectors of a connector must be connected to the same constituents.
3) Connectors must be connected to at least two different constituents.
4) Weld layouts cannot be connected to more than two constituents.
5) If a constraint block exists in the scene, then for each member there must be at least one chain of constituents continuously connected, connecting the member (first constituent in the chain) to the constraint block (final constituent in the chain).
6) If the jnode of the renode is hierarchical, then for each slave member there must be at least one chain of constituents continuously connected, connecting the slave member (first constituent in the chain) to the master member (final constituent in the chain).
7) If the jnode of the renode is central, the same condition valid for hierarchical jnodes must be checked, considering one of the members as master.
8) All the constituents must appear into at least one chain.

These conditions can all be checked automatically once the scene has been set up. In order to do that, a first step is to create a vector having as many elements as the total number of connectors in the scene (number of bolt layout plus number of weld layout, if only bolts and welds are used as connectors). Each element of the array is a data structure named CONNECTORDATA, and embeds:

1) the unique identifier of the connector, e.g. a number in the scene 3D object array
2) a flag indicating the type of the connector (weld layout or bolt layout)
3) the multiplicity n, of the connector, i.e. the number of constituents connected
4) a vector of connected object identifiers, having n elements
5) a vector of connected object types (member or non-member), having n elements
6) a vector indicating the connected objects subpart, having n elements
7) a vector indicating the connector extremity, having n elements.

Assuming that a connector cannot connect more than 20 constituents, CONNECTORDATA might look as follows, in C/C + + :

```
typedef struct tagCONNECTORDATA
{
        DWORD theobject;                 // number of the connector in
                                            the scene vect
        DWORD theobjectkind;             // kind of the connector
        BYTE multiplicity;               // multiplicity of the
                                            connector > = 2
        DWORD obj [20] ;                 // objects connected
        WORD obj_piece [20] ;            // marker of different pieces
                                            of one object
        DWORD joiner_extremity [20] ; // conventional connector
                                            extremity (1, 2, 3..)
} CONNECTORDATA;
```

CONNECTORDATA is filled for each connector, automatically detecting the constituents connected by the connector by means of the geometrical rules already summarized (in Sections 4.5.2 and 4.5.3).

However, this vector is not enough in order to check all the scene compliance criteria. To do that the *chains* must be found, using the CONNECTORDATA vector described.

4.7.2 Chains

The logic of a well-designed renode is that the force packets must flow from a slave member to the master member in hierarchical jnodes, or from a member to the constraint block in constraint jnodes. As has already been pointed out, a central jnode can always be seen, under this particular aspect, as a special hierarchical jnode. Cuspidal and tangent jnodes can always be avoided and so will not be mentioned.

The flow of the forces is assured by a number of *chains*, which are basically as follows:

...Constituent*Connector*Constituent*Connector*Constituent...

It always starts and ends with a constituent (member, force transferrer, stiffener, or constraint block).

If a chain meets a connector having a multiplicity higher than two, the chain is split into two or more chains: specifically $(n - 1)$ if the connector has multiplicity n. For a multiplicity 3 connector this means:

*Constituent.....

.....Constituent*Connector*Constituent* Connector

*Constituent.....

Clearly, if the logic of the connection is complex, finding all the chains might be quite difficult (see Figure 4.21 for a relatively simple example).

The physical meaning of the chains is apparent: *they are load paths.* Connectors may imply the selective transferring of only some components of forces (such as in shear only bolt layouts or in pins), but the set of all the available chains must be able to carry the applied loads, delivering them from one constituent to another.

If a constituent is not part of any chains, it cannot be considered connected. Chains referring to stiffeners are closed loops or broken chains.

Chains can be unconnected to anything, closed loops or broken. These chains have no effect on the renode (Figure 4.22).

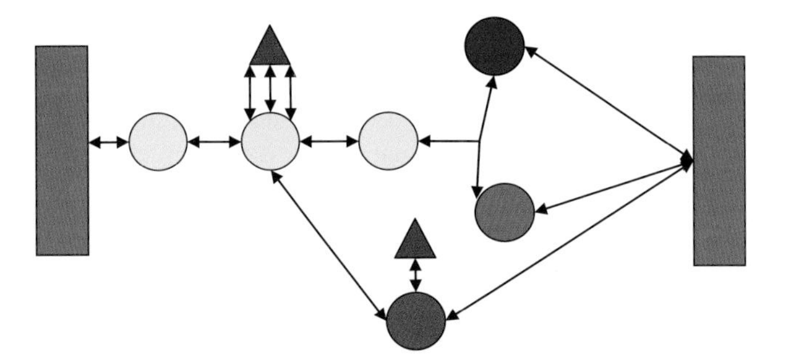

Figure 4.21 Structured arrangements imply more chains, here five (two broken and three complete).

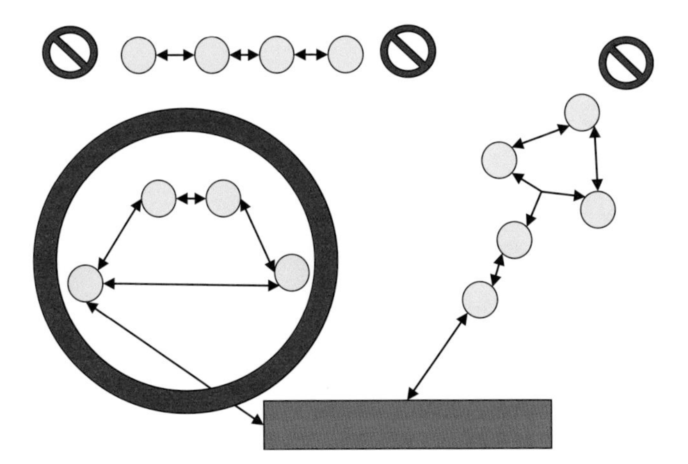

Figure 4.22 Invalid chains: unconnected (top left), loop (bottom left), broken (top right).

It is clear that finding all the existing chains and classifying them is a needed step in order to automatically ensure that a renode is properly conceived. If a renode is not properly conceived, there is no hope that it could work effectively. A bad layout would imply some lack of constraint for some constituent, which is clearly unacceptable. However, finding the chains, and ensuring that they do indeed exist, is not enough in order to assess that a connection is well designed. For instance, a connector may undergo forces that it is unable to resist, and this might lead to failure, not simply because the level of the forces applied is too high, but because the geometry of the connection is ill-conceived, for instance if a moment resisting beam-end is connected only by means of a short weld along the web center.

In the next section we will explain how the chains can be found by an automatic algorithm.

4.7.3 Finding Chains

Finding the chains automatically is a quite difficult task. It will be assumed that the CONNECTORDATA vector has been set up, detecting the existing connections on the basis of the geometry of the scene (as explained). So, for each connector, all the constituents connected, at each extremity, are known. Also, the vector of the existing constituents, members, force transferrers, stiffeners, and a possible constraint block, is also known, as well as the nature of the jnode of the renode at hand.

The algorithm accepts a constituent S as start, and another constituent E as end. All the chains connecting S to E will be found, and added to a list of existing chains. Chains are added only if they are different from all the other ones previously added. A chain from S to E can be synthetically written as (S-E). If more than one chain exists, each chain can be identified by an index, like this: $(S\text{-}E)_i$.

The algorithm is repeated (*nmembers* – 1) times for a hierarchical or central jnode, and *nmembers* times for a constraint. Precisely:

- Hierarchical and central: (*nmembers* – 1) run of the algorithm with $S = m_i$ and $E = m_1$, with $i \neq 1$. The chains will be $(m_i - m_1)$.
- Constraints: (*nmembers*) run of the algorithm with $S = m_i$ and $E = cb$, with cb = constraint block. The chains will be $(m_i - cb)$.

The algorithm is described by the function *DoPaths*, which accepts the following parameters in its prototype:

1) the number of connectors in the renode, *njoin*
2) the vector of CONNECTORDATA, *pjd*[]
3) the number of chains (by reference, i.e. will be modified), *nchains*
4) the chain vector (by reference, i.e. will be modified), *pchains*[]
5) the start constituent, *objstart,* and its subconstituent number, *cobjstart*
6) the end constituent, *objend*
7) the maximum number of objects in a chain, *maxnobjchain*
8) a flag to decide if broken chains must be added, or not, *deletebrokenchains*

Realistically, the maximum object in a single chain can be set to 30. However, there are situations where a chain might have more elements. The number 100 can be considered a comfortable upper bound.

Chains are stored in an appropriate data structure, CHAIN, which embeds the following data for a single chain (that is an object *pchains*[i]):

- number of objects in the chain, *nobject*
- chain-objects identifier vector, *obj*[100]
- chain-objects subconstituent-number vector, *obj_comp*[100]
- a vector of chain-objects selection flags, *selection*[100]

We will described the algorithm to find all the chains of a hierarchical renode. Extension to constraints is straightforward.

At the beginning, it sets up a loop on the slave members, and for each member all its subconstituents (compound sections), calling the function *DoPaths*:

```
nchains = 0;
pchains= NULL VECTOR;
nmaster = 1; // usually the master member is the member number 1
For(each slave Member i) // i > 1
{
        For(each subconstituent icomp of Member i)
        {
                // TRUE means that broken chains will be discarded
                DoPaths(njoin, pjd, nchains, pchains, i, icomp,
                nmaster, maxnobjchain, TRUE);
        }
}
```

The function *DoPaths* has the following layout, in pseudocode:

```
DoPaths(njoin, pjd, nchains, pchains, objstart, cobjstart,
objend, maxnobjchain, deletebc)
{
        // Add a new empty chain to the vector
        AddNewEmptyChain();
        // Initializes the new chain
        LastChain = GetLastChain(nchain, pchain);
        LastChain.nobject = 1;
        LastChain.obj[0] = objstart;
        LastChain.obj_comp = cobjstart;
        // Next function is described below: repeatedly adds a
        constituent to the chain
        ret = StepForward(njoin, pjd, nchains, pchains, nchains,
        objend, maxnobjchain);
        If(deletebc = True) // delete broken chains
        {
                For(each chain i)
                {
                        chain = GetChain(pchain, i); // gets the
                        current chain from array
```

```
                    If(chain.obj[chain.nobject-1] != objend)
                    // does not end where it should
                    {
                          DeleteChain(pchain, i); // remove
                          the current chain from array
                    }
              }
        }
  }
```

The function *StepForward* is used to add a constituent to the current chain, and it is the kernel of the algorithm: it calls itself with a recursive call. As it is complex, it will be described in parts.

```
StepForward(njoin, pjd, nchains, pchains, thischain, objtarget,
maxnobjchain)
{
      // get the last object of the current chain
      CurrentChain = GetChain(pchains, thischain);
      obj = CurrentChain.obj[CurrentChain.nobject-1];
      cobj = CurrentChain.obj_comp[CurrentChain.nobject-1];
      If(ItIsAConnector(obj)) execute Part 1
      Else                    execute Part 2
}
```

Part 1 is executed if the last object in the chain is a connector and it is as follows:

```
// Get the JOINERDATA of the current object, a connector
jdata = GetJData(obj);
countexit = 0; // a useful counter
SaveChain = CurrentChain;
For(i=0; i < jdata.multiplicity; i++) // i is an extremity of the
connector
{
      currentobj = jdata.obj[i];
      currentcobj = jdata.obj_piece[i];
      If(IsThisObjectOneBeforeInChain(SaveChain, currentobj,
      currentcobj)) Do Nothing;
      Else If(IsThisObjectAlreadyInChain(SaveChain, currentobj,
      currentcobj)) Do Nothing;
      Else
      {
            countexit = countexit + 1;
            If(countexit == 1) // the first time we get out of the
            connector, same chain
            {
                  execute Part 3: add the new object to the
                  current chain
            }
```

```
        Else // exit number > 1: we need a new chain (chain splitting)
        {
            execute Part 4: add a new chain = to the current, and add
            new object to it
        }
}
}
```

Part 3 adds to the current chain the new object, and calls *StepForward* again:

```
//adds this new object to the current chain
CurrentChain.nobject = CurrentChain.nobject + 1;
CurrentChain.obj [CurrentChain.nobject-1] = currentobj;
CurrentChain.obj_comp [CurrentChain.nobject-1] = currentcobj;
If (CurrentChain.nobject = = maxnobjchain) return;
Else If ((currentobj = = objtarget)
{
        return; // we arrived at destination
}
// Now we move forward
StepForward(njoin, pjd, nchains, pchains, thischain, objtarget,
maxnobjchain);
```

Part 4 adds a new chain, as the connector has more than two extremities. Then this new chain is made equal to the current one, and finally the new object is added and *StepForward* called again:

```
AddChain(nchains, pchains); // this also increments nchains by 1
NewChain = GetLastChain(nchains, pchains);
NewChain = SaveChain; // so new chain is identical to current
NewChain.nobject = NewChain.nobject + 1;
NewChain.obj [NewChain.nobject-1] = currentobj;
NewChain.obj_comp [NewChain.nobject-1] = currentcobj;
If (NewChain.nobject = = maxnobjchain) return;
Else If ((currentobj = = objtarget)
{
        return; // we arrived at destination
}
// Now we move forward
StepForward(njoin, pjd, nchains, pchains, nchains, objtarget,
maxnobjchain);
```

Part 2 is executed if the last object in the chain is not a connector and it is as follows:

```
countexit = 0; // a useful counter
SaveChain = CurrentChain;
For(i=0; i < njoin; i) // i is the generic connector
{
        // Get the JOINERDATA of the current connector
        jdata = GetJData(i);
```

```
        found = FALSE; // a Boolean flag
        For(j=0; j < jdata.multiplicity; j++) // j is an extremity
        of the connector
        {
                currentobj = jdata.obj[j];
                currentcobj = jdata.obj_piece[j];
                // obj,cobj are the object and subconstituent
                number of the last obj in chain
                If((currentobj == obj) AND (currentcobj == cobj))
                //curr = last of current chain
                {
                        Found = TRUE;
                }
                If(found == TRUE)
                {
                        currentobj = jdata.theobject; //
                        connector id
                        currentcobj = 0; // no subconstituent for a
                        connector
                        // Do nothing for first two ifs.
                        If(IsThisObjectOneBeforeInChain
                        (SaveChain, currentobj, currentcobj));
                        Else If(IsThisObjectAlreadyInChain
                        (SaveChain, currentobj, currentcobj));
                        Else
                        {
                                countexit = countexit + 1;
                                If(countexit == 1) // the first time we
                                get out of the connector
                                {
                                        execute Part 3
                                }
                                Else // exit number > 1: we need a new
                                chain (chain splitting)
                                {
                                        execute Part 4
                                }
                        }
                }
        }
}
```

With this algorithm a quite important goal is achieved: the automatic detection of the logic of a renode, and so the ability to automatically control it.

Chains can be transformed easily into strings, if each object numerical identifier is replaced by its unique alphanumerical identifier. An example will better explain this.

The example of Figure 4.23 shows a three member renode using a *double* angle T1 (so having two subconstituents) and a plate P1 as force transferrers. There are four bolt layouts, and two weld layouts, a total number of six connectors, connecting objects according to the following connectivity list:

1) B1 connects T1, first and second subconstituent, and m2
2) B2 connects m3 and P1
3) B3 connects T1 second subconstituent and m1
4) B4 connects T1, first subconstituent, and m1
5) W1 connects P1 and T1, second subconstituent
6) W2 connects P1 and T1, second subconstituent

The chains found by the automatic algorithm are the following, grouped by slave member:

```
CHAIN 1 m2 :(B1):T1(2):(B3):m1
CHAIN 2 m2 :(B1):T1:(B4):m1

CHAIN 1 m3 :(B2):P1*(W1)*T1(2):(B1):T1:(B4):m1
CHAIN 2 m3 :(B2):P1*(W1)*T1(2):(B3):m1
CHAIN 3 m3 :(B2):P1*(W2)*T1(2):(B1):T1:(B4):m1
CHAIN 4 m3 :(B2):P1*(W2)*T1(2):(B3):m1
```

The strings are generated taking into account the subconstituent number and using the "∗()∗" symbols for the weld layouts, and the ":():" symbols, for the bolt layouts.

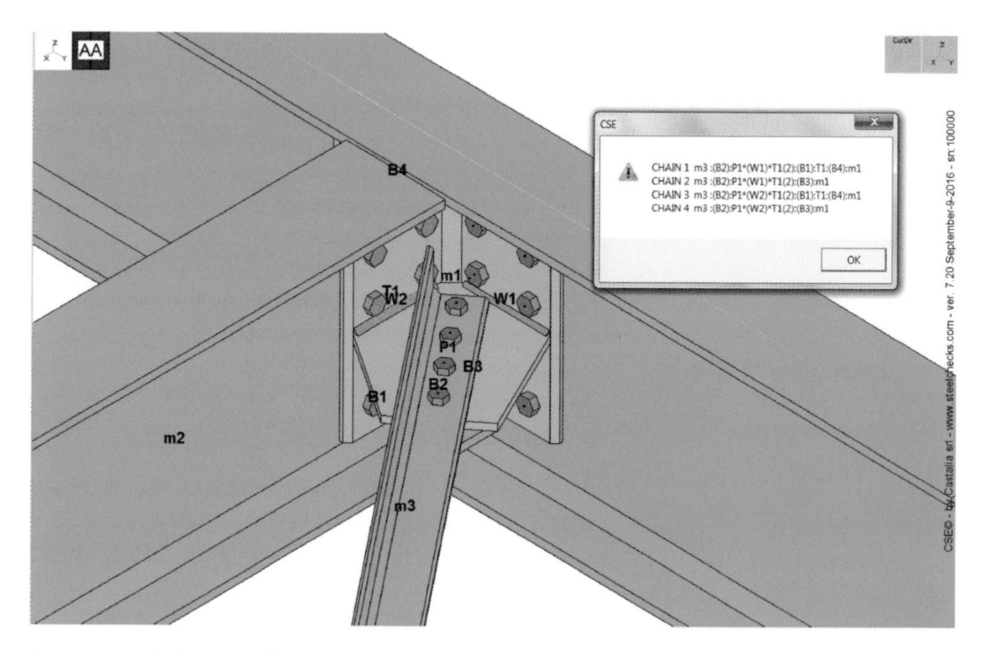

Figure 4.23 A three member connection using T1 and P1 as force transferrers. Some of the chains automatically found are listed (the model was prepared using software CSE by AMSIS srl, Rovato, Italy).

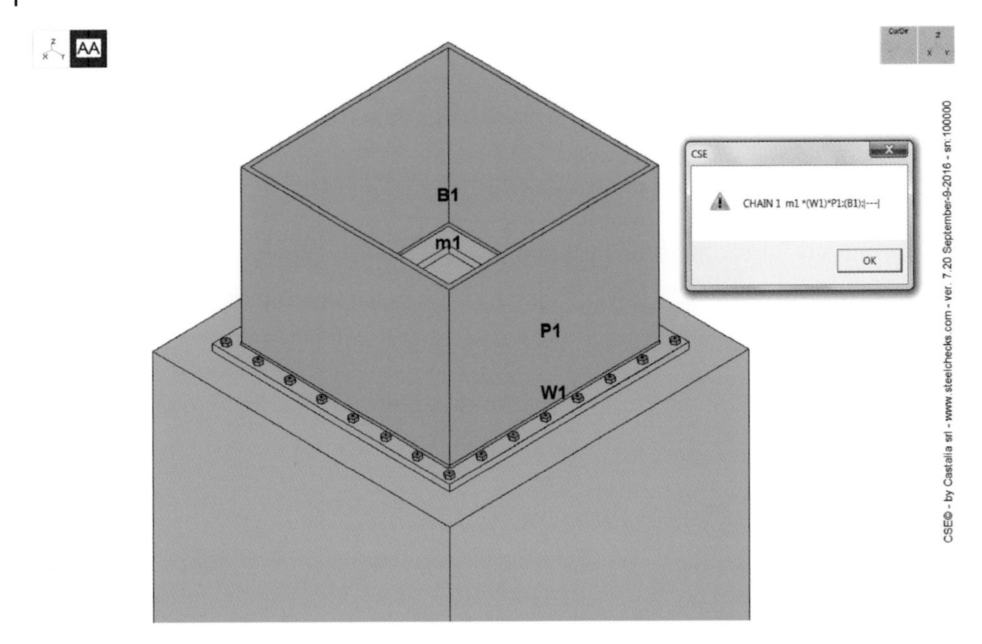

Figure 4.24 A base node isoconnected (the model was prepared using software CSE by AMSIS srl, Rovato, Italy).

As all the slave members are connected to the master, and as there is no constituent unconnected, the renode is capable of transferring the forces from the slave members to the master.

The example of Figure 4.24 shows a base node having one member m1 only, and using a constraint block (named "|---|"), a plate P1, a weld layout W1, and a bolt layout B1. The chain is unique and it is as follows:

```
CHAIN 1 m1 *(W1)*P1:(B1):|---|
```

Chains can be hundreds. – in a complex renode it is relatively easy to find several hundred chains. If the allowed maximum length of chains is too low, chains may be broken before getting to the target.

4.8 Prenodes

Scenes can be automatically generated for special families of renodes, which have been parameterized. They will be called *prenodes* (parameterized real nodes). For instance, a prenode might assume that a rectangular base plate *P*1 of a column *m*1 made by a rolled H is created by adding a plate having width equal to the width of the H cross-section plus 150 mm, height equal to the depth of the H cross-section plus 90 mm, and thickness equal 2 times the thickness of the H cross-section flange. This can be easily described as follows:

$$P1.b = m1.b + 150mm$$

$$P1.h = m1.h + 90mm$$

$$P1.t = 2^{*} m1.tf$$

Each object initially has a well defined geometry, which can be identified by a limited number of parameters. So, if the parameters of a new object are defined as a function of the parameters of the already existing 3D objects, new objects can be added to the scene in a parameterized way. At the beginning of the operations, the scene contains only the members, and so only their parameters will be used.

New variables can be defined in a very general way, and using the internal forces flowing at the extremities of the members, identified by variables like $m2.N$, or $m3.M3$, also special checking formulas can be defined. For instance, in this way, to check for net cross-section:

$$m1.N < (m1.A - B1.nb^{*} B1.Dh^{*} m1.tw)^{*} m1.fu/gamma.M2$$

where

$m1.N$	is the axial force in the member m1
$m1.A$	is the cross-section area
$B1.nb$	is the number of bolts in a bolt layout
$B1.Dh$	is the diameter of the hole related to the bolts of bolt layout B1
$m1.tw$	is the thickness of the web of the cross-section of m1.
$m1.fu$	the ultimate stress of member m1 material.
$gamma.M2$	the partial safety factor for material rupture

Parameterization will not be directly described in this work, but much work has been done by the author in this area and almost 800 families of renodes have been parameterized (it is acknowledged that the extensive work of parameterization has been done by Ing. Marco Croci). Using the predefined variables it is possible to teach a procedure how to do specific checking jobs.

4.9 After Scene Creation

This chapter has briefly summarized how the scene is created and manipulated, by means of analytic geometry. The final goal is to store the positions, orientations, sizes, and exact shapes of all the constituents used in order to connect a number of members to themselves, or to a reference (the constraint block).

Just using the geometry of the objects, it is also possible to detect the chains, and so to inspect connections and the general coherence of a renode, a preliminar step to get to the checks.

Scenes can be quite complex, and the aim of the analysis should be to analyze all the connections at a time. Up to now, it has been shown how to deal with the geometrical part of the problem. The following chapters will get into the detail of the analytical methods available.

5

Pillars of Connection Analysis

5.1 Equilibrium

5.1.1 Generality

When considering the effects of the loads in all the constituents of a renode, a first necessary condition to be satisfied is that all the constituents must be in equilibrium. This may appear obvious, but there are ways in which subtle violation of this basic principle may arrive to the final design.

Neglecting some of the forces exchanged at the constituents interface is a way to violate the equilibrium, as these neglected forces will be unbalanced.

Also applying the forces in the wrong points of application, or in points of application which are not coherent in the different stages of the analysis, will indirectly lead to unbalanced constituents.

A typical way to apply forces at the wrong points of application is neglecting eccentricities. Setting these eccentricities to zero definitely means positioning the forces onto a different line of action, and this, in turn, implies an equilibrium violation (see Figures 5.1 and 5.2). There is no way to assess, a priori, that the sum of all these apparently minor violations may not lead to a severe underestimate of the utilization.

Consider, for instance, a bolt layout with three bolts of diameter d, connecting two plates, loaded by a shear force F equal to k times the limit shear for the bolts. So $F = 3kF_{lim,bolt}$. The value of the limit force in bolts can be set equal to the effective area times a limit stress: $F_{lim,bolt} = Af_v$. Let the effective area be the gross area of the shaft of diameter d: $A = 0.25\ \pi d^2$. If the thickness of the plates is t, and F is applied at mid-thickness, the additional moment at the plate interface is $M = 0.5Ft$.

Two possible models can be assumed to balance the additional moment.

The first uses tensile and compressive forces in the bolt shafts (Figure 5.3). A variation of this model assigns the compression of the bolts as pressures exchanged at the interface between the plates.

The second model assigns the additional moment to the bolt shafts.

Consider the first model.

Assuming a distance between the bolts of $3d$, the axial force N in the external bolts can be evaluated as

$$N = \frac{M}{6d} = \frac{3 \cdot k \cdot f_v 0.25\pi d^2 0.5t}{6d} \approx 0.2kdtf_v$$

Steel Connection Analysis, First Edition. Paolo Rugarli.
© 2018 John Wiley & Sons Ltd. Published 2018 by John Wiley & Sons Ltd.

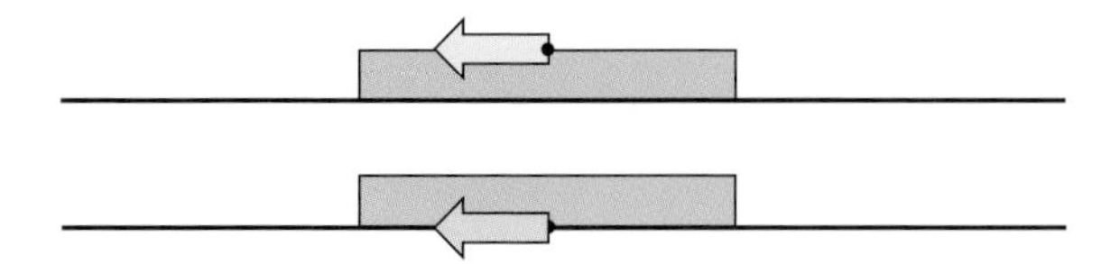

Figure 5.1 In this example an eccentricity has been neglected.

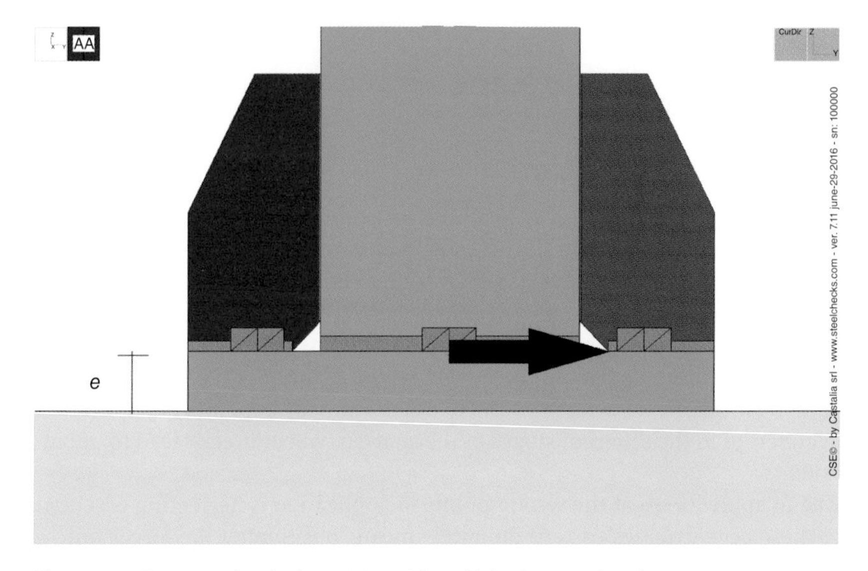

Figure 5.2 An example of a base joint with a thick plate, under shear.

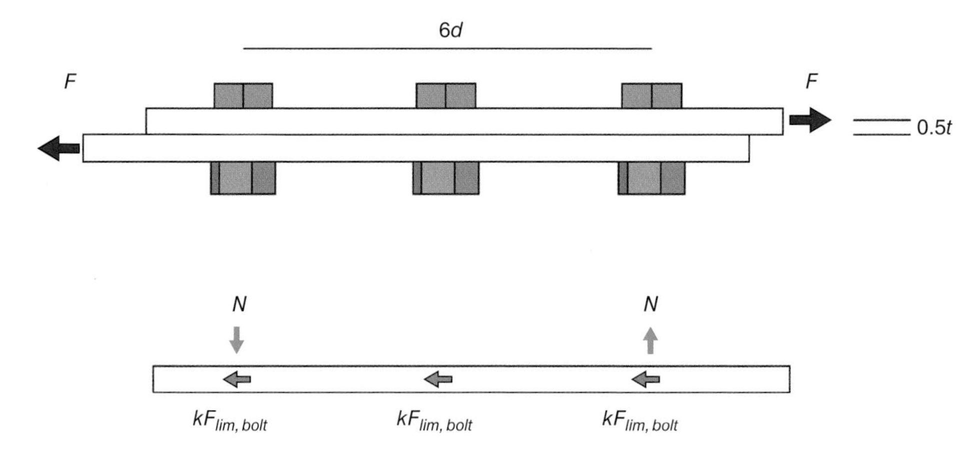

Figure 5.3 First model to balance the transport moment: extra axial forces in the bolt shafts.

If bolts class 8.8 are used, and the rules of Eurocode 3 are applied, $f_v = 384$ MPa. The limit stress of the bolts in tension is $f_t = 576$ MPa. The checking formula for the bolt would be (neglecting for simplicity the fact that in tensile force the threaded area must be used, which would strengthen the remark)

$$k + 1.4\frac{\frac{3 \cdot k \cdot f_v 0.25\pi d^2 0.5t}{6d}}{0.25\pi d^2 f_t} = k + 1.4 \cdot 0.25\frac{f_v}{f_t}\frac{kt}{d} = k + 1.4 \cdot 0.25\frac{384}{576}\frac{kt}{d} = k\left(1 + 0.23\frac{t}{d}\right) \leq 1$$

The second term of the first member of the inequality is usually neglected. As the two plates are in contact, it is often assumed that the moment arising by transferring the line of action of the forces from the mid-thickness to the interface between plates is absorbed by contact pressures. However, contact pressures alone would not be in equilibrium, as the plate would displace in the direction of the bolt axes. A tensile force in the bolt shaft must develop, to balance the pressures at contact. Although the amount of this axial force can be evaluated in several different ways, the simple model outlined predicts the order of magnitude of the force. The second term of the inequality, indeed, is not negligible. If $d = 2\,t$, the increment of utilization is 11.5%; if $d = t$, this is 23%.

A second possible model would divide the moment into three additional moments in the bolt shafts (Figure 5.4). Each bolt would carry $M/3$ which would lead to an additional tensile stress in the shaft equal to

$$\sigma = \frac{M}{3W} = \frac{32M}{3\pi d^3} = \frac{32 \cdot 3k \cdot f_v \cdot 0.25\pi d^2 0.5t}{3\pi d^3} = 4kf_v\frac{t}{d}$$

Once more, this stress is not negligible; indeed it is quite high.

Many methods currently set up in order to precisely check the stress state of the constituents of a renode imply a significant amount of computational effort. A first preliminary question arising is: can the violations of the equilibrium due to the currently accepted simplifications, lead to errors of importance higher than that of the errors avoided by refined computational approaches?

The answer is: in general, yes. Violations of equilibrium may lead to severe errors.

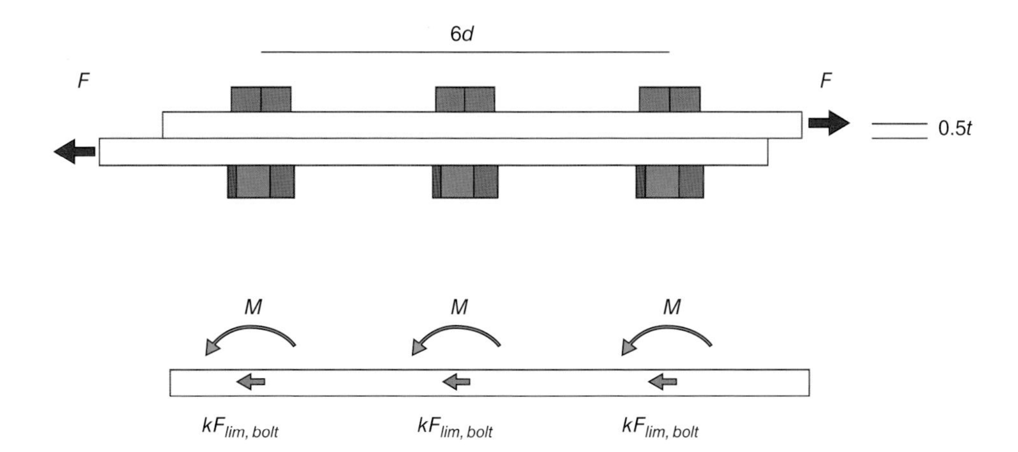

Figure 5.4 Second model to balance the moment of transport: additional moments in the bolt shafts.

On the other hand, the exact computation of equilibrium conditions in 3D is not an easy task using by hand or by spreadsheet computations. Typical renodes may have several inclined members, carrying six internal generalized forces at their extremities (three forces and three moments), and tens other constituents. Even if the equilibrium conditions to be satisfied are simple, as they involve simple vector operations, their number can be high, and the correct evaluation of all of them may become almost impossible to do by hand. The points of application of the forces in 3D space will differ from constituent to constituent, and the number of elementary operations to be done increases abruptly. Moreover, there can be tens of such renodes to be computed.

5.1.2 Statics of Free Rigid Bodies

When considering the problem of equilibrium, each constituent of the *scene* – that is, of the set of all the constituents that are needed to properly connect members, as positioned in 3D space – can be considered as a free rigid body. So, the set of all the forces and moments applied to it must satisfy the rules of rigid body equilibrium (Figure 5.5).

In this book the term *generalized forces* will refer to the forces and the moments. Considering the forces and moments applied to a point **P**, the two vectors $\mathbf{f}_\mathbf{P}$ and $\mathbf{m}_\mathbf{P}$ can be formed. If the symbols F_x, F_y, and F_z are used for the three components of the force vector relative to the global reference system, then

$$\mathbf{f}^\mathrm{T} = \left[F_x \;\; F_y \;\; F_z\right]$$

In the very same way, the vector $\mathbf{m}_\mathbf{P}$ is formed using the three components of the moment vector with reference to the global axes (X, Y, Z):

$$\mathbf{m}^\mathrm{T} = \left[M_x \;\; M_y \;\; M_z\right]$$

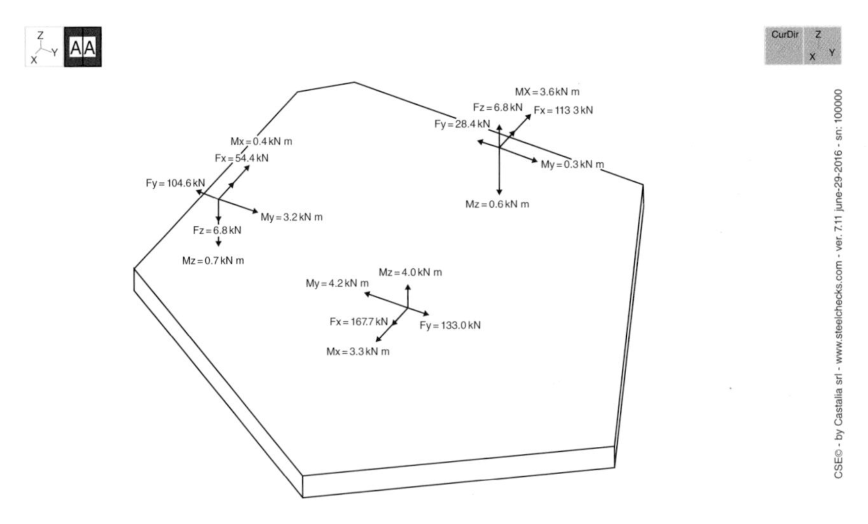

Figure 5.5 A constituent in space under the effect of three force packets.

Considering the vector having order 6 obtained by listing first the components of vector **f** and then the components of vector **m**, the *generalized force vector* **s** can be introduced:

$$\mathbf{s}^{\mathbf{T}} = \left[F_x \;\; F_y \;\; F_z \;\mid\; M_x \;\; M_y \;\; M_z\right]$$

This notation is much more compact and will be the preferred one.

The *elementary generalized force packet*, or briefly the *force packet*, is a vector **s** applied to a point **P** of a solid, and will be identified by the symbol $\mathbf{s_P}$. Although the moments have no application point, the forces do, because is the application point is changed, moments arise.

When considering the equilibrium of a rigid body (Figure 5.6), the generalized forces $\mathbf{s_P}$ can be transferred to a different point **Q**, becoming $\mathbf{s_Q}$, with no modification of the conditions of equilibrium, provided that the necessary transports are considered.

Considering the vector **f** acting at point **P**, if this vector is transferred to point **Q**, a moment of transport will arise. If x_{QP}, y_{QP}, z_{QP} are the components of the vector (**Q** – **P**), going from **P** to **Q**, the three components of the moment vector **m** can be written:

$$m_x = F_z y_{QP} - F_y z_{QP}$$

$$m_y = F_x z_{QP} - F_z x_{QP}$$

$$m_z = F_y x_{QP} - F_x y_{QP}$$

This means that

$$\mathbf{m_Q} = (\mathbf{Q} - \mathbf{P}) \wedge \mathbf{f_P} = \mathbf{f_P} \wedge (\mathbf{P} - \mathbf{Q})$$

The subscript "Q" applied to vector **m** merely recalls that these moments are due to the displacement of vector $\mathbf{f_P}$ to point **Q**.

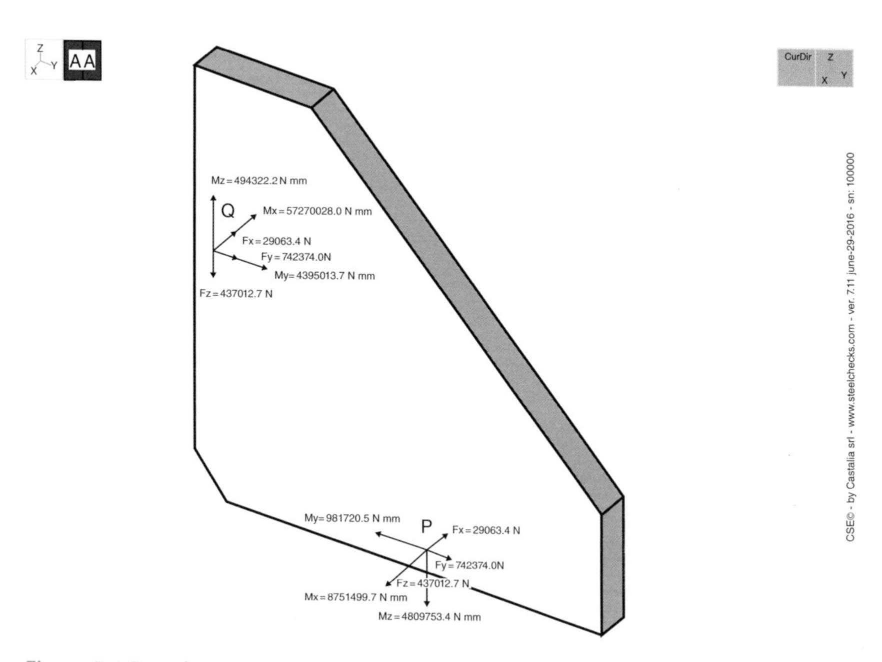

Figure 5.6 Transferring of a force packet from P to Q.

The previous relations can be written synthetically in matrix form as

$$\mathbf{m_Q} = \mathbf{T_{3PQ}f_P}$$

where the 3×3 square matrix $\mathbf{T_{3PQ}}$ can be defined as

$$\mathbf{T_{3PQ}} = \begin{bmatrix} 0 & -z_{QP} & y_{QP} \\ z_{QP} & 0 & -x_{QP} \\ -y_{QP} & x_{QP} & 0 \end{bmatrix}$$

By definition

$$x_{QP} = x_Q - x_P$$

$$y_{QP} = y_Q - y_P$$

$$z_{QP} = z_Q - z_P$$

so it can be easily seen that

$$\mathbf{T_{3PQ}} + \mathbf{T_{3QP}} = \mathbf{0}$$

Now if the generalized force packet **s** applied to point **P**, $\mathbf{s_P}$, is applied to point **Q**, then

$$\mathbf{s_Q} = \mathbf{T_{PQ}s_P}$$

where the 6×6 square matrix $\mathbf{T_{PQ}}$ can be written (using $4 - 3 \times 3$ blocks)

$$\mathbf{T}_{PQ} = \left[\begin{array}{c|c} \mathbf{I} & \mathbf{0} \\ \hline \mathbf{T}_{3PQ} & \mathbf{I} \end{array}\right]$$

which means

$$\mathbf{f_Q} = \mathbf{f_P}$$

$$\mathbf{m_Q} = \mathbf{T_{3PQ}f_P} + \mathbf{m_P}$$

It can be easily seen that for the order 6 matrix $\mathbf{T_{PQ}}$

$$\mathbf{T_{QP}T_{PQ}} = \mathbf{T_{PQ}T_{QP}} = \mathbf{I}$$

The proof follows:

$$\mathbf{T}_{QP}\mathbf{T}_{PQ} = \left[\begin{array}{c|c} \mathbf{I} & \mathbf{0} \\ \hline \mathbf{T}_{3QP} & \mathbf{I} \end{array}\right] \cdot \left[\begin{array}{c|c} \mathbf{I} & \mathbf{0} \\ \hline \mathbf{T}_{3PQ} & \mathbf{I} \end{array}\right] = \left[\begin{array}{c|c} \mathbf{I} & \mathbf{0} \\ \hline \mathbf{T}_{3QP} + \mathbf{T}_{3PQ} & \mathbf{I} \end{array}\right] = \left[\begin{array}{c|c} \mathbf{I} & \mathbf{0} \\ \hline \mathbf{0} & \mathbf{I} \end{array}\right]$$

This merely means that transferring a packet **s** from a point **P** to a point **Q** and then back from point **Q** to point **P**, the original packet is found.

If at a point **P** of a body A, a packet of generalized forces **s** is exchanged with another body B, the packet applied at point **P** of the body A, due to its interaction with the body B, will be identified by the symbol $\mathbf{s_{PB}}$, meaning "forces at **P** due to the interaction with body B".

Given a rigid body in space loaded at a number of n points $\mathbf{P_1}, \mathbf{P_2}, \mathbf{P_i}, \ldots, \mathbf{P_n}$ by the elementary generalized force packets $\mathbf{s_{P1}}, \mathbf{s_{P2}}, \mathbf{s_{Pi}}, \ldots, \mathbf{s_{Pn}}$ the necessary equilibrium conditions can be written assuming that all the force packets are transported to a single point **O**, and writing the following equilibrium condition:

$$\sum_{i=1}^{n} \mathbf{s}_{\mathrm{O_i}} = \mathbf{0} \tag{5.1}$$

As the force packets in point **O** can be obtained by the force packets in point $\mathbf{P_i}$, the equilibrium condition can also be written as

$$\sum_{i=1}^{n} \mathbf{T}_{\mathbf{PiO}} \mathbf{s}_{\mathbf{Pi}} = \mathbf{0}$$

The previous matrix equation states that summing up all the forces and moments transported at a single point O, the result must be null. Also, if $n = 1$ then the generalized forces applied at a single point must all be null. If this is not the case, the body cannot be in equilibrium.

If it is assumed that some force packets are unknown, while others not, the equilibrium condition is an equation. If the number of unknown packets is m, the following considerations apply.

It can be easily seen that if $m = 1$ the equation can be solved, expressing the force packet at the point of the unknown packet $\mathbf{P_k}$, as a function of the force packets at all other points $\mathbf{P_i}$ with $\mathrm{i} \neq \mathrm{k}$.

$$\mathbf{s}_{\mathbf{P}k} = - \sum_{i=1, i \neq k}^{n} \mathbf{T}_{\mathbf{PiPk}} \mathbf{s}_{\mathbf{Pi}}$$

Generally speaking, if $m > 1$ there are infinite possible solutions to the equations. If this happens, the "exact" solution can only be found by considering the bodies as deformable.

5.2 Action Reaction Principle

In his famous work *Philosophiae Naturalis Principia Mathematica* (Newton 1687), Sir Isaac Newton stated as the third law of motion the action reaction principle, which stands as a rock, unmodified after centuries (Figure 5.7):

> *Actioni contrariam semper & aequalem esse reactionem : sive corporum duorum actiones in se mutuo semper esse aequales & in partes contrarias dirigi.*

In the revised translation made by Ian Bruce (Newton 1687) to better convey the exact words delivered by Isaac Newton, this is

> *The action is always opposite and equal to the reaction: or the actions exchanged by two bodies are always mutually equal and directed in opposite directions.*

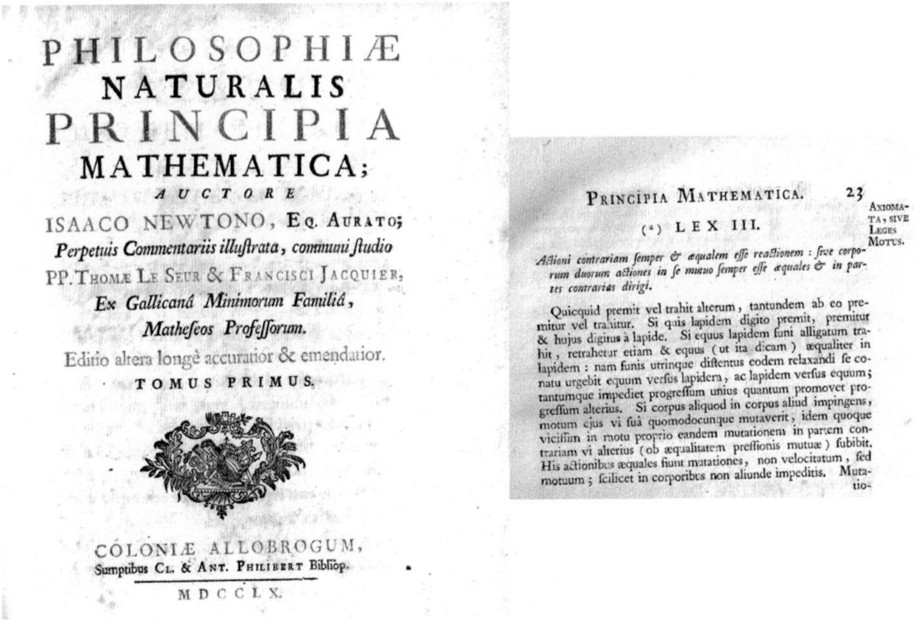

PHILOSOPHIÆ
NATURALIS
PRINCIPIA
MATHEMATICA;
AUCTORE
ISAACO NEWTONO, EQ. AURATO;
Perpetuis Commentariis illustrata, communi studio
PP. THOMÆ LE SEUR & FRANCISCI JACQUIER,
Ex Gallicanâ Minimorum Familiâ,
Matheseos Professorum.
Editio altera longè accuratior & emendatior.
TOMUS PRIMUS.

COLONIÆ ALLOBROGUM,
Sumptibus CL. & ANT. PHILIBERT Bibliop.
MDCCLX.

PRINCIPIA MATHEMATICA. 23

AXIOMATA, SIVE LEGES MOTUS.

(a) LEX III.

Actioni contrariam semper & æqualem esse reactionem : sive corporum duorum actiones in se mutuo semper esse æquales & in partes contrarias dirigi.

Quicquid premit vel trahit alterum, tantundem ab eo premitur vel trahitur. Si quis lapidem digito premit, premitur & hujus digitus à lapide. Si equus lapidem funi alligatum trahit, retrahetur etiam & equus (ut ita dicam) æqualiter in lapidem : nam funis utrinque distentus eodem relaxandi se conatu urgebit equum versus lapidem, ac lapidem versus equum; tantumque impediet progressum unius quantum promovet progressum alterius. Si corpus aliquod in corpus aliud impingens, motum ejus vi suâ quomodocunque mutaverit, idem quoque vicissim in motu proprio eandem mutationem in partem contrariam vi alterius (ob æqualitatem pressionis mutuæ) subibit. His actionibus æquales fiunt mutationes, non velocitatum, sed motuum; scilicet in corporibus non aliunde impeditis. Mutatio-

Figure 5.7 The 1760 2nd Jesuitical edition of *Philosophiae Naturalis Principia Mathematica* by Isaac Newton. On the right the Third Law, at page 23 of the first volume (courtesy of the author's friend Giorgio Nieri, owner of the volumes).

To better explain, Newton proposed several examples, which are still of interest:

> *Anything pressing or pulling another, by that is pressed or pulled just as much. If anyone presses a stone with a finger, the finger of this person is pressed by the stone. If a horse pulls a stone tied to a rope, and also the horse (as thus I may say) is drawn back equally by the stone: for the rope stretched the same on both sided requiring itself to be loosened will draw upon the horse towards the stone, and the stone towards the horse; and yet it may impede the progress of the one as much as it advances the progress of the other.*

So, if two constituents are connected, they must exchange the same force packet. As in the present model of connections the connection is realized by connectors, according to the third law the force packet loading each connector at each extremity, is equal and opposite to the force packet loading the connected constituent (member, force transferrer, stiffener, or constraint block) at that extremity. With the notation introduced, this means that between two constituents A and B (one of the two is necessarily a connector), at a point **P** (which will be one of the extremities of the connector)

$$\mathbf{s}_{\mathbf{P}A} = -\mathbf{s}_{\mathbf{P}B}$$

$$\mathbf{s}_{\mathbf{P}B} = -\mathbf{s}_{\mathbf{P}A}$$

The action reaction principle is often violated in current connection analysis. A violation of the action reaction principle will, sooner or later, lead to a violation of equilibrium, somewhere.

The possible interfaces where the third law must be applied, according to the model that is used in this work, are:

- member to node
- member to connector
- force transferrer to connector
- stiffener to connector
- constraint block to connector.

These interfaces may be reduced to two:

- member to node (external interface)
- constituent to connector (internal interface).

At the member-to-node interface the third law is often violated in current connection analysis practice. This is done, for instance, when after having analyzed a 3D BFEM, checking the members for six internal forces (which are all needed to ensure the equilibrium of nodes), getting to the connection analysis a number of these internal forces are thrown away, often assuming that they flow to other constituents nearby (typically floors: the simplification is sometimes seen also when no floor at all is available). If N is then the node of the BFEM and E is the ideal extremity of a member with no eccentricity, this means

$$\mathbf{s}_{PN} \neq -\mathbf{s}_{PE}$$

More precisely, if torque M_1, one shear V_2, and one bending M_3 are neglected, what happens is as follows (see also Appendix 1 for the notation referring to the internal forces of members).

The internal forces at the extremity E of a member are a vector $\boldsymbol{\sigma}$ expressed in local reference system of the member, axes (1, 2, 3).

Usually these internal forces are one of the main results of the BFEM, and are used to check members for resistance and stability.

These internal forces are used to form the $\boldsymbol{\sigma}$ vector in this way:

$$\boldsymbol{\sigma}^{\mathrm{T}} = [N \quad V_2 \quad V_3 \mid M_1 \quad M_2 \quad M_3]$$

If the local reference system (1, 2, 3) of the member has orientation matrix $\mathbf{T}$, then

$$\mathbf{s}_{PE} = \mathbf{T}^{\mathrm{T}}\boldsymbol{\sigma}$$

Where $\mathbf{s}_{PE}$ are the forces applied to the member extremity in global reference system.

At the node, the BFEM analysis ensures that:

$$\mathbf{s}_{PN} = -\mathbf{T}^{\mathrm{T}}\boldsymbol{\sigma}$$

while at the extremity of the member, switching to connection analysis it is assumed

$$\mathbf{s}_{PE} = \mathbf{T}^{\mathrm{T}}\mathbf{C}\boldsymbol{\sigma} = \mathbf{T}^{\mathrm{T}} \cdot \left[\begin{array}{ccc|ccc} 1 & 0 & 0 & 0 & 0 & 0 \\ 0 & 0 & 0 & 0 & 0 & 0 \\ 0 & 0 & 1 & 0 & 0 & 0 \\ \hline 0 & 0 & 0 & 0 & 0 & 0 \\ 0 & 0 & 0 & 0 & 1 & 0 \\ 0 & 0 & 0 & 0 & 0 & 0 \end{array}\right] \left[\begin{array}{c} N \\ V_2 \\ V_3 \\ M_1 \\ M_2 \\ M_3 \end{array}\right] = \mathbf{T}^{\mathrm{T}} \left[\begin{array}{c} N \\ 0 \\ V_3 \\ \hline 0 \\ M_2 \\ 0 \end{array}\right] = \mathbf{T}^{\mathrm{T}} \cdot \boldsymbol{\sigma}'$$

So

$$\mathbf{T}^{\mathrm{T}}\boldsymbol{\sigma}' - \mathbf{T}^{\mathrm{T}}\boldsymbol{\sigma} \neq \mathbf{0}$$

The introduction of the diagonal matrix **C** is the equivalent of assuming null V_2, M_1, and M_3.

This kind of violation can be applied to the problem almost without notice. The unbalanced part will be the node, if $\boldsymbol{\sigma}'$ is used coherently, but this equilibrium is of no interest for the connection analyst once the BFEM has already been run. Moreover, the node in the BFEM is indeed in equilibrium, because *there* the internal forces components are not neglected. In this way, forces get lost.

To apply such simplifications coherently, probably some end release would have to be applied to the beam elements simulating the members, which in turn would possibly have led to higher lateral displacements (globally and locally), and increased P-Δ effects: all definitely unwanted by the analyst. With this incorrect procedure a double advantage is apparently obtained, at no cost.

A second interface where the forces are modified in such a way as to violate the third law is often the interface between connectors and connected constituents, i.e. the internal interfaces.

The third law would require that the forces acting over a connector (i.e. over all the connector subconstituents, one by one) should be applied with sign reversed, at each connector extremity, to the connected constituent. However, when getting to the checking of constituents, the forces flowing from the connectors are sometimes taken not fully coherently with that acting over the subconstituents of the same connectors.

For instance, considering a shear-only bolt layout under in plane eccentric shear (say, a simple beam shear), a shear force vector having different modulus and direction loads each bolt. However, this is not always properly taken into consideration, and the forces acting over the constituent are made all parallel (neglecting torque). This is often seen, for instance, in block tear check, where the elementary forces adopted to check, for instance, a coped beam end, are not exactly the ones adopted to check the bolts (see Chapter 8).

Similar considerations would apply to fillet weld layouts, which are checked against a stress distribution that is not then applied, with sign reversed, to the welded constituents.

The third law not only states that the forces exchanged must be equal and with sign reversed, but it also states that the points where these forces are exchanged must be the same in space. Otherwise, due to the additional moments related to transport, it's easy to see that equilibrium is violated.

The correct modeling of the points of application of elementary generalized stresses coming from connectors may have an importance in constituent checks, as the flow of stresses in the connected constituents may differ significantly. For instance, applying a single fillet weld having a length of 100 mm, is not the same as applying two fillet welds having a length of 30 mm, as the forces are more concentrated, and are exchanged in different parts of the connected constituent. However, when checking the constituents, this is often forgotten, assuming a perfect transmission and spreading of the stresses over the whole available length. This model is not coherent and implicitly violates the third law, as the forces used to check the constituent are not the ones used to check the welds, with sign reversed.

5.3 Statics of Connections

5.3.1 Equilibrium of Members in Renodes: Proper and Dual Models

Getting back to the model of connections outlined in Chapter 3, it can be assumed that the *scene* of a real node is made by members, a possible constraint block, force transferrers, stiffeners, and connectors. All these bodies must be in equilibrium under the applied loads.

The region where the members meet in the BFEM node, with all the constituents needed to physically realize the connections, will be named *nodal region*. The list of all the constituents with their geometrical position and orientation defines the *scene* (Figure 5.8).

When considering the surroundings of the renode, members can be cut at a distance from the ideal extremities, defining the extent of the nodal zone. In this way, the nodal region is isolated in space from the rest of the world and must be in equilibrium under the applied loads.

The length of the member stumps is decided in such a way that the internal stress state at the cross-sections obtained by cutting the members is coherent with the beam theory, that is, it is far enough from the connected end.

Recalling the De Saint Venant principle, it can be assumed that the length of the member stump L is obtained multiplying the maximum depth D of the cross-section of the member times a suitable multiplier, of the order of 3 to 5. The length of the stump L must be computed considering the force transferrers and the stiffeners connected to each member, and getting to a sufficient distance from the farthest constituent to the connected extremity, named **N** (near) (Figure 5.9). In this way, local effects should be considered expired, so that the stress state at the new extremity **F** (far) can be computed by

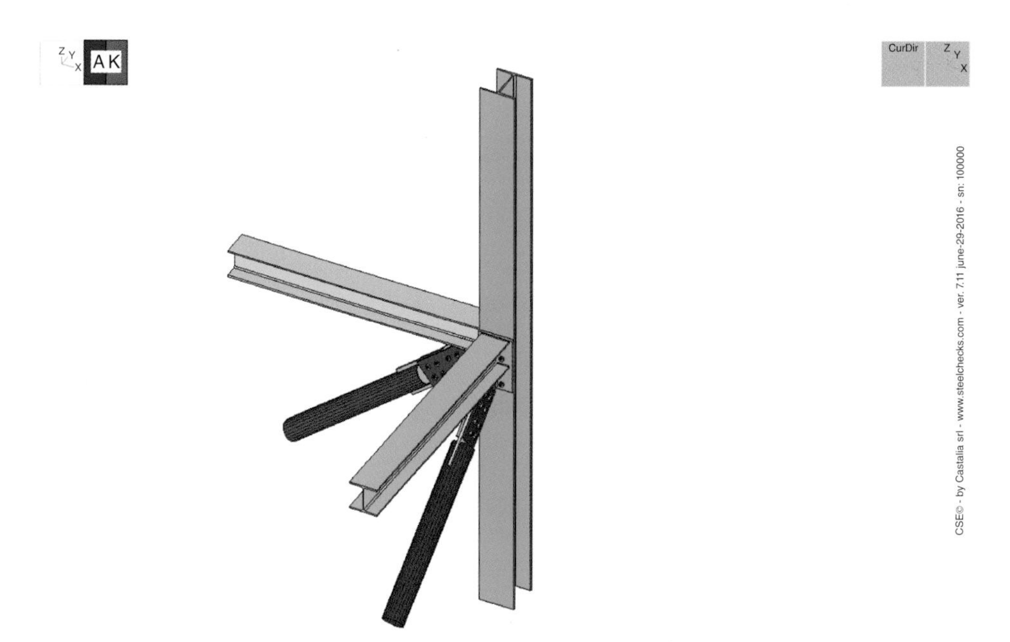

Figure 5.8 Nodal zone for a five member real node.

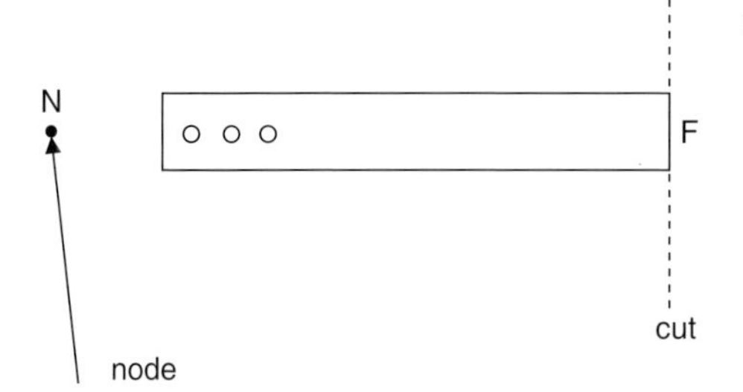

Figure 5.9 Member stump extremity N, and the new one F.

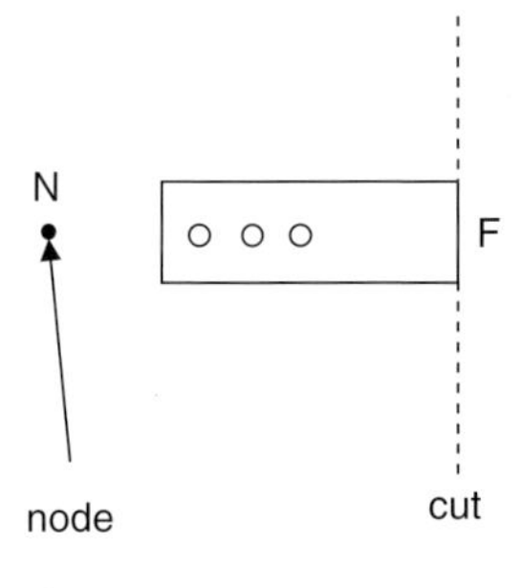

Figure 5.10 Incorrect member stump definition: too short.

simple beam theory. Otherwise the member stump is too short (Figure 5.10). Point **F** is the point of the member axis at distance L to the ideal extremity **N**, that is, to the ideally connected extremity of the member.

The force packet applied to each member at the new extremity **F** obtained by cutting it, can be computed by means of equilibrium conditions, considering the member alone, and reverting to the underlying BFEM.

If the effect of external loads applied to a member stump can be neglected, the force-packet at the new extremity **F** can be obtained easily by imposing equilibrium. The forces acting at the extremity **N**, that is, the one connected, are assumed to be known in the local reference system of the member: they are the internal forces $\boldsymbol{\sigma}$ acting at the connected member extremity. Usually they are one of the main results of the BFEM and are used to check members for resistance and stability. If the BFEM is not available, they can be set by design.

These internal forces form the $\boldsymbol{\sigma}$ vector in this way:

$$\boldsymbol{\sigma}^{\mathrm{T}} = [N \;\; V_2 \;\; V_3 \;\; | \;\; M_1 \;\; M_2 \;\; M_3]$$

If the local reference system (1, 2, 3) of the member has orientation matrix $\mathbf{T}$, then

$$\mathbf{s}_{\mathrm{N}} = \mathbf{T}^{\mathrm{T}}\boldsymbol{\sigma}_{\mathrm{N}}$$

Transferring the packet $\mathbf{s}_{\mathrm{N}}$ from **N** to **F**, and imposing equilibrium, we can find the value of the unknown packet $\mathbf{s}_{\mathrm{F}}$ (Figure 5.11):

$$\mathbf{s}_{\mathrm{F}} = -\mathbf{T}_{\mathrm{NF}}\mathbf{T}^{\mathrm{T}}\boldsymbol{\sigma}_{\mathrm{N}}$$

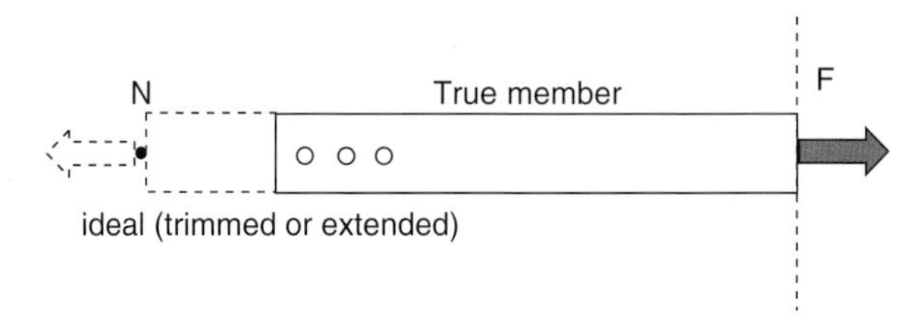

Figure 5.11 Forces at far extremity F (pulled bar). Equilibrium is obtained from the forces exerted by the bolts over the member. In the BFEM, those forces are equivalent to $\mathbf{s}_{\mathrm{N}}$ applied at extremity N.

So far, the effects of the external loads directly applied to the member stumps have been neglected. Usually these externally applied loads refer to beam members and are distributed forces.

If the effect of the external forces applied to the member stump cannot be neglected (Figure 5.12), then these forces will be equivalent to a number l of *known* force packets $\mathbf{s_i}$, applied at some *known* points $\mathbf{P_l}$ of the member stump. The equilibrium condition must then be modified accordingly:

$$\mathbf{s_F} = -\mathbf{T_{NF}}\mathbf{T}^{\mathrm{T}}\boldsymbol{\sigma}_{\mathbf{N}} - \sum_{i=1}^{l}\mathbf{T_{PiF}}\mathbf{s_i}$$

If the underlying BFEM is fully known (including the elementary loads applied in each load case), then $\mathbf{s_F}$ can also be obtained more easily from

$$\mathbf{s_F} = \mathbf{T}^{\mathrm{T}}\boldsymbol{\sigma}_{\mathbf{F}}$$

Considering the nodal zone isolated from the rest of the structure, the packets applied to the extremities **F** of the members can be considered external loads. These forces will flow into the constituents of the renode in such a way that all constituents will be in equilibrium.

When considering the problem of equilibrium, the nodal zone can be seen in a different, dual way (Figure 5.14). Instead of considering the packets at the far extremities **F** of the member stumps and the external loads directly applied to members (Figure 5.12), members can be considered loaded only by those packets ($\mathbf{s_F}$ and the l $\mathbf{s_i}$) transferred back to the connected extremities **N** of the members. In turn, this will indeed mean applying the original packet $\mathbf{s_N}$ to the extremity **N** of each member, with the sign reversed: $-\mathbf{s_N}$ (see Figures 5.13 and following).

The dual model is valid for computing equilibrium, but if the stress state of the member is to be assessed, then all the forces must be applied where they are: at the far extremity **F**, at the points where external loads are applied, and at the connector extremities. These latter forces are equivalent to the packet $-\mathbf{s_N}$ applied at the **N** end.

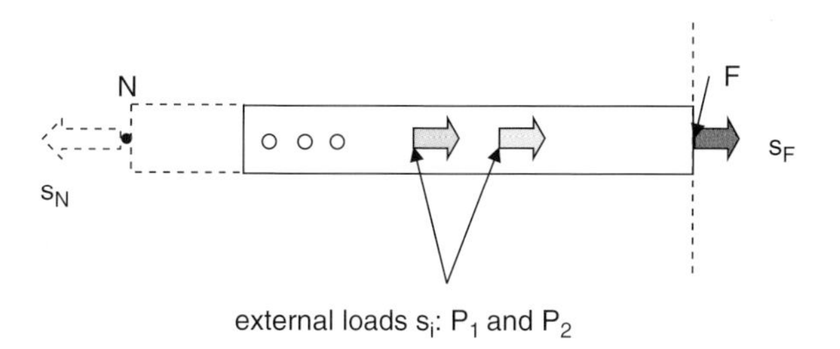

Figure 5.12 Forces at far extremity F, if external loads are applied. Equilibrium is obtained from the forces exerted by the bolts over the true member. In the BFEM, those forces are equivalent to s_N applied at extremity N.

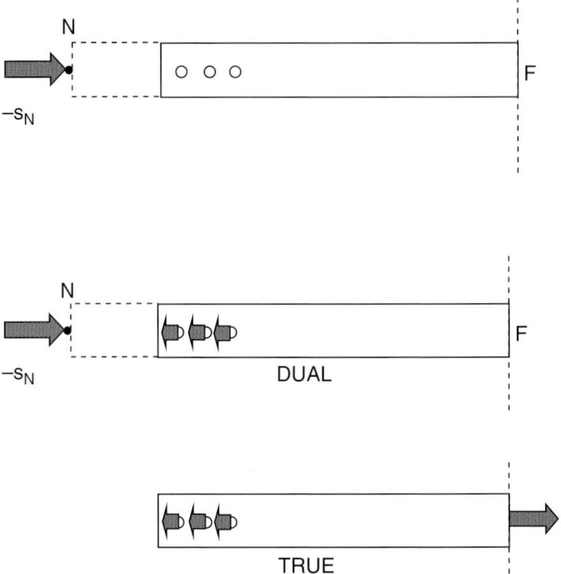

Figure 5.13 Dual model for pulled bar (Figure 5.11). Equilibrium is obtained from the forces exerted by the bolts over the member.

Figure 5.14 Dual and true systems under balanced packets (no external load).

If all the members have no rigid offsets, then they all meet at the nodal point, and the extremities $\mathbf{N_i}$ coincide with the nodal point **N** of the BFEM (here N means “node”). In this case, the BFEM solution guarantees that if the number of connected members is *nmembers*

$$\sum_{i=1}^{nmembers} -\mathbf{s_{Ni}} = \mathbf{0}$$

that is, it is certain that the nodes are balanced (note that the sum is done in the global reference system).

If no BFEM solution is available, the *nmembers* packets $\mathbf{s_{Ni}}$ *must be balanced* or otherwise the scene will not be in equilibrium and the fundamental equilibrium condition will be violated. This must be kept well in mind when fixing the values of the design internal forces by using envelopes (which violate equilibrium), or by using simplified member sets of internal forces.

If one or more members have rigid offsets, their extremities $\mathbf{N_i}$ will not be coincident with the nodal point **N** of the BFEM, but will however be well defined points in space. Transferring the packets from the extremities $\mathbf{N_i}$ to the nodal point **N**, once more a final balanced configuration will have to be found.

So now the problem of equilibrium assessment is stated in these two alternative ways:

> *Given the nodal zone separated from the rest of the structure, loaded by the known force packets $\mathbf{s_{Fi}}$ applied at all the far extremities F_i of the members, find the force packets flowing into each constituent of the scene.*

Given the nodal zone separated from the rest of the structure, loaded by the known force packets $-\mathbf{s}_{Ni}$ *applied at all the near extremities* $\mathbf{N}_i$ *of the members, find the force packets flowing into each constituent of the scene.*

Up to now it has been considered a jnode with no constraint. If a constraint block is also included in the scene, the previous considerations do still hold true, but among the known packets applied it will be necessary to include also the constraint reaction packet **r**. For exactly the same reasons applicable to the members, the packet **r** (which are the generalized forces acting over the structure from the constraint) will be applied to the nodal point **N**.

5.3.2 Force Packets for Compound Members

Sometimes the BFEM of members use compound cross-sections simulating the existence of members made up of two or more subparts. Typical examples are side to side angles, or back to back channels used as diagonals or chords for lattice structures. However, more generally, a cross-section within a BFEM can be obtained considering more simple cross-sections acting all together in an organized manner.

Within the BFEM the underlying fundamental hypothesis is that all the submembers act as a single constituent thanks to some device which will guarantee the connection between submembers. The organization between submembers takes the form of a single linear strain variation over the compound domain A, obtained by considering the union of all the elementary cross-sections domains A_i. This simplification can be safely assumed as far as the connecting devices between the submembers are effective enough. In turn, this is assessed by means of specific checks that must ensure the stiffness and the resistance of the connecting devices themselves.

This model is quite useful as it avoids the need to model each submember and the connecting devices in the BFEM, but it has a related drawback when the problem of connections is considered in a realistic 3D scene.

On the one hand in the 3D scene the submembers are now disjoint from one another, and the connecting devices will have to be explicitly modeled and checked. On the other hand the force packets flowing into each submember are not directly known, as the result of the BFEM stress analysis is the overall generalized forces $\boldsymbol{\sigma}$ applied to the center of gravity of the compound domain.

The first problem is related to connection analysis and must be considered. However, in order to do that, the second problem must be solved, finding the elementary force packets flowing into each submember, and applying each force packet at the submember cross-section center.

Let $\boldsymbol{\sigma}$ be the internal forces applied at the **N** extremity of a composed member, in the local reference system of the composed member, and let $\boldsymbol{\sigma}_i$ be the internal forces applied at each submember i cross-section center $\mathbf{G}_i$. The **N** extremity is coincident with the compound domain gravity center **G**, as by definition the finite elements of the BFEM are referred to the center lines.

It must be understood that the principal axes orientation of the compound cross-section **T** is not the same of the orientation of the elementary cross-section of each submember, $\mathbf{T}_i$ (Figure 5.15). So, because by definition the internal forces flowing into a

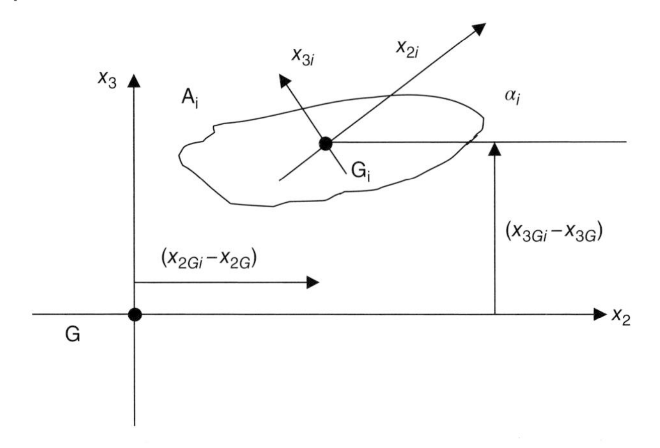

Figure 5.15 Compound reference system and submember cross-section reference system.

member (or submember) are expressed with reference to the local system of the member (or submember), in general $\boldsymbol{\sigma}$ and $\boldsymbol{\sigma}_i$ will not be referring to the same coordinate system. The submember internal forces $\boldsymbol{\sigma}_i$ can be expressed in the global reference system pre-multiplying by the transpose of the submember orientation matrix $\mathbf{T}^T{}_i$. In order to transform a vector from the global system to the compound member reference system $\mathbf{T}$ it is necessary to multiply it by $\mathbf{T}$. So the internal forces of the submember i expressed in the compound member reference system are

$\mathrm{T}\,\mathrm{T}^T{}_i\,\sigma_i$

As the equilibrium must be satisfied, the following condition must be met, expressed in the global reference system:

$$\mathbf{T}^T\boldsymbol{\sigma} = \sum_{i=1}^{n} \mathbf{T}_{GiG}\mathbf{T}_i^T\boldsymbol{\sigma}_i \tag{5.2}$$

The first member of the equation is the internal forces flowing into the compound member $\boldsymbol{\sigma}$, expressed in the global reference system, $\mathbf{T}^T\boldsymbol{\sigma}$. The second member is the sum of n terms, where n is the total number of submembers. Each term is obtained by considering the internal force packet of the submember i $\boldsymbol{\sigma}_i$, unknown, transformed into global system (by pre-multiplication by $\mathbf{T}^T{}_i$), and then transferred from the submember center $\mathbf{G}_i$ to the compound center $\mathbf{G}$, thanks to transfer matrix $\mathbf{T}_{GiG}$.

Equation 5.2 cannot be solved if $n > 1$ which is always true. So, in order to evaluate the force packets $\boldsymbol{\sigma}_i$ from the force packet $\boldsymbol{\sigma}$, a different path must be followed.

Assuming a linear strain field over the compound cross-section A, and referring to the principal system (2, 3) of the compound domain, with origin in $\mathbf{G}$, we can write that the strain ε:

$$\varepsilon = ax_2 + bx_3 + c$$

If a linear constitutive law is assumed, the normal stress is

$$\sigma = -\frac{M_3}{J_3}x_2 + \frac{M_2}{J_2}x_3 + \frac{N}{A} \tag{5.3}$$

where A, J_2, and J_3 are the area and second area moment of the compound domain A, with reference to its principal system (2, 3), and N, M_2, M_3 are the axial force and the bending moments (see also Appendix 1). Clearly the strain ε is

$$\varepsilon = -\frac{M_3}{EJ_3}x_2 + \frac{M_2}{EJ_2}x_3 + \frac{N}{EA}$$

where E is Young's modulus.

The generalized forces N_i, M_{2i}, M_{3i} flowing into each subconstituent i, and expressed in the subconstituent principal system $(2, 3)_i$, can be obtained by expressing the variables (x_2, x_3) as a function of (x_{2i}, x_{3i}), that is, the principal coordinates of the submember cross-section domain i. Let α_i be the angle between the x_{2i} axis and the x_2 axis. This is known by the definition of the compound cross-section. The following relations hold true:

$$\begin{aligned} x_{2i} &= +(x_2 - x_{2Gi})\cos\alpha_i + (x_3 - x_{3Gi})\sin\alpha_i \\ x_{3i} &= -(x_2 - x_{2Gi})\sin\alpha_i + (x_3 - x_{3Gi})\cos\alpha_i \end{aligned}$$

Theses equations can be inverted, leading to

$$\begin{aligned} x_2 &= x_{2i}\cos\alpha_i - x_{3i}\sin\alpha_i + x_{2Gi} \\ x_3 &= x_{2i}\sin\alpha_i + x_{3i}\cos\alpha_i + x_{3Gi} \end{aligned} \tag{5.4}$$

Replacing the expressions of Equation 5.4 into Equation 5.3, the stress field in the local reference system of cross-section i is found:

$$\sigma_i = -\frac{M_3}{J_3}\left(x_{2i}\cos\alpha_i - x_{3i}\sin\alpha_i + x_{2Gi}\right) + \frac{M_2}{J_2}\left(x_{2i}\sin\alpha_i + x_{3i}\cos\alpha_i + x_{3Gi}\right) + \frac{N}{A}$$

Now, the axial force and bending moments flowing into cross-section i, and expressed in local reference system of cross-section i, can be found using their definitions:

$$\begin{aligned} N_i &= \int_{A_i} \sigma_i dA_i \\ M_{2i} &= \int_{A_i} \sigma_i x_{3i} dA_i \\ M_{3i} &= -\int_{A_i} \sigma_i x_{2i} dA_i \end{aligned}$$

and considering that by the definition of $\mathbf{G_i}$ and of principal axes

$$\begin{aligned} &\int_{A_i} x_{2i} dA_i = 0 \\ &\int_{A_i} x_{3i} dA_i = 0 \\ &\int_{A_i} x_{2i}^2 dA_i = J_{3i} \\ &\int_{A_i} x_{3i}^2 dA_i = J_{2i} \\ &\int_{Ai} x_{2i} x_{3i} dA_i = J_{23i} = 0 \end{aligned}$$

It can be now be easily seen that the following relationships hold true:

$$N_i = -\frac{M_3}{J_3}x_{2Gi}A_i + \frac{M_2}{J_2}x_{3Gi}A_i + N\frac{A_i}{A}$$

$$M_{2i} = M_3\frac{J_{2i}}{J_3}\sin\alpha_i + M_2\frac{J_{2i}}{J_2}\cos\alpha_i$$

$$M_{3i} = M_3\frac{J_{3i}}{J_3}\cos\alpha_i - M_2\frac{J_{3i}}{J_2}\sin\alpha_i$$

These can be written in matrix form introducing the distribution matrix $\mathbf{D_{iNM}}$, as follows:

$$\begin{vmatrix} N_i \\ M_{2i} \\ M_{3i} \end{vmatrix} = \begin{bmatrix} \frac{A_i}{A} & \frac{A_i x_{3Gi}}{J_2} & -\frac{A_i x_{2Gi}}{J_3} \\ 0 & \frac{J_{2i}}{J_2}\cos\alpha_i & \frac{J_{2i}}{J_3}\sin\alpha_i \\ 0 & -\frac{J_{3i}}{J_2}\sin\alpha_i & \frac{J_{3i}}{J_3}\cos\alpha_i \end{bmatrix} \cdot \begin{vmatrix} N \\ M_2 \\ M_3 \end{vmatrix} \equiv \mathbf{D_{iMN}} \cdot \begin{vmatrix} N \\ M_2 \\ M_3 \end{vmatrix}$$

Clearly, if $n = 1$ then $\alpha_i = 0$ $x_{2Gi} = x_{3Gi} = 0$ and $\mathbf{D_{iNM}}$ becomes equal to identity matrix $\mathbf{I}$.

Up to now only normal stress has been considered. When dealing with shear stress (and so with torque M_1 and shears V), there is no unique distribution over the compound cross-section. It depends on the cross-section type. However, remembering that

$$\frac{dM_2}{dx_1} = V_3$$

$$\frac{dM_3}{dx_1} = V_2$$

the following relationships can be obtained for submember shears (expressed in the submember i reference system. It should be noted that x_1 does not change for compound member and submembers):

$$V_{2i} = \frac{dM_{3i}}{dx_1} = V_2\frac{J_{3i}}{J_3}\cos\alpha_i - V_3\frac{J_{3i}}{J_2}\sin\alpha_i$$

$$V_{3i} = \frac{dM_{2i}}{dx_1} = V_2\frac{J_{2i}}{J_3}\sin\alpha_i + V_3\frac{J_{2i}}{J_2}\cos\alpha_i$$

The same forces, expressed in the compound member principal system, would be written as follows, in matrix form ("compound" is applied to distinguish these latter):

$$\begin{vmatrix} V_{2i,\,compound} \\ V_{3i,\,compound} \end{vmatrix} = \begin{bmatrix} \cos\alpha_i & -\sin\alpha_i \\ \sin\alpha_i & \cos\alpha_i \end{bmatrix} \cdot \begin{bmatrix} \frac{J_{3i}}{J_3}\cos\alpha_i & -\frac{J_{3i}}{J_2}\sin\alpha_i \\ \frac{J_{2i}}{J_3}\sin\alpha_i & \frac{J_{2i}}{J_2}\cos\alpha_i \end{bmatrix} \cdot \begin{vmatrix} V_2 \\ V_3 \end{vmatrix}$$

Executing the matrix product, the following matrix conversion is found (for brevity $\cos\alpha_i$ and $\sin\alpha_i$ are replaced by "c_i" and "s_i"):

$$\begin{vmatrix} V_{2i,compound} \\ V_{3i,compound} \end{vmatrix} = \begin{bmatrix} \frac{J_{3i}}{J_3}c_i^2 - \frac{J_{2i}}{J_3}s_i^2 & -\frac{J_{3i}}{J_2}s_i c_i - \frac{J_{2i}}{J_2}s_i c_i \\ \frac{J_{3i}}{J_3}s_i c_i + \frac{J_{2i}}{J_3}s_i c_i & -\frac{J_{3i}}{J_2}s_i^2 + \frac{J_{2i}}{J_2}c_i^2 \end{bmatrix} \cdot \begin{vmatrix} V_2 \\ V_3 \end{vmatrix}$$

Referring to torque, its submember value can be obtained by considering the relative torsional stiffness of the submember, and re-adding the effect of the submember shears shifted from **G** to $\mathbf{G_i}$:

$$M_{1i} = M_1\frac{J_{1i}}{J_1} + V_{2i,compound}(x_{3Gi} - x_{3G}) - V_{3,compund}(x_{2Gi} - x_{2G})$$

And so, finally

$$M_{1i} = M_1\frac{J_{1i}}{J_1} + \left[\left(\frac{J_{3i}}{J_3}c_i^2 - \frac{J_{2i}}{J_3}s_i^2\right)V_2 - \left(\frac{J_{3i}}{J_2}s_i c_i + \frac{J_{2i}}{J_2}s_i c_i\right)V_3\right](x_{3Gi} - x_{3G}) +$$
$$-\left[\left(\frac{J_{3i}}{J_3}s_i c_i + \frac{J_{2i}}{J_3}s_i c_i\right)V_2 + \left(-\frac{J_{3i}}{J_2}s_i^2 + \frac{J_{2i}}{J_2}c_i^2\right)V_3\right](x_{2Gi} - x_{2G})$$

It is now possible to define the distribution matrix $\mathbf{D_i}$ (6 × 6), as follows:

$$\boldsymbol{\sigma}_\mathbf{i} = \mathbf{D_i}\boldsymbol{\sigma}$$

where

$$\boldsymbol{\sigma}_\mathbf{i} = |N_i \ \ V_{2i} \ \ V_{3i} \ \ M_{1i} \ \ M_{2i} \ \ M_{3i}|^\mathrm{T}$$
$$\boldsymbol{\sigma} = |N \ \ V_2 \ \ V_3 \ \ M_1 \ \ M_2 \ \ M_3|^\mathrm{T}$$

$$\mathbf{D_i} = \begin{bmatrix} \frac{A_i}{A} & 0 & 0 & 0 & \frac{A_i x_{3Gi}}{J_2} & -\frac{A_i x_{2Gi}}{J_3} \\ 0 & \frac{J_{3i}}{J_3}c_i & -\frac{J_{3i}}{J_2}s_i & 0 & 0 & 0 \\ 0 & \frac{J_{2i}}{J_3}s_i & \frac{J_{2i}}{J_2}c_i & 0 & 0 & 0 \\ 0 & \begin{matrix}\left(\frac{J_{3i}}{J_3}c_i^2 - \frac{J_{2i}}{J_3}s_i^2\right)(x_{3Gi}-x_{3G}) + \\ -\left(\frac{J_{3i}}{J_3}s_i c_i + \frac{J_{2i}}{J_3}s_i c_i\right)(x_{2Gi}-x_{2G})\end{matrix} & \begin{matrix}-\left(\frac{J_{3i}}{J_2}s_i c_i + \frac{J_{2i}}{J_2}s_i c_i\right)(x_{3Gi}-x_{3G}) + \\ +\left(+\frac{J_{3i}}{J_2}s_i^2 - \frac{J_{2i}}{J_2}c_i^2\right)(x_{2Gi}-x_{2G})\end{matrix} & \frac{J_{1i}}{J_1} & 0 & 0 \\ 0 & 0 & 0 & 0 & \frac{J_{2i}}{J_2}c_i & \frac{J_{2i}}{J_3}s_i \\ 0 & 0 & 0 & 0 & -\frac{J_{3i}}{J_2}s_i & \frac{J_{3i}}{J_3}c_i \end{bmatrix}$$

Thanks to this matrix, it is easily possible to get the force packet flowing into each submember, as a function of the force packet of the compound member and of the elementary cross-section data of the submember.

5.3.3 Primary Unknowns: Iso-, Hypo-, and Hyperconnectivity

The equilibrium assessment of the nodal zone can be better understood once the primary unknowns of the problem are introduced.

In Chapter 3, we proposed a model for the connectors that uses the concept of multiplicity of connection, and that introduces the extremities of the connectors as the ideal points where the force packets are exchanged with the connected constituents.

If a single connector i is considered (usually a layout), at each extremity e of the connector, a packet of forces will be exchanged, $\mathbf{c_{ie}}$.

Three indeterminacies can be defined, when dealing with steel connection analysis:

1) indeterminacy related to the evaluation of member forces in the BFEM: it will here be called *structural indeterminacy*
2) indeterminacy related to the evaluation of the force packets flowing into each connector of a renode: here called *renode* or *external indeterminacy*
3) indeterminacy related to the distribution of elementary forces between the subconnectors of a connector, at an equal force packet taken by the connector at a given extremity: here called the *connector* or *internal indeterminacy.*

The adjectives "internal" and "external" have meaning related to the context. In the context of the analysis of a single renode, they refer to the renode and to the connector indeterminacy, respectively. Using "renode indeterminacy" and "connector indeterminacy" avoids any possible misunderstanding.

The structural indeterminacy is solved in the BFEM. At the renode level, member forces are considered to be already known.

The packets applied *to* the connectors at their extremities can be introduced as primary unknowns of the problem at hand, that is, the renode indeterminacy is the first problem to be solved. The internal static indeterminacy related to the distribution of the forces *between the subconnectors* (connector indeterminacy) can be considered, and it is usually removed by the introduction of suitable displacement hypotheses relative to the displacement field of the subconnectors, or relative to their force distribution, for example plastic or elastic (see Section 3.3 and Chapters 6 and 7).

In fact, as members, force transferrers and stiffeners are connected by definition *only by means of connectors*, and as the action reaction principle must hold true, the force packets exerted by the connectors *to* the connected constituents are simply $-\mathbf{c_{ie}}$, for each connector i and each connector extremity e.

A force transferrer or a stiffener will have to be in equilibrium under a set of packets $-\mathbf{c_{ie}}$ exchanged with the set of connectors *connected to it* (at least two).

A member stump will have to be in equilibrium under the packet $-\mathbf{s_N}$ applied at point $\mathbf{N}$, and under the packets $-\mathbf{c_{ie}}$ exchanged with the set of connectors *connected to it.*

A connector i will have to be in equilibrium under all the packets $\mathbf{c_{ie}}$ applied at all its extremity points $\mathbf{E_e}$. So, considering as reference point the first extremity of the connector $\mathbf{E_1}$, and assuming that the number of extremities is cm (connector multiplicity)

$$\mathbf{c_{i1}} + \sum_{e=2}^{cm} \mathbf{T_{EeE1}} \mathbf{c_{ie}} = \mathbf{0}$$

No other unknown packets are needed.

If a scene is such that:

- nm is the total number of members
- nc is the total number of connectors
- ng is the total number of force transferrers
- ns is the total number of stiffeners
- nx is the number of constraint blocks (1 or 0)

then the total number ne of equilibrium conditions of the type of Equation 5.1 is

$$ne = nm - 1 + nc + ng + ns + nx$$

As the nodal zone as a whole is already in equilibrium in space, the number of available equations for the member stumps is $(nm - 1)$ and not nm. All the other constituents imply a single equilibrium equation.

If the total multiplicity of the connectors is tm, there will be tm packets unknown. Clearly

$$tm = \sum_{i=1}^{nc} cm_i$$

If the number of force packet unknown is equal to the available equations, the renode is said to be *isoconnected.*

If the number of force packet unknown is higher than the available equations, the renode is said to be *hyperconnected.*

If the number of force packet unknown is lower than the available equations, the renode is said to be *hyponnected.*

The *degree of hyperconnectivity dh* is defined as

$$dh = tm - ne = tm - nm - nc - ng - ns - nx + 1$$

A negative degree of hyperconnectivity means that some constituents still possess some degree of freedom.

If the renode is isoconnected, all the force packets flowing into the constituents can be computed by means of simple matrix algebra.

It must be underlined that a high proportion of renodes are isoconnected. The typical subsystem, member–simple connector–force transferrer–simple connector–master member, is intrinsically isoconnected as the master member does not enter into the unknown count $(nm - 1)$, and the subsystem has total multiplicity 4 (2 + 2) and four constituents, so that 4 = 4, see graph in Figure 5.16, top. This connection graph is suitable for a number of typical connections.

If two connectors with multiplicity 3 are used, and two force transferrers, the number of unknowns is six and the number of equations is five, so this subsystem is one time hyperconnected (Figure 5.16 bottom).

Just as for normal structures, simply counting unknowns and equations may hide more complex situations where a subpart is hyperconnected, while another is hypoconnected. For example in Figure 5.17 an apparently isoconnected scheme can be found. However, it can be seen that the dark force transferrer is hypoconnected (unconnected at all) and can be in equilibrium only if totally unloaded. The number of unknowns is six, the number of

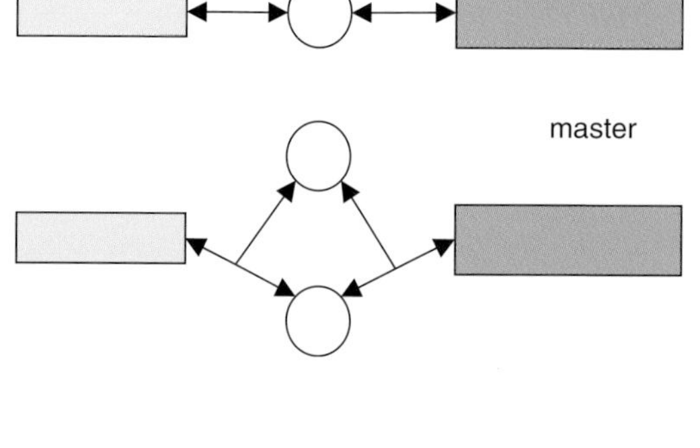

Figure 5.16 Graphs for typical subsystems: isoconnected (top); one time hyperconnected (bottom).

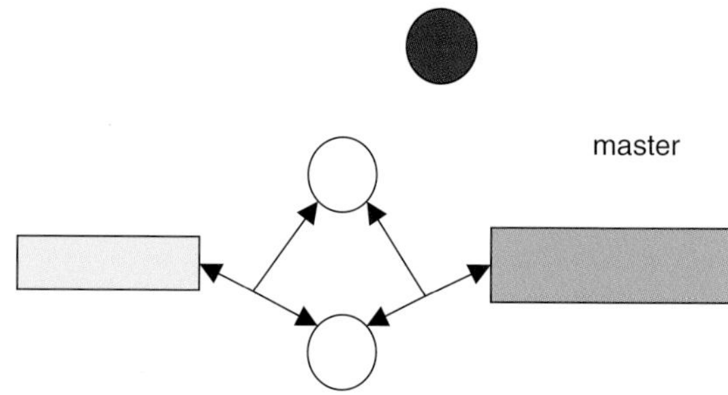

Figure 5.17 Apparently isoconnected system, hiding a hypoconnected subsystem and a hyperconnected surrounding.

equations is $1 + 3 + 2 = 6$, and $dh = 0$. But there is a subsystem having $dh = -1$, and another having $dh = +1$.

So although the number of unknowns and equations is a very important index, a deeper analysis must be done to assess the coherence and degree of hyperconnectivity of a renode. As already mentioned in Chapter 4 this is done by analyzing chains.

For hyperconnected renodes, the force packets exchanged by the connectors are indeterminate and, as mentioned, this indeterminacy in the context of a renode analysis will be named *renode* or *external indeterminacy*. This is not the only indeterminacy, as the distribution of forces among a single connector subconstituents is also unknown. As already said, this latter indeterminacy will be named *connector* or *internal indeterminacy* and it is normally solved by assuming a special distribution of forces between subconstituents (linear, parabolic, plastic, etc.).

In a hyperconnected renode, an infinite number of balanced solutions exist, each with a different flow of forces into the connectors. As the connector extremities are shared with the connected constituents, and the third law holds true, depending on the distribution of forces chosen to solve the renode indeterminacy the constituents will be loaded differently. So, both their plastic and buckling behavior will be affected by the choice of the distribution of forces between connectors.

Example

Consider the renode of Figure 4.23.

The hyperconnectivity count is as follows:

- total multiplicity of connectors: $3 + 2 + 2 + 2 + 2 + 2 = 13$
- number of members: 3
- number of force transferrers (double angle has two subconstituents): $1 + 2$

- number of connectors: 6
- number of stiffeners: 0
- hyperconnectivity: 13 – [(3 – 1) + 3 + 6] = 2

There are indeed two questions about the force flows that cannot be answered by equilibrium only:

1) Which part of the forces flowing into member m2 will get in the first subconstituent of the double angle, and which part will get into the second one.
2) Which part of the forces flowing into member m3 will get in the weld layout W1, and which part will get into the weld layout W2.

5.4 Static Theorem of Limit Analysis

The design of steel structures is implicitly based on the plastic properties of steel.

As brilliantly explained in Jacques Heyman's masterpiece *The Stone Skeleton* (Heyman 1995) there is no hope of exactly predicting the stresses and strains according to elasticity theory, in real systems, as there are a number of variables that may indeed affect elastic results and that are not and cannot be known in detail. This is also true for the connections between members (see also below, Section 5.6.1.6).

Moreover, the stress peaks predicted by elasticity theory are not experimentally found because, due to the plasticity of the material, the stresses are "redistributed" to the still not yielded surroundings.

When the loads are increased, plasticity gradually spreads through the system, or in single parts of the system, up to a point where equilibrium cannot be assured.

As will be better seen in the following sections, a number of different phenomena may interact with the spread of plasticity, stopping it: buckling, fracture, slip, and fatigue. However, in this section, they will not be considered.

There are several possible constitutive models that can be adopted in plastic static or pseudostatic analyses.

Probably the most common is that of "perfect plasticity" which means that the single material grain, when stressed to the yield point, cannot bear increasing stresses and so increases its strain with no residual stiffness. The increase in strain is the reason why the surrounding material grains, still elastic, are called to help and increase their stress state possibly to the yield, thus distributing the effort to a region wider than the elastic overstressed one.

No matter how wide the plastic region is, if the virtual work done by a small increase of the load applied is still positive, also locally, the system is able to resist. A *plastic mechanism* is reached when there is no possible increase of all the externally applied loads, no matter what the amplitude of the displacements (and of the strains) are.

The perfect plasticity model is usually integrated by some limitation on the strain amplitude, so that when a strain bound is reached, a state of internal fracture is notionally assigned to the grain. The non-linear analysis may then be arrested or it may go on, neglecting the effect of the fractured grains. Clearly the fracture of one grain (usually a Gauss/Lobatto integration point, as will be seen in Chapter 9) implies a stress redistribution and unbalanced loads that will have to be considered in the next iterations of the analysis.

A second very useful model, nearer to the experimental results, assumes that the stiffness after the yield does not abruptly drop to zero but keeps a low but significant value. An increase of the stresses is then possible, and the stiffness does not vanish once the yield is reached. When this model is assumed, the failure of the system is usually signaled by the reaching of the "ultimate stress" in some grains, which in turn is notionally related to fracture. Post fracture analyses are also possible, but they are not presently common and not very meaningful in the context of safe designs.

The reach of a plastic mechanism is usually explained with reference to frame systems, that is, systems made up of beam elements. In this specific context, the concept of "plastic hinge" has been widely used (and still is), by means of several possible numerical methods.

This model is quite simply not applicable in the context of steel connection analysis, as it relies on a basically wireframe model. Indeed, plastic connection analysis means in some way looking inside the "plastic hinge" in order to see exactly what happens: which plates reach plasticity and if and why the "plastic rotation" can develop.

The workhorse of connection analysis, as will be better seen in Chapter 9, is the plate–shell element, not the beam element.

Despite this fundamental difference, connection analysis and frame analysis in the plastic range do still have some features in common.

In both cases, to reach the "exact" answer, a step-by-step non-linear analysis should be performed, in order to carefully assess the spreading of plasticity (first here, later there) and to carefully arrive at the "mechanism", that is, to an external load level related to high displacements and loss of convergence. By the way, as will further explained in Chapter 9, the "loss of convergence" and "high displacements" conditions must be defined numerically in some practical way, which introduces some arbitrary but welcome engineering judgments into the procedure.

If the aim is to assess the load level related to a plastic mechanism, in the context of static or pseudostatic analyses, it is of no practical use to walk through the so-called "softening branch", which can be followed by decreasing the externally applied loads and using special numerical techniques such as *line search* and siblings. Instead, this ability, and the ability to compute cycles of loading and unloading, is fundamental to seismic truly dynamic analyses.

True loads are intrinsically dynamic, but the commonly adopted model is that of slow quasistatic increase, also for the very important special case of seismic loading. Truly dynamic analyses, however, are quickly gaining an important role in structural systems design, and it can be foreseen that in the coming years it will become the common tool for checking for seismic loads.

Although static, a true non-linear step-by-step analysis of a plastic system can be, or be considered to be, expensive. Indeed, every passing year it is less expensive, and today it can be done in a limited computational time. However, if quicker analyses can deliver a reasonable and safe estimate of the external load level related to the plastic mechanism, this will be welcome.

It is for this very reason that the so-called static or "safe" theorem of limit (plastic) analysis is of great interest in connection analysis and, more generally, in structural design.

It must be underlined that in the context of steel connection analysis as treated in this book, the "system" is a renode, with all the members and maybe tens of constituents, including connectors, and the externally applied loads related to all the load combinations to be checked. Among them there may also be notional load combinations, which are added to study specific design situations such as the over strength required by capacity design.

It has been said that in the context of steel connection analysis the force packets exchanged by the connectors can be seen as the primary unknowns of the problem, and that for hyperconnected system there may be infinite distributions of force packets balanced with the externally applied loads.

A generic distribution of force packets balanced with the externally applied loads is defined as a statically admissible force packet distribution.

For isoconnected renodes, the force packet distribution is unique and there is no possible choice to be done, just like for a cantilever of length L loaded at the tip by a transverse force P the moment at the clamp, where the unique plastic hinge develops, must always be $M = PL$.

For hyperconnected renodes, a statically admissible force packet distribution is not necessarily also kinematically admissible. This means that the displaced shapes of the renode constituents, under the effect of this statically admissible force packet distribution, do not match one another. This also happens in frame analysis and is expected when a statically admissible force distribution is used, in lieu of the "exact" one, which is at the same time statically and kinematically admissible.

The choice of a statically admissible force packet distribution removes the *renode indeterminacy* related to hyperconnected renodes, but it does not solve the *connector indeterminacy* related to the infinite possible distributions of forces between the subconstituents of the connector, exchanging a given external force packet $\mathbf{c}_{ie}$.

Also, this kind of indeterminacy must be solved, and this is usually done assuming a simple model (linear or plastic) for the internal subconstituents force or strain distribution.

Once more, this statically admissible force distribution is not, in general, kinematically admissible. The "exact" distribution of forces among the subconstituents of a connector (bolt layout or weld layout) strongly depends on the connected constituents' local stiffness, in the elastic and then in the plastic range.

Removing the external and internal indeterminacies using an external (i.e. renode) force packet distribution, and several internal subconnector force distributions is a quick way to have a set of known forces to be checked. These forces, by definition, are balanced with a set of factored loads, amplified by a scalar α_{L1}.

The static theorem of limit analysis, also applicable to the problem at hand, states that considering all constituents as being elastic–plastic, and neglecting other phenomena which may stop the load-increase process (fracture, buckling, fatigue, and slip), the "exact" load multiplier α_{L2} that would be obtained using the "exact" external and internal force distributions, is higher than the one assumed, α_{L1}. The only necessary condition is that the distributions are statically admissible and that nowhere are the stresses higher than the plastic ones, that is, nowhere is the plastic stress locus crossed or violated.

So if no other phenomena such as the one mentioned are triggered, using the load multiplier related to a generic statically admissible distribution is on the safe side, as this load multiplier is lower than the "exact" one.

It must once more be underlined that the deformed shapes of the constituents obtained by this method are not kinematically admissible. However, the load level predicted is lower than the “exact” one, which can be obtained only by following the loading path with a step-by-step procedure.

Although it is ideally preferable to use statically admissible distributions not too different from the “exact” ones, the theorem allows the use of generic statically admissible distributions. If one distribution does not fit (because the load multiplier is too low or because one constituent fails), then it is always possible to try another, in order to improve the estimate.

5.5 The Unsaid of the Engineering Simplified Methods

Maybe it was not clear to 19th- and 20th-century designers, who were using old-fashioned simple models and formulae to check connections, but their effectiveness was largely founded on undiscovered principles of the theory of plasticity (see Heyman 1995, and Heyman 1998).

Using balanced internal force configurations has been one of the key methods of getting to simple checking models and formulae, and still is.

However, it is only relatively recently, during the 1950s, that the theory of plasticity explained why such methods, basically arbitrary, eventually will sometimes work. Simple rules such as strut and tie models (quite useful both for steel connection analysis and for reinforced concrete design), 30° degrees (or whatever other angle) stress spreading, rules like “the bending moment flows into the flanges, the shear to the web”, and also deciding the rotation center of an end plate bolted connections are all rules that are not strictly justified if not invoking the static theorem of limit analysis.

Searching for “the” elastic “exact” solution is meaningless, as this single solution is influenced by a huge number of unpredictable initial conditions and loads. But using “a” balanced configuration can have a meaning only in the shelter of the static theorem of limit analysis, which, in turn, holds true under well defined conditions and in the absence of other phenomena, able to lead constituents to failure.

The simple rules based on the static theorem sometimes fail to predict the true “limit” load of connections, as they do not properly take into account other possible phenomena, different from plastic flow, which may also drive a constituent to failure. For these other phenomena, no simple rules such as equilibrium conditions, action reaction principle, or static theorem of limit analysis are available.

In the next section they will be considered one by one.

5.6 Missing Pillars of Connection Analysis

As already said, the shelter provided by the static theorem is not fully comprehensive. Some failure modes may be triggered by loads lower than those considered by the static theorem.

This means that these failure modes must be checked, or otherwise the design will be unsafe.

There are several failure modes that are not related to the formation of a mechanism in the classic perfect plasticity meaning. Generally speaking they are related to at least four different physical phenomena: buckling, fracture, slip and fatigue.

Each constituent or subconstituent of a renode, embedding a set of connections, should be checked for these failure mechanisms, which are named differently according to the constituents involved, but that are basically always related to the very same problems.

In the following sections the issues related to such failure mechanisms will be considered.

5.6.1 Buckling

5.6.1.1 Generality

Buckling is related to the change of configuration triggered by compressive forces in relatively flexible constituents. As plasticity lowers the stiffness, buckling is affected by it, and it is therefore necessary to take into account how the two phenomena interact in order to properly check a constituent. So, if the load multiplier, not taking into account the buckling, is α_L, then the load multiplier also taking into account buckling, α_R, may be a small fraction of it.

There are indeed formulations which, in the frame of the theory of elasticity, predict the buckling loads of a number of typical structural problems, such as Euler's prismatic column, thin plates, and cylindrical shells, but these brilliant formulations, which have had a central role in past decades, are now partly losing their effectiveness. The reason is that the buckling load level depends on all the following features of the problem at hand, with no exceptions:

1) The geometry of the structure, which is nowadays more and more not as regular and clean as the closed formulae require.
2) The load path, i.e. the sequence of the applied loads. This is because the instability is triggered by compressive stresses, which depend on the externally applied load distribution, level, and sequence. Assuming, as it is often done, that all the loads of a loading combination are increased by a unique load factor, this neglects the load-path effect. This also holds true for the plastic response of a system, which is, in theory, path dependent.
3) The distribution of active forces over the structure or constituent. Due to the need for checking many load combinations, these distributions of forces can be many and varied.
4) The distribution and nature of the constraints which are almost never as perfect, clean, and well posed as expected by the theoretical formulations.
5) The nature, level, and distribution of the imperfections, which lead to buckling loads lower than those predicted by the elastic methods.
6) The interaction of the buckling with plasticity and fracture, which can dramatically change the problem to be solved, modifying the stiffness of the loaded parts.

Considering a typical renode, where n members are connected by means of tens constituents (stiffeners, gusset plates, end-plates, haunches, etc., and bolts and welds), with specific geometry, and connectors acting as constraints, under the effect of dozens or hundreds of load combinations, it is clear that no traditional closed formulae approach can really cope. Each plate, stiffener, or part of a member may buckle, and there might be global buckling modes involving more constituents, and these possibilities have to be checked.

Traditionally the problem of the buckling of parts under the loads applied has been tackled by rules of good practice, which are still an invaluable, unavoidable means of getting a proper design. For instance, the thickness of the stiffeners has been related to the thickness of the stabilized parts; and the distribution of such stiffeners, which have in connection design a role comparable to that of reinforcing in reinforced concrete structure, has been carefully described for typical problems in the textbooks.

However, stiffeners, usually welded, have a cost, and the tendency has been to make them thinner and to apply them as little as possible. Moreover, understanding when, where, and how to apply stiffeners is still a sort of "art", like reinforcing design in concrete, and the lack of a general method for checking them is definitely a severe limitation.

Also, if the stability of the stiffeners and of the stiffened part is not checked for all load combinations, there is no formal proof of their effectiveness.

So, are these traditional approaches to buckling checks still sufficient?

The problem is the tradeoff between weight and cost, on one side, and safety on the other. Traditional methods are probably able to protect against unwanted buckling of parts, but if the cost has to be reduced (as apparently the more recent standards require) then there is a real problem.

As will be better seen in the next chapters, there are general tools available, which can greatly help to solve the buckling check problems in a general way. All the issues enumerated above can be tackled, and solved, the possible exception being issue 2 in current practice, which requires a tremendous computational effort because from one single load combination stem several different possible loading paths.

There is still much work to be done in this area, in order to find clever bounds at reasonable computational cost.

5.6.1.2 Buckling Checks Methodology

It is clear that instability interacts with plasticity, as plasticity weakens the constituents, decreasing their stiffness and increasing their displacements.

There are basically two methods of checking for buckling of structures, that are also used in checking the buckling of renode constituents, under the effect of tens or maybe hundreds loading combinations.

1) A first method uses elastic critical buckling loads, and computes a critical load multiplier α_{cr}. If the problem at hand is not covered by exact formulations, this method usually implies a finite element model and an eigenvalue analysis. Once the plastic load multiplier α_L and the elastic critical load multiplier α_{cr} are known, the "real" or better expected load multiplier α_R is computed by means of a reduction factor χ of α_L, which takes into account the imperfections.
2) A second method (which will be described in Chapter 8) uses a non-linear step-by-step finite element analysis, considering both sources of non-linearity at the same time: material and geometrical. To take into account imperfections, initial correction to geometries are possibly applied. If the analysis reaches a convergence under the loads applied, with no mechanism, then the system is able to resist. To get the real load multiplier, loads can be increased indefinitely, up to the first mechanism.

5.6.1.2.1 Elastic Critical Load and the General Method

Instability and plasticity may occur at very different load levels, at least when instability is predicted using an eigenvalue analysis. To do so, the following eigenvalue problem is usually set up:

$$(\mathbf{K} - \alpha_{cr}\mathbf{K_G}) \cdot \mathbf{u} = \mathbf{0} \tag{5.5}$$

where $\mathbf{K}$ (with no subscripts) is the elastic stiffness matrix, α_{cr} is the critical load multiplier, the eigenvalue searched, $\mathbf{K_G}$ is the geometrical stiffness matrix taking into account the geometrical effects, and $\mathbf{u}$ is the critical nodal displacements vector (eigenvector). The geometrical stiffness matrix $\mathbf{K_G}$, in general, depends on the stress level, and in Equation 5.5 it is coherent with the stress level related to $\alpha = 1$. Computing the critical load multiplier α_{cr} in this way implicitly assumes that during the load increase the stiffness matrix $\mathbf{K}$ would remain elastic, and so unchanged, and that the stresses used to fill the geometrical stiffness matrix would remain linearly dependent on α, so that the geometrical stiffness matrix related to the load level α, would simply be computed as $\alpha\mathbf{K_G}$. If during the load increase from 1 to α no part gets into the plastic range, the assumption is acceptable, but if some constituent goes into plasticity, possibly triggered by imperfections, then the assumption is not correct.

Two multipliers have thus been computed, for each load combination under analysis:

1) The first is the limit load multiplier, α_L, computed assuming a plastic behavior for the material, and totally neglecting the geometrical effects, thus neglecting the buckling phenomena.
2) The second is the critical load multiplier, α_{cr}, computed basically with an elastic constitutive law, and neglecting plasticity.

As is well known, the elastic critical multiplier is not a reliable measure of the danger of instability. This is due to a number of conditions which, in real problems, change the terms of the problem at hand, and in practice lead to lower buckling loads: imperfections in the geometry and in the loads, initial stresses, interaction of plasticity and buckling, and so on.

In Eurocode 3, the problem of finding reliable measures of the buckling danger, from the two aforementioned multipliers is tackled by the introduction of appropriate "stability curves", named "a_0", "a", "b", "c", "d", which allow us to compute a reduction factor χ, which, applied to the limit multiplier α_L, leads to the final expected multiplier α_R:

$$\alpha_R = \chi\alpha_L \tag{5.6}$$

This method using stability curves was introduced first for columns, and then extended to very different structural subsystems, by Eurocode 3, so that now it is also known as a "general method" or "general approach" (e.g. see Bernuzzi and Cordova 2016 or Rugarli 2008). Considering Eurocode 3, this method is used directly or indirectly at least in these clauses:

- EN 1993.1.1 § 6.3.4 — flexural buckling and lateral sway of beams
- EN 1993.1.3 §4.2(4) — cold formed structural elements
- EN 1993.1.3 §4.3.1(7) — cold formed structural elements
- EN 1993.1.3 §4.3.4.2(9) — cold formed structural elements
- EN 1993.1.3 §4.3.4.3(10) — cold formed structural elements

- EN 1993.1.3 §5.8.P(6) cold formed structural elements
- EN 1993.1.3 §6.3.P(1) cold formed structural elements
- EN 1993.1.5 § 10.(3) plated structural elements
- EN 1993.1.5 §B.1 plated structural elements
- EN 1993.1.5 §D.2.2 plated structural elements
- EN 1993.1.6 §8.5.2 shells
- EN 1993.2 §5.2.4.3(3)c bridges
- EN 1993.2 §5.5.2.2(3) bridges
- EN 1993.2 §5.5.4.3.1(1)c bridges

As is well known, for the column flexural buckling problem the reduction factor χ is related to *slenderness* λ, $\chi = \chi(\lambda)$, which in turn is computed as the ratio of the buckling length to the radius of gyration of the cross-section of the column, for the buckling-bending axis considered. The buckling length is a function of the length of the column and of its constraints. The choice of the buckling curve depends on the level of imperfection, so that, at equal slenderness λ, buckling curve "d" leads to a lower χ value than buckling curve "c", and curve "c" to lower χ value than "b", and so on.

The general method uses the same stability curves named "a_0", "a", "b", "c", "d" for all the possible problems, but changes the slenderness definition into a "non-dimensional" slenderness $\bar{\lambda}$ defined as

$$\bar{\lambda} = \sqrt{\frac{\alpha_L}{\alpha_{cr}}} \tag{5.7}$$

The "non-dimensional" classic attribute is somewhat misleading, because the normal slenderness is also non-dimensional. However, it refers to the normalization of the slenderness with respect to the limit slenderness λ_1 which acts as a notional bound, for Euler columns, between collapse for instability ($\lambda > \lambda_1$) and collapse for plasticity ($\lambda < \lambda_1$):

$$\lambda_1 = \pi\sqrt{\frac{E}{f_y}}$$

where E is Young's modulus, and f_y the yield stress of the steel.

In the general context of the stability analysis of a renode, no such normalization is needed as no "buckling length" or radius of gyration can be found. The only needed parameters of crucial importance are a limit load multiplier α_L, and a critical load multiplier α_{cr}. From them the "non-dimensional slenderness" is immediately computed.

For the Euler column problem, the term "slenderness" is reasonable, as the radius of gyration is related to the size of the cross-section, and the buckling length to the actual length of the column. Their ratio relates well to the "thinness" or "slenderness" of the column. For general problems, where no such parameters can be obtained or have meaning, and where critical modes assume strange, unpredictable shapes, the term seems less justified. For this reason instead of "non-dimensional slenderness", it has been proposed to name it "criticalness" (Rugarli 2008).

The stability curves and the general method are only fully justified from the theoretical point of view for Euler columns. Stability curves have been derived by extended experimental tests carried out in the past decades on such systems. Their applicability to different problems is questionable. However, the possibility of choosing, by means of an "imperfection factor", the severity of the reduction carried out by $\chi(\bar{\lambda})$, makes the method flexible and quite useful for tackling problems that would otherwise need specific tests.

It is immediately clear that the general method may have some problems. For the column axially loaded or for the bending of a slender beam, the plastic and critically displaced are effectively related, in the context of a general system, this is not necessarily true.

It may happen, for instance, that the first plastic mechanism is placed in a different region of the system (the renode) which is not related to the region where the critical mode can be seen. It would probably be better not to always use the first critical multiplier, that is, the lowest α_{cr}, in order to evaluate the interaction between plasticity and buckling, but to use the first which implies a critical shape referring to, or at least, also to, the region where the first plastic mechanism is expected. However, as can be readily seen by the formula defining the non-dimensional slenderness (criticalness), using the lowest α_{cr} is on the safe side, as it overestimates the criticalness, which, in turn, overestimates the reduction related to the χ factor.

According to Eurocode 3, the following expressions must be used in order to compute the correction factor χ (we will use the symbol χ_{EC} to identify the reduction factor according to Eurocode 3):

$$\begin{aligned}
&\chi_{EC} = 1 \quad \text{if } \bar{\lambda} < 0.2, \quad \text{else} \\
&\chi_{EC} = \frac{1}{\phi + \sqrt{\phi^2 - \bar{\lambda}^2}} = \frac{\phi - \sqrt{\phi^2 - \bar{\lambda}^2}}{\bar{\lambda}^2} \\
&\phi = 0.5 \cdot \left[1 + k(\bar{\lambda} - 0.2) + \bar{\lambda}^2\right]
\end{aligned} \tag{5.8}$$

The imperfection factor k has the values shown in Table 5.1, depending on the stability curve.

It can be easily seen that for each value of the non-dimensional slenderness the reduction factor χ_{EC} is lower than $(1/\bar{\lambda}^2)$:

$$\begin{aligned}
&\chi_{EC} = \frac{\phi - \sqrt{\phi^2 - \bar{\lambda}^2}}{\bar{\lambda}^2} < \frac{1}{\bar{\lambda}^2} \quad \Rightarrow \\
&\phi - \sqrt{\phi^2 - \bar{\lambda}^2} < 1 \quad \Rightarrow \\
&\sqrt{\phi^2 - \bar{\lambda}^2} > \phi - 1 \quad \Rightarrow \\
&\phi^2 - \bar{\lambda}^2 > \phi^2 - 2\phi + 1 \quad \Rightarrow \\
&\bar{\lambda}^2 < 2\phi - 1 = 1 + k \cdot (\bar{\lambda} - 0.2) + \bar{\lambda}^2 - 1 \quad \Rightarrow \\
&k \cdot (\bar{\lambda} - 0.2) > 0, \quad \text{always}
\end{aligned}$$

Table 5.1 Imperfection factors for the stability curves according to Eurocode 3.

Stability curve	a_0	a	b	c	d
Imperfection factor k	0.13	0.21	0.34	0.49	0.76

This property of χ can also be written as

$$\chi < \frac{1}{\bar{\lambda}^2} = \frac{\alpha_{cr}}{\alpha_L}$$

This means that

$$\alpha_R = \alpha_L \chi < \alpha_{cr}$$

that is, the expected load multiplier can never be higher than the critical one, a remark which is particularly meaningful for systems having an high criticalness λ: $\alpha_{cr} \ll \alpha_L$.

The reduction factor χ_{EC} should be computed with the complex formula only if $\bar{\lambda} > 0.2$; otherwise it must be considered equal to 1, meaning that no reduction due to instability needs to be applied to limit multiplier α_L. This implies

$$\alpha_{cr} > 25\alpha_L$$

An early evaluation of the reduction factor χ is that of the Merchant–Rankine formula, which sets

$$\chi_{MR} = \frac{\alpha_{cr}}{\alpha_L + \alpha_{cr}} = \frac{1}{1 + \bar{\lambda}^2} \tag{5.9}$$

Also the reduction factor computed with the Merchant–Rankine formula satisfies the condition

$$\chi_{MR} = \frac{1}{1 + \bar{\lambda}^2} < \frac{1}{\bar{\lambda}^2}$$

5.6.1.2.2 Non-Linear Elastic Plastic Analysis with Geometrical Effects Included

If a non-linear analysis is performed including the geometrical effects, depending on the features of the system at hand, several different possible situations might arise.

If the criticalness of the system under the applied loads is low, or very low (say lower than 0.2, which means $\alpha_{cr} > 25\alpha_L$), the system will be able to reach its plastic mechanism having a null interaction with geometrical effects.

If the criticalness of the system is high or very high (say higher than 3, which means $\alpha_{cr} < \alpha_L/9$), then the system will probably buckle much before getting into the plastic range. Buckling will immediately lead to an increase of the stresses due to P-Δ effects, and plasticity will appear, quickly causing a mechanism, and so failure, or a post-buckling configuration quite different from the initial one, but still able to carry increasing loads. While post buckling configurations may indeed carry increasing loads, buckling is typically unwanted in actual structures, and so it is rare that this effect is taken into account (one classic example of buckling is the out-of-plane buckling of the web panel in a beam-to-column connection). Only specific, well delimited, and well studied structural subparts are designed to buckle for design loads, and are proportioned by means of simplified approaches.

If the criticalness is confined between these values, geometrical effects and plasticity effects will interact, and only a true step-by-step analysis is able to "exactly" predict how. Internal constituents, lightly loaded but very critical (local criticalness of the subsystem 1.5 or 2), might buckle before those that are highly stressed but stocky. Geometrical effects in buckled constituents will always trigger higher stresses and yielding, which, implying a local decrease in the stiffness, will indeed cause increasing displacements, in a self-feeding loop. The maximum level reachable of the externally applied loads, α_R, is the equivalent of the "expected" or "real" load multiplier obtained with the general method.

Step-by-step analysis with geometrical effects is influenced by imperfections and by residual stresses, which are never known with exactitude.

Steel connections are usually designed in such a way as to ensure that a plastic mechanism can be obtained, with limited effects related to buckling. In turn, this means ensuring that the thicknesses and sizes of the constituents are such that the criticalness is pushed to low values, avoiding or reducing the interaction between the two phenomena.

However, this need clashes with the need for lighter and simpler structures, so the ability to correctly tackle the issue without exceeding in the weight is important nowadays.

Moreover, it is not always simple to have a good estimate of the bounding thickness to be used to avoid buckling, in the total absence of numerical analyses.

5.6.1.3 A Fundamental Question

When considering the limit load in light of the static theorem, and the buckling problem, a fundamental question arises.

Let D_1 be a balanced distribution of internal forces, implying a choice for the mutual ratios of the redundant force packets, a choice of internal force distribution for the single connectors subconstituents, and a specific plastic stress distribution related to a plastic mechanism, and let α_{L1} be the limit load multiplier related to it. If under the effect of D_1 all constituents are stressed by a distribution that does not violate the plastic limit at any point, and if, under the effect of D_1 the final load multiplier α_{R1} (obtained by the general method or by a step-by-step non-linear analysis considering plasticity and geometrical effects) is higher than 1, can we assume that all the constituents in the renode are checked both for resistance *and stability*?

The uncomfortable answer to this fundamental question is: no, we cannot.

Had we assumed a different distribution D_2, related to a different plastic mechanism and load multiplier α_{L2}, we would have found a different stress distribution for the force packets and/or for the internal forces in the subconstituents. So we would have done a different buckling check (due to the issue 3 above: we would have used a different plastic stress distribution and a different plastic mechanism), finding a different expected load multiplier α_{R2}.

In general, the distribution D_1 may be obtained by starting with an elastic stiffness matrix different from the other, so the two critical multipliers obtained by eigenvalue analysis are not identical; and also different are the initial slopes of the load–displacement curve, for any given displacement component (Figure 5.18). The difference in the elastic stiffness may be due to different simplifying models, e.g. using secant or tangent stiffness, or considering a load level as one reference or another, or one contact condition or another in contact non-linearity problems.

Is the buckling check done with D_1 leading to α_{R1} lower than 1, enough to avoid the buckling check with D_2? In principle it is not, we have tested a different *distribution* of forces, which may overload different constituents or different parts of the same constituents (Figure 5.18).

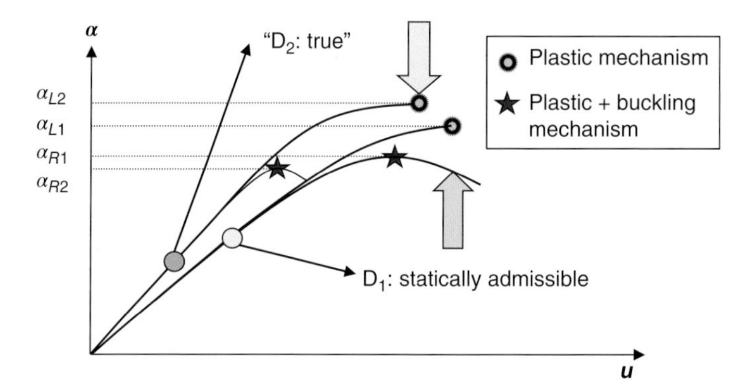

Figure 5.18 Static "safe" theorem of limit analysis does not provide protection from buckling.

Let's assume that D_2 is the "exact" plastic distribution, i.e. the one both statically and kinematically admissible. We will discuss later if in the context of the connection analysis that is searching this "exact" solution is meaningful and possible.

If the distribution D_2 is the "exact" one, according to the theory of plasticity, then certainly,

$$\alpha_{L2} \geq \alpha_{L1}$$

but we cannot assume, in general, that the following condition also holds true:

$$\alpha_{R2} > \alpha_{R1}$$

If the "exact" plastic stress distribution at limit can be computed, then it will also be possible to assess if this distribution can be reached without buckling or other failures. If this "exact" stress distribution cannot be computed, or if it is not possible to find an "exact" distribution, then we will need to guard against buckling and other phenomena by means of a design able to resist plastically without other failures to a wide set of possible distributions, not only the chosen statically admissible ones.

Jacques Heyman, one of the leading scholars in the field of plastic design, did perceptively warn about this very important issue in his classic text *Structural Analysis A Historical Approach* (Heyman 1998, §10.4).

It is wise to have a closer look at the instability issue, in order to understand the importance of the problem described, and its possible solutions.

5.6.1.4 The "General Method" and the Static Theorem

If the general method is used to check the renode for buckling, or some parts of it, then an important result can be obtained, considering the properties of the static theorem, and those of the reduction factor χ.

If it is assumed that a statically admissible plastic mechanism for the renode has been considered, this would imply a limit load multiplier α_{L1}, lower that related to the "true" plastic mechanism according to the theory of plasticity, α_{L2}.

Considering the elastic stress distribution, and executing an eigenvalue analysis, let us assume that a unique lowest critical elastic load multiplier α_{cr} can be found, which does not depend on the plastic mechanisms; this implies that a reasonably correct elastic stress distribution is known, and that the plastic mechanisms differ in how this elastic stress distribution has been abandoned, gradually increasing the load.

In general, a statically admissible plastic distribution and the "exact" one are related to different original elastic distributions, that is, to a different way of solving the indeterminacy also in the elastic range. Of course, if the renode is isoconnected, there is no renode indeterminacy in the force packets, but there may still be one in the internal distribution of forces between subconnectors, which is locally affected by the stiffnesses of the connected constituents, point by point. Any simplification of this problem would lead to an elastic stiffness matrix different from the "exact" one.

The expected load multipliers, in the two cases would be

$$\alpha_{R1} = \alpha_{L1}\chi_1(\bar{\lambda}_1)$$
$$\alpha_{R2} = \alpha_{L2}\chi_2(\bar{\lambda}_2)$$

The solution obtained with the first, statically admissible plastic mechanism, is safe if and only if

$$\alpha_{R1} \le \alpha_{R2}$$

What is relevant is the product of the two factors α_L and χ, and so being $\alpha_{L2} > \alpha_{L1}$ is not enough to be on the safe side. As it has been assumed that the critical load multiplier is the same, the criticalness $\bar{\lambda}_2$ related to the "true" plastic mechanism is higher than $\bar{\lambda}_1$, and so the reduction factor χ_2 is lower than χ_1, which means that we cannot be sure of which is the worst effect.

However, it can be proved that using the expressions proposed by the standards, it is always true that

$$\alpha_{R1} \le \alpha_{R2}$$

which has very important consequences.

Considering first the Merchant–Rankine correction, the proof follows:

$$\alpha_{R1} = \frac{\alpha_{L1}\alpha_{cr}}{\alpha_{L1} + \alpha_{cr}} \le \alpha_{R2} = \frac{\alpha_{L2}\alpha_{cr}}{\alpha_{L2} + \alpha_{cr}} \quad \Rightarrow$$
$$\alpha_{L1}\cdot(\alpha_{L2} + \alpha_{cr}) \le \alpha_{L2}\cdot(\alpha_{L1} + \alpha_{cr}) \quad \Rightarrow$$
$$\alpha_{L1}\alpha_{cr} \le \alpha_{L2}\alpha_{cr} \quad \Rightarrow$$
$$\alpha_{L1} \le \alpha_{L2} \quad \text{always}$$

Now, considering the Eurocode 3, in Figure 5.19 the curves $\alpha_R = \alpha_R(\alpha_L)$ are plotted for several values of α_{cr}, and using buckling curve "b".

For each curve, a value of α_{cr} is fixed, and a range of α_L values is considered. From Equation 5.7 a value of criticalness can be obtained for each of the values of α_L at given α_{cr}, and using equations 5.8 and Table 5.1 χ is computed.

Finally, by Equation 5.6, the value of α_R for each α_L and for a given α_{cr} and imperfection factor k, is obtained.

Next, Figures 5.20, 5.21 and 5.22 refer to buckling curve "b", "c", and "d", respectively.

Considering these curves, it is clear that for each value of α_L, the first derivative of α_R to α_L is always positive. This means that using the general method to take into account the stability issue, at equal α_{cr} between mechanisms, it is always on the safe side to use a statically admissible expected plastic multiplier α_{R1}, instead of the "exact" one, α_{R2}, which is certainly greater than α_{R1}; the expected load multiplier related to the true plastic

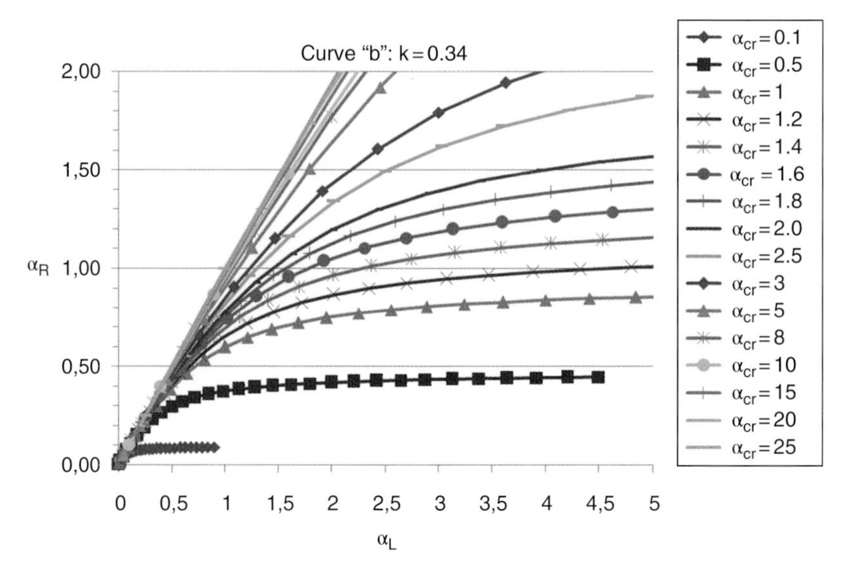

Figure 5.19 The value of α_R as a function of α_L, for different α_{cr}, and for buckling curve "b". If $\alpha_{L1} < \alpha_{L2}$, then it is always $\alpha_{R1} < \alpha_{R2}$, for every value of α_{cr}.

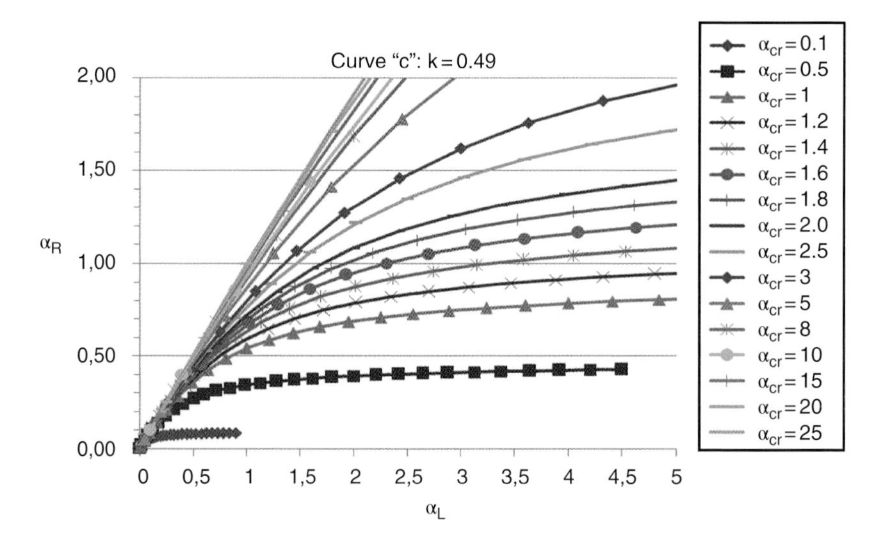

Figure 5.20 The value of α_R as a function of α_L, for different α_{cr}, and for buckling curve "c". If $\alpha_{L1} < \alpha_{L2}$, then it is always $\alpha_{R1} < \alpha_{R2}$, for every value of α_{cr}.

mechanism α_{R2}, is always greater than that obtained by using the statically admissible mechanism, α_{R1} (Figure 5.22).

All the curves have these important features:

- When α_{cr} gets much higher than 1, the curves tend to become a straight line, and the plastic multiplier tends to be unaffected by buckling, so the expected load multiplier gets to the plastic one. If plastic design is wanted, with $\alpha_R \approx \alpha_L$, then high critical multipliers are a necessary precondition.

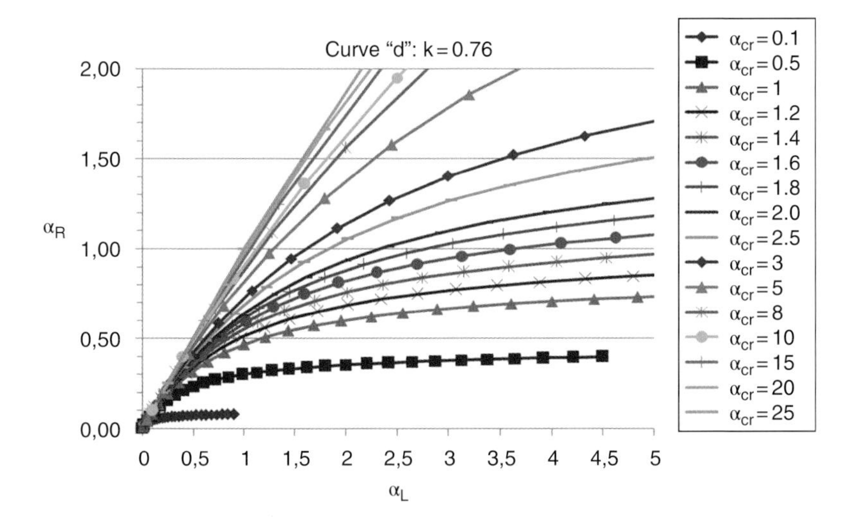

Figure 5.21 The value of α_R as a function of α_L, for different α_{cr}, and for buckling curve "d". If $\alpha_{L1} < \alpha_{L2}$, then it is always $\alpha_{R1} < \alpha_{R2}$, for every value of α_{cr}.

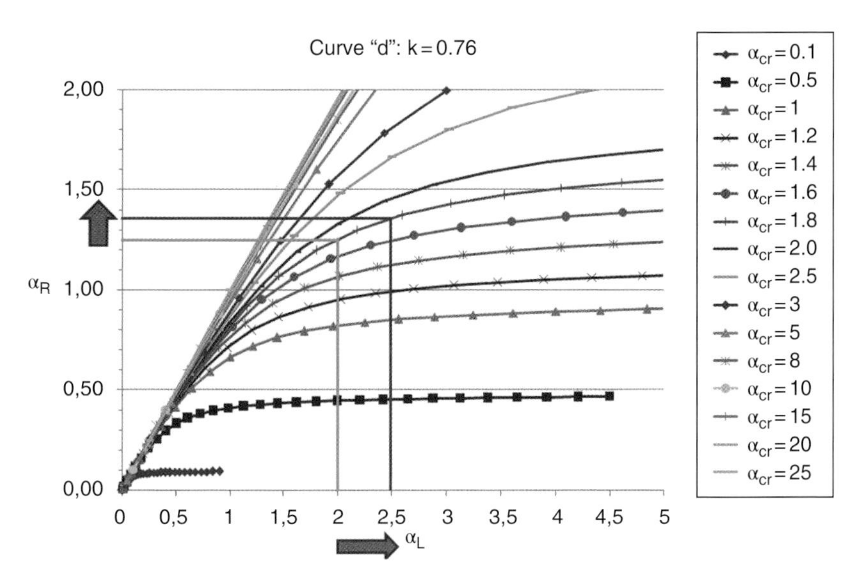

Figure 5.22 Example of application of the static theorem of limit analysis using the general method. Statically admissible plastic multiplier $\alpha_{L1} = 2$, "true" plastic multiplier $\alpha_{L2} = 2.5$, $\alpha_{cr} = 1.8$.

- If $\alpha_L > 1$ and $\alpha_{cr} > 1$ this does not necessarily imply $\alpha_R > 1$. Considering each buckling curve, there is a well defined minimum value of α_L for each α_{cr}, related to $\alpha_R = 1$ (consider the horizontal straight line $\alpha_R = 1$). Lower plastic multipliers α_L, albeit greater than 1, imply $\alpha_R < 1$ which is an unsafe design.
- The underestimate of α_R due to the use of a statically admissible plastic multiplier tends to increase with α_{cr}. For low α_{cr} (say <2) the underestimate tends to lose importance.

American standard AISC 360-10, does not use or mention the "general approach" of Eurocode 3. However, although different formulations are used for the different buckling problems, these formulations always take into account the imperfections and the interaction of stability with plasticity. Considering for instance the formulations adopted for the compressed members in AISC 360-10, and using the symbols introduced here, the following relationships are found:

$$\alpha_R = 0.658^{\bar{\lambda}^2} \alpha_L \qquad \text{for } \bar{\lambda} < 1.5$$

$$\alpha_R = \frac{0.877}{\bar{\lambda}^2} \alpha_L = 0.877 \alpha_{\text{cr}} \qquad \text{for } \bar{\lambda} > 1.5$$

The second one, valid for high criticalness, is not dependent on α_L. The first one can be found by taking the first partial derivative of α_L, giving

$$\frac{\partial \alpha_R}{\partial \alpha_L} = \frac{\partial 0.658^{\frac{\alpha_L}{\alpha_{cr}}} \cdot \alpha_L}{\partial \alpha_L} = 0.658^{\bar{\lambda}^2} \frac{\ln 0.658}{\alpha_{cr}} + 0.658^{\bar{\lambda}^2} = 0.658^{\bar{\lambda}^2} \left(1 - \frac{0.419}{\alpha_{cr}}\right) > 0$$

So, increasing α_L, α_R will always increase in the meaningful range of values of α.

Another example is Canadian Standard S16-14, where the formulation is

$$\alpha_R = \frac{0.9 \alpha_L}{\left(1 + \bar{\lambda}^{2n}\right)^{\frac{1}{n}}} \quad \text{with } n = 1.34 \text{ or } n = 2.24$$

Using two values of α_L, $\alpha_{L1} < \alpha_{L2}$, it can be proved also that $\alpha_{R1} < \alpha_{R2}$:

$$\frac{0.9\alpha_{L1}}{(1 + \bar{\lambda}_1{}^{2n})^{\frac{1}{n}}} \le \frac{0.9\alpha_{L2}}{(1 + \bar{\lambda}_2{}^{2n})^{\frac{1}{n}}} \Rightarrow \frac{1 + \bar{\lambda}_2{}^{2n}}{1 + \bar{\lambda}_1{}^{2n}} \le \left(\frac{\alpha_{L2}}{\alpha_{L1}}\right)^n \Rightarrow$$

$$\frac{1 + \left(\frac{\alpha_{L2}}{\alpha_{cr}}\right)^n}{1 + \left(\frac{\alpha_{L1}}{\alpha_{cr}}\right)^n} \le \left(\frac{\alpha_{L2}}{\alpha_{L1}}\right)^n \Rightarrow \frac{\alpha_{cr}^n + \alpha_{L2}^n}{\alpha_{cr}^n + \alpha_{L1}^n} \le \frac{\alpha_{L2}^n}{\alpha_{L1}^n} \Rightarrow$$

$$\alpha_{cr}^n \alpha_{L1}^n \le \alpha_{cr}^n \alpha_{L2}^n \quad \text{always}$$

Similar results can also be obtained for other standards.

Up to now it has been assumed that the *elastic* external and internal distribution of the force packets are known and exact. If this is not rigorously the case, as when using simplifying methods that neglect the exact stiffness related to each single connector subconstituent, strictly speaking nothing can be said, in general, about the existing relationship between α_{cr1} and a_{cr2}. In this case, proving that the statically admissible plastic mechanism is not limited by buckling is not enough to prove that buckling is avoided for the "true" plastic mechanism. In fact, α_{cr1} is computed using *an* elastic distribution, not *the* elastic distribution, and assuming $\alpha_{cr1} = a_{cr2}$ is not allowed.

The buckling checks made by the general method or by a step-by-step analysis, however, will allow us to have a measure of the safety against buckling for one of the statically admissible distributions. Ideally more distributions would have to be checked. For each constituent, the safest distribution is the one that maximizes the buckling effects over it.

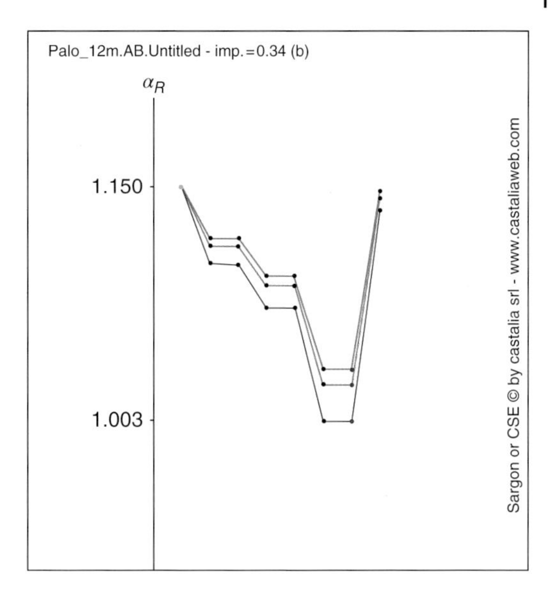

Figure 5.23 The general method applied to an eight-combination model (constant limit load lower bound $\alpha_L = 1.15$ used). Curve "b" has been used. The minimum value is $\alpha_R = 1.003 > 1.0$. The first three critical multipliers have been used, so three curves can be distinguished.

If the check is repeated for a number of different loading combinations (see Figure 5.23), using a single statically admissible force distribution for each combination, implying different stress distributions and stress levels in all the constituents, and if the critical multiplier is always sufficiently high to assume that $\alpha_R > 1$, then it may be indirectly assumed that the constituents are well designed. This approach, testing by finite elements a large number of loading combinations, is better than the traditional approaches neglecting part of the forces, part of the geometry, and evaluating only a restricted number if not a single loading configuration.

5.6.1.5 Isomorphically Loaded Constituents

Consider a force transferrer or a stiffener inside a hyperconnected renode. Generally speaking this constituent is loaded by a number n of force packets delivered by its connectors, $-\mathbf{c}_i$ and it is self-balanced in space under the effect of these loads. Depending on how the external indeterminacy has been solved, these force packets will be different, so there will be a set $-\mathbf{c}_{i1}$ related to distribution D_1 and a set $-\mathbf{c}_{i2}$ related to another distribution D_2.

If the two distributions are different only due to a scalar, that is, if d is a numerical value, for each i

$$-\mathbf{c}_{i2} = -d\mathbf{c}_{i1}$$

then the constituent is said to be isomorphically loaded. Isomorphically loaded constituents are constituents that, no matter what the external distribution of force packets is, at the renode level, they are basically loaded internally in the same way, differing only by the intensity, from one distribution D_1 to another distribution D_2 at equal externally applied loads. Many constituents exhibit such a behavior, like for instance internal stiffeners or haunches: these constituents are loaded depending basically on their connectors, at different intensities.

Let us consider one isomorphically loaded constituent. For the distribution D_1, its elastic stress state can be obtained by the following finite element analysis:

$$\mathbf{Ku} = \sum_{i=1}^{n} -\mathbf{A_i}(\mathbf{c_i})$$

where $\mathbf{K}$ is the stiffness matrix, $\mathbf{A_i}$ is an assembly matrix transferring to the nodes of the finite element model, $\mathbf{u}$ is the displacement vector and $\mathbf{c_i}$ are the force packets, acting as loads.

Using the stress state computed, a limit multiplier α_L and a critical multiplier α_{cr} can be obtained.

If now a second distribution of force packets D_2 is used, the constituent will have to be solved for the following finite element analysis:

$$\mathbf{Ku} = d\sum_{i=1}^{n} -\mathbf{A_i}(\mathbf{c_i})$$

and so the limit load multiplier will be α_L/d and the critical load multiplier will be α_{cr}/d. It can then be concluded that

$$\bar{\lambda}_1 = \sqrt{\frac{\alpha_L}{\alpha_{cr}}} = \bar{\lambda}_2 = \sqrt{\frac{d\alpha_L}{d\alpha_{cr}}}$$

If the general method is used to check for resistance and buckling

$$\alpha_{R1} = \chi\alpha_{L1}$$

$$\alpha_{R2} = \chi\alpha_{L2} = \chi\alpha_{L1}/d = \frac{\alpha_{R1}}{d}$$

So if the expected load multiplier α_{R1} has been computed for the constituent under a distribution D_1, and if it is expected that other distributions might imply a loading d times that related to D_1, then the expected multiplier for those unknown distributions might be obtained by multiplying α_{R1} times d.

For instance, if a stiffener is checked for resistance and stability under a distribution D_1 with a load multiplier $\alpha_{R1} = 3$, and if it is expected that the load could be between 0.5 and 2 times that related to the distribution D_1, then the expected multiplier would be between 6 and 1.5, and the utilization between 1/6 and 2/3.

It is clear that this property is of great help, once the expected multipliers α_R under a distribution D_1 are known.

Sometimes, several identical constituents act in parallel in order to carry some load (Figure 5.24), and the exact force level carried by each is not known. A finite element buckling check will automatically consider the worst constituent and will also be indirectly valid for the others. The only need will be to evaluate the maximum d related to the worst loaded constituent in order to get a final safety measure.

Isomorphically loaded constituents can be preliminarily designed in such a way as to avoid relevant interaction between plasticity and buckling. This is why the standards sometimes ask that some width-to-thickness ratios are respected for specific connections-constituent kinds.

A wide set of isomorphically loaded constituents are the identical force transferrers isoconnected (Figure 5.25).

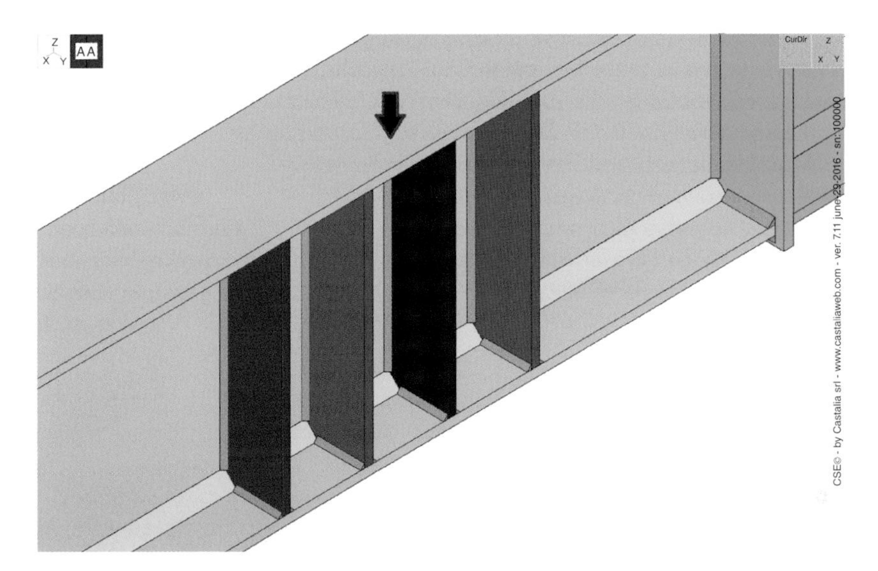

Figure 5.24 Identical iso-loaded constituents forming a set.

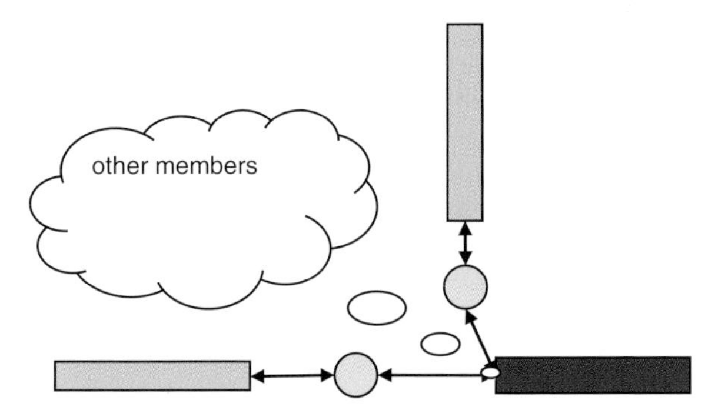

Figure 5.25 Isomorphically loaded constituents.

Consider that:

- the force transferrer is isoconnected, by one connector to a constituent and by another connector to another constituent
- by equilibrium the forces in one connector are determined by the forces in the other
- it has been assumed that the internal force *distribution* in the connector subconstituents is fixed (linear, plastic, or others)
- if the member can only bear specific force-packets (like axial force in trusses, or simple shears in beams) every statically admissible distribution of forces in the total renode, D_i, will locally have to comply with the transfer of similar forces.

So it is apparent that the forces flowing into the force transferrer will not depend on the distribution assumed at renode level, D_i. Locally, the distribution will always have to be the same.

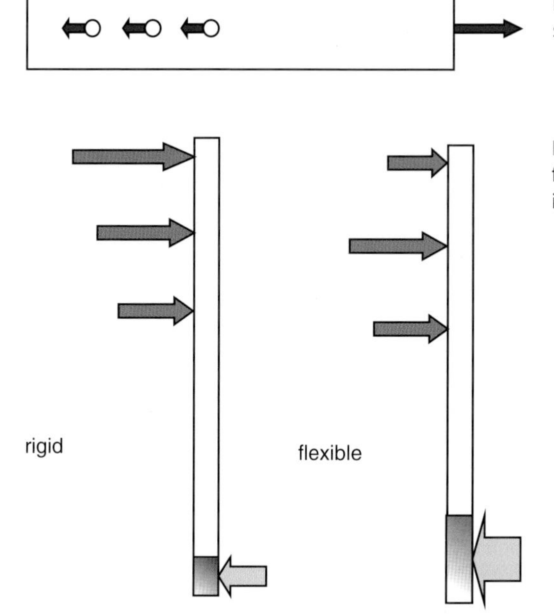

Figure 5.26 Equal force distribution for shear applied to a connector.

Figure 5.27 Internal distribution of the forces in a moment connector is not immediately determinable.

Also the members, that are connected to another member or to a constraint block only by means of a single isoconnected force transferrer, are in the same situation.

It has been assumed that the internal force distribution in the connector force packets joining isoconnected parts does not depend on the external distribution D_i. This is particularly true for shear connectors (Figure 5.26), as local plasticization is basically independent of the geometry of the connected parts, and will soon lead to equal forces in the connector subconstituents (see Chapter 7 and the example in Chapter 10). Also, assuming a linear or a plastic distribution for the local shear related to torque, the equal distribution hypothesis can be extended to connectors loaded by shear and torque.

The equal distribution hypothesis is not very sound in the context of moment or axially loaded connections (Figure 5.27), as the distribution of forces in the connector subconstituents strongly depends on the plasticization of the connected plates in bending, and this in turn also depends on the different possible hypotheses about the contact between the parts (contact non-linearity); the different possible distributions to be controlled in light of the static theorem are here basically the internal distribution of forces between the connector subconstituents.

However, there is a huge set of constituents that can be considered isomorphically loaded at equal internal force distributions, and this can be very helpful in the checking phase.

5.6.1.6 The "True" or "Exact" Plastic Mechanism

Given a system under an external loading, the correct way to assess the "exact" behavior of this system is to perform a non-linear step-by-step elastic–plastic analysis. As some phenomena may interact with plasticity and stop the load increase before it reaches the theoretical value of the limit load in the sense of the theory of plasticity, these phenomena must also be taken into account. This poses a tough problem to be solved, as the non-linear procedure must take into account contact, buckling, fracture, and slip.

Following this ideal path would in theory allow us to compute the "exact" distribution of forces, both externally (distribution of force packets between the connectors in hyper-connected renodes) and internally (distribution of the forces between the bolts of the bolt layouts, or the weld seams of the weld layouts of a renode).

This is however only true in theory. The knowledge of the system is always notional and never perfect; trains of decimal digits are basically an invention.

This holds true for the governing parameters when considering the contact non-linearity, which is strongly affected by plays, lack of planarity, tolerances in thicknesses, holes and widths, and by imperfections in general.

Also, buckling is strongly affected by imperfections, and by contact non-linearity itself (which might locally overload some constituent), and these are never known in full detail. The best example are prying forces, where a small difference in displacement may lead to strong differences in the stress state. Moreover, residual stresses, which depend on rolling, cold forming, and cooling, play an important role in triggering buckling, because they may locally lower the stiffness.

Friction coefficients, and "ultimate strains" are usually nominal values, as they are known only statistically.

So the ideal path cannot be realistically followed for true systems, and some kind of envelope must be found. For this reason the word "exact" has been used between inverted commas.

What must be done therefore is to design a system resilient to possible minor changes, and that tends to behave in a clear, predictable way. Generally this can be achieved by correct sizing and by sound layouts.

5.6.1.7 Conclusions

The ability to check buckling of a whole renode – as will be seen – by finite element analyses, may be very important as a direct index of good design, but it cannot formally prove that a statically admissible force distribution, safe according to the theory of plasticity, is also safe when considering the buckling effects.

A tradeoff between the computational effort needed to run *a number* of fully non-linear analyses, trying to correctly estimate the "true" system, and a generalized over-sizing of the constituents is using buckling checks and the general method for a high number of loading combinations, referring to statically admissible external and internal force distributions.

5.6.2 Fracture

Given a steel grain, this will begin to fracture if its strain exceeds a given limit. Fracture in a structural part is a highly unwanted behavior, as it can lead to a sudden collapse. Also, even if unloading occurs, the damage is still unrecoverable.

If fracture happens in a wide region involved in a plastic mechanism which has already developed its ultimate load, it might be considered the unavoidable end of each monotonically increasing loading. Constituents are designed for loads lower than those leading to such failure, and so its importance is relatively low.

However, fracture may be also triggered by local peaks of the stresses and strains which are not related to any mechanism. If this happens, the fracture of one particle, possibly near to a defect, or of a sharp corner, may trigger a crack growth, which can suddenly lead to global collapse. Classic examples are the Comet airplanes or the WW II Liberty Ships.

It is also possible, as in block tearing when a bolt layout is applied to a steel plate, that several regions highly stressed, one near the other, interact with one another leading to a globally organized pattern of cracks, able to detach a part of the plate from the remainder. This is a special failure mode, which needs specific methods to address it, as will be seen in Chapter 8. A bolt layout with a given pattern of holes can be applied differently to a steel plate, in different regions and with different distances from free edges. So, good rules of design may greatly help.

Just as the static theorem of limit analysis cannot prove that a different distribution of internal forces leading to a different mechanism will not experience buckling before attaining the limit load, it cannot prove that a different distribution will not lead to local stress peaks able to generate a crack.

Indeed, the resistance of the constituents to this kind of failure is not proved by a necessarily limited set of loading combinations to check, but indirectly by the way the constituent is designed.

Local stress peaks are not usually dangerous as they immediately imply a local plasticization which will redistribute the stresses in the region in the vicinity (thermal effects must be considered when the construction is exposed at very low temperature). However, there are special cases that may indeed trigger a crack, as even low stress levels can be related to quite high peaks.

This happens near sharp corners, related to cuts or to holes that have not been properly rounded (Figure 5.28). Several books report the stress intensity factors for different geometries and different loading conditions. In the field of steel connection design, dangerous geometries should always be avoided: cuts should always have rounded corners, and edges may need to meet at sufficiently high angles.

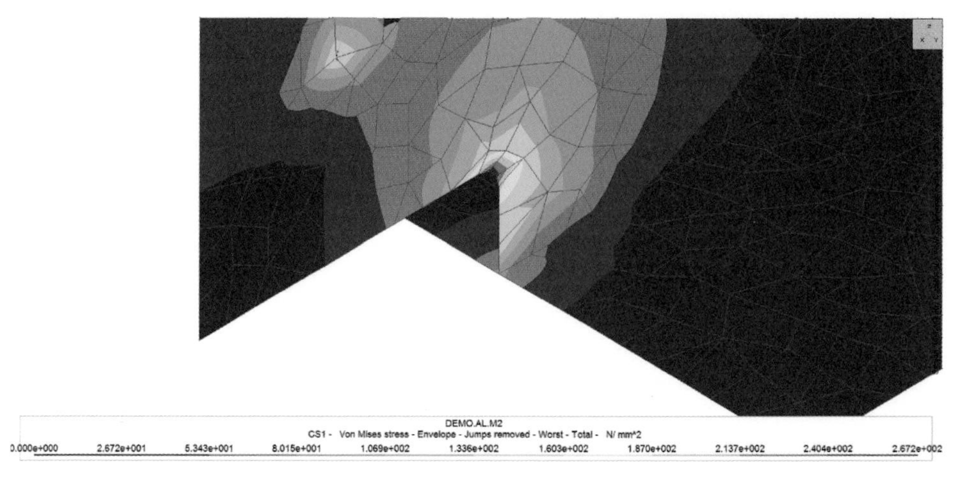

Figure 5.28 Local stress peak near to a rounded corner (coarse mesh).

Considering the stress obtained by simple algebra, dividing the force by the area, it is possible to define the stress concentration factor, which increases the notional stress depending on the geometry of the corners and of the holes.

For instance, a pulled plate with a circular hole (Figure 5.29) has the following stress concentration factor k (Young 1989):

$$\sigma_{nom} = \frac{P}{t(D-2r)}$$

$$\sigma_{\max} = k\sigma_{nom}$$

$$k = 3 - 3.13\left(\frac{2r}{D}\right) + 3.66\left(\frac{2r}{D}\right)^2 - 1.53\left(\frac{2r}{D}\right)^3$$

where P is the force applied, D the width of the plate, r the radius of the hole, and t the thickness of the plate. Plotting the functions, the curves are as in Figure 5.30. So, the stress may increase by a factor 3.

Similar results can also be obtained for other geometries and loads.

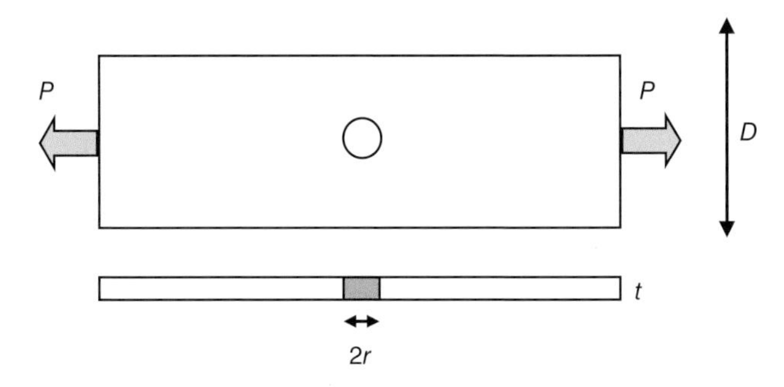

Figure 5.29 Plate with a circular hole, in tension.

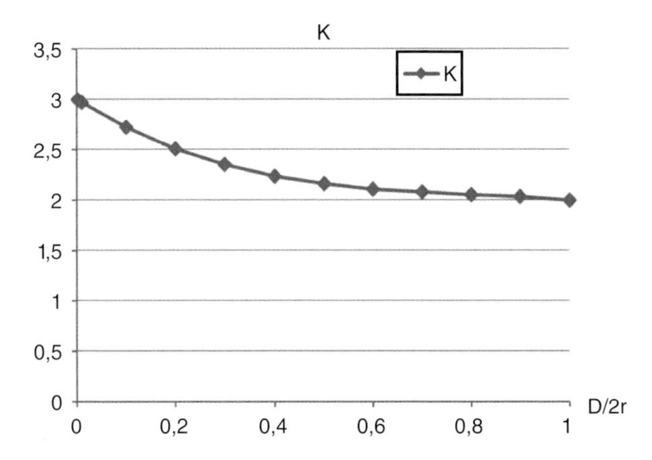

Figure 5.30 Stress concentration factor for a pulled plate with a circular hole.

The results obtained for isomorphically loaded constituents are also qualitatively valid when considering fracture failure.

If no stress peak able to lead to fracture has been observed in a constituent with a distribution D_1, it is not possible for a stress peak to be observed in that isomorphically loaded constituent, with a distribution D_2. The stresses in the constituent will only differ by a factor *d*. If the spreading of plasticity will occur before fracture for a given limit load level α_{L1}, it will do exactly the same for the load level $\alpha_{L2} = \alpha_{L1}/d$, as the two patterns of forces resulting, at limit, will be the same.

On the other hand, for isomorphically loaded constituents, if fracture is reached before the spreading of plasticity can lead to a mechanism, then this will also happen if the starting load is different.

Other constituents, not isomorphically loaded, will in general be loaded differently, depending on the force distribution D. However, if there is no region where a concentration of stress might occur (no corners or holes), it is not expected that fracture could be the governing failure mode, before reaching plastic mechanism.

When considering the problem of cracking, a special problem is posed by welding defects, that may trigger fracture and collapse. To avoid such defects, welding must be subjected to a number of checks, ensuring that no inclusions or empty small volumes have been left in the welded material (for an extended and rich discussion see Bruneau et al. 2011).

5.6.3 Slip

There are two types of no-slip (bolted) connections: those which must resist with no slip for service loads and those which must resist with no slip at ultimate loads. In the first case once the slip of the connection has occurred, the resisting system relies on bolt bearing and all other normal resistance modes.

Usually, no-slip connections are locally isoconnected, that is, they do not act in parallel with other type of connections (Figure 5.31 top left). If this is true (Figure 5.31 right),

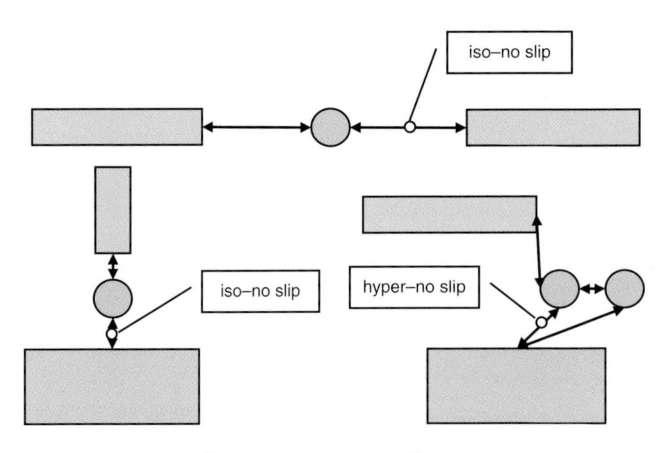

Figure 5.31 Iso and hyperconnected no-slip connectors.

most of the standards do presently oblige us to consider only one of the parallel systems as totally resisting.

No-slip connections are typically used to connect a member to a chord in truss works, by means of a gusset plate, or to connect a column to a concrete slab or block. It is rare that a no-slip connection is under the effect of force packets which cannot be directly determined by equilibrium. An important exception is bolted splice joints of I or H cross-sections, but usually all the bolt layouts are preloaded, and so the relative stiffness can be evaluated correctly. This means that the problem of considering different force packets distributions, D_i, in order to prove that a different distribution does not imply a slip, is not usually existing, as all the distributions D_i will always imply the same force packets in the no-slip connector. On the other hand, if more than one connector acts in parallel, and one of these is a no-slip connector, then it will be necessary to decide: if the no-slip connection is going to take all the load, then the problem vanishes, reverting to the previous one. Otherwise, the other connectors will have to be able to resist the total loads applied.

If the external indeterminacy does not usually exist or can be solved safely, what can be said about the internal one? Under the shelter of the static theorem of limit analysis, several possible internal force distributions can be considered, and it is important to assess if, provided no slip occurs under an internal force distribution D_i, this will also hold true for another internal distribution, D_j.

If the connector is under the effect of shear, the single subconnectors are usually considered loaded in the very same way, albeit the "true" loads flowing into each bolt would have to be computed considering the stiffness of the parts in contact. This effect is usually neglected. An increased shear in one bolt, might trigger a local failure of the no-slip condition, but will not be able to trigger global slip, if the no-slip check is satisfied using the components of the force packet absorbed by the connector as a whole.

If the connector is under the effect of a pure torque, and axial force pre-stress, the forces flowing in the single bolts are also in this case related to the stiffness of the connected parts, but also in this case the effect is normally neglected. The forces flowing into each bolt are then computed using some simple internal distributions (typically polar), and imposing equilibrium. What if the shear induced by torque will locally exceed the limit related to the existing local compression? The exceeding force will be distributed to the less stressed parts, having a lower lever, by means of elastic distortion in the body in contact, until a proper balance is found, if any. Assuming that all pre-stresses are equal in the bolts, the same shear will at limit be allowable. So, the "plastic" distribution is the correct one at limit.

Axial force and bending moments acting over the connector lead to different axial forces in the bolt shafts, depending on the flexibility of the connected part, and their contact. So, in general, a wide set of possible distributions, all balanced, can be computed.

However, a bending moment applied to the layout does not imply a change in the resultant axial force. If a part is detached, other parts will be over-pressed.

So from the no-slip check point of view, once it is ensured that the no-slip condition is satisfied with the resulting shear, axial force, and pre-stress force, and that a contact surface may still exist with the axial force and bending moment applied, the distributions can be different, but all with no slip. This depends on the fact that the resisting friction force does not depend on the contact-surface amount, but just on the normal force and the friction coefficient.

5.6.4 Fatigue

There are two kinds of fatigue that must be taken into account:

- high number of cycles fatigue
- low number of cycles fatigue

The first typically has to be checked when connections are subjected to a high number of repeated relatively low loadings. This happens typically to parts under rotation, or under the continuous effect of relevant wind actions, or for example to structures placed over ships.

If a high number of cycles is expected, despite the compliance of the load peaks to the requirements of plastic and buckling static analyses, it must be checked that the details are not subjected to rupture due to high cycle fracture. High cycle fatigue is typically observed in the elastic range, for those details that undergo stresses often much lower than yield. Fatigue rupture is more probable if the variation of the stresses is higher relative to the constant stress level, and for this reason bolt preloading often has very useful effects on reducing fatigue. In fact, using preloading (see Chapter 7), the true variation of the stresses due to the externally applied variable loads is a much lower fraction of the constant stresses.

High cycle fatigue requires the evaluation of the stress variations, and to this end, detailed finite element models like those described in Chapters 9 and 10 are very useful, as they can help to predict the stress levels taking into account the true geometry, load, and constraints of the detail under examination.

The second type of fatigue is completely different and it is typically observed in parts subjected to high plastic strains, and to repeated inverted cycles of loading. This is the typical condition of dissipative zones of earthquake resistant structures.

In order to take this problem into account, which is designed only for special parts of dissipative structures in seismic design, care must be taken to ensure that the degradation of the resisting mechanisms due to the residual accumulated plastic strains, is not such as to limit the number of sustainable plastic cycles to too low a number.

If this happens, the detail is able to resist only part of the time history signal of the earthquake, and failures due to low cycles fatigue after some seconds from the beginning of the earthquake shock may directly lead to an immediate collapse. The problem is made more difficult by the fact that the limit number of cycles is a function of the maximum plastic strain, and this latter strongly depends on the frequency content of the seismic signal (which imposes sign reversals), which is hardly predictable. Also for this reason, a high number of signals should ideally be analyzed.

High plastic strains are favored by buckling (made easier by the P-δ effect triggered by residual strains), so for this reason buckling of plastic parts should be avoided. The study of details able to be repeatedly loaded and unloaded in the plastic range with no loss of stiffness and dissipation capacity is one of the areas where presently (and for the past few decades) researchers are working more. They were not only using experimental approaches valid for assessing the reliability of specific structural solutions, but they were also trying to tune reliable models of the degradation that can be used in non-linear dynamic solvers, so as to model the typical behavior of dissipative structures under earthquake.

5.7 Analysis of Connections: General Path

It is now possible to describe a general path that may be followed in order to analyze a renode having a generic number of members connected one another or to a constraint block:

1) For each renode, define its nodal region, cutting all the members at proper far extremities **F**.
2) Verify that each constituent (members, force transferrers, stiffeners, and constraint blocks) is correctly connected, that is, that no constituent is hypoconnected.
3) Verify that each connector is properly connected to a number of constituents, so that all its subconstituents are connected to the same constituents.
4) Introduce, as primary unknowns of the problem at hand, the force packets $\mathbf{c}_{ie}$ exchanged at each connector i extremity e.
5) By means of a proper computational model, compute a balanced distribution of force packets $\mathbf{c}_{ie}$, that is, a distribution D_1 of force packets satisfying equilibrium equations, solving the external and internal indeterminacy.
6) Check the connectors for resistance (including slip) against the computed force packets, properly unpacked to each single connector subconstituent k (single welds and bolts, force subpackets $\mathbf{c}_{iek}$).
7) Apply the computed force packets with sign reversed $-\mathbf{c}_{ie}$ to all the connected constituents. Unpack the packets so that each single connector subconstituent k delivers its own force subpacket $-\mathbf{c}_{iek}$ at a specific set of points of the constituent (e.g. along the single weld projected throat surface, or at each bolted mid-thickness). The proper modeling of the point of application of the forces is a fundamental requirement for saving the equilibrium.
8) For each non-member constituent under the action of the set of force packets delivered by the connectors, properly distributed among the subconstituents as $-\mathbf{c}_{iek}$, check all the relevant failure modes, and ensure that the constituent is able to carry the force subpackets applied. The set of the force subpackets applied to the constituent must be self-balanced, as no other force packet or subpacket loads the constituent. As the generic constituent is self-balanced, and isolated in space, dummy constraints can be added, to analyze it if needed, exerting no constraint reactions.
9) For each member, consider the force packets delivered by the connectors connected to the member, properly distributed among the subconnectors as $-\mathbf{c}_{iek}$, the force packet applied at the far end **F** (obtained by properly cutting the members) and if needed the force packets directly applied to member by externally applied loads. Check the member loaded by all these force subpackets and packets, against all the relevant failure modes of the member. The set of all the force packets and subpackets applied to the member should be self-balanced, so once more, dummy constraints can be safely applied at the far extremity.
10) If some constituent is not checked for some failure modes, choose one of the following two, or both:
    a) Review the design of the constituent increasing its strength or stiffness.
    b) Find a different set of force packet distribution $\mathbf{c}_{ie}$ and start back from step 6.
11) Be sure that buckling is checked or by means of simple truly enveloping rules, and by means of sufficiently stiff proportions, or by means of a wide set of buckling analyses

(linear or non-linear) related to many load combinations, ensuring that the real multiplier is always greater than 1.

12) Check design for fatigue when fatigue conditions are expected.
13) If all the constituents are checked, their "true" utilization will be higher than the one computed, and the design can be accepted.

This analysis path is general and can be applied to every renode. No limitation of geometry, layout, number or nature of members or cross-sections, loads applied, or internal forces distribution has been required. The approach is general, and can be applied to every renode of every structure.

The following chapters will get into the details in order to explain how this path can be implemented.

References

Bernuzzi B. & Cordova B. 2016, *Structural Steel Design to Eurocode 3 and AISC Specifications*, Wiley Blackwell.

Bruneau, M., Uang, C.M. & Sabelli, S. 2011, *Ductile Design of Steel Structures*, McGraw Hill.

Canadian Standard Association Group 2014, Design of Steel Structures, *S16-14*, CSA Group.

Heyman, J. 1995, *The Stone Skeleton*, Cambridge University Press.

Heyman, J. 1998, *Structural Analysis – A Historical Approach*, Cambridge University Press.

Newton, I. 1687, *Philosophiae Naturalis Principia Matematica*, trans. I. Bruce, 2015, http://www.17centurymaths.com.

Rugarli, P. 2008, *Calcolo di Strutture in Acciaio*, EPC Libri.

Young, C. 1989, *Roark's Formulas for Stress & Strain*, McGraw Hill.

6

Connectors: Weld Layouts

6.1 Introduction

This chapter and the next aim to discuss the analytical procedures that can be used in order to find the forces and stresses flowing in the subparts of a single connector, once the overall forces acting at different connector extremities are known (i.e. the overall three forces and three moments acting at each extremity of the connector). They also aim to discuss how the intrinsic stiffness of a connector might be evaluated, and which geometrical limitations have to be verified according to different standards.

The discussion of the failure modes checks that are strictly related to the connectors (basically their resistance, or no-slip for friction connectors) is in Chapter 8.

Also, it is not within the scope of these chapters to discuss the experimental results related to connectors, or to discuss the very important design considerations related to good practice in using connectors. These very important goals are already covered in depth by excellent texts, such as those by Bruneau et al. 2011 or Tamboli 2010.

Three families of connectors will be considered: weld layouts, bolt layouts, and contact. The last does not need the existence of any fastener, and it's an immaterial connector: it works when two surfaces are in contact and acts asymmetrically in the direction normal to the plane of contact (the *faying surface*). In tangential direction, the maximum force exerted is a function of normal force and friction coefficient.

In considering the behavior of connections, very many different problems are found, and there are a number of uncertain physical values that might affect the detail of the connector behavior (plays, residual stresses, micro defects, tolerances, yield levels, lack of planarity, contact, friction, etc.). For this reason, taking also into account the typical uncertainty of the physical actions related to civil structures (wind, snow, and especially earthquake), it seems meaningless to try to find an out of scale precision, using too many significant digits (Rugarli 2014).

What is relevant is to get methods that are reliable and safe, which might help designers to tackle generic problems and give them confidence that the solution found is safe and possibly also economical. It is also needed to avoid neglecting important issues, while perhaps deepening too much others. The depth of our analyses should possibly be coherent, as it is not very logical to search a four-digit precision using data having two digits precision. We must consider all the relevant facts, and not neglect them. One of those facts is the geometrical position of the constituents, and the effects of eccentricities, or the existence of six generalized forces in members.

Steel Connection Analysis, First Edition. Paolo Rugarli.
© 2018 John Wiley & Sons Ltd. Published 2018 by John Wiley & Sons Ltd.

As discussed in Chapter 5, the shelter offered by the static theorem of limit analysis is not comprehensive, and designs that were believed to be safe have shown significant weaknesses during earthquakes. It is very important to avoid such problems in the future, and so again, rather than four digits precision, it is the soundness of the design that must be given priority.

The need for *coherence* might imply the use of complex math, but the complex math should not be used at the expense of the very goal of engineering analyses, and forgetting the unavoidable intrinsic lack of detailed precision of the engineering methods. This means that the goal of the present work is not to push precision beyond two significant digits (if it were ever reachable), but sometimes, in order to save those digits and coherence, some complex math will have to be used. It is important to underline this aspect, so as to avoid possible misunderstandings: in no way is the aim to push "precision" beyond reasonable limits.

6.2 Considerations of Stiffness Matrix of Connectors

The stiffness matrix of connectors is needed in order to properly consider their effect on the other constituents in the scene. Two stiffness matrices are of interest: the stiffness matrix of the connector as a whole, and that of single connector subconstituents, that is, single weld seams or weld seams segments and single bolts (there is no subconstituent for contact).

When considering the shear forces acting over a connector, it is commonly assumed, for good reasons, that the distribution of forces between the subconstituents is plastic, not elastic. This means that we consider the redistribution of forces between the subconnectors, due to the plasticization of the overloaded ones. This rule has some limitations, especially for very long joints, as the displacements needed to redistribute the force applied to all the subconstituents may be too high.

So, when considering for instance a bolt row loaded in shear, it is common to assume that the bolts are equally loaded (for experimental results showing this behavior see Kulak et al. 2001), albeit the elastic distribution would be quite different.

If the determination of the forces is according to the theory of elasticity, then at increasing load the initially most loaded subconnectors are predicted to fail, with no redistribution.

Considering first the stiffness of the connector as a whole, the following important considerations can be made.

If two members are isoconnected (see Chapter 5), the force packets flowing into their unique connectors (e.g. Figure 3.1) are by definition computable regardless of the evaluation of the stiffness: they are defined by equilibrium conditions. So, in these very frequent cases, if the aim is to get the forces flowing into the connectors, the stiffness evaluation is only needed to get a reasonable amount of displacement. However, having the ability to use a stiffness matrix is very important, since using the standard finite element methods, the forces flowing into the connectors will be evaluated precisely, taking into account all the internal forces, the exact position of the constituents, and all the eccentricities. This means being freed of a huge amount of work, and with no errors.

If two members are hyperconnected, more than one connector will share the forces, and the relative forces flowing into each one will depend on the ratio of the stiffnesses. It must be underlined that to assess the relative proportion of forces it is not important only the stiffness of the connector, but also the stiffness of the connected parts, in the surrounding region.

For example, in a row of sheared bolts, it is not only the stiffness of the bolt shaft that it is important, but also the stiffness of the plates connected, and the local plasticization due to bolt bearing.

These plasticizations lead to a sharp drop in stiffness, so that the resulting equivalent stiffness of the bolt in shear is much lower. As the distribution of forces between the bolts depends on the ratio of the stiffness of the plate to the stiffness of the bolts, decreasing the bolt stiffness implies a redistribution of forces between the bolts, which somehow mimics the true plastic redistribution (lowering the stiffness of the bolts, the plate tends to behave as if it were rigid). This has been shown with a very simple model in Figure 6.1 by decreasing the diameter of the bolts at equal plate stiffness: the distribution of forces tends to be constant. More refined models taking into account this effect will be described in Chapter 7.

If the stiffness of the surrounding part is explicitly modeled by a finite element model, correctly taking into account the thicknesses, widths, positions, and mutual constraints of all the subplates of all the constituents, then it will be the model to drive the correct forces to the connectors. In practice, this means having modeled the whole renode, or some wide parts of it, by plate–shell elements or by solid elements. If this is true, then the single connector subconstituents will probably be modeled one by one by means of suitable elements. This approach is the most promising, and will probably be used as the standard one in the near future. This is considered in Chapter 10 and it is named *pure fem* in this book.

But if the model used to evaluate the forces flowing in the connectors does not use a detailed modeling of the connected parts, then the following considerations apply.

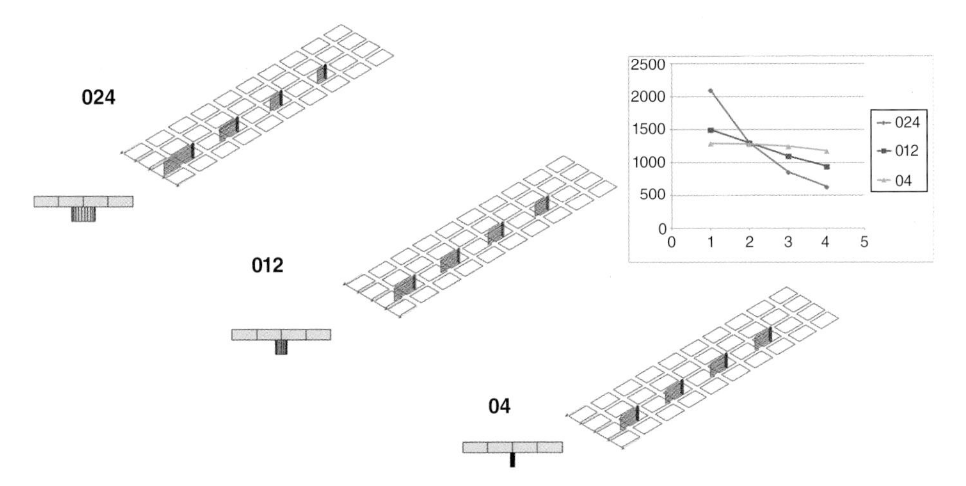

Figure 6.1 Progressive redistribution of elastic forces obtained by lowering the bolt stiffness. Plates are pulled toward the left. Bolt cross-sections are circles of decreasing diameter (24 mm, 12 mm, 4 mm). This simulates the effect of the bolt bearing plasticization effect.

Some parts of the surrounding stiffness can be evaluated locally, and might have important effects on the distribution of forces. The typical example has already been made, and it's the hole–shaft interaction and the related plasticity.

However, other parts may depend on the global size, position, orientation, and layout of the constituents, and cannot easily be modeled in general. Typically, for normal stresses the flexibility of the plates, their sizes and constraints, and the possible contact with other parts are important.

If typical connections, with well defined geometries and specific loads – that is, specific components of internal forces – are considered, then some methods are available in order to model the surrounding of the connectors, so as to set up a reasonable model of the stiffnesses (e.g. the *component method* of Eurocode 3, Jaspart 1991, or the semi-rigid connections stiffness estimate made by interpolation of the experimental curves, e.g. see Chen, 2000). In Eurocode 3, for example, in Section 6, *Structural Joints Connecting H or I Sections*, clearly a subset of all possible connections, the following typical situations are mapped for the stiffness:

1) column web panel in shear
2) column web in transverse compression
3) column web in transverse tension
4) column flange in bending
5) end plate in bending
6) flange cleat in bending
7) beam or column flange and web in compression
8) beam web in tension
9) plate in tension or compression
10) bolts in tension
11) bolts in shear
12) bolts in bearing
13) concrete in compression including grout
14) base plate in bending under compression
15) base plate in bending under tension
16) anchor bolts in tension
17) anchor bolts in shear: no information provided
18) anchor bolts in bearing: no information provided
19) welds
20) haunched beams

Several of these stiffnesses are sometimes considered infinite, such as when stiffeners are applied; this means assuming the hypothesis of rigid bodies. For instance, the stiffnesses 1, 2, 3 are considered infinite when stiffeners are applied (indeed, this is a frequent condition of good design practice), and so 11 and 12 if preload is applied, and 19, welds, always. Other stiffnesses are quite local, and can be easily taken into account, such as 10, 11, 12, 16.

However, the use of *all* such stiffnesses in the *general* context of generic steel connections seems hard. This is so not only because the steel elements do not use only H or I cross-sections, but more because six internal forces components act at the same time with unknown signs (torsion, shears, weak and strong axis bending, and axial force), several possible work processes can be applied to the ideal constituents modifying their shapes, bolt "rows" may well not exist, and so on.

However, for typical, well delimited steel connections – perhaps the most frequently used – under well defined loading conditions, these methods are fit. Also, for local behavior of connectors (e.g. bolt in bearing and in axial force and shear) these formulations are very useful. One legitimate question is whether, because to apply them standard connections must be used, the USA approach of pre-qualified, standard connections should be preferred, avoiding many difficult calculations. In such cases, the answer seems to be yes. However, this book does not deal only with typical connections, but rather it aims to find a general method that can be used to tackle every connection, getting to usable engineering results. So, the frame is necessarily wider.

Evaluating the stiffness of the surroundings is not strictly necessary if the aim is to compute the forces flowing into connectors of hyperconnected members, because what is relevant for force flow is the ratio of the stiffnesses.

If the evaluation of the value of the stiffness is important (e.g. to tune the semi-rigid stiffness models of frames), then experimental methods (for industrial productions, such as racks) or numerical simulations by finite elements as those described in Chapter 10, seem to be more general and even quicker, at least for non-standard joints. Besides, it must be underlined that the search of "precise" stiffness evaluations is frustrated by the fact that these stiffnesses change depending on the loads, and on a number of uncertain parameters, and that, on the other hand, all the available methods to assess the loads and to check for global stability are always affected by a non-negligible systematic error, of the order of 30%, not less. So, once more, an excess search for precision appears to be illusory. More than precision, engineers need safe and economical approximations. This has always been the kernel of engineering.

Evaluating the ratios of the stiffnesses does not necessarily mean evaluating exactly the stiffnesses themselves. Starting from an initial local estimate for the stiffness, it will be possible to modify the ratio of the stiffnesses by increasing or decreasing one term of the ratio, for each ratio, so also taking into account the effect of the surrounding regions. This can be accomplished by the introduction of the *flexibility index*, i.e. a real number used to tune the stiffness of each connector, if needed. It can be easily seen considering two springs in parallel, having initial stiffnesses K_1 and K_2. By assigning a flexibility index f initially equal to 1 to both, the fraction of force flowing into spring number 1 is

$$\frac{\dfrac{K_1}{f_1}}{\dfrac{K_1}{f_1}+\dfrac{K_2}{f_2}} = \frac{f_2 K_1}{f_2 K_1 + f_1 K_2}$$

which tends to 1 when f_2 tends to infinity *or* f_1 to zero, and to 0 when f_2 tends to 0 *or* f_1 to infinity. By changing f_1, or f_2, the forces flowing into the springs can be modified.

All the distribution of forces between the connectors connecting two hyperconnected members are statically admissible, as they all satisfy equilibrium equations. So, they are all under the shelter of the static theorem of limit analysis, as long as we only consider plasticity. However, as we saw in Chapter 5, we are also interested in considering realistic distributions, so as to reduce the probability of missing some important failure mode, related to buckling and fracture – albeit the "exact" force flow evaluation is unreachable

(Heyman 1995), and so the true protection against these failure modes must be accomplished by good design rules. For this reason, the global distribution of forces between connectors should be evaluated realistically, if possible. This specific goal does not need the exact evaluation of the stiffnesses but a reasonable evaluation of the ratio of the stiffnesses.

Once the global forces acting on a connector are known, the forces acting on single subconnectors can be evaluated by plastic methods, that is, assuming a redistribution between connectors. This is the counterpart of what is already done for member checking, when the internal forces are computed by means of elastic (or plastic) models, while the cross-section checks are nowadays always carried on assuming plastic moduli, if the sections are compact (AISC) or of class 1 or 2 (Eurocode 3), and so using a redistribution of stresses between the different fibers of the member cross-section.

Why is there a need for pertinent stiffness matrices of single subconstituents of connectors? The aim is to allow an evaluation of the forces flowing into each subconstituent, not by means of distributions assumed a priori, but by means of finite element analyses, which must take into account the flexibility and possible plasticization of the connected constituents, as well of the connectors. Elastic or plastic analyses can be run, and ideally the stiffness matrices of single connectors should be available for both. In other words, this would solve the connector indeterminacy (see Chapter 5).

In elastic analyses, we need to avoid overloading some subconstituents: it must somehow be taken into account the redistribution of forces and the possible plasticization.

In plastic analyses, the single subconnectors might still be modeled as elastic or they can be modeled as elastic-plastic. Clearly, this choice might affect the distribution of stresses in the connected parts. Usually the connectors themselves need to be sheltered, as their failure is not ductile but rather brittle.

6.3 Introduction to Weld Layouts

In this book, weld layouts always connect two constituents. Sometimes, elsewhere, weld layouts are considered as unique groups, despite the fact that some weld seams of the group weld two constituents say A-B, while some other weld seams of the same group, other constituents, say A-C; it is sufficient that the first object welded A (O_1 – see Chapter 4 for a definition of O_1 and O_2) is common to all weld seams, so that their first fusion faces all share the same plane (an example of this approach is given, for example, in Jaspart and Weynand, 2016, §6.3.5).

This is not what is assumed in this book. Here, all the weld seams of the same weld layout will weld the same two constituents, O_1, and O_2. So if a set of weld seams weld A to B, and another set welds A to C, two layouts will be needed.

Welds can be classified in quite a number of different ways, which refer to the way the welds are applied, the geometrical position of the constituents to be connected, and so on.

Here, weld layouts are divided into two categories: penetration weld layouts (groove welds) and fillet weld layouts. Plug weld layouts (slotted weld layouts are a special case of plug ones) can be considered an extension of fillet weld layouts.

Within the two families, fillet welds are quite commonly used, penetration welds are also often used, while plug welds are rarer.

The main difference between these welds is related to the stress flow within the welds.

In penetration welds, the stresses can flow in a manner not too different from that obtained by assuming the two constituents welded as a unique, continuous body. As the weld is placed in an empty space usually obtained by applying proper bevels to the original thicknesses, the stresses can follow more or less straight paths.

In fillet and slot welds this is just not possible (Figure 6.2). The stresses must follow a path external to the thicknesses, and this leads to a higher deformability (Figure 6.3) and to a lower resistance. The stress state inside the fillet is much complex, and special checking rules must be used.

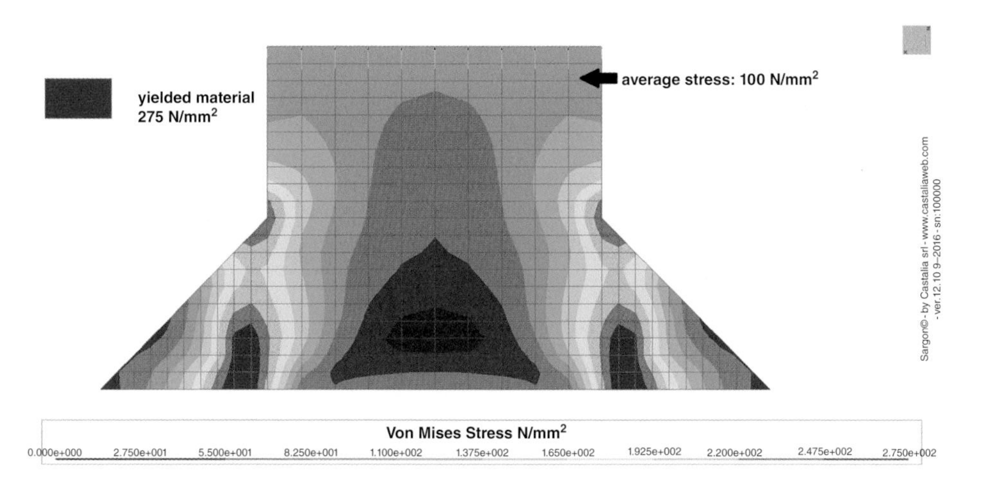

Figure 6.2 A pulled plate fixed by two fillet welds. Total throat is 7.07 mm, plate thickness is 10 mm. The normal stress in the vertical plate is 100 N/mm^2. The Von Mises stress in the welds is much higher than $100 \times 10/7 = 141$ N/mm^2, and reaches the yield stress of 275 N/mm^2 in extended parts of the fillet welds. Brick and wedge elements are used. Perfect plasticity behavior is assumed.

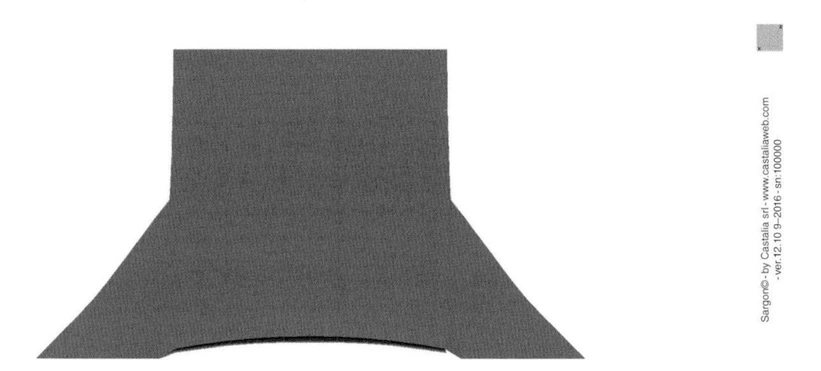

Figure 6.3 Deformed view of the pulled plate (amplification 100×). Maximum Z displacement is 0.0095 mm. Weld and plate length: 50 mm. Plate thickness: 10 mm. Weld seams thickness: 5 mm. Throat size of single weld: 3.53 mm. Force applied: 50 kN.

6.4 Reference Systems and Stresses for Welds

As explained in Chapter 4, a weld layout is inserted in the scene at a point **I**, using three orthogonal local axes (1, 2, 3). Axis 3 is perpendicular to the common plane of the weld seams (object O_1, the *faying plane*), while axes 1 and 2 lie in that plane, with no specific properties. Usually axis 1 is parallel to one of the chosen-face sides, since the weld seams use those sides to be defined (see Chapter 4).

Considering the weld seams, their position will be identified by coordinates (x_1, x_2), relative to insertion point **I** (Figure 6.4).

Each weld seam i has its own local axes, defined by the longitudinal axis $\mathbf{u_i}$, the transverse axis $\mathbf{v_i}$, and the normal axis $\mathbf{w_i}$, which is parallel to axis 3 of the weld layout, for all seams. These axes ($\mathbf{w_i}$, $\mathbf{u_i}$, $\mathbf{v_i}$), in this order,[1] are referred to the center of the weld seam effective surface: the point $\mathbf{G_i}$. The angle between axis $\mathbf{u_i}$ and axis 1 is α_i. These axes are also principal axes for the single weld seam, i. The weld position is identified thanks to two points, $\mathbf{P_1}$ and $\mathbf{P_2}$, which have to belong to a side of the chosen face (belonging to object O_2 – see Figures 4.8 and 4.10). Depending on the weld layout type (penetration or fillet), the weld seam will be on the left or on the right of the segment $\mathbf{P_1}$-$\mathbf{P_2}$, as explained in Chapter 4. The effective area of the weld seam will be:

- the whole weld area laying over the faying surface for partial or full penetration weld seams
- the throat area projected on the faying surface for fillet welds.

The throat a of the weld seam, Figure 6.5, can be defined as the height of the inscribed triangle, and its length depends on the thickness t and on the angle between active faces γ (the fusion faces of the weld seam).

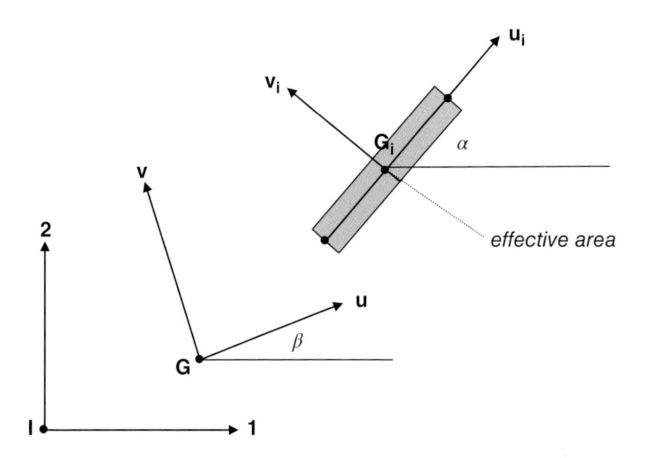

Figure 6.4 Reference systems for a weld layout and single weld seam.

1 The reason for ordering the axes so that **wi** is the first one is to consider the axial force of the finite elements in first row of the stiffness matrix, as is usually done.

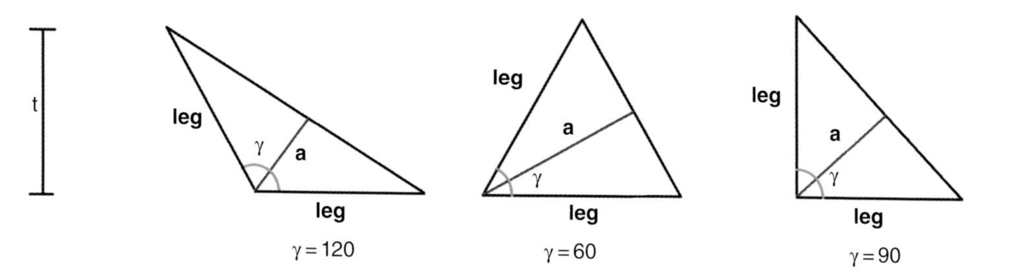

Figure 6.5 Throat, leg and thickness definition.

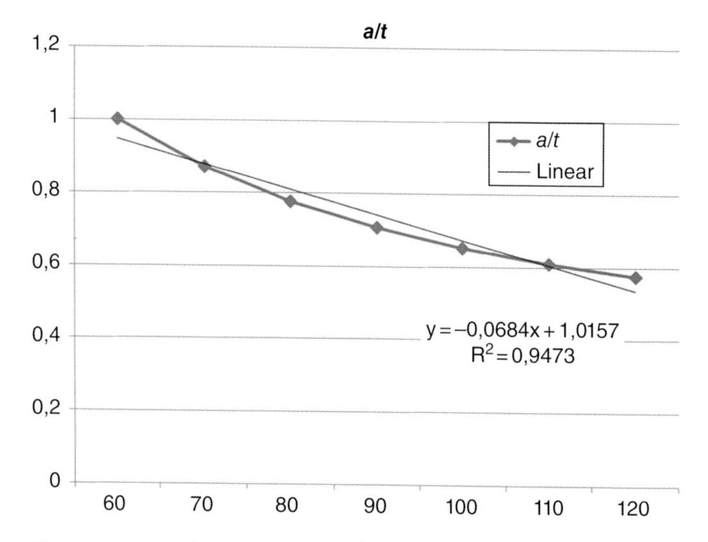

Figure 6.6 Ratio between throat thickness a and thickness t, depending on the angle between active faces γ, in degrees. An interpolation formula is also displayed, with the related R^2.

The relationship between the thickness t and the throat a is (see Figure 6.6)

$$a = t\frac{\sin\left(\frac{\pi-\gamma}{2}\right)}{\sin(\pi-\gamma)} \tag{6.1}$$

while the leg length is

$$leg = \frac{t}{\cos\left(\frac{\pi}{2}-\gamma\right)} \tag{6.2}$$

The weld layout *as a whole* has a center **G**. Its principal axes (their properties will be seen later) are named **u** and **v**. In general their orientation is not that of any single weld seam, or of axes 1 or 2. The third axis, **w**, is parallel to axis 3.

The coordinate transformation between the system (**1**, **2**) and the system (**u**, **v**) are as follows, using the well known coordinate transformation rules:

$$u = (x_1 - x_{1G})\cos\beta + (x_2 - x_{2G})\sin\beta$$
$$v = -(x_1 - x_{1G})\sin\beta + (x_2 - x_{2G})\cos\beta \quad (6.3)$$

and β is the angle between the axis **u** and the axis 1.

A similar coordinate transformation also holds true for the conversion between the system ($\mathbf{u_i}$, $\mathbf{v_i}$) and the system (**u**, **v**).

$$u_i = (u - u_{Gi})\cos(\alpha_i - \beta) + (v - v_{Gi})\sin(\alpha_i - \beta)$$
$$v_i = -(u - u_{Gi})\sin(\alpha_i - \beta) + (v - v_{Gi})\cos(\alpha_i - \beta) \quad (6.4)$$

Displacements will be identified by the letter "*d*", and the subscript referring to the axis (e.g. "1", "u", "ui", as in d_1, d_u, d_{ui}).

Rotations (in radians) will be identified by the letter "*r*" and the subscript referring to the axis, using the right hand rule to set positive values.

The stresses acting over a single weld will be generally projected onto the plane common to all weld seams (the faying plane).

The symbols "*n*" and "*t*" will be used, to refer to normal and tangential stress acting on the projected plane. The subscripts "par" and "per" will refer to the direction of axes **ui** and **vi**, respectively. Alternatively, we will also used the subscripts "ui" and "vi", to refer to the reference axes of the weld seam *i*.

So, for each weld seam, three stresses should be considered, as shown in Figure 6.7:

- normal stress n_i, acting in direction $\mathbf{w_i}$, i.e. that of axis 3 of the weld layout
- tangential stress t_{pari}, acting in the direction of the single weld seam axis ($\mathbf{P_{2i}}$-$\mathbf{P_{1i}}$), $\mathbf{u_i}$
- tangential stress t_{peri}, acting in the direction perpendicular to the weld seam, obtained by the vector product $\mathbf{w_i}$ ^ ($\mathbf{P_{2i}}$-$\mathbf{P_{1i}}$), i.e. the direction of local axis $\mathbf{v_i}$ of weld seam
- normal stresses n_{pari} and n_{peri} are negligible or can be neglected.

For penetration welds, the stresses projected over the common plane, the faying plane, are the usual normal stress σ and tangential stress τ. The reason why they will be referred to by letters *n* and *t* is for uniqueness of notation with fillet welds to avoid confusion.

For fillet welds (Figure 6.8) it would be more correct to refer to the stresses acting on the unprojected throat plane, named σ, τ_{per}, and τ_{par}. If the legs are equal, the unprojected throat plane is inclined by $\gamma/2$ to the faying plane (see red lines of Figure 6.5).

The studies of fillet welds carried on in the past few decades, have in fact shown that these are the relevant stresses for fillet weld resistance checks. However, computing them

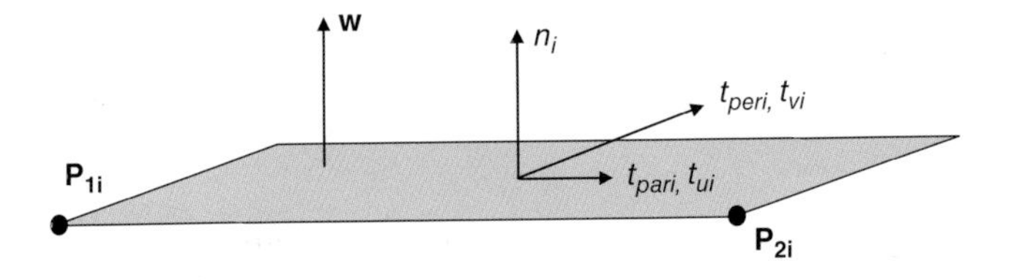

Figure 6.7 The stresses in a weld seam: active face on object O_1.

is not quick, so they are not used as much as the projected ones. The relationships between the projected stresses and the unprojected ones are

$$\tau_{par} = t_{par}$$
$$\tau_{per} = +t_{per}\cos\left(\frac{\gamma}{2}\right) + n\sin\left(\frac{\gamma}{2}\right) \tag{6.5}$$
$$\sigma = -t_{per}\sin\left(\frac{\gamma}{2}\right) + n\cos\left(\frac{\gamma}{2}\right)$$

and the inverse ones are

$$t_{par} = \tau_{par}$$
$$t_{per} = \tau_{per}\cos\left(\frac{\gamma}{2}\right) - \sigma\sin\left(\frac{\gamma}{2}\right) \tag{6.6}$$
$$n = \tau_{per}\sin\left(\frac{\gamma}{2}\right) + \sigma\cos\left(\frac{\gamma}{2}\right)$$

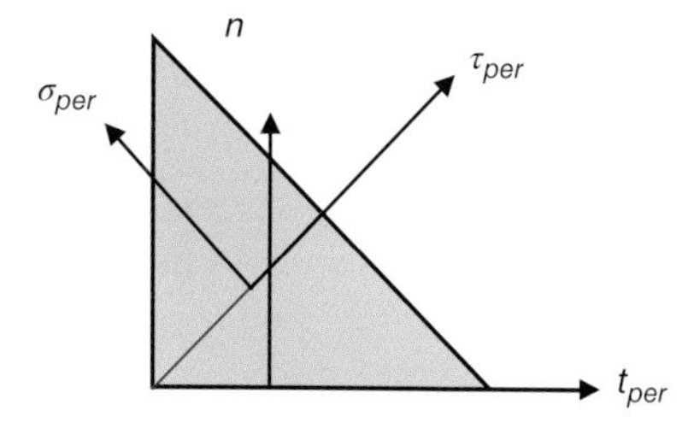

Figure 6.8 Stresses in fillet weld: on the unprojected throat plane (σ, τ) and on a projected throat plane (n, t). Both t_{par} and τ_{par} are normal to the paper, positive toward the reader.

For standard $\gamma = 90°$ fillet welds, the unprojected stresses are obtained by the following rules:

$$\tau_{par} = t_{par}$$
$$\tau_{per} = \frac{\sqrt{2}}{2}\left(n + t_{per}\right) \tag{6.7}$$
$$\sigma = \frac{\sqrt{2}}{2}\left(n - t_{per}\right)$$

6.5 Geometrical Limitations

6.5.1 Penetration Weld Layouts

6.5.1.1 Eurocode 3 (EN 1993-1.8)

Intermittent penetration welds are not allowed.

The thickness of the welded parts will be more than 4 mm, otherwise specific rules applicable to cold-formed sheetings apply (EN 1993-1.3).

Single sided partial penetration welds (Figure 6.9) should possibly be avoided as they lead to local eccentricity. This does not hold true if the single weld is part of a layout *globally* able to absorb the moment applied. If the root of the weld is in tension the eccentricity should be taken into account (the root of the weld is the part of the weld made by the first weld pass).

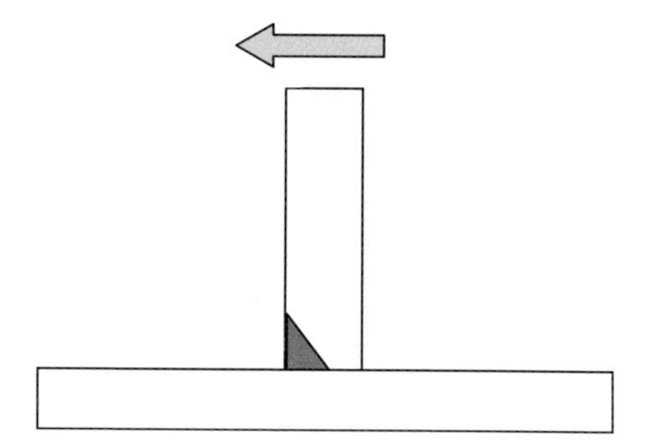

Figure 6.9 Single-sided partial penetration weld under the effect of an eccentric force is under bending, with tension at the root of the weld.

6.5.1.2 AISC 360-10

The partial penetration weld thickness a must not be lower than a function of the lower thickness welded t, according to the following limits:

- $0 \le t \le 6$ mm $a \ge 3$ mm
- $6 < t \le 13$ mm $a \ge 5$ mm
- $13 < t \le 19$ mm $a \ge 6$ mm
- $19 < t \le 38$ mm $a \ge 8$ mm
- $38 < t \le 57$ mm $a \ge 10$ mm
- $57 < t \le 150$ mm $a \ge 13$ mm
- $150 < t$ $a \ge 16$ mm

6.5.2 Fillet Weld Layouts

6.5.2.1 Eurocode 3

The angle between fusion (active) faces should be between 60° and 120°. If the angle is lower than 60°, then the fillet weld should be considered as a partial penetration weld.

For angles greater than 120° the resistance should be determined by testing.

Intermittent fillet welds are allowed provided that they are not used in corrosive environment and provided that the gap between two subsequent welds is lower than 16 times the minimum thickness welded or 200 mm for tension, or 12 times the minimum thickness welded or 200 mm for compression.

For single sided fillet welds the same considerations that are valid for partial penetration welds also hold true.

Throat length a is defined by the code as the height of the largest triangle inscribed in the weld cross-section, measured perpendicular to the outer (not fusion) face. This must not be lower than 3 mm ($a \ge 3$ mm).

Fillet welds must have an effective length higher than the largest of 30 mm or six times the throat thickness.

6.5.2.2 AISC 360-10

The *minimum size* of fillet welds (i.e. the minimum leg, w), is a function of the thickness welded according to the following list:

- $0 \le t \le 6$ mm $w \ge 3$ mm
- $6 < t \le 13$ mm $w \ge 5$ mm
- $13 < t \le 19$ mm $w \ge 6$ mm
- $19 < t$ $w \ge 8$ mm

The *maximum size* w of fillet weld is a function of the thickness welded, according to the following list:

- $0 \le t \le 6$ mm $w \le t$
- $6 < t \le 13$ mm $w \le (t - 2\text{ mm})$

The *minimum length* of fillet welds, l, will be four times the nominal weld size.

If the length is less than or equal to 100 times the weld size w, the effective length l_{eff} can be considered equal to the length of the weld. Otherwise, a reduction factor β has to be applied to l in order to get l_{eff}, computed as

$$\beta = 1.2 - 0.002(l/w) \leq 1.0$$

otherwise

$$l_{eff} = 180w \ \text{ if } \ (l/w) > 300$$

Intermittent fillet welds are allowed, provided that each individual length $l > 4w$, and provided that $l > 38$ mm.

The angle γ between fusion faces should be between 60° and 135°.

6.6 Penetration-Weld Layouts (Groove Welds)

6.6.1 Generality

These weld layouts are often used as full penetration in order to restore the full resistance of welded parts. In plastic design, partial penetration weld layouts should not be used, because by definition they are not able to resist the plastic forces of the connected parts.

When considering full penetration weld layouts, care must be taken to completely restore the connected constituents' cross-sections, also considering the cross-sections' fillets, which may play a non-negligible role.

From an analytical point of view, it is also possible to define full penetration weld seams not extending along all the available plate lengths: this clearly implies a partial resistance WL (Figure 6.10). However, both Eurocode 3 and AISC 360-10 explicitly forbid this kind of welding (intermittent penetration welds).

The generalized forces acting on a weld layout extremity can be written as a vector $\boldsymbol{\sigma}$, defined as $\boldsymbol{\sigma} = \{N, V_u, V_v, M_w, M_u, M_v\}^T$, that is, one normal force N (normal to the faying

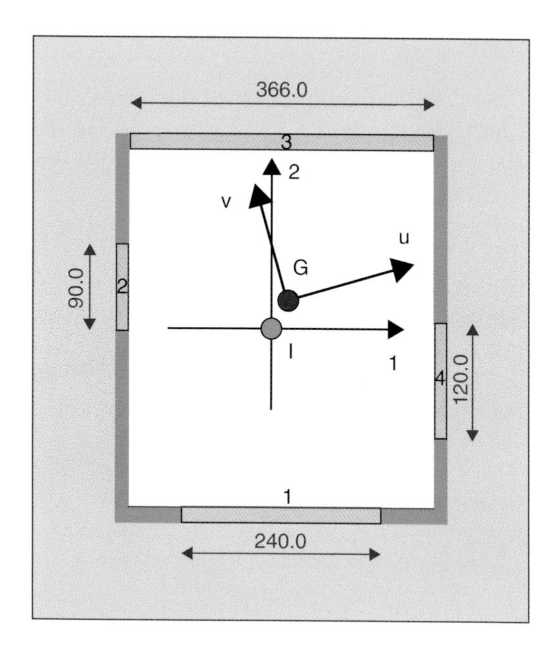

Figure 6.10 Full penetration weld layout, having seams not extending along all the plate length. Center G and principal axes (**u**, **v**) are displayed, as well as insertion point I and reference axes 1 and 2.

plane), two shears V_u and V_v, one torque M_w, and two bending moments M_u and M_v, all referred to principal axes (**w**, **u**, **v**) of the layout.

Practically all the standards consulted (including but not limited to Eurocode 3 and AISC 360-10), allow an elastic computation of the stresses in the welds, starting from the forces and moments externally applied to the weld layout. This holds true for both shear and torque, leading to shear stresses, and for axial force and moments, leading to normal stresses.

Indeed, plastic distributions are also allowed, in special cases, and the instantaneous center of rotation method is also considered by AISC for fillet welds (and bolts). This method takes into account that the force to displacement relationship in single subconnectors (welds but also bolts) is non-linear, and acts accordingly. It is more precise but also more complex (it will be considered in Section 6.7.4).

However, there is no doubt that the most widely diffused method for computing stresses in welds is an "elastic" distribution, that is, a *linear* one.

In turn, this means that most of the methods used to compute stresses in single welds closely resemble the method used to compute the stresses in beam cross-sections according to simple beam theory. Indeed, there is no difference.

So, using the symbols introduced in the previous section, in each point **P** of the weld layout effective area, necessarily belonging to a weld seam i, the normal stress n, and shear stresses t_{par} and t_{per} may be computed according to the following.

6.6.2 Simple Methods to Evaluate the Stresses

The normal stress n will be computed by

$$n = \frac{N}{A} + \frac{M_u}{J_u}v - \frac{M_v}{J_v}u$$

where:

- (**u**,**v**) are the principal axes of the whole weld layout, that is, those with null mixed second area moment J_{uv}, and with origin in the weld layout effective area center
- J_u and J_v are the second area moment to axis **u**, and **v**, and A is the weld layout effective area
- n is the normal stress on the faying surface.

The shear stresses in point **P** are the result of different contributions.

It is commonly accepted that the shear stress due to shear forces is averagely distributed on the resisting area A, as plasticization and consequent redistribution is expected, also for low forces. This means that the stress distribution commonly assumed is not globally "elastic", but a mix of elastic and plastic.

So

$$t_u = \frac{V_u}{A}$$

$$t_v = \frac{V_v}{A}$$

These are the tangential stresses parallel to the principal axes of the weld layout. They must then be rotated, for each weld seam i, to the direction of the axis of the weld seam itself, so:

$$t_{pari} = t_u \cos(\alpha_i - \beta) + t_v \sin(\alpha_i - \beta)$$
$$t_{peri} = -t_u \sin(\alpha_i - \beta) + t_v \cos(\alpha_i - \beta)$$

The shear stress due to torsion is computed by the elastic method, assuming that the tangential stress at a point **P** is perpendicular to the straight line connecting the weld layout center **G** to the point **P**, and proportional to the length r_P of the segment **GP**, so that at each point (Figure 6.11)

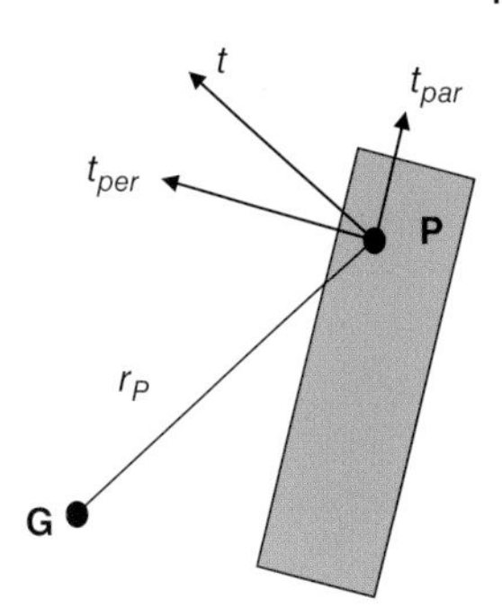

Figure 6.11 Shear stress due to torsion, elastic method.

$$t = \frac{M_w}{J_t} r_P$$

The torsional second area moment is the polar second area moment to axis **w** of the whole weld layout, and can be easily computed as:

$$J_t = J_p = J_w = J_u + J_v$$

If the point **P** has the coordinates x_{1P} and x_{2P} with reference to the insertion point **I**, and if the weld layout center has coordinates x_{1G} and x_{2G}, then (Figure 6.11)

$$r_P = \sqrt{(x_{1P} - x_{1G})^2 + (x_{2P} - x_{2G})^2}$$
$$t_{1,P} = -\frac{M_w}{J_t}(x_{2P} - x_{2G})$$
$$t_{2,P} = \frac{M_w}{J_t}(x_{1P} - x_{1G})$$
$$t_{par,P} = t_1 \cos(\alpha_i) + t_2 \sin(\alpha_i)$$
$$t_{per,P} = -t_1 \sin(\alpha_i) + t_2 \cos(\alpha_i)$$

If the principal axes are used instead, then

$$r_P = \sqrt{u_P^2 + v_P^2}$$
$$t_{u,P} = -\frac{M_w}{J_t} v$$
$$t_{v,P} = \frac{M_w}{J_t} u$$
$$t_{par,P} = t_u \cos(\alpha_i - \beta) + t_v \sin(\alpha_i - \beta)$$
$$t_{per,P} = -t_u \sin(\alpha_i - \beta) + t_v \cos(\alpha_i - \beta)$$

In the previous equations it has been assumed that the point **P** belongs to the weld seam i, and the shear stresses have first been computed referred to the reference axes (1, 2), which have been later rotated to axis $\mathbf{u_i}$ and $\mathbf{v_i}$. These new t_{par} and t_{per} caused by torsion should be added to those due to the shears, also considering signs.

These sets of equations, used to compute n and t, are well known to every structural engineer, and have the main advantage of being simple. They have been used for many years due to their ease of use and because of the implicit shelter offered by the static theorem of limit analysis.

To use these equations, it is necessary to consider a unique "cross-section" W obtained by summing the contributions of all welds seams, that is, a number n_w or rectangles of width a_i, length l_i, and angle of inclination to the reference axis 1 α_i, and to the principal axis **u**, $(\alpha_i - \beta)$.

The next section explicitly lists the necessary steps for computing all the relevant data of the weld layout "cross-section" W.

6.6.3 Weld Layout Cross-Section Data

The principal axes of the weld layout can be computed considering a section obtained by summing all the weld seam contributions. Each weld seam is a rectangle of proper center $\mathbf{G_i}$, length l_i, thickness a_i, and angle α_i relative to weld layout reference axis 1.

The following formulae can be used to derive all the necessary data, where:

- n_w is the number of weld seams, each having two end points $\mathbf{P_1}$ and $\mathbf{P_2}$
- x_1 and x_2 are the coordinates of a point relative to insertion point **I**, that is, the origin of axes (1, 2, 3)
- x_{1P1i} is the coordinate x_1 of the point $\mathbf{P_1}$ of weld i, x_{2P1i} is the coordinate x_2 of the point $\mathbf{P_1}$ of weld i, x_{1P2i} is the coordinate x_1 of the point $\mathbf{P_2}$ of weld i, and x_{2P2i} is the coordinate x_2 of the point $\mathbf{P_2}$ of weld i
- **G** is the center of the weld layout, and $\mathbf{G_i}$ the center of each weld seam
- A is the effective area of the weld layout
- J_{ui} is the second area moment of the weld seam i to its principal axis **ui**
- J_{vi} is the second area moment of the weld seam i to its principal axis **vi**
- J_{1Gi} is the second area moment of the weld seam i to the axis parallel to axis 1, and passing through its center, $\mathbf{G_i}$
- J_{2Gi} is the second area moment of the weld seam i to the axis parallel to axis 2 and passing through its center, $\mathbf{G_i}$
- J_{12Gi} is the mixed second area moment, as the two previous ones
- x_{1Gi} and x_{2Gi} are the coordinates of the center of the weld seam i
- x_{1G} and x_{2G} are the coordinates of the weld layout center
- J_{1G}, J_{2G}, and J_{12G} are the second area moment of the whole weld layout relative to the axes parallel to axes 1 and 2, and passing through the weld layout center, **G**
- β is the angle between principal axis **u**, and reference axis 1.

With these definitions, the coordinates of the penetration weld-seam i effective-area center are (weld to the right of **P1-P2**):

$$\begin{aligned} x_{1Gi} &= 0.5\cdot(x_{1P1i} + x_{1P2i}) - 0.5a_i \sin\alpha_i \\ x_{2Gi} &= 0.5\cdot(x_{2P1i} + x_{2P2i}) + 0.5a_i \cos\alpha_i \end{aligned} \tag{6.8}$$

For fillet welds, they would have been (weld to the left of P1-P2)

$$\begin{aligned} x_{1Gi} &= 0.5\cdot(x_{1P1i} + x_{1P2i}) + 0.5a_i \sin\alpha_i \\ x_{2Gi} &= 0.5\cdot(x_{2P1i} + x_{2P2i}) - 0.5a_i \cos\alpha_i \end{aligned} \tag{6.9}$$

The total area A is

$$A = \sum_{i=1}^{nw} a_i l_i \tag{6.10}$$

The coordinates of the weld layout center are found as

$$\begin{aligned} x_{1G} &= \frac{\sum_{i=}^{nw} x_{1Gi} a_i l_i}{A} \\ x_{2G} &= \frac{\sum_{i=}^{nw} x_{2Gi} a_i l_i}{A} \end{aligned} \tag{6.11}$$

The area and second area moment of the weld seam to its principal axes are

$$\begin{aligned} A_i &= a_i l_i \\ J_{ui} &= \frac{l_i\, a_i^3}{12} \\ J_{vi} &= \frac{l_i^3 a_i}{12} \end{aligned} \tag{6.12}$$

The second area moment relative to the axes parallel to 1 and 2 and passing through the weld seam center $\mathbf{G_i}$, are

$$\begin{aligned} J_{1Gi} &= J_{ui}\cos^2(\alpha_i) + J_{vi}\sin^2(\alpha_i) \\ J_{2Gi} &= J_{ui}\sin^2(\alpha_i) + J_{vi}\cos^2(\alpha_i) \\ J_{12Gi} &= (J_{vi} - J_{ui})\sin\alpha_i \cos\alpha_i \end{aligned} \tag{6.13}$$

while the second area moment to the axes 1 and 2 of the whole weld layout are

$$\begin{aligned} J_{1G} &= \sum_{i=1}^{nw} \left[J_{1Gi} + l_i a_i (x_{2Gi} - x_{2G})^2\right] \\ J_{2G} &= \sum_{i=1}^{nw} \left[J_{2Gi} + l_i a_i (x_{1Gi} - x_{1G})^2\right] \\ J_{12G} &= \sum_{i=1}^{nw} \left[J_{12Gi} + l_i a_i (x_{1Gi} - x_{1G}) \cdot (x_{2Gi} - x_{2G})\right] \end{aligned} \tag{6.14}$$

Now, imposing a null mixed second area moment, the angle β of the principal axis $\mathbf{u}$ to the reference axis 1 is found as

$$\beta = \frac{1}{2}\arctan\frac{2J_{12G}}{J_{2G} - J_{1G}} \tag{6.15}$$

The angle α_i can be computed easily once the coordinates of the extremities of the weld seam on the chosen face side, $\mathbf{P_{1i}}$ and $\mathbf{P_{2i}}$, are known:

$$\alpha_i = \arctan\left(\frac{x_{2P2i} - x_{2P1i}}{x_{1P2i} - x_{1P1i}}\right)$$

The second area moment to the principal axes (**u**, **v**) of the weld layout are eventually obtained from the following well-known formulae:

$$\begin{aligned} J_u &= J_{2G}\sin^2\beta + J_{1G}\cos^2\beta - J_{12G}\sin(2\beta) \\ J_v &= J_{1G}\sin^2\beta + J_{2G}\cos^2\beta + J_{12G}\sin(2\beta) \end{aligned} \tag{6.16}$$

The torsional second area moment of the weld layout J_t, is now easily obtained according to

$$J_t = J_u + J_v \tag{6.17}$$

This definition uses the polar inertia (as the distance from origin $r^2 = u^2 + v^2$).

6.6.4 Stiffness Matrix

The next step is straightforward: it is assumed that the behavior of a penetration weld layout (or, as will be seen later, of a fillet weld layout) can be globally modeled in the scene by a single finite element of proper stiffness and proper principal axes (**w**, **u**, **v**).

There is nothing new and nothing different in this assumption, nothing that it is not already intrinsically accepted by assuming a linear displacement field and a linear variation for the normal stress due to bending, or the shear stress due to torque.

Clearly, here, some important assumptions have previously been made, and are implicitly used by all the world standards: that the weld seams displacements are *organized*, that is, that the weld seams do not behave independently in the scene, but that their displacement is related to that of the siblings by a linear law. Indeed, this might be true for stiff constituents, almost acting as rigid bodies at the scale of connection design, but for very flexible ones this assumption can lead to distributions that are not so realistic. Connected parts in the region of the connection are normally very stiff, to restore continuity of displacements, but there are designs that rely on semi-rigid connections, or that connect weak parts, and in those contexts the rigid-like, linear behavior might be questionable.

A pale reflection of this important problem can be found, for instance, in the clauses of Eurocode 3 part 1.8 referring to the "effective length" of welds applied to "unstiffened flanges" (§4.10). Here the effective length of the weld is much reduced (Figure 6.12) and limited to the nearby length of the flange's unique stiffener (a web, usually). In this way, the resistance of the more flexible parts is neglected, and the width of the weld is limited to the stiff parts connected.

This issue is a fundamental hint suggesting that the traditional methods used to compute welds are potentially flawed.

As it will better seen in Chapter 10, a full modeling of the renode would implicitly tackle this issue, as no organization is assumed between the displacements of the single seams of a weld layout. However, when dealing with simplified models, not explicitly modeling the renode scene using plate–shells, care will have to be taken in order to neglect the effect of too flexible parts.

From a computational point of view, the first extremity $\mathbf{E_1}$ of a penetration weld layout is the geometrical center of the welding seams **G**, placed over the plane of object O_1, that is, over the faying plane.

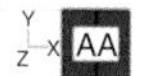

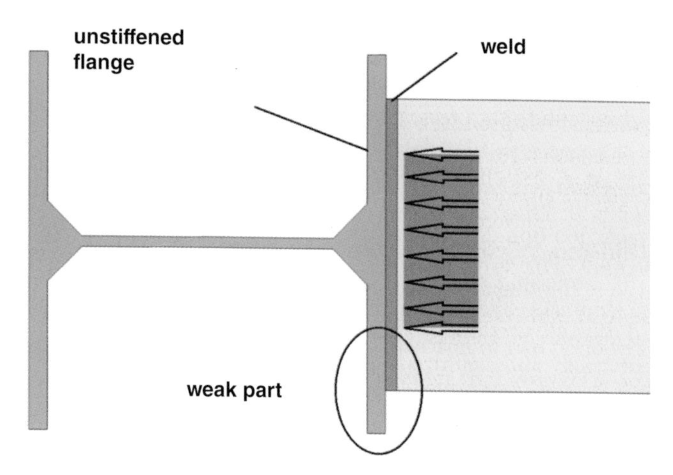

Figure 6.12 Weld of unstiffened flange.

The two extremities of the weld layout $\mathbf{E}_1$ and $\mathbf{E}_2$ are obtained as

$$\begin{aligned} \mathbf{E}_1 &= \mathbf{G} \\ \mathbf{E}_2 &= \mathbf{G} + L \cdot \mathbf{v}_3 \end{aligned} \tag{6.18}$$

where L is a suitable distance and $\mathbf{v}_3$ is the unit vector directed such as the axis 3 of the weld layout (i.e. like the opposite $-\mathbf{n}$ of the normal $\mathbf{n}$ of the chosen face defined in Chapter 4).

If the thicknesses of the weld seams are different, as the definition of the second extremity is notional, two possible choices for L are:

$$L = \frac{\sum_{i=1}^{nw} t_i l_i}{\sum_{i=1}^{nw} l_i}$$

$$L = \frac{\sum_{i=1}^{nw} t_i}{nw}$$

It should be observed that the thickness t of a penetration weld is here by definition measured normal to the faying plane, no matter what the inclination of the connected parts. If all the thicknesses t of the weld seams are equal, then $t_i = t$ and $L = t$.

There are several beam elements available in the most frequently used software package. The two reference beam elements are the Bernoulli beam, which works well for slender elements, and the Timoshenko beam, which is fit for stocky members. The latter will here be considered, as the average thickness of a weld layout is quite short, and as this element has useful properties that will be later discussed.

Assuming that the first extremity $\mathbf{E_1}$ is the first node of the element, and the second extremity $\mathbf{E_2}$ is the second node of the element, the stiffness matrix of the element can be written, relating the overall forces $\boldsymbol{\sigma}$ and the vector of generalized displacement $\mathbf{d}$ at the extremities (nodes) $\mathbf{E_1}$ and $\mathbf{E_2}$:

$$\boldsymbol{\sigma} = \mathbf{Kd}$$

where

$$\mathbf{d} = \{d_{w1} \; d_{u1} \; d_{v1} \; r_{w1} \; r_{u1} \; r_{v1} \; d_{w2} \; d_{u2} \; d_{v2} \; r_{w2} \; r_{u2} \; r_{v2}\}^T$$

is the vector of the nodal displacements vector in the principal system of the weld layout, and

$$\sigma = \{N_1 \; V_{u1} \; V_{v1} \; M_{w1} \; M_{u1} \; M_{v1} \; N_2 \; V_{u2} \; V_{v2} \; M_{w2} \; M_{u2} \; M_{v2}\}^T$$

is the vector of the nodal forces in the principal system of the weld layout. In partitioned form, using blocks of size (3, 1), they can be written as

$$\mathbf{d} = \begin{vmatrix} \mathbf{d}_1 \\ \mathbf{r}_1 \\ \hline \mathbf{d}_2 \\ \mathbf{r}_2 \end{vmatrix}$$

$$\boldsymbol{\sigma} = \begin{vmatrix} \mathbf{F}_1 \\ \mathbf{M}_1 \\ \hline \mathbf{F}_2 \\ \mathbf{M}_2 \end{vmatrix}$$

$\mathbf{K}$ might be partitioned in (3 × 3) blocks as in Equation 6.19:

$$\mathbf{K} = \begin{vmatrix} \mathbf{A} & \mathbf{B} & -\mathbf{A} & \mathbf{B} \\ \mathbf{B}^{\mathrm{T}} & \mathbf{D} & -\mathbf{B}^{\mathrm{T}} & \mathbf{E} \\ -\mathbf{A} & -\mathbf{B} & \mathbf{A} & -\mathbf{B} \\ \mathbf{B}^{\mathrm{T}} & \mathbf{E} & -\mathbf{B}^{\mathrm{T}} & \mathbf{D} \end{vmatrix} \tag{6.19}$$

The blocks **A**, **B**, **D**, **E** are defined in this way:

A is the term related to direct translational stiffnesses:

$$\mathbf{A} = \begin{vmatrix} \frac{EA}{L} & 0 & 0 \\ 0 & \frac{GA_u}{L} & 0 \\ 0 & 0 & \frac{GA_v}{L} \end{vmatrix} \equiv \begin{vmatrix} k_N & 0 & 0 \\ 0 & k_{Vu} & 0 \\ 0 & 0 & k_{Vv} \end{vmatrix} \tag{6.20}$$

E is the Young's modulus, and G the shear modulus. A is the cross-sectional area. The terms A_u and A_v are the shear area related to axes **u** and **v**. *In general, they can be different.*

B is related to the coupling of translational and rotational degrees of freedom:

$$\mathbf{B} = \begin{vmatrix} 0 & 0 & 0 \\ 0 & 0 & \dfrac{GA_u}{2} \\ 0 & -\dfrac{GA_v}{2} & 0 \end{vmatrix} = \begin{vmatrix} 0 & 0 & 0 \\ 0 & 0 & \dfrac{k_{Vu}L}{2} \\ 0 & -\dfrac{k_{Vv}L}{2} & 0 \end{vmatrix} \tag{6.21}$$

D is related to the direct rotational stiffnesses:

$$\mathbf{D} = \begin{vmatrix} \dfrac{GJ_t}{L} & 0 & 0 \\ 0 & R_u & 0 \\ 0 & 0 & R_v \end{vmatrix} \equiv \begin{vmatrix} k_T & 0 & 0 \\ 0 & R_u & 0 \\ 0 & 0 & R_v \end{vmatrix} \tag{6.22}$$

J_t is the torsional constant, and R_u and R_v are defined as

$$\begin{aligned} R_u &= \frac{EJ_u}{L} + \frac{GA_vL}{3} \equiv k_{Mu} + \frac{k_{Vv}L^2}{3} \\ R_v &= \frac{EJ_v}{L} + \frac{GA_uL}{3} \equiv k_{Mv} + \frac{k_{Vu}L^2}{3} \end{aligned} \tag{6.23}$$

J_u and J_v are the second area moments to the principal axes **u**, **v**.

Finally, the block **E** is related to the indirect stiffness (rotational and translational):

$$\mathbf{E} = \begin{vmatrix} -\dfrac{GJ_t}{L} & 0 & 0 \\ 0 & K_u & 0 \\ 0 & 0 & K_v \end{vmatrix} = \begin{vmatrix} -k_T & 0 & 0 \\ 0 & K_u & 0 \\ 0 & 0 & K_v \end{vmatrix} \tag{6.24}$$

where

$$\begin{aligned} K_u &= -\frac{EJ_u}{L} + \frac{GA_vL}{6} = -k_{Mu} + \frac{k_{Vv}L^2}{6} \\ K_v &= -\frac{EJ_v}{L} + \frac{GA_uL}{6} = -k_{Mv} + \frac{k_{Vu}L^2}{6} \end{aligned} \tag{6.25}$$

It should be observed that the submatrices **A**, **D**, and **E** are diagonal, so they are equal to their transpose.

If no specific difference is assumed in longitudinal and transverse direction, then it can be assumed that $A_u = A_v = A$. As a constant shear stress distribution is assumed, the shear factor is set to 1. Then, for the whole weld layout, it would result that

$$A_u = A_v = A_s = A$$

The expressions written refer to the stiffness matrix of a generic Timoshenko beam element, and can be applied to the weld layout when seen as the assembly of single weld seams, each one behaving like a single Timoshenko beam, once the proper constraint for the displacement field is applied, that is, when it is considered that the weld seams *displacement field* is assumed organized and linear. This is also compliant with the instantaneous center of rotation method, where it is assumed the motion is that of a rigid body.

This can be shown as follows, with some matrix manipulation that will also be useful later.

A generic weld seam i has its own displacements and rotations, in its reference system, at its center $\mathbf{G_i}$. These displacements are related to the displacement of the whole layout, because the layout is assumed to behave in an organized manner. So, the following relationships might be written, linking the displacements $\mathbf{d_i}$ of the single weld seam i center $\mathbf{G_i}$ to the displacements $\mathbf{d}$ of the whole weld layout:

$$
\begin{aligned}
d_{wi} &= d_w + r_u v_{Gi} - r_v u_{Gi} \\
d_{ui} &= (d_u - r_w v_{Gi})\cos(\alpha_i - \beta) + (d_v + r_w u_{Gi})\sin(\alpha_i - \beta) \\
d_{vi} &= -(d_u - r_w v_{Gi})\sin(\alpha_i - \beta) + (d_v + r_w u_{Gi})\cos(\alpha_i - \beta) \\
r_{wi} &= r_w \\
r_{ui} &= r_u \cos(\alpha_i - \beta) + r_v \sin(\alpha_i - \beta) \\
r_{vi} &= -r_u \sin(\alpha_i - \beta) + r_v \cos(\alpha_i - \beta)
\end{aligned}
\tag{6.26}
$$

where:

- d_{wi}, d_{ui}...r_{vi} are the displacements and rotations of the center $\mathbf{G_i}$ of the single weld seam i, defined in its local reference system (**ui, vi, wi**) – see Figure 6.13
- u_{Gi}, v_{Gi} are the coordinates of the weld i center $\mathbf{G_i}$, in the principal system of the weld layout.

For brevity we will also use

$$
\begin{aligned}
c_i &= \cos(\alpha_i - \beta) \\
s_i &= \sin(\alpha_i - \beta)
\end{aligned}
$$

In matrix form, considering the displacements of both the nodes of the single weld i, $\mathbf{E_{1i}}$, $\mathbf{E_{2i}}$, this can be written as

$$
\begin{aligned}
\mathbf{d_i} &= \{ d_{wi1} \;\; d_{ui1} \;\; d_{vi1} \;\; r_{wi1} \;\; r_{ui1} \;\; r_{vi1} \;\; | \;\; d_{wi2} \;\; d_{ui2} \;\; d_{vi2} \;\; r_{wi2} \;\; r_{ui2} \;\; r_{vi2} \}^T \\
\mathbf{d_i} &= \mathbf{Q_i d}
\end{aligned}
\tag{6.27}
$$

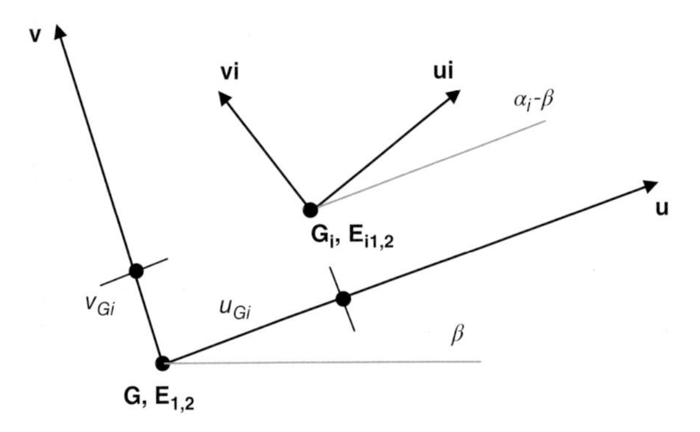

Figure 6.13 Nodal displacements of weld i and of the whole weld layout.

The transfer matrix $\mathbf{Q_i}$ can be expressed in partitioned form as

$$\mathbf{Q_i} = \left|\begin{array}{c|c|c|c} \mathbf{T_i} & \mathbf{S_i} & \mathbf{0} & \mathbf{0} \\ \hline \mathbf{0} & \mathbf{T_i} & \mathbf{0} & \mathbf{0} \\ \hline \mathbf{0} & \mathbf{0} & \mathbf{T_i} & \mathbf{S_i} \\ \hline \mathbf{0} & \mathbf{0} & \mathbf{0} & \mathbf{T_i} \end{array}\right| \tag{6.28}$$

where

$$\mathbf{T_i} = \begin{vmatrix} 1 & 0 & 0 \\ 0 & c_i & s_i \\ 0 & -s_i & c_i \end{vmatrix}$$
$$\mathbf{S_i} = \begin{vmatrix} 0 & v_{Gi} & -u_{Gi} \\ Y_i & 0 & 0 \\ X_i & 0 & 0 \end{vmatrix} \tag{6.29}$$

with

$$X_i = v_{Gi}s_i + u_{Gi}c_i$$
$$Y_i = -v_{Gi}c_i + u_{Gi}s_i$$

The terms X_i and Y_i are such that if r_i is the distance between $\mathbf{G_i}$ and $\mathbf{G}$, then

$$r_i^2 = X_i^2 + Y_i^2 \tag{6.30}$$

Also, as will be useful later,

$$s_iY_i + c_iX_i = u_{Gi}$$
$$-c_iY_i + s_iX_i = v_{Gi}$$

Considering the single weld seam, its local internal forces, in its local reference system, are

$$\boldsymbol{\sigma}_\mathbf{i} = \{N_{i1} \;\; V_{ui1} \;\; V_{vi1} \mid M_{wi1} \;\; M_{ui1} \;\; M_{vi1} \mid N_{i2} \;\; V_{ui2} \;\; V_{vi2} \mid M_{wi2} \;\; M_{ui2} \;\; M_{vi2}\}^T$$

Introducing the stiffness matrix of the single weld seam $\mathbf{K_i}$, defined in its local reference system,

$$\boldsymbol{\sigma}_\mathbf{i} = \mathbf{K_i d_i} \tag{6.31}$$

This stiffness matrix is quite similar to the global one already seen, but it uses the following list of local quantities (the expressions will also be useful later, and it is assumed that the shear areas A_{ui} and A_{vi} are not equal):

$$\begin{aligned} &A_i = a_il_i \\ &J_{ui} = \frac{1}{12}l_ia_i^3 \\ &J_{vi} = \frac{1}{12}a_il_i^3 \\ &J_{ti} = J_{ui} + J_{vi} \\ &L_i \\ &A_{ui} \\ &A_{vi} \end{aligned} \tag{6.32}$$

Of course, this matrix $\mathbf{K_i}$ can be partitioned like the global one resulting in

$$\mathbf{K}_i = \left[\begin{array}{c|c|c|c} \mathbf{A}_i & \mathbf{B}_i & -\mathbf{A}_i & \mathbf{B}_i \\ \hline \mathbf{B}_i^T & \mathbf{D}_i & -\mathbf{B}_i^T & \mathbf{E}_i \\ \hline -\mathbf{A}_i & -\mathbf{B}_i & \mathbf{A}_i & -\mathbf{B}_i \\ \hline \mathbf{B}_i^T & \mathbf{E}_i & -\mathbf{B}_i^T & \mathbf{D}_i \end{array}\right] \tag{6.33}$$

It is now possible to express the stiffness matrix of a single weld seam $\mathbf{K_i}$ in the global layout reference system, $\mathbf{K_{il}}$, in the standard manner (for an introductory explanation of the finite element method, see for example Rugarli 2010)

$$\mathbf{K_{il}} = \mathbf{Q_i^T K_i Q_i} \tag{6.34}$$

The stiffness matrix of the whole weld layout is obtained by assembly, which is now merely

$$\mathbf{K_l} = \sum_{i=1}^{nw} \mathbf{K_{il}} = \sum_{i=1}^{nw} \mathbf{Q_i^T K_i Q_i} \tag{6.35}$$

Considering the partitioned form both of the stiffness matrix and the transfer matrix, the following final expression can be obtained for the stiffness matrix of the weld layout, seen as the assembly of the stiffnesses of single weld seams:

$$\mathbf{K}_l = \sum_{i=1}^{nw} \left[\begin{array}{c|c|c|c} \mathbf{T}_i^T\mathbf{A}_i\mathbf{T}_i & \mathbf{T}_i^T\mathbf{A}_i\mathbf{S}_i + \mathbf{T}_i^T\mathbf{B}_i\mathbf{T}_i & -\mathbf{T}_i^T\mathbf{A}_i\mathbf{T}_i & -\mathbf{T}_i^T\mathbf{A}_i\mathbf{S}_i + \mathbf{T}_i^T\mathbf{B}_i\mathbf{T}_i \\ \hline \mathbf{S}_i^T\mathbf{A}_i\mathbf{T}_i + \mathbf{T}_i^T\mathbf{B}_i^T\mathbf{T}_i & \begin{array}{c}\mathbf{S}_i^T\mathbf{A}_i\mathbf{S}_i + \mathbf{S}_i^T\mathbf{B}_i\mathbf{T}_i \\ + \mathbf{T}_i^T\mathbf{B}_i^T\mathbf{S}_i + \mathbf{T}_i^T\mathbf{D}_i\mathbf{T}_i\end{array} & -\mathbf{S}_i^T\mathbf{A}_i\mathbf{T}_i - \mathbf{T}_i^T\mathbf{B}_i^T\mathbf{T}_i & \begin{array}{c}-\mathbf{T}_i^T\mathbf{B}_i^T\mathbf{S}_i - \mathbf{S}_i^T\mathbf{A}_i\mathbf{S}_i \\ + \mathbf{T}_i^T\mathbf{E}_i\mathbf{T}_i + \mathbf{S}_i^T\mathbf{B}_i\mathbf{T}_i\end{array} \\ \hline -\mathbf{T}_i^T\mathbf{A}_i\mathbf{T}_i & -\mathbf{T}_i^T\mathbf{A}_i\mathbf{S}_i - \mathbf{T}_i^T\mathbf{B}_i\mathbf{T}_i & \mathbf{T}_i^T\mathbf{A}_i\mathbf{T}_i & \mathbf{T}_i^T\mathbf{A}_i\mathbf{S}_i - \mathbf{T}_i^T\mathbf{B}_i\mathbf{T}_i \\ \hline -\mathbf{S}_i^T\mathbf{A}_i\mathbf{T}_i + \mathbf{T}_i^T\mathbf{B}_i^T\mathbf{T}_i & \begin{array}{c}-\mathbf{S}_i^T\mathbf{A}_i\mathbf{S}_i - \mathbf{S}_i^T\mathbf{B}_i\mathbf{T}_i \\ + \mathbf{T}_i^T\mathbf{E}_i\mathbf{T}_i + \mathbf{T}_i^T\mathbf{B}_i^T\mathbf{S}_i\end{array} & \mathbf{S}_i^T\mathbf{A}_i\mathbf{T}_i - \mathbf{T}_i^T\mathbf{B}_i^T\mathbf{T}_i & \begin{array}{c}\mathbf{S}_i^T\mathbf{A}_i\mathbf{S}_i - \mathbf{S}_i^T\mathbf{B}_i\mathbf{T}_i \\ + \mathbf{T}_i^T\mathbf{D}_i\mathbf{T}_i - \mathbf{T}_i^T\mathbf{B}_i^T\mathbf{S}_i\end{array} \end{array}\right] \tag{6.36}$$

The matrix *is symmetric* and it depends only on a few terms – just seven, all (3 × 3) matrices, which are here listed:

- Block number 1: $\mathbf{T_i^T A_i T_i}$
- Block number 2: $\mathbf{T_i^T A_i S_i}$
- Block number 3: $\mathbf{T_i^T B_i T_i}$
- Block number 4: $\mathbf{S_i^T A_i S_i}$
- Block number 5: $\mathbf{S_i^T B_i T_i}$
- Block number 6: $\mathbf{T_i^T D_i T_i}$
- Block number 7: $\mathbf{T_i^T E_i T_i}$

It should be noted that some matrix terms are the transpose of others, for instance:

$$\mathbf{S_i^T A_i T_i} = \mathbf{T_i^T A_i S_i}$$

Considering in particular the term $\mathbf{T_i^T A_i T_i}$ (block number 1), it can be seen that it is responsible for the direct translational stiffnesses, and can be used to determine the shear area of the assembled weld layout.

So, block number 1:

$$\mathbf{T_i}^{\mathbf{T}}\mathbf{A_i}\mathbf{T_i} = \begin{vmatrix} 1 & 0 & 0 \\ 0 & c_i & -s_i \\ 0 & s_i & c_i \end{vmatrix} \begin{vmatrix} \frac{EA_i}{L_i} & 0 & 0 \\ 0 & \frac{GA_{ui}}{L_i} & 0 \\ 0 & 0 & \frac{GA_{vi}}{L_i} \end{vmatrix} \begin{vmatrix} 1 & 0 & 0 \\ 0 & c_i & s_i \\ 0 & -s_i & c_i \end{vmatrix}$$
$$= \begin{vmatrix} \frac{EA_i}{L_i} & 0 & 0 \\ 0 & \frac{GA_{ui}}{L_i}c_i^2 + \frac{GA_{vi}}{L_i}s_i^2 & \left(\frac{GA_{ui}}{L_i} - \frac{GA_{vi}}{L_i}\right)c_i s_i \\ 0 & \left(\frac{GA_{ui}}{L_i} - \frac{GA_{vi}}{L_i}\right)c_i s_i & \frac{GA_{ui}}{L_i}s_i^2 + \frac{GA_{vi}}{L_i}c_i^2 \end{vmatrix}_i \quad (6.37)$$

block number 2:

$$\mathbf{T_i}^{\mathbf{T}}\mathbf{A_i}\mathbf{S_i} = \begin{vmatrix} 1 & 0 & 0 \\ 0 & c_i & -s_i \\ 0 & s_i & c_i \end{vmatrix} \begin{vmatrix} \frac{EA_i}{L_i} & 0 & 0 \\ 0 & \frac{GA_{ui}}{L_i} & 0 \\ 0 & 0 & \frac{GA_{vi}}{L_i} \end{vmatrix} \begin{vmatrix} 0 & v_{Gi} & -u_{Gi} \\ Y_i & 0 & 0 \\ X_i & 0 & 0 \end{vmatrix}$$
$$= \begin{vmatrix} 0 & \frac{EA_i}{L_i}v_{Gi} & -\frac{EA_i}{L_i}u_{Gi} \\ \frac{GA_{ui}Y_i c_i - GA_{vi}X_i s_i}{L_i} & 0 & 0 \\ \frac{GA_{ui}Y_i s_i + GA_{vi}X_i c_i}{L_i} & 0 & 0 \end{vmatrix} \quad (6.38)$$

block number 3:

$$\mathbf{T_i}^{\mathbf{T}}\mathbf{B_i}\mathbf{T_i} = \begin{vmatrix} 1 & 0 & 0 \\ 0 & c_i & -s_i \\ 0 & s_i & c_i \end{vmatrix} \cdot \begin{vmatrix} 0 & 0 & 0 \\ 0 & 0 & \frac{GA_{ui}}{2} \\ 0 & -\frac{GA_{vi}}{2} & 0 \end{vmatrix} \cdot \begin{vmatrix} 1 & 0 & 0 \\ 0 & c_i & s_i \\ 0 & -s_i & c_i \end{vmatrix}$$
$$= \begin{vmatrix} 0 & 0 & 0 \\ 0 & \left(\frac{-GA_{ui} + GA_{vi}}{2}\right)c_i s_i & \left(\frac{GA_{ui}\,c_i^2 + GA_{vi}\,s_i^2}{2}\right) \\ 0 & \left(\frac{-GA_{ui}\,s_i^2 - GA_{vi}\,c_i^2}{2}\right) & \left(\frac{GA_{ui} - GA_{vi}}{2}\right)c_i s_i \end{vmatrix} \quad (6.39)$$

block number 4:

$$\mathbf{S_i}^{\mathbf{T}}\mathbf{A_i}\mathbf{S_i} = \begin{vmatrix} 0 & Y_i & X_i \\ v_{Gi} & 0 & 0 \\ -u_{Gi} & 0 & 0 \end{vmatrix} \begin{vmatrix} \frac{EA_i}{L_i} & 0 & 0 \\ 0 & \frac{GA_{ui}}{L_i} & 0 \\ 0 & 0 & \frac{GA_{vi}}{L_i} \end{vmatrix} \begin{vmatrix} 0 & v_{Gi} & -u_{Gi} \\ Y_i & 0 & 0 \\ X_i & 0 & 0 \end{vmatrix}$$
$$= \begin{vmatrix} \frac{GA_{ui}\,Y_i^2 + GA_{vi}\,X_i^2}{L_i} & 0 & 0 \\ 0 & \frac{EA_i}{L_i}v_{Gi}^2 & -\frac{EA_i}{L_i}u_{Gi}v_{Gi} \\ 0 & -\frac{EA_i}{L_i}u_{Gi}v_{Gi} & \frac{EA_i}{L_i}u_{Gi}^2 \end{vmatrix} \tag{6.40}$$

block number 5:

$$\mathbf{S_i}^{\mathbf{T}}\mathbf{B_i}\mathbf{T_i} = \begin{vmatrix} 0 & Y_i & X_i \\ v_{Gi} & 0 & 0 \\ -u_{Gi} & 0 & 0 \end{vmatrix} \cdot \begin{vmatrix} 0 & 0 & 0 \\ 0 & 0 & \frac{GA_{ui}}{2} \\ 0 & -\frac{GA_{vi}}{2} & 0 \end{vmatrix} \cdot \begin{vmatrix} 1 & 0 & 0 \\ 0 & c_i & s_i \\ 0 & -s_i & c_i \end{vmatrix}$$
$$= \begin{vmatrix} 0 & \frac{-GA_{ui}Y_is_i - GA_{vi}X_ic_i}{2} & \frac{GA_{ui}Y_ic_i - GA_{vi}X_is_i}{2} \\ 0 & 0 & 0 \\ 0 & 0 & 0 \end{vmatrix} \tag{6.41}$$

block number 6:

$$\mathbf{T_i}^{\mathbf{T}}\mathbf{D_i}\mathbf{T_i} = \begin{vmatrix} 1 & 0 & 0 \\ 0 & c_i & -s_i \\ 0 & s_i & c_i \end{vmatrix} \cdot \begin{vmatrix} \frac{GJ_{ti}}{L_i} & 0 & 0 \\ 0 & R_{ui} & 0 \\ 0 & 0 & R_{vi} \end{vmatrix} \cdot \begin{vmatrix} 1 & 0 & 0 \\ 0 & c_i & s_i \\ 0 & -s_i & c_i \end{vmatrix}$$
$$= \begin{vmatrix} \frac{GJ_{ti}}{L_i} & 0 & 0 \\ 0 & R_{ui}c_i^2 + R_{vi}s_i^2 & R_{ui}c_is_i - R_{vi}c_is_i \\ 0 & R_{ui}c_is_i - R_{vi}c_is_i & R_{ui}s_i^2 + R_{vi}c_i^2 \end{vmatrix} \tag{6.42}$$

and finally block 7:

$$\mathbf{T_i}^{\mathbf{T}}\mathbf{E_i}\mathbf{T_i} = \begin{vmatrix} 1 & 0 & 0 \\ 0 & c_i & -s_i \\ 0 & s_i & c_i \end{vmatrix} \cdot \begin{vmatrix} -\dfrac{GJ_{ti}}{L_i} & 0 & 0 \\ 0 & K_{ui} & 0 \\ 0 & 0 & K_{vi} \end{vmatrix} \cdot \begin{vmatrix} 1 & 0 & 0 \\ 0 & c_i & s_i \\ 0 & -s_i & c_i \end{vmatrix}$$
$$= \begin{vmatrix} -\dfrac{GJ_{ti}}{L_i} & 0 & 0 \\ 0 & K_{ui}c_i^2 + K_{vi}s_i^2 & K_{ui}c_is_i - K_{vi}c_is_i \\ 0 & K_{ui}c_is_i - K_{vi}c_is_i & K_{ui}s_i^2 + K_{vi}c_i^2 \end{vmatrix} \tag{6.43}$$

The blocks that have been computed and summed up lead to the assembled matrix of the weld layout **K**, in the form of Equation 6.35.

On assembling, some terms disappear, if the principal axes of the weld layout (**u**, **v**) are used and equal lengths $L_i = L$, namely:

block 2, terms (1, 2) and (1,3):

$$\sum_{i=1}^{nw} \frac{EA_i}{L} v_{Gi} = \sum_{i=1}^{nw} \frac{EA_i}{L} u_{Gi} = 0$$

block 4, terms (2, 3) and (3, 2):

$$\sum_{i=1}^{nw} \frac{EA_i}{L} u_{Gi} v_{Gi} = \sum_{i=1}^{nw} \frac{EA_i}{L} u_{Gi} v_{Gi} = 0$$

If $A_{vi} = A_{ui} = A_v$ and $L_i = L$ then further relevant simplifications can be obtained, namely

$$\sum_{i=1}^{nw} \mathbf{T_i^T A_i S_i} = \mathbf{0}$$
$$\sum_{i=1}^{nw} \mathbf{S_i^T B_i T_i} = \mathbf{0} \tag{6.44}$$

Block 1, $\mathbf{T_i}^{\mathbf{T}}\mathbf{A_i}\mathbf{T_i}$, terms (2, 3) and (3, 2) are null

$$\left(\frac{GA_{ui} - GA_{vi}}{L_i}\right) = 0$$

Block 3, $\mathbf{T_i}^{\mathbf{T}}\mathbf{B_i}\mathbf{T_i}$, terms (2, 2) and (3, 3) are null

$$\left(\frac{GA_{ui} - GA_{vi}}{2}\right) = 0$$

However, if $A_u \neq A_v$, the coupling terms *do not vanish*, in general, and they must be taken into account in computing the forces flowing into a single weld seam and then computing the stresses.

In the special case of $A_{vi} = A_{ui}$ and $L_i = L$, it is also apparent that the form of the assembled matrix **K** closely resembles the general form of Equations 6.20–6.25, provided that the cross-sectional data used is that of the whole weld layout cross-sectional data W, i.e. the total area, the total shear area, the torsional constant, and the second area moment to the principal axes (**u**, **v**). The assembly takes the form of integral sums, that is, the assembled constants are the same constant that would have been obtained by evaluating the properties of the weld layout section W, as the sum of n_w rectangles having length l_i, width a_i, and **ui** axis inclination α_i to the axis 1.

For instance, considering the diagonal terms (block 1 and block 6), we get

$$K_{11} = \sum_{i=1}^{nw} \frac{EA_i}{L_i} = \frac{EA}{L}$$

$$K_{22} = K_{33} = \sum_{i=1}^{nw} \frac{GA_i}{L_i} = \frac{GA}{L}$$

$$K_{44} = \sum_{i=1}^{nw} \left(\frac{GA_i}{L_i} r_i^2 + \frac{GJ_{ti}}{L_i} \right) = \frac{GJ_t}{L}$$

$$K_{55} = \sum_{i=1}^{nw} \left(\frac{EA_i}{L} v_{Gi}^2 + R_{ui} c_i^2 + R_{vi} s_i^2 \right) = \sum_{i=1}^{nw} \left(\frac{EA_i}{L} v_{Gi}^2 + \frac{EJ_{ui}\, c_i^2}{L_i} + \frac{GA_i L_i\, c_i^2}{3} + \frac{EJ_{vi}\, s_i^2}{L_i} + \frac{GA_i L_i\, s_i^2}{3} \right) = \frac{EJ_u}{L} + \frac{GAL}{3}$$

$$K_{66} = \sum_{i=1}^{nw} \left(\frac{EA_i}{L} u_{Gi}^2 + R_{ui} s_i^2 + R_{vi} c_i^2 \right) = \sum_{i=1}^{nw} \left(\frac{EA_i}{L} u_{Gi}^2 + \frac{EJ_{ui}\, s_i^2}{L_i} + \frac{GA_i L_i\, s_i^2}{3} + \frac{EJ_{vi}\, c_i^2}{L_i} + \frac{GA_i L_i\, c_i^2}{3} \right) = \frac{EJ_v}{L} + \frac{GAL}{3}$$

The following definitions have been used:

$$A = \sum_{i=1}^{nw} A_i = \sum_{i=1}^{nw} a_i l_i$$

$$A_u = A_v = \sum_{i=1}^{nw} A_i$$

$$J_t = \sum_{i=1}^{nw} J_{ti} + A_i r_i^2 = \sum_{i=1}^{nw} \left[\frac{1}{12} l_i a_i^3 + \frac{1}{12} a_i l_i^3 + A_i \left(u_{Gi}^2 + v_{Gi}^2 \right) \right] \tag{6.45}$$

$$J_u = \sum_{i=1}^{nw} \left(\frac{1}{12} l_i a_i^3 c_i^2 + \frac{1}{12} a_i l_i^3 s_i^2 + A_i v_{Gi}^2 \right) = \sum_{i=1}^{nw} \left(J_{ui} c_i^2 + J_{vi} s_i^2 + A_i v_{Gi}^2 \right)$$

$$J_v = \sum_{i=1}^{nw} \left(\frac{1}{12} l_i a_i^3 s_i^2 + \frac{1}{12} a_i l_i^3 c_i^2 + A_i u_{Gi}^2 \right) = \sum_{i=1}^{nw} \left(J_{ui} s_i^2 + J_{vi} c_i^2 + A_i v_{Gi}^2 \right)$$

So, clearly this closely resembles the methods used to get the weld-layout cross-section W properties.

The stiffness matrix of a weld layout can be obtained by assembling the stiffness matrices of the single weld seams, and the result is the same as would result from considering a single layout element and the area properties of W. This is a special property of the Timoshenko beam element.

This consideration opens the path to the modeling of single welds in complex finite element models (those of *pure fem*); and in turn, as will be better seen in Chapter 10, each weld seam may indeed be modeled by a set of Timoshenko-like beam elements at equal distance one to another. As there will no longer be any predefined organization between the displacements of the welds belonging to a single weld layout, the distribution of forces and stresses will change, and will depend on the local stiffness of the surrounding parts.

The stresses in single welds can be computed using the global method already recalled in Section 6.6.2 and using $\boldsymbol{\sigma}$, that is, the internal forces acting over the whole layout, in the principal reference system (**w**, **u**, **v**).

As an equivalent but more general alternative (usable also when coupling occurs, i.e. when $A_u \neq A_v$, and for special formulations, as will be seen in next section), the disassembly of single weld seam elements can be used to get the stresses flowing into each weld.

If a short Timoshenko beam element is loaded at a free node, say node 1, when the other is clamped, the displacements at the free tip, $\{\mathbf{d_1}, \mathbf{r_1}\}$, a (6×1) vector, can be obtained from the matrix equation

$$\mathbf{K_{11}} \cdot \begin{vmatrix} \mathbf{d_1} \\ \mathbf{r_1} \end{vmatrix} = \boldsymbol{\sigma_1}$$

or writing explicitly the (3×3) blocks of $\mathbf{K_{11}}$ and $\boldsymbol{\sigma_1}$

$$\left[\sum_{i=1}^{nw} \begin{vmatrix} \mathbf{T_i^T A_i T_i} & \mathbf{T_i^T A_i S_i} + \mathbf{T_i^T B_i T_i} \\ \mathbf{S_i^T A_i T_i} + \mathbf{T_i^T B_i^T T_i} & \mathbf{S_i^T A_i S_i} + \mathbf{S_i^T B_i T_i} + \mathbf{T_i^T B_i^T S_i} + \mathbf{T_i^T D_i T_i} \end{vmatrix} \right] \cdot \begin{vmatrix} \mathbf{d_1} \\ \mathbf{r_1} \end{vmatrix} = \begin{vmatrix} \mathbf{F_1} \\ \mathbf{M_1} \end{vmatrix}$$

These displacements, however, are not suitable for getting the stresses because part of them is due to the finite length L of the element. For instance, a pure shear V_u would imply a translation d_u but also an unwanted rotation r_v. That rotation would in fact lead to displacements $d_{wi} \neq 0$ in welds. So, a modified stiffness must be used, obtained by considering

$$L \to 0$$

If this is done, the terms

$$(\cdot)$$

$$(\cdot) \cdot L$$

$$(\cdot) \cdot L^2$$

in $\mathbf{A_i}$, $\mathbf{B_i}$, $\mathbf{D_i}$, would vanish when compared to the terms

$$(\cdot) \cdot \frac{1}{L}$$

So:

- deleting the terms (2, 3) and (3, 2) in $\mathbf{B_i}$, it vanishes
- deleting the terms $GA_{vi}L_i/3$ in R_{ui} and $GA_{ui}L_i/3$ in R_{vi}, R_{ui} tends to EJ_{ui}/L_i, and R_{vi} tends to EJ_{vi}/L_i in matrix $\mathbf{D_i}$
- matrix $\mathbf{E_i}$ becomes equal to $-\mathbf{D_i}$, as K_{ui} tends to $-EJ_{ui}/L_i$ and K_{vi} tends to EJ_{vi}/L_i.

The blocks $\mathbf{T_i}^\mathbf{T}\mathbf{B_i}\mathbf{T_i}$ and $\mathbf{S_i}^\mathbf{T}\mathbf{B_i}\mathbf{T_i}$ disappear and $\mathbf{T_i}^\mathbf{T}\mathbf{D_i}\mathbf{T_i} = -\mathbf{T_i}^\mathbf{T}\mathbf{E_i}\mathbf{T_i}$.

In this way, the modified assembled stiffness matrix $\mathbf{K_{11,mod}}$, is found. Now the local stresses in a weld seam i can be obtained as follows.

First, the modified *assembled* displacements are obtained by

$$\mathbf{K}_{11,mod}\cdot\begin{vmatrix}\mathbf{d}_1\\ \mathbf{r}_1\end{vmatrix}=\left[\sum_{i=1}^{nw}\begin{vmatrix}\mathbf{T}_i^T\mathbf{A}_i\mathbf{T}_i & \mathbf{T}_i^T\mathbf{A}_i\mathbf{S}_i\\ \mathbf{S}_i^T\mathbf{A}_i\mathbf{T}_i & \mathbf{S}_i^T\mathbf{A}_i\mathbf{S}_i+\mathbf{T}_i^T\mathbf{D}_{i,mod}\mathbf{T}_i\end{vmatrix}\right]\cdot\begin{vmatrix}\mathbf{d}_1\\ \mathbf{r}_1\end{vmatrix}=\begin{vmatrix}\mathbf{F}_1\\ \mathbf{M}_1\end{vmatrix}$$

$$\begin{vmatrix}\mathbf{d}_1\\ \mathbf{r}_1\end{vmatrix}=\mathbf{K}_{11,mod}^{-1}\begin{vmatrix}\mathbf{F}_1\\ \mathbf{M}_1\end{vmatrix} \tag{6.46}$$

Then, the displacements of the generic weld seam i center are obtained by applying the kinematic organization, which forces the welds to behave like a unique set (an assumption which can be relaxed in pure fem models):

$$\begin{vmatrix}\mathbf{d}_{i1}\\ \mathbf{r}_{i1}\end{vmatrix}=\begin{vmatrix}\mathbf{T}_i & \mathbf{S}_i\\ 0 & \mathbf{T}_i\end{vmatrix}\cdot\begin{vmatrix}\mathbf{d}_1\\ \mathbf{r}_1\end{vmatrix} \tag{6.47}$$

Finally, the forces loading the single weld seam are obtained by

$$\boldsymbol{\sigma}_i=\begin{vmatrix}\mathbf{F}_{i1}\\ \mathbf{M}_{i1}\end{vmatrix}=\mathbf{K}_{i11,mod}\begin{vmatrix}\mathbf{d}_{i1}\\ \mathbf{r}_{i1}\end{vmatrix} \tag{6.48}$$

These are the forces and moments acting over a single weld seam, in its reference system (**wi, ui, vi**). Usually the stresses are evaluated along the local axis **ui** (mid throat thickness of weld seam), at the boundaries of the weld seam length:

$$u_i=\pm\frac{l_i}{2}$$

If so, no normal stress n is related to M_{ui}, and no tangential stress t_{par} to M_{ti}, so that

$$\begin{aligned}
n_i&=\frac{N_i}{A_i}\mp\frac{6M_{vi}}{a_i\,l_i^2}\\
t_{par,i}&=\frac{V_{ui}}{a_il_i}\\
t_{per,i}&=\frac{V_{vi}}{a_il_i}\pm\frac{6M_{ti}}{a_i\,l_i^2+a_i^3}
\end{aligned} \tag{6.49}$$

Summarizing, the following results have been found:

1) The method commonly accepted to compute the stresses in single welds from the global internal forces flowing into a weld layout is exactly the same as a beam element, and depends on assumed linear displacement fields and planar rotations of the weld layout projected “cross-section”.

2) The displacements of single weld seams are constrained to follow an ad hoc distribution (linear), which is not generally verified but is used anyway to compute stresses under the shelter of static theorem of limit analysis, albeit with the limitation discussed in Chapter 5. So, the welds displacements are not free but organized.
3) These very same hypotheses, with no addition, allow us to consider the weld layout *intrinsic* stiffness as that of a short Timoshenko beam element.
4) This latter hypothesis is equivalent to considering each weld seam as a single Timoshenko beam element, and forcing the nodal displacements of the single weld seam $\mathbf{E}_{i1}$ and $\mathbf{E}_{i2}$ to depend on the nodal displacement of the weld layout extremities, $\mathbf{E}_1$ and $\mathbf{E}_2$, considered as the nodes of a weld layout beam-like element.
5) Relaxing the hypothesis about the organization of the weld seams, each weld seam may be considered as a set of weld segments, each modeled by an independent Timoshenko beam, whose displacements have no relationship with those of the other weld segments. In turn, this means reducing the constraints related to linear kinematic behavior to the one single weld seam segment, allowing for non-linear weld layout displacements in finite element models.

6.6.5 Special Models

6.6.5.1 Shear Only, No Shear, Longitudinal Shear Only

In normal practice, and using the shelter offered by the static theorem of limit analysis, it is common to distribute the forces between the weld seams (or between the bolts) so as to neglect the effect of some seams, while considering only some of the others.

For instance, the shear force may be distributed between the weld seams so as to load the seams only longitudinally (i.e. along **ui**). Another instance is to use some seams to carry the normal stresses (axial force and bending) while others carry only the shear stresses (shear and torque).

There is no general law that we can use, in order to choose, but only the static theorem of limit analysis. Depending on the context, some subconnectors may be less efficient, and can be neglected (also to avoid an overestimate of their contribution). This heuristic approach can be very useful.

Several "recipes" are available, in literature.

If on the one hand these recipes are sheltered by the static theorem, on the other hand it must be recalled that the static theorem is not a complete shelter. Weld elements whose existence is neglected may well fail under the *true* loads, and this may, in principle, dynamically trigger more failures: the sudden failure of some subelements of the connector may imply loads higher than those computed by static equilibrium in the others. Fortunately, there is some ductility available in welds, so redistribution would occur before failure.

The idea of neglecting the forces acting transversely on the weld seams has good motivation. As it will be better seen in Section 6.7.1, fillet welds loaded transversely are stiffer and much less ductile than when loaded longitudinally.

From the analytical point of view, it would then be possible to declare some weld seam as alternatively:

- fully reactive (shear and normal stress), abbreviation: FR
- no shear (only normal stresses), abbreviation: NS
- shear only (only shear stresses due to torque and shear forces), abbreviation: SO.

Also, independently, it would be possible to assign these other two alternative flags to each weld seam:

- all shears, abbreviation: AS
- only longitudinal shear, abbreviation: LO

Clearly, these choices imply a difference in the evaluation of the stiffness matrix of the whole layout.

First of all it must be stressed that it would ideally be possible to consider all the weld seams as shear only, or as no shear, thus making some stiffness null. Also, if all the welds are loaded transversely, and the flag LO is assigned to them all, then no stiffness in that direction would be available.

This does not, in general, mean that a hypostatic configuration is met. If the two constituents connected, A and B, are hyperconnected, then more weld layouts (or other connectors) are available, and are possibly able to resist the loads. The choice of neglecting some stiffness of one connector may then have a good reason: *to force the generalized forces to flow into another connector.*

So these special conditions must in general be allowed.

However, it is also possible that the connection between A and B is isoconnected, or also that the other connectors in a hyperconnected connection lack of the necessary stiffness (and also, maybe, due to a mistake in the analysis).

Setting the stiffness to zero would then imply a null determinant in the assembled matrix, and an unsolvable system of equations.

More intelligent than zeroing the stiffness is to set it to a very low value. In this way, the forces will continue to flow elsewhere, if needed, but if no other connector is available, then the low-stiffness connector will do. This comes at a cost: *very high displacements.* The analyst will then detect the problem not by an excess of stresses, but by an excess of displacements.

How must the definition of the stiffness matrix be changed?

The easiest way to get the assembled matrix is to simply sum the contributions of each weld seam using Equation 6.35. The assumptions about the displacement field are still the same.

The stiffness contribution of each weld seam will be different depending on the flags applied to it.

Three values will be used, μ_n, μ_t, and μ_l, defined as follows:

- μ is a suitable small value (of the order of 10^{-3})
- μ_n is equal to μ for SO welds and equal to 1 for the others
- μ_t is equal to μ for NS welds and equal to 1 for the others
- μ_l is equal to μ for LO welds and equal to 1 for the others.

The center and principal axes of the weld layout can still be computed in the usual way, provided that the contribution of each weld seam is weighted by a factor μ_{ni}, equal to 1 if the weld seam is FR or NS, and equal to μ if the weld seam is shear only, SO.

Equations (6.45) are needed with the proper addition of the weight factor μ_{ni}:

$$A = \sum_{i=1}^{nw} \mu_{ni} a_i l_i$$

$$x_{1G} = \frac{\sum_{i=}^{nw} x_{1Gi} \mu_{ni} a_i l_i}{A}$$

$$x_{2G} = \frac{\sum_{i=}^{nw} x_{2Gi} \mu_{ni} a_i l_i}{A}$$

$$J_{ui} = \mu_{ni} \frac{l_i\, a_i^3}{12}$$

$$J_{vi} = \mu_{ni} \frac{l_i^3 a_i}{12}$$

$$J_{1Gi} = J_{ui} \cos^2(\alpha_i) + J_{vi} \sin^2(\alpha_i)$$

$$J_{2Gi} = J_{ui} \sin^2(\alpha_i) + J_{vi} \cos^2(\alpha_i)$$

$$J_{12Gi} = (J_{vi} - J_{ui}) \sin\alpha \cos\alpha$$

$$J_{1G} = \sum_{i=1}^{nw} \left[J_{1i} + \mu_{ni} l_i a_i (x_{2Gi} - x_{2G})^2 \right]$$

$$J_{2G} = \sum_{i=1}^{nw} \left[J_{2i} + \mu_{ni} l_i a_i (x_{1Gi} - x_{1G})^2 \right]$$

$$J_{12G} = \sum_{i=1}^{nw} \left[J_{12i} + \mu_{ni} l_i a_i (x_{1Gi} - x_{1G}) \cdot (x_{2Gi} - x_{2G}) \right]$$

$$\beta = \frac{1}{2} tan^{-1} \frac{2J_{12G}}{J_{2G} - J_{1G}}$$

$$J_u = J_{2G} \sin^2\beta + J_{1G} \cos^2\beta - J_{12G} \sin(2\beta)$$

$$J_v = J_{1G} \sin^2\beta + J_{2G} \cos^2\beta + J_{12G} \sin(2\beta)$$

The new cross-sectional global terms x_{1G}, x_{2G}, β, A, J_u, and J_v also embed the values of μ_{ni}, and so are in general different from the ones computed normally. Similar modified quantities can also be computed for shear areas A_u, A_v, and for torsional constant J_t, but if a difference of μ_{li} is applied between the weld seams (if some weld seam is longitudinal only), directions **u** and **v** are coupled, and the simple expressions can no longer be used.

If the weld seams are all SO, then no difference is found in the center and principal axes position, when compared to an all-FR weld layout.

Having defined the three μ, μ_n, μ_t, and μ_l, the following expressions should be inserted in the definitions of the submatrices, $\mathbf{A}_i$, $\mathbf{B}_i$, $\mathbf{D}_i$, $\mathbf{E}_i$:

$$EA_i \Rightarrow E\mu_{ni} a_i l_i$$

$$GA_{ui} \Rightarrow G\mu_{ti} a_i l_i$$

$$GA_{vi} \Rightarrow G\mu_{li} \mu_{ti} a_i l_i$$

$$GJ_{ti} \Rightarrow G\mu_{ti} \left[(1/12)\, l_i a_i^3 + \mu_{li} (1/12)\, a_i l_i^3 \right]$$

$$EJ_{ui} \Rightarrow E\mu_{ni} (1/12)\, l_i a_i^3$$

$$EJ_{vi} \Rightarrow E\mu_{ni} (1/12)\, a_i l_i^3$$

These can be thought of as modifications applied to the elastic properties of the material, E and G, so that the same strain leads to much lower stresses. However, from the computational point of view it is totally equivalent to computing the area properties applying the same factors, as explained above.

Once the stiffness matrix of weld seam i is evaluated, it can be assembled in the usual way. The assembled matrix can be later used to compute the internal forces in the layout, $\boldsymbol{\sigma}$. From the forces flowing into the layout the forces flowing in the single seam are computed using Equations (6.46–6.48) and from these, the stresses n, t_{par}, and t_{per} at relevant points are computed using Equations (6.49) that is:

$$n_i = \frac{N_i}{A_i} \mp \frac{6M_{vi}}{a_i\, l_i^2}$$

$$t_{par,i} = \frac{V_{ui}}{a_i l_i}$$

$$t_{per,i} = \frac{V_{vi}}{a_i l_i} \pm \frac{6M_{ti}}{a_i\, l_i^2 + a_i^3}$$

with no modifications.

6.6.5.2 Flexibility Index

It has been mentioned that the stiffness of a connector can be modified in order to decrease or increase its value, so as to get different forces flowing into connectors working in parallel. Also, in practice, the *shear-only, no-shear*, or *longitudinal-shear only* flags accomplish the task of tuning the stiffnesses. However, it might also be useful to reduce or increase all the stiffnesses of a connector by a common factor, i.e. the flexibility index, f.

This value f can be related to each weld layout, and can be used as a factor applied to the "length" L of the weld seams' Timoshenko beam elements. So, it is sufficient to replace L by $f{\cdot}L$ in the stiffness matrix, to get the needed stiffness tuning. The approach is general, and it can be used to modify the forces flowing into the connectors.

Initially, the flexibility index is set to 1. Different values in the real-numbers range can be set, however. It must be underlined that the stiffness of the connectors does not depend only on the intrinsic stiffness of the connector itself, but also, in a renode scene, on the absolute position of a connector, that is, by its distance from the forces to be transferred.

6.6.6 Example

A full penetration weld layout made by eight welds is considered. Six welds have $t = 7.5$ mm and are used for the flanges. Two welds have $t = 4.5$ mm and are used for the web. The member cross-section thickness is filled with two welds. The cross-section is HEB 200 ($h = 200$ mm, $b = 200$ mm, $t_w = 9$, $t_f = 15$ mm, see Figure 6.14).

The weld group is loaded by six different loading conditions, each referring to one of the principal axes, and each with a unique component applied:

1) $N = 183{,}488$ N
2) $V_v = 72{,}248$ N

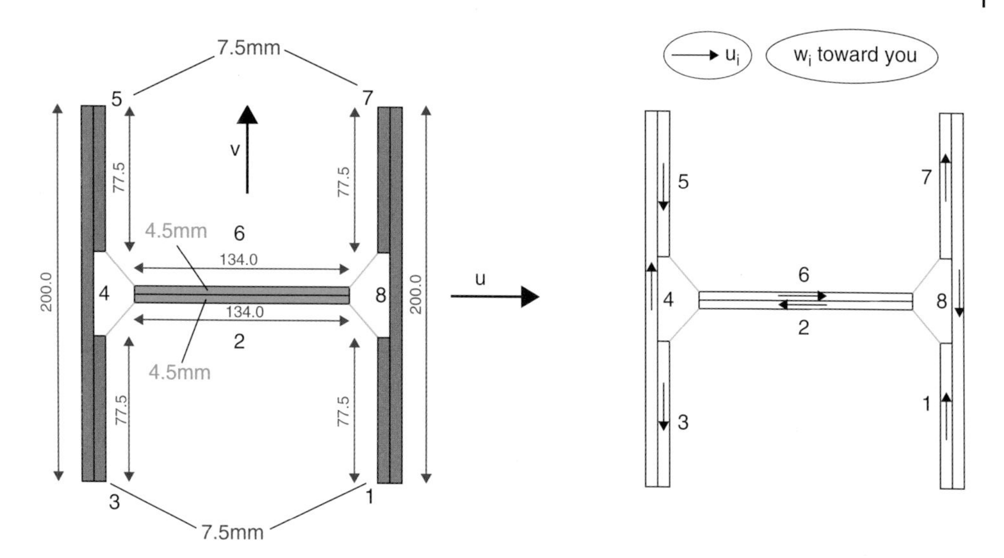

Figure 6.14 Example of weld layout (full penetration welds). Eight welds, four having $t = 7.5$ mm and the two web ones having $t = 4.5$ mm. Principal axes of weld layout (**u**, **v**) are also drawn, as well as $\mathbf{u_i}$ direction for all welds.

3) $V_u = -33{,}689$ N
4) $M_w = 536{,}197$ Nmm
5) $M_v = 13{,}386{,}070$ Nmm
6) $M_u = -4{,}707{,}990$ Nmm

These are 10% of the elastic limits of the cross-section (axial force, shears, torque, and bending) using a material S235.

The first computation assumes that all the welds are FR (fully reactive). The stresses are listed in Table 6.1.

For instance, axial force is balanced as for load case 1

$$(2 \times 200 \times 210.7 + 4 \times 77.5 \times 210.7 + 2 \times 134 \times 126.4) = 183{,}472\text{N}$$

which is equal to 183,488 N, allowing for rounding errors. Similar considerations can be done for the other load cases.

The second computation assumes that all the flange welds are FR, while the web welds (numbers 2 and 6) are shear only. The stresses are listed in Table 6.2.

Comparing the normal stress in load case 1, for instance, the value 34.5 instead of 28.1 is found. In load case 5, strong axis bending, the maximum normal stress changes from 26.9 to 27.9 MPa. There is no difference in load cases 2, 3, and 4 (shears and torque).

The third computation is like the second (2 and 6 are shear only, the other welds FR), but all the welds are longitudinally sheared only. The stresses are listed in Table 6.3.

It can be seen that t_{per} is always null. Load 2 is resisted only by flange welds, while load 3 only by web welds. The torque is balanced as

$$2 \times 8.4 \times 200 \times (100 - 7.5/2) + 4 \times 7.7 \times 77.5 \times (100 - 15 + 7.5/2) + 2 \times 0.1 \times 134 \times 4.5/2 = 535{,}306$$

Table 6.1 Stresses in the welds for six load cases, MPa, N/mm. All welds FR. Stresses n, $t_{par} = t_u$, $t_{per} = t_v$ at extremity "ext", and force per unit length. Only the worst stressed extremity is listed, for each weld.

Load	Weld seam	n_{Per}	t_{Par}	t_{Per}	Force	Ext
1	1	28.1	0.0	0.0	210.7	1
1	2	28.1	0.0	0.0	126.4	1
1	3	28.1	0.0	0.0	210.7	1
1	4	28.1	0.0	0.0	210.7	1
1	5	28.1	0.0	0.0	210.7	1
1	6	28.1	0.0	0.0	126.4	1
1	7	28.1	0.0	0.0	210.7	1
1	8	28.1	0.0	0.0	210.7	1
2	1	−0.0	−11.1	0.0	83.0	1
2	2	−0.0	0.0	11.1	49.8	1
2	3	−0.0	11.1	0.0	83.0	1
2	4	0.0	−11.1	0.0	83.0	1
2	5	0.0	11.1	0.0	83.0	1
2	6	0.0	0.0	−11.1	49.8	1
2	7	0.0	−11.1	0.0	83.0	1
2	8	−0.0	11.1	0.0	83.0	1
3	1	0.0	−0.0	−5.2	38.7	1
3	2	−0.0	−5.2	0.0	23.2	1
3	3	−0.0	−0.0	5.2	38.7	1
3	4	−0.0	−0.0	−5.2	38.7	1
3	5	−0.0	−0.0	5.2	38.7	1
3	6	0.0	5.2	0.0	23.2	1
3	7	0.0	−0.0	−5.2	38.7	1
3	8	0.0	−0.0	5.2	38.7	1
4	1	0.0	−0.7	0.8	7.9	2
4	2	0.0	0.0	−0.5	2.4	1
4	3	0.0	−0.7	−0.8	7.9	1
4	4	0.0	0.8	−0.8	8.2	1
4	5	0.0	−0.7	0.8	7.9	2
4	6	0.0	0.0	−0.5	2.4	1
4	7	0.0	−0.7	−0.8	7.9	1
4	8	0.0	0.8	−0.8	8.2	1
5	1	−24.8	0.0	0.0	185.9	1
5	2	18.7	0.0	−0.0	84.2	1
5	3	24.8	0.0	−0.0	185.9	1
5	4	26.9	0.0	0.0	201.6	1

Table 6.1 (Continued)

Load	Weld seam	n_{Per}	t_{Par}	t_{Per}	Force	Ext
5	5	24.8	0.0	−0.0	185.9	1
5	6	−18.7	−0.0	−0.0	84.2	1
5	7	−24.8	0.0	0.0	185.9	1
5	8	−26.9	0.0	−0.0	201.6	1
6	1	23.7	0.0	−0.0	177.5	2
6	2	0.5	−0.0	−0.0	2.4	1
6	3	23.7	−0.0	−0.0	177.5	1
6	4	−23.7	0.0	−0.0	177.5	1
6	5	−23.7	−0.0	−0.0	177.5	2
6	6	−0.5	−0.0	0.0	2.4	2
6	7	−23.7	0.0	−0.0	177.5	1
6	8	23.7	−0.0	−0.0	177.5	1

Table 6.2 Stresses in the welds for six load cases. All welds FR, but the web welds, 2 and 6, that are shear only (rows shaded). Stresses n, $t_{par} = t_u$, $t_{per} = t_v$ at extremity "ext". Only the worst stressed extremity is listed, for each weld.

Load	Weld seam	n_{Per}	t_{Par}	t_{Per}	Force	Ext
1	1	34.5	−0.0	0.0	258.4	1
1	2	0.0	0.0	0.0	0.2	1
1	3	34.5	0.0	0.0	258.4	1
1	4	34.5	0.0	0.0	258.4	1
1	5	34.5	0.0	0.0	258.4	1
1	6	0.0	0.0	0.0	0.2	1
1	7	34.5	0.0	0.0	258.4	1
1	8	34.5	0.0	0.0	258.4	1
2	1	−0.0	−11.1	0.0	83.0	1
2	2	−0.0	−0.0	11.1	49.8	1
2	3	−0.0	11.1	0.0	83.0	1
2	4	0.0	−11.1	0.0	83.0	1
2	5	0.0	11.1	0.0	83.0	1
2	6	0.0	0.0	−11.1	49.8	1
2	7	0.0	−11.1	0.0	83.0	1
2	8	−0.0	11.1	0.0	83.0	1

(Continued)

Table 6.2 (Continued)

Load	Weld seam	n_{Per}	t_{Par}	t_{Per}	Force	Ext
3	1	−0.0	−0.0	−5.2	38.7	1
3	2	0.0	−5.2	0.0	23.2	1
3	3	0.0	−0.0	5.2	38.7	1
3	4	0.0	−0.0	−5.2	38.7	1
3	5	0.0	−0.0	5.2	38.7	1
3	6	−0.0	5.2	0.0	23.2	1
3	7	−0.0	−0.0	−5.2	38.7	1
3	8	−0.0	−0.0	5.2	38.7	1
4	1	0.0	−0.7	0.8	7.9	2
4	2	0.0	0.0	−0.5	2.4	1
4	3	0.0	−0.7	−0.8	7.9	1
4	4	0.0	0.8	−0.8	8.2	1
4	5	0.0	−0.7	0.8	7.9	2
4	6	0.0	0.0	−0.5	2.4	1
4	7	0.0	−0.7	−0.8	7.9	1
4	8	0.0	0.8	−0.8	8.2	1
5	1	−25.8	−0.0	−0.0	193.1	1
5	2	0.0	−0.0	0.0	0.1	1
5	3	25.8	−0.0	0.0	193.1	1
5	4	27.9	−0.0	−0.0	209.5	1
5	5	25.8	−0.0	0.0	193.1	1
5	6	−0.0	0.0	0.0	0.1	1
5	7	−25.8	−0.0	−0.0	193.1	1
5	8	−27.9	−0.0	0.0	209.5	1
6	1	23.7	−0.0	0.0	177.6	2
6	2	0.0	−0.0	0.0	0.0	2
6	3	23.7	0.0	0.0	177.6	1
6	4	−23.7	−0.0	0.0	177.6	1
6	5	−23.7	0.0	0.0	177.6	2
6	6	−0.0	0.0	−0.0	0.0	2
6	7	−23.7	−0.0	0.0	177.6	1
6	8	23.7	0.0	0.0	177.6	1

It is equal to 536,197 Nmm, allowing for rounding errors. This torque is carried only by longitudinal shears, as expected.

Finally, as a last case, some welds have been removed, to have an asymmetric weld over a symmetric cross-section. The forces acting over the (column) cross-section are the

Table 6.3 Stresses in the welds for six load cases. All welds LO (longitudinally sheared only), flange welds are FR, and web welds, 2 and 6, are shear only (rows shaded). Stresses n, $t_{par} = t_u$, $t_{per} = t_v$ at extremity "ext". Only the worst stressed extremity is listed, for each weld. As all the welds are LO, t_{per} is always null.

Load	Weld seam	n_{Per}	t_{Par}	t_{Per}	Force	Ext
1	1	34.5	0.0	0.0	258.4	1
1	2	0.0	0.0	0.0	0.2	1
1	3	34.5	0.0	0.0	258.4	1
1	4	34.5	0.0	0.0	258.4	1
1	5	34.5	0.0	0.0	258.4	1
1	6	0.0	0.0	0.0	0.2	1
1	7	34.5	0.0	0.0	258.4	1
2	1	−0.0	−13.6	0.0	101.7	1
2	2	−0.0	0.0	0.0	0.1	1
2	3	−0.0	13.6	−0.0	101.7	1
2	4	0.0	−13.6	0.0	101.7	1
2	5	0.0	13.6	0.0	101.7	1
2	6	0.0	0.0	−0.0	0.1	1
2	7	0.0	−13.6	0.0	101.7	1
3	1	0.0	−0.0	−0.0	0.2	1
3	2	−0.0	−27.8	−0.0	125.2	1
3	3	−0.0	−0.0	0.0	0.2	1
3	4	−0.0	−0.0	−0.0	0.2	1
3	5	−0.0	−0.0	0.0	0.2	1
3	6	0.0	27.8	0.0	125.2	1
3	7	0.0	−0.0	−0.0	0.2	1
3	8	0.0	−0.0	0.0	0.2	1
4	1	0.0	−1.0	0.0	7.7	2
4	2	0.0	0.0	−0.0	0.1	1
4	3	0.0	−1.0	−0.0	7.7	1
4	4	0.0	1.1	−0.0	8.4	1
4	5	0.0	−1.0	0.0	7.7	2
4	6	0.0	0.0	−0.0	0.1	1
4	7	0.0	−1.0	−0.0	7.7	1
4	8	0.0	1.1	−0.0	8.4	1
5	1	−25.8	−0.0	−0.0	193.1	1
5	2	0.0	−0.0	−0.0	0.1	1
5	3	25.8	−0.0	0.0	193.1	1

(Continued)

Table 6.3 (Continued)

Load	Weld seam	n_{Per}	t_{Par}	t_{Per}	Force	Ext
5	4	27.9	−0.0	−0.0	209.5	1
5	5	25.8	−0.0	0.0	193.1	1
5	6	−0.0	0.0	0.0	0.1	1
5	7	−25.8	−0.0	−0.0	193.1	1
5	8	−27.9	−0.0	0.0	209.5	1
6	1	23.7	0.0	−0.0	177.6	2
6	2	0.0	−0.0	−0.0	0.0	2
6	3	23.7	−0.0	0.0	177.6	1
6	4	−23.7	0.0	−0.0	177.6	1
6	5	−23.7	−0.0	−0.0	177.6	2
6	6	−0.0	−0.0	0.0	0.0	2
6	7	−23.7	0.0	−0.0	177.6	1
6	8	23.7	−0.0	0.0	177.6	1

same as the previous cases and are aligned with its principal axes but are transformed to the principal axes of the weld layout which have a different origin and a different axis **u** (Figure 6.15). The stresses are listed in Table 6.4.

Considering load 4, torque, it can first be seen that the force resultant is null:

$$0.2 \times 134 - 0.2 \times 134 = 0$$

$$14.1 \times 200 - 9.4 \times 77.5 - 10.4 \times 200 = 11.5 \cong 0$$

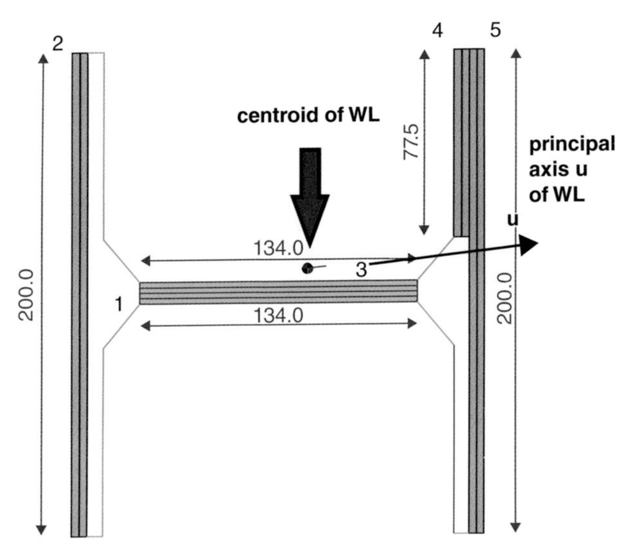

Figure 6.15 Asymmetric weld of a symmetric cross-section. Welds are now numbered differently. The flange welds are FR, the web welds are shear only. All welds are LO. The loads applied to the weld layout are the same of the other cases.

Table 6.4 Stresses in the welds for six load cases. All welds LO (longitudinally sheared only), flange welds are FR, and web welds, 1 and 3, are shear only (rows shaded). Stresses n, $t_{par} = t_u$, $t_{per} = t_v$ at extremity "ext". Only the worst stressed extremity is listed, for each weld. As all the welds are LO, t_{per} is always null.

Load	Weld seam	n_{Per}	t_{Par}	t_{Per}	Force	Ext
1	1	0.1	0.0	0.0	0.3	1
1	2	74.0	−0.0	0.0	555.1	2
1	3	0.1	−0.0	−0.0	0.3	2
1	4	44.2	−0.0	0.0	331.2	2
1	5	60.1	0.0	−0.0	450.9	1
2	1	0.0	−0.1	0.0	0.4	1
2	2	0.0	−23.8	−0.0	178.5	2
2	3	0.0	−0.1	−0.0	0.3	2
2	4	−0.0	−17.7	0.0	132.9	1
2	5	0.0	17.5	0.0	131.1	1
3	1	−0.0	−27.9	0.0	125.3	1
3	2	−0.0	−0.0	−0.0	0.2	2
3	3	−0.0	27.9	−0.0	125.3	2
3	4	−0.0	0.0	−0.0	0.2	2
3	5	−0.0	−0.0	0.0	0.2	1
4	1	0.0	0.0	−0.0	0.2	1
4	2	0.0	1.9	0.0	14.1	2
4	3	0.0	0.0	0.0	0.2	2
4	4	0.0	−1.3	−0.0	9.4	1
4	5	0.0	1.4	−0.0	10.4	1
5	1	0.0	0.0	0.0	0.2	1
5	2	56.2	−0.0	0.0	421.1	1
5	3	0.0	−0.0	−0.0	0.2	2
5	4	−30.9	−0.0	0.0	231.4	2
5	5	−45.6	0.0	−0.0	342.1	1
6	1	0.0	0.0	−0.0	0.0	2
6	2	39.8	−0.0	−0.0	298.8	2
6	3	0.0	−0.0	0.0	0.0	1
6	4	−33.2	0.0	0.0	248.9	1
6	5	46.2	−0.0	0.0	346.6	1

It must be noted that a precision of 0.05 N/mm for a weld 200 mm long means an expected loss of precision of 10 N, here the out-of-balance load is 11.5 N.

Then it can be seen that the torque applied is balanced by that of the cross-section:

$$14.1 \times 200 \times (100 - 7.5/2) + 9.4 \times 77.5 \times (85 + 7.5/2) + 10.4 \times 200 \times (100 - 7.5/2) = 536{,}279$$

The contribution due to the web has been neglected. The difference with the applied torque is due to the cut-off and rounding, but the difference is negligible.

6.7 Fillet-Welds Weld Layouts

6.7.1 The Behavior of Fillet Welds

Fillet weld layouts behave in a different manner from penetration weld layouts. The flow of the stresses is not direct, and this makes the connection more flexible and weaker.

Many studies have been carried on in the past few decades, referring to fillet weld layouts, and limit domains have been drawn, leading to approximate checking formulae used by standards all over the world (for an excellent review, see Ballio and Mazzolani, 1983).

A first difference with penetration weld layouts is that the effective size of the fillet, useful for checks, is the so-called *throat thickness*. This is related to the legs and the thickness of the fillet weld, and to the angle between the fusion faces, by the Equations 6.1 and 6.2.

From the static point of view, fillet welds behave in very different manner when loaded by longitudinal and transverse shears (i.e. longitudinal V_{par} and transverse V_{per}). In the transverse direction, fillet welds are stronger, stiffer, and less ductile, while in the longitudinal direction they are weaker, more flexible, and ductile. According to AISC 360-10, the resistance of fillet welds loaded transversely is 1.5 greater than the resistance of fillet welds loaded longitudinally. This difference of behavior poses a real problem when modeling the fillet welds.

Often this difference of behavior in transverse and longitudinal direction is neglected, using both the resistance and the stiffness of the longitudinally loaded fillets. This model will be called isotropic. If the model adopted uses different resistances and stiffnesses for the longitudinal and transverse direction, it is called orthotropic.

The aim of this section is to get to an engineering model of fillet welds and of fillet weld layouts, which can be used to evaluate deformations and stresses. Once more, it must be underlined that these are intrinsic stiffness, that is, they do not consider the surroundings.

The reference model used will be that of a Timoshenko beam for the single fillet welds, with modifications. The stiffness of the weld layout will be evaluated by assembling the contributions of the single weld seams, as already done for penetration weld layouts, and adding the linear displacement hypothesis for the displacement field of the weld layout. For fillet weld layouts, the considerations already made about the existing computational methods used to get the stresses from the layout generalized forces, still hold true. The linear model is still accepted by all the standards.

The cross-section of the weld layout, W, will be that of the throat surfaces projected onto the faying surface. The extremities of the elements will be defined as for the penetration weld layouts (Equation 6.18), but now the thickness t will in general be different from weld seam to weld seam. The length of the elements L, will be considered equal to half the average thickness of the weld seams. This length is notional. Numerical tests have shown that in the linear range this value of length predicts well the displacement of the fillet welds, using a properly tuned Timoshenko beam element.

Experimental tests carried out by many researchers have shown that the force–displacement curve of a fillet weld is strongly non-linear, and also depends on the direction of loading. The angle θ is used, and it is 0 for longitudinally loaded fillet welds, and 90° for transversely loaded ones. Note that a single weld seam under "axial force", that is, a force perpendicular to the faying plane, directed as **wi**, is under the effect of a $\theta = 90°$ force, as well as when under a force directed as **vi**.

Lesik and Kennedy (1990) proposed a load–displacement curve that was later adopted by the AISC standard, both for weld groups under plane eccentric loading (shear plus torsion) and for weld layouts under out-of-plane eccentric loading (axial force plus bending). These load–displacement curves can be seen in Figure 6.16. The curves have the following analytical form:

$$
\begin{aligned}
R &= R_0 \cdot \left(1 + 0,5 \cdot \sin^{1.5}\theta\right) \cdot \left[p \cdot (1.9 - 0.9p)\right]^{0.3} = R_0 \cdot q(\theta) \cdot h(p) \\
p &= \frac{d}{d_{\max}} \\
d_{\max} &= 0.209 \cdot (\theta + 2)^{-0.32} \cdot t \\
d_{ult} &= 1.087 \cdot (\theta + 6)^{-0.65} \cdot t
\end{aligned}
\qquad (6.50)
$$

where:

- R is the force applied to the fillet weld
- p is a strain measure
- R_0 *is the resistance of a fillet weld loaded longitudinally,* $R_0 = 0.6 \cdot f_{uw} \cdot l \cdot a$ according to AISC, where f_u is the ultimate stress of the weld, l is the weld length, and a the throat size
- θ is the angle of the force to the weld axis, in degrees (between 0° and 90°)
- d is the displacement
- t is the weld thickness
- d_{max} is the displacement at maximum force
- d_{ult} is the ultimate displacement, i.e. the fracture displacement
- $q(\theta)$ is a function of θ between 1 and 1.5
- $h(p)$ is a function of p between 0 and 1.

Transversely loaded fillet welds experience a higher load at equal displacement, and are less ductile. At equal displacement, when a transversely loaded fillet weld reaches its maximum, a longitudinally loaded fillet weld of equal thickness is at 83% of its resistance.[2] Higher loads would soon imply a fracture in the transversely loaded fillet welds, and these would not be able to carry more loads, even if applied longitudinally. This is quite a dangerous condition. Traditionally, sometimes this has been avoided considering

2 The value that can be obtained by the equations in AISC 360-10 is slightly lower (0.80), and the standard itself uses the value 0.85. However, the value of 0.83 is often found in the literature regarding this issue, e.g. Tamboli, 2010. The difference is probably due to the rounding of the experimental curve fitting.

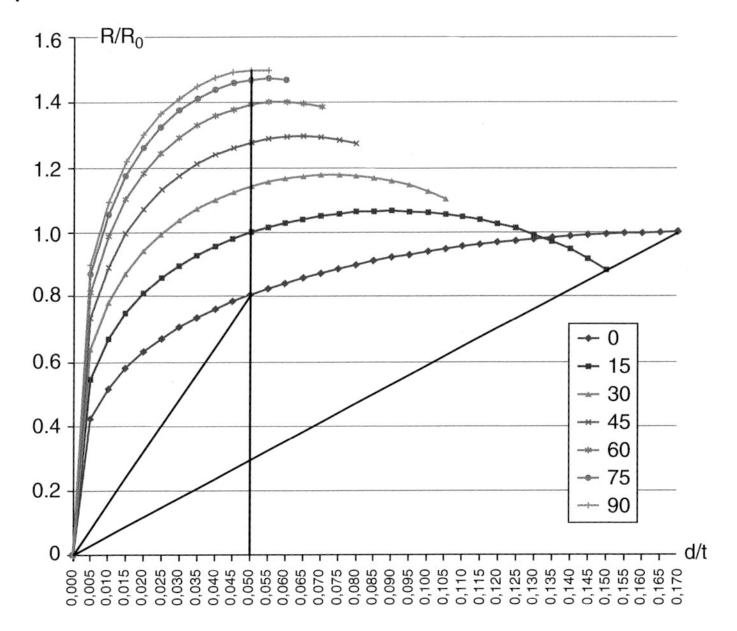

Figure 6.16 Load–displacement curves for fillet welds loaded at different angles θ. The horizontal axis is the ratio of displacement to weld thickness. The vertical axis maps the ratio of the force to the ultimate force of longitudinally loaded fillet welds (R_0, $\theta = 0$). The curves plotted are those of the AISC method. The two straight lines are related to the secant stiffness and modified secant stiffness for $\theta = 0$.

only the longitudinally loaded fillet welds, and neglecting the resistance of the transversely loaded ones. However, it can be easily proved that this method is not always on the safe side.

Let A_t be the projected throat area of transversely loaded fillet welds, and A_l the projected throat area of longitudinally loaded fillet welds having all the same thickness. It will be assumed that the existing fillet welds are loaded only at $\theta = 0$ or $\theta = 90°$, as in a normal I or H weld, under principal axis shears. The loading capacity is

$$V = 0.83R_0A_l + 1.5R_0A_t$$

Using the simple method of neglecting the transversely loaded fillet welds gives

$$V^* = R_0A_l$$

The method is safe as far as

$$R_0A_l < 0.83R_0A_l + 1.5R_0A_t$$

that is

$$A_l < 8.82A_t$$

The extended experimental work carried on by Gomez et al. (2008), an excellent work where many questions find an answer, has shown that the methods used by AISC when applied to out-of-plane eccentric loading are too much on the safe side, as the test-to-predicted ratios are on average 1.75 with a coefficient of variation equal to 0.25. The reason is that the bent welded plate finds a bearing in the other constituent welded, after an initial rotation has occurred, and this (assumed stiff) bearing, shifts the neutral axis, leading to an increase of strength. In the work by Gomez et al., several methods are proposed in order to take into account this effect.

At the moment, the standards do not explicitly mention this issue. If the component fillet welded is in direct contact with the other, and if it exchanges a strong compression, then contact and friction may carry the loads, and the fillet welds stay unloaded. It is on the safe side to assume that the forces do always flow into the fillet welds, no matter what the direction of the loading, as if a gap between the connected parts were present. However, more refined mechanical models, taking into account contact and friction can be set up: the issue will be covered in Section 6.7.6.

In the remaining part of this chapter, with the exception of Section 6.7.6, the contact plus friction effect is not taken into account.

In Eurocode 3, no direct mention is apparently paid to these issues, even if the *directional method* implicitly considers the difference of resistance related to the direction of loading (transverse and longitudinal, see Section 8.4.3.1).

The mechanical model for fillet welds not using contact and friction is as follows.

When $p = 1$, that is, the force has reached its maximum, the point reached depends on the angle θ according to the following relations:

$$d = d_{max} = 0.209 \cdot (\theta + 2)^{-0.32} \cdot t$$
$$R = R_0 \cdot \left(1 + 0.5 \cdot \sin^{1.5}\theta\right)$$

So, a secant stiffness K_s can be defined, depending on the angle θ:

$$K_s = \frac{R_0 \cdot \left(1 + 0.5 \cdot \sin^{1.5}\theta\right)}{t \cdot 0.209 \cdot (\theta + 2)^{-0.32}} = 2.87 \frac{laf_{uw}}{t} \cdot \frac{\left(1 + 0.5 \cdot \sin^{1.5}\theta\right)}{(\theta + 2)^{-0.32}}$$

The normalized secant stiffness K_{sn} is obtained by

$$K_{sn} = \frac{K_s}{\left(\dfrac{laf_{uw}}{t}\right)}$$

This means:

- for $\theta = 0$ $K_{sn,0} = 3.58$
- for $\theta = 90°$ $K_{sn,90} = 18.3$.

Plotting the curve $K_{sn} = K_{sn}(\theta)$ the results are in Figure 6.17. If a Timoshenko beam element is used for the weld seam, only two stiffnesses can be specified: one for $\theta = 0$, K_{s0}, longitudinal loading, and one for $\theta = 90°$, K_{s90}, transverse loading. For a force inclined at an angle θ, the stiffness would be

$$K_s^* = K_{s0} \cdot \cos^2\theta + K_{s90} \cdot \sin^2\theta$$

This can be seen well considering the term (2, 2) of submatrix $\mathbf{T_i}^{\mathbf{T}}\mathbf{A_i}\mathbf{T_i}$, Equation 6.37.

In Figure 6.17 the comparison between $K_{sn}(\theta)$ and $K_{sn}{*}(\theta)$ is presented. As can be seen, the agreement is good. So, a Timoshenko beam element well predicts the secant stiffness of a weld seam also for inclined loading.

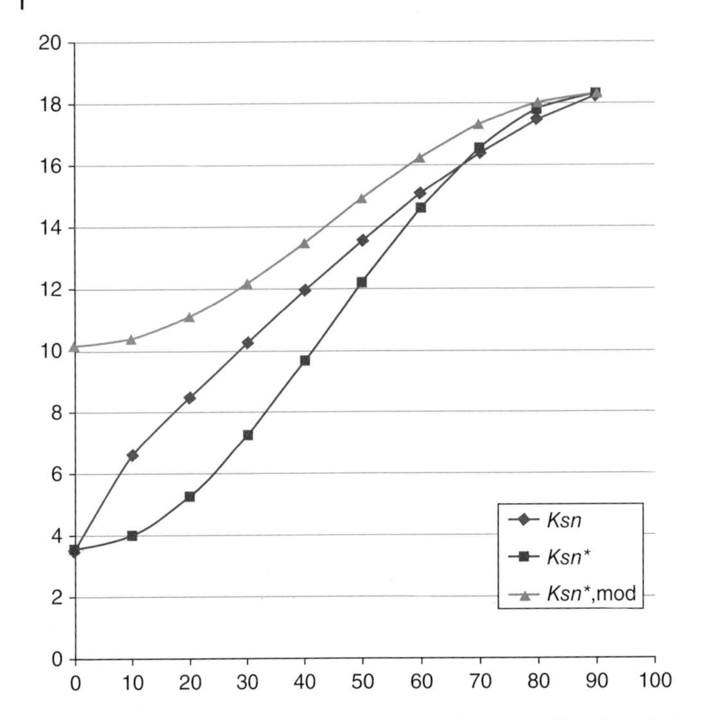

Figure 6.17 Normalized secant stiffness as a function of angle θ. K_{sn} is the normalized secant stiffness curve predicted by the force–displacement curve by AISC. $K_{sn}*$ is the curve predicted by a Timoshenko beam element using only the stiffnesses valid for $\theta = 0$ and $\theta = 90°$.

Using a secant stiffness is a good compromise between the need for quick evaluation of the forces, and a reasonable evaluation of the stiffness. Clearly, a more rigorous method would imply a non-linear force–displacement curve. The assumed linear curve, however, is tuned to correctly predict the maximum force of the weld seam and its related displacement. For loads lower than the maximum one, the displacements will be overestimated.

When considering a weld layout under the effect of a concentric force, loading some welds transversely and some others longitudinally, a drawback of the secant stiffness approach is that the forces flowing into the longitudinally loaded weld seams are underestimated when the transversely loaded welds reach their maximum. For instance, if a weld layout of a typical I or H cross-section is considered, and assuming that all the welds have the same thickness, the following considerations apply. Under the effect of a pure shear parallel to the web or to the flanges, some welds are loaded longitudinally and some others transversely. If the secant stiffness is used for the welds, when the transversely loaded welds reach their maximum resistance, 1.5 R_0, the longitudinally loaded weld seams will receive a force $R*$, much lower than 0.83 R_0:

$$R^* = K_s(0°)\cdot d_{\max}(90°) = \frac{2.87\cdot laf_{uw}}{t\cdot 2^{-0.32}}\cdot 0.209\cdot 92^{-0.32}t = 0.176\cdot laf_{uw}$$

$$\frac{R^*}{R_0} = \frac{0.176}{0.6} = 0.293$$

So, the maximum shear force predicted would be

$$V = 1.5R_0A_t + 0.29\cdot R_0A_l < 1.5R_0A_t + 0.83\cdot R_0A_l$$

In order to get a limiting force equal to that predicted by the experimental tests (and by the instantaneous center of rotation method in pure shear), an increased secant stiffness can be used for $\theta = 0$. The needed secant stiffness can be computed as

$$K_{s,\,\mathrm{mod},0} = \frac{0.83\cdot R_0}{d_{\max}(90^\circ)} = \frac{0.83\cdot 0.6\cdot laf_{uw}}{0.209\cdot 92^{-0.32}t} = \frac{0.498\cdot laf_{uw}}{0.049174\cdot t} = 10.14\frac{laf_{uw}}{t}$$

If this modification is applied to the stiffness related to longitudinally loaded welds, the stiffness of the Timoshenko beam element at angles different by 90° will be different by the previously defined one. Namely it will be

$$K^*_{s,\,\mathrm{mod}} = K_{s,\,\mathrm{mod},0}\cdot\cos^2\theta + K_{s90}\cdot\sin^2\theta$$

The curve $K_{s,mod}*(\theta)$ is now different by $K_s(\theta)$, as the stiffnesses are higher (Figure 6.17). However, the forces flowing into the welds having $\theta \neq 90^\circ$ are much nearer to the exact ones, and are exact for the longitudinally loaded welds, having $\theta = 0^\circ$.

The secant stiffness can be computed also for torsion and for bending of a single weld. The numerical factors are similar for torsion and bending, as the fillet is loaded transversely.

Considering torsion, at the extremities of the weld $p = 1$ and $d = d_{max}$. If half weld is considered, and the result doubled, the maximum resistance torque would be obtained by

$$M = 2\int_0^{l/2} x\cdot 1.5\cdot 0.6\cdot f_{uw}\cdot[p\cdot(1.9-0.9p)]^{0.3}\, adx$$

The abscissa x and the strain p are related by

$$p = \frac{2x}{l}$$

So, the integral can be rewritten as

$$M = \frac{al^2}{2}0.9\cdot f_{uw}\cdot\int_0^1 p[p\cdot(1.9-0.9p)]^{0.3}dp = \frac{al^2}{2}0.9\cdot f_{uw}\cdot 0.466 = 0.2097\cdot al^2 f_{uw}$$

It must be noted that this moment is much higher than that which would be predicted by a linear stress–strain curve and assuming $1.5\cdot 0.6\cdot f_u$ as maximum stress. That would be

$$M_{elastic} = \frac{al^2}{6}\cdot 1.5\cdot 0.6\cdot f_{uw} = 0.15\cdot al^2 f_{uw}$$

and it is 1.43 times lower. This depends on the shape of the curve $h(p)$, which is strongly non-linear.

Now, dividing the moment by the rotation at limit, the secant stiffness gives

$$K_{rw} = \frac{0.2097\cdot al^2 f_{uw}}{\dfrac{0.049174\cdot t}{l/2}} = \frac{0.2097}{2\cdot 0.049174}\cdot\frac{al^3 f_{uw}}{t} = 2.13\cdot\frac{al^3 f_{uw}}{t}$$

This is the secant stiffness of the weld seam considered as a whole. If the weld seam is divided into segments, each having the transverse and normal secant stiffness

$$K_v = K_w = 18.3 \cdot \frac{alf_{uw}}{t}$$

the bending secant stiffness found would have been simply

$$K_{rw} = K_w \cdot \frac{l^2}{12} = 1.52 \cdot \frac{al^3 f_{uw}}{t}$$

Up to now the secant stiffnesses of a single fillet weld have been found. It is also useful to consider the *initial* stiffness, that is, the stiffnesses at the very beginning of the loading, when the displacements are null or very low. As can be clearly seen, this is infinite as the curve

$$h(p) = [p(1.9 - 0.9p)]^{0.3}$$

has a derivative

$$h'(p) = 0.3[p(1.9 - 0.9p)]^{-0.7}(1.9 - 1.8p)$$

which is infinite for $p = 0$. The issue can be solved assuming a very small variation of p. For $p = 0.001$, $d = d_{max}/1000$. The following two stiffnesses K_{ini} can be found for $\theta = 0°$ and for $\theta = 90°$:

$$K_{ini,0} = \frac{h(0.001) \cdot 0.6 \cdot f_{uw} \cdot la}{0.001 \cdot 0.1674 \cdot t} = \frac{0.1526 \cdot 0.6 \cdot f_{uw} \cdot la}{0.001 \cdot 0.1674 \cdot t} = 546 \cdot \frac{laf_{uw}}{t}$$

$$K_{ini,90} = \frac{h(0.001) \cdot 1.5 \cdot 0.6 \cdot f_{uw} \cdot la}{0.001 \cdot 0.049174 \cdot t} = \frac{0.1526 \cdot 1.5 \cdot 0.6 \cdot f_{uw} \cdot la}{0.001 \cdot 0.049174 \cdot t} = 2793 \cdot \frac{laf_{uw}}{t}$$

If in the initial range $p \leq 0.001$ a linear behavior is assumed (i.e. $h(p) = 152.6 \cdot p$), then the bending and torsional stiffnesses of a single fillet weld can be obtained by multiplying the translational $\theta = 90°$ stiffness by $l^2/12$.

It is interesting to compare the utilization ratios that would be obtained by using a secant stiffness model with modified longitudinal stiffness, with those obtained by the exact analysis, considering a layout of welds made of

- "flange" welds of area A_f and thickness t_f
- "web" welds of area A_w and thickness t_w

and assuming that the web welds are perpendicular to the flange welds and that $t_f \geq t_w$. The aim is to demonstrate that the secant stiffness with modification almost always predicts safe utilizations, that is, higher than the correct ones. This has some importance in linear pure fem approaches, when the welds are modeled as segments using a linearized secant stiffness. If the weld layout is computed as a whole by means of the (intrinsically non-linear) instantaneous center of rotation method (Section 6.7.4.1), the utilization is instead predicted correctly.

Three loading conditions will be analyzed:

1) force perpendicular to the faying plane, i.e. acting at $\theta = 90°$ for all welds
2) force parallel to the flanges, i.e. acting with $\theta = 0°$ for the flange welds and with $\theta = 90°$ for the web welds
3) force parallel to the web, i.e. acting with $\theta = 90°$ for the flange welds and with $\theta = 0°$ for the web welds

Force normal to the faying plane – Let F_{lim} be the maximum force allowed. It is assumed that the web welds are at the limit first (as $t_f > t_w$). Then

$$p_w = 1 = \frac{d}{0.049 \cdot t_w}$$

$$p_f = \frac{d}{0.049 \cdot t_f} = \frac{0.049 \cdot t_w}{0.049 \cdot t_f} = \frac{t_w}{t_f}$$

$$h_w = 1$$

$$h_f = \left[p_f \cdot \left(1.9 - 0.9 p_f\right)\right]^{0.3}$$

$$F_{\lim} = 0.6 \cdot 1.5 \cdot f_{uw} \cdot \left(A_w + h_f A_f\right)$$

The secant stiffness of the web fillets is

$$K_w = \frac{18.3 A_w f_{uw}}{t_w}$$

and that of the flange fillets is

$$K_f = \frac{18.3 A_f f_{uw}}{t_f}$$

If F is the total force applied, the force flowing in the web fillets is

$$F_w = \frac{A_w t_f}{A_w t_f + A_f t_w} F$$

The correct utilization ratio is

$$U = \frac{F}{F_{\lim}}$$

while the predicted one would be

$$U_p = \frac{F_w}{0.6 \cdot 1.5 \cdot A_w \cdot f_{uw}}$$

If the ratio between the areas α is introduced,

$$\alpha = \frac{A_w}{A_f}$$

and the ratio between the thicknesses, Δ

$$\Delta = \frac{t_w}{t_f} \le 1$$

then it turns out that

$$r_N = \frac{U_p}{U} = \frac{\alpha + h(\Delta)}{\alpha + \Delta}$$

The curve $r_N(\alpha)$ for several Δ is plotted in Figure 6.18. As can be seen, the ratio is always greater than 1, and not so far from 1 if Δ is about 1.

Force parallel to flanges – Here we have

$$p_w = 1 = \frac{d}{0.049 \cdot t_w}$$

$$p_f = \frac{0.049 \cdot t_w}{0.1674 \cdot t_f} = 0.2927 \cdot \Delta$$

$$h_w = 1$$

$$h_f = \left[p_f \cdot \left(1.9 - 0.9 p_f\right)\right]^{0.3}$$

$$F_{\lim} = 0.6 \cdot 1.5 \cdot f_{uw} \cdot A_w + 0.6 \cdot f_{uw} \cdot h\left(p_f\right) \cdot A_f$$

The secant stiffness of the web fillets is

$$K_w = \frac{18.3 A_w f_{uw}}{t_w}$$

and that of the flange fillets is

$$K_f = \frac{10.14 \cdot A_f f_{uw}}{t_f}$$

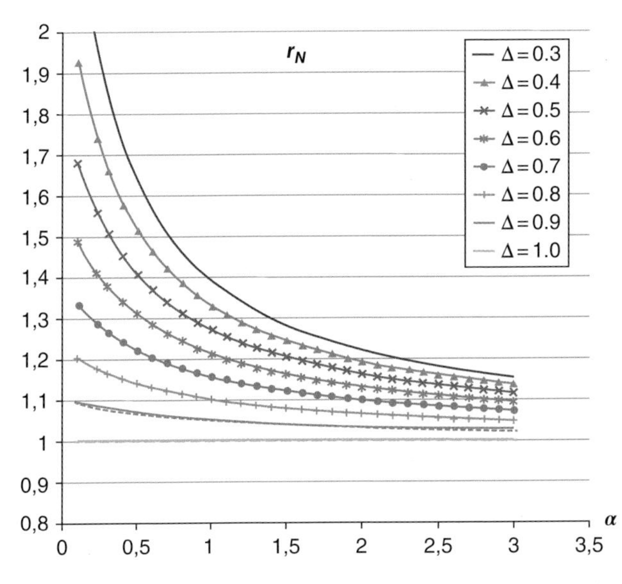

Figure 6.18 Ratio predicted to exact utilization ratios, using a linearized modified secant stiffness for the welds. Load perpendicular to faying plane.

If F is the total force applied, the force flowing in the web fillets is

$$F_w = \frac{18.3 \cdot A_w t_f}{18.3 \cdot A_w t_f + 10.14 \cdot A_f t_w} F = \frac{18.3}{18.3 + 10.14 \alpha \Delta} F$$

The correct utilization ratio is

$$U = \frac{F}{F_{\text{lim}}}$$

while the predicted one would be

$$U_p = \frac{F_w}{0.6 \cdot 1.5 \cdot A_w \cdot f_{uw}}$$

so

$$r_{par,f} = \frac{U_p}{U} = \frac{18.3\alpha + 12.2 h_f}{18.3\alpha + 10.14\Delta}$$

The curve $r_{par,f}(\alpha)$ for several Δ is plotted in Figure 6.19. As can be seen, the ratio is always greater than 1 if $\Delta < 1$, and not so far from 1 if Δ is about 1.

Force parallel to web – Here it is not always certain that the limit load will be reached first in the web welds. This happens if

$$t_w < \frac{0.049}{0.164} t_f \cong 0.3 \cdot t_f$$

which seems a rare condition. Analyzing then the case when the limit load is reached first in the flange welds (loaded transversely)

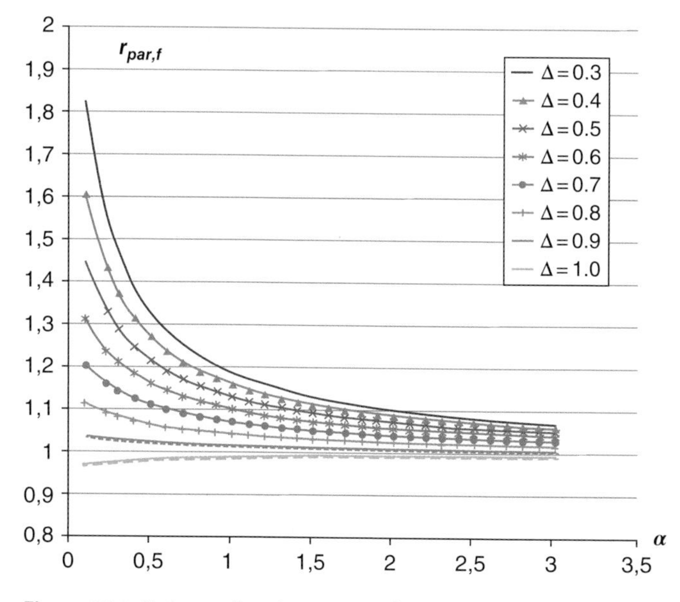

Figure 6.19 Ratio predicted to exact utilization ratios, using a linearized modified secant stiffness for the welds. Load parallel to flange.

$$p_f = 1 = \frac{d}{0.049 \cdot t_f}$$

$$p_w = \frac{0.049 \cdot t_f}{0.1674 \cdot t_w} = 0.2927 \cdot \frac{1}{\Delta}$$

$$h_f = 1$$

$$h_w = [p_w \cdot (1.9 - 0.9 p_w)]^{0.3}$$

$$F_{\text{lim}} = 0.6 \cdot 1.5 \cdot f_{uw} \cdot A_f + 0.6 \cdot f_{uw} \cdot h(p_w) \cdot A_w$$

The secant stiffness of the flange fillets is

$$K_f = \frac{18.3 A_f f_{uw}}{t_f}$$

and that of the web fillets is

$$K_w = \frac{10.14 \cdot A_w f_{uw}}{t_w}$$

If F is the total force applied, the force flowing in the web fillets (where the predicted utilization will be highest) is

$$F_w = \frac{10.14 \cdot A_w t_f}{18.3 \cdot A_f t_w + 10.14 \cdot A_w t_f} F = \frac{10.14}{18.3 \frac{\Delta}{\alpha} + 10.14} F$$

The correct utilization ratio is

$$U = \frac{F}{F_{\text{lim}}}$$

while the predicted one would be

$$U_p = \frac{F_w}{0.6 \cdot A_w \cdot f_{uw}}$$

so

$$r_{par,w} = \frac{U_p}{U} = \frac{\frac{15.21}{\alpha} + 10.14 \cdot h_w}{18.3 \cdot \frac{\Delta}{\alpha} + 10.14} = \frac{15.21 + 10.14 \cdot \alpha \cdot h_w}{18.3 \cdot \Delta + 10.14 \cdot \alpha}$$

The curve $r_{par,w}(\alpha)$ for several Δ is plotted in Figure 6.20. As can be seen, the ratio is not always greater than 1: only if $\Delta < 0.7$ is the ratio almost always around 1. If Δ is 1 the ratio of the utilizations is about 0.8, which means that the utilization ratio would be under predicted by 20%. This is the worst case.

This kind of utilization would be predicted by fillet welds individually modeled using weld segments in a pure fem (PFEM) model, in linear range, assuming the same displacement field (i.e. rigid) implicitly assumed by the instantaneous center of rotation method. The issue will be covered in Chapter 10.

The results presented briefly in this section will be useful in order to set up a computational model for welds.

The experimental works have been carried out only on fillet welds having an angle between the active faces of 90° (angle γ). The author is not aware of studies related to

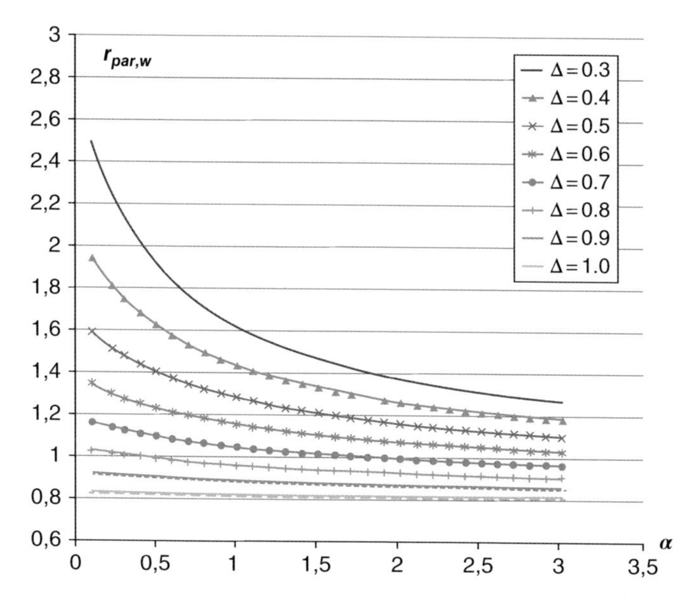

Figure 6.20 Ratio predicted to exact utilization ratios, using a linearized modified secant stiffness for the welds. Load parallel to web.

the variation of stiffness as a function of the angle γ. For this reason, a number of computational models have been set up, with the aim for estimating the variation of the stiffness of fillet welds with the angle between the fusion faces.

The results obtained in the linear range are then extended to the secant stiffnesses. In this way, a complete model able to predict the stiffness of weld seams as a function of θ and γ will be available, and can be used to set the stiffness matrix of equivalent Timoshenko-like elements, related to single weld seams.

In the next section these numerical tests will be presented briefly.

6.7.2 Numerical Tests of Fillet Welds in the Linear Range

A set of different fillet welds has been modeled under typical loading. The thickness, length, and angle between active faces have been considered in a wide range, covering most of the applications. The following values have been used:

- •Thickness: $t = 5, 7.5, 10$ mm
- Ratio between length of fillets and thickness: $r = 5, 10, 15, 20, 40$
- Angle between active faces: $\gamma = 60°, 75°, 90°, 105°, 120°$

So, the total number of models run is $3 \times 5 \times 5 = 75$.

The models use brick eight noded elements and wedge six nodes elements (Figure 6.21). Two symmetrical fillet welds have been considered, to avoid asymmetries. The loads are applied by means of a stiff plate always 10 mm thick, also modeled by brick elements. In this way, no effects related to the deformation of the plate are embedded in the results. The nodes laying on the (horizontal) faying plane are clamped. A row of elements is deleted from the plate, to avoid the contact of the stiff plate with the clamped

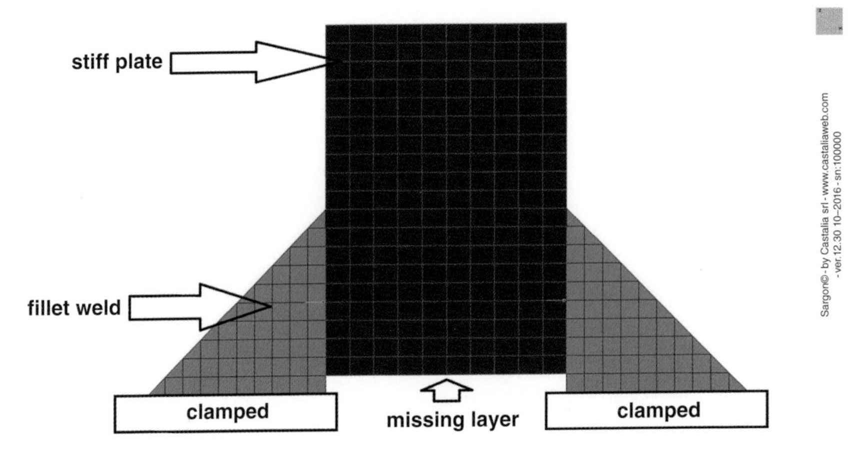

Figure 6.21 Typical cross-section of models ($\gamma = 90°$).

nodes, which would nullify the stresses. The number of degrees of freedom (dof) is the same for all the models, and it's 19530 (rotational dofs are constrained). The number of nodes is 6978.

The material used is S235, having $E = 210{,}000$ MPa, and Poisson's ratio 0.3. Typical models are shown in Figures 6.22 and 6.23. Models are identified by the following names:

2 W_t_r_T_γ

Where:

- 2 W is an identification string, fixed
- t is the thickness of fillet welds in mm
- r is the ratio between length and thickness t
- T is the thickness of the plate in mm (10)
- γ is the angle between active faces in degrees.

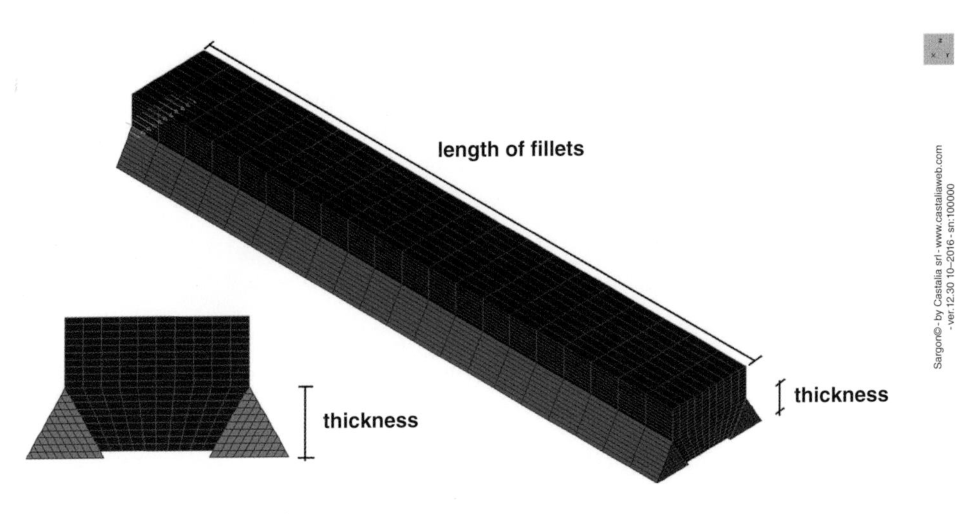

Figure 6.22 Typical model for $\gamma < 90°$.

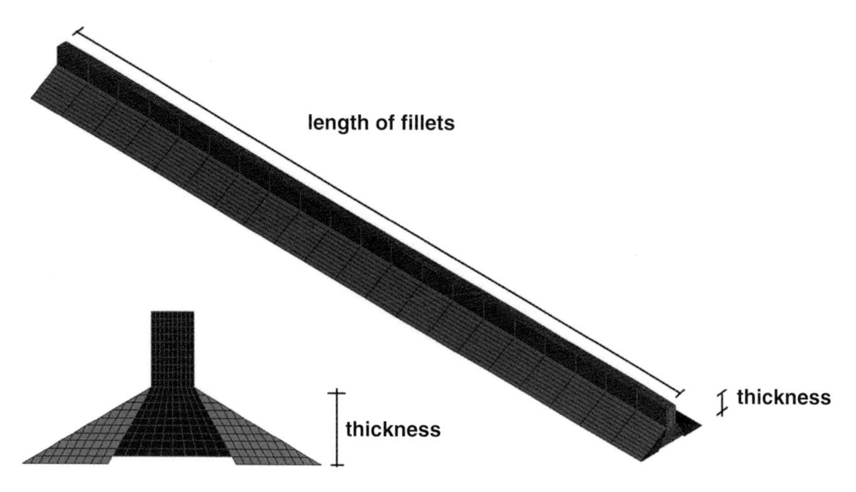

Figure 6.23 Typical model for $\gamma > 90°$.

In order to study the stiffness of the fillets, two loading conditions have been applied. The loads have been distributed evenly in order to avoid local effects. The very high stiffness of the plate imposes a rigid body deformation. So, the typical deformation of the welds can be studied by:

- a force normal to the faying plane, to simulate the "axial" behavior of the Timoshenko beam element; this is also valid for transverse loading ($\theta = 90°$)
- a longitudinal force, applied at nodes placed at an height equal to $t/2$, i.e. 2.5, 3.75, or 5 mm depending on the models.

All the forces are nodal forces. The reason why the forces have been applied at the height $t/2$ of the plate is that the length of the equivalent Timoshenko beam, L, has been set equal to $t/2$. This should be intended as a reasonable amount, at mid height of the fillet.

In order to evaluate the stiffness, and to identify the relevant term of the Timoshenko stiffness matrix, the nodal displacements of pertinent nodes have been read and processed. The values identified are A, A_u, and A_v. So, the equivalent beam properties EA/L, GA_u/L, G_vA_v/L have been identified. As expected, these quantities depend on a number of parameters, including the thickness t, the length to thickness ratio r, and the angle γ between the active faces.

In the following subsections, results will be presented one by one, and discussed.

6.7.2.1 Axial Behavior

This is studied by applying a vertical force to the plate, having resultant of 1 N. The vertical displacement d_w is read in the model (Figure 6.24), and by this a stiffness

$$k_N = \frac{EA}{L} = \frac{1}{d_w}$$

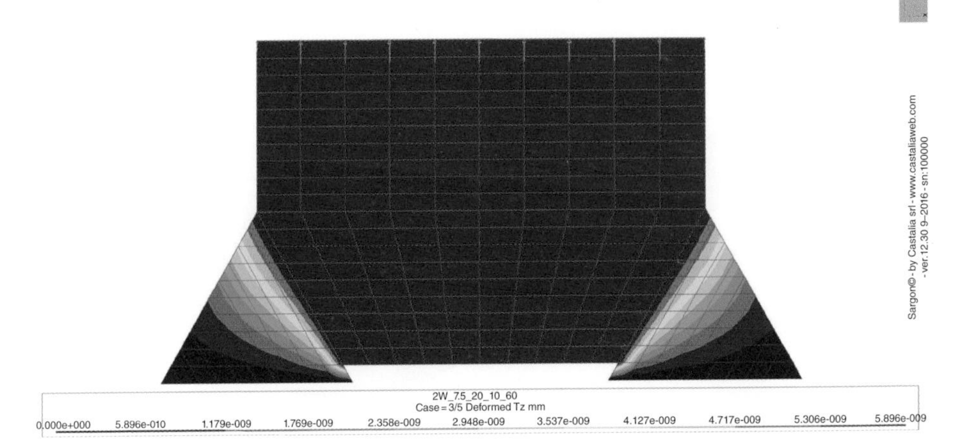

Figure 6.24 A typical vertical displacement map. Model 2W_7.5_20_10_60.

This stiffness is then compared to the Timoshenko stiffness of

$$k_{Nn} = \frac{EA_n}{L}$$

where A_n is the nominal area and, considering that there are two fillets,

$$E = 210,000\,\mathrm{MPa}$$

$$A_n = 2 \cdot a \cdot r \cdot t$$

$$L = 0.5 \cdot t$$

the throat size a is computed using Equation 6.1. The comparison between the stiffnesses is done by computing the ratio

$$q = \frac{A}{A_n}$$

so that for each model, the target axial modulus A, can be computed as

$$A = qA_n$$

Results are independent of the thickness of the stiff plate T (and numerical tests have confirmed). At fixed γ, they are also independent of the thickness t, and the ratio r. Considering the curve $q = q(\gamma)$ (Figure 6.25), a clear linear law can be drawn, which enables us to find a general formula for q.

In fact, the following interpolation formula can be used with very good agreement:

$$A = [2 - 1.13 \cdot \gamma / 100] \cdot A_n$$

where γ is measured in degrees and it's normally between 60 and 120.

Now if the correction factor C_w to be applied to the axial stiffness valid for $\gamma = 90°$ is needed, this can be evaluated as

$$C_w = C_v = \frac{[2 - 1.13 \cdot \gamma / 100]}{0.983} = 2.035 - 1.15 \cdot \gamma / 100$$

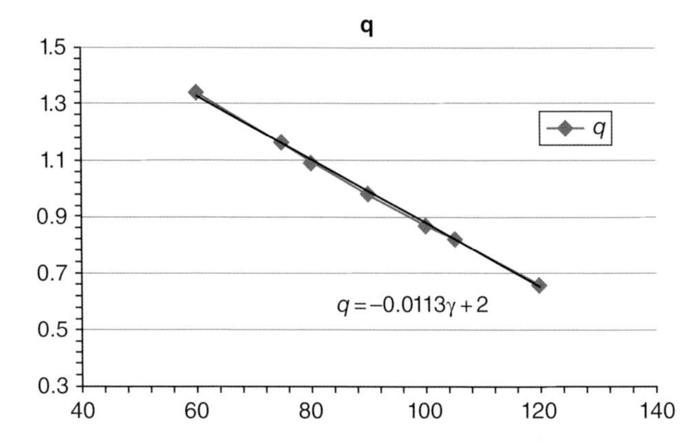

Figure 6.25 Curve $q = q(\gamma)$: a clear linear law can be detected. Also the points at $\gamma = 80$ and $\gamma = 100$ have been used.

The same factor, named C_v, can also be applied to the transverse stiffness, that is, to the "GA_v/L" stiffness, as $\theta = 90°$.

6.7.2.2 Longitudinal Shear Behavior

Considering a Timoshenko beam element clamped at the first node extremity, and loaded by a transverse force at the free tip, second node, the following expressions can be obtained for the displacement d_u and rotation r_v at the tip:

$$F = \frac{GA_u}{L} d_u - \frac{GA_u}{2} r_v$$

$$0 = -\frac{GA_u}{2} d_u + \left(\frac{EJ_v}{L} + \frac{GA_u L}{3}\right) \cdot r_v$$

From these relationships, the following stiffness can be obtained, setting $F = 1$:

$$\frac{GA_u}{L} = \frac{1}{\left(d_u - \frac{L}{2} r_v\right)}$$

$$\frac{EJ_v}{L} = \frac{GA_u}{L} \cdot \left(\frac{L d_u}{2 r_v} - \frac{L^2}{3}\right)$$

Once the displacement d_u and the rotation r_u are read from the models, an appropriate estimate of the stiffness can be obtained, particularly of A_u and J_v. The longitudinal displacement d_u is read in a node of the plate, on the symmetry plane, at $t/2$ from the faying plane. The rotation r_u is computed considering the vertical displacement at the end of the plate (in the longitudinal direction), and dividing by $0.5rt$, that is, half the fillets length.

The nominal values of the area and second area moment are

$$A_{un} = 2 \cdot a \cdot r \cdot t$$

$$J_{vn} = \frac{2}{12} a (r \cdot t)^3$$

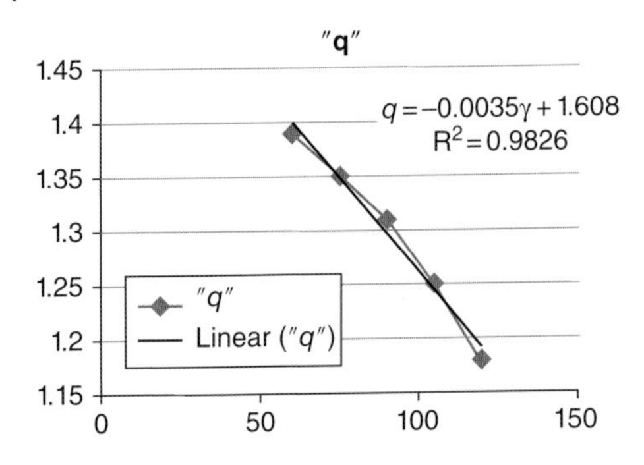

Figure 6.26 Curve $q = q(\gamma)$ for longitudinally sheared fillet welds. The tendency linear line equation is also plotted, with the R^2 value.

It is now possible to define the ratio

$$q = \frac{A_u}{A_{un}}$$

and consider its variation depending on the thickness t, the ratio between fillet length, and thickness r and angle between the fusion faces γ.

Analyzing the results, it turns out that q is not very dependent on the thickness t, and that it is slightly dependent on the ratio r between fillet length and fillet thickness t. For a high r, 40, a slight increase in the values of q is noted. The maximum variation of the ratio is about 2.5% for q. So, within reasonable limit of precision, it can be assumed that q is constant with r. The most important variation is related to γ, the angle between the fusion faces. The stiffness decreases when the angle increases, so the fillet welds having $\gamma = 60°$ are much more rigid than those having $\gamma = 120°$.

Plotting the average q values as a function of γ, the curve shown in Figure 6.26. is found. This curve may well be assumed linear, with slight errors (the R^2 value is also plotted, according to the Excel^{TM} rules).

So, it can be set as

$$A_u = [1.61 - 0.35 \cdot \gamma / 100] \cdot A_{un}$$

If once more the factor C_u to be applied to the 90° stiffness is needed, this can be evaluated as

$$C_u = \frac{[1.61 - 0.35 \cdot \gamma / 100]}{1.295} = 1.243 - 0.27 \cdot \gamma / 100$$

6.7.3 The Stiffness Matrix of a Single Fillet Weld

6.7.3.1 Orthotropic Model

Taking into account what was explained in the previous sections, the secant stiffness matrix of a single fillet weld can be expressed using the expressions already used for the penetration weld layouts. However, the following substitutions will have to be used, if the orthotropic behavior of fillets is considered:

$$\begin{aligned}
&\frac{EA_i}{L_i} \Rightarrow k_{Ni} \Rightarrow 18.3\frac{A_i f_u}{t_i} C_{wi}\mu_{ni} \\
&\frac{GA_{ui}}{L_i} \Rightarrow k_{Vui} \Rightarrow \frac{3.58 \cdot A_i f_u}{t_i} C_{ui}\mu_{ti} \\
&\frac{GA_{vi}}{L_i} \Rightarrow k_{Vvi} \Rightarrow \frac{18.3 \cdot A_i f_u}{t_i} C_{vi}\mu_{ti}\mu_{li} \\
&\frac{EJ_{ui}}{L_i} \Rightarrow k_{Mui} \Rightarrow \frac{1.52 \cdot A_i\, a_i^2 f_u}{t_i} C_{wi}\mu_{ni} \\
&\frac{EJ_{vi}}{L_i} \Rightarrow k_{Mvi} \Rightarrow \frac{1.52 \cdot A_i\, l_i^2 f_u}{t_i} C_{wi}\mu_{ni} \\
&\frac{GJ_{ti}}{L_i} \Rightarrow k_{Ti} \Rightarrow 1.52 \cdot A_i\, \frac{l_i^2 f_u}{t_i} C_{vi}\mu_{ti}\mu_{li}
\end{aligned} \tag{6.51}$$

If the modified secant stiffness is to be used for the longitudinal shears, then

$$\frac{GA_{ui}}{L_i} \Rightarrow k_{Vui} \Rightarrow \frac{10.14 \cdot A_i f_u}{t_i} C_{ui}\mu_{ti}$$

These stiffnesses also take into account the factors μ_n, μ_t, and μ_l needed for the special flags SO, NS, and LO.

If the initial stiffness is needed, it can be obtained by using these different values:

$$\begin{aligned}
&\frac{EA_i}{L_i} \Rightarrow k_{Ni} \Rightarrow 2793\frac{A_i f_u}{t_i} C_{wi}\mu_{ni} \\
&\frac{GA_{ui}}{L_i} \Rightarrow k_{Vui} \Rightarrow \frac{546 \cdot A_i f_u}{t_i} C_{ui}\mu_{ti} \\
&\frac{GA_{vi}}{L_i} \Rightarrow k_{Vvi} \Rightarrow \frac{2793 \cdot A_i f_u}{t_i} C_{vi}\mu_{ti}\mu_{li} \\
&\frac{EJ_{ui}}{L_i} \Rightarrow k_{Mui} \Rightarrow \frac{232 \cdot A_i\, a_i^2 f_u}{t_i} C_{wi}\mu_{ni} \\
&\frac{EJ_{vi}}{L_i} \Rightarrow k_{Mvi} \Rightarrow \frac{232 \cdot A_i\, l_i^2 f_u}{t_i} C_{wi}\mu_{ni} \\
&\frac{GJ_{ti}}{L_i} \Rightarrow k_{Ti} \Rightarrow 232 \cdot A_i\, \frac{l_i^2 f_u}{t_i} C_{vi}\mu_{ti}\mu_{li}
\end{aligned} \tag{6.52}$$

By using this method, the stiffness matrix of a single weld seam can be created and later assembled in the usual way.

6.7.3.2 Isotropic Model

If instead of the orthotropic model we use the isotropic model, then the following substitutions will have to be used (0.298 = 3.58/12):

$$
\begin{aligned}
&\frac{EA_i}{L_i} \Rightarrow k_{Ni} \Rightarrow 3.58\frac{A_i f_u}{t_i} C_{wi}\mu_{ni} \\
&\frac{GA_{ui}}{L_i} \Rightarrow k_{Vui} \Rightarrow \frac{3.58 \cdot A_i f_u}{t_i} C_{ui}\mu_{ti} \\
&\frac{GA_{vi}}{L_i} \Rightarrow k_{Vvi} \Rightarrow \frac{3.58 \cdot A_i f_u}{t_i} C_{vi}\mu_{ti}\mu_{li} \\
&\frac{EJ_{ui}}{L_i} \Rightarrow k_{Mui} \Rightarrow \frac{0.298 \cdot A_i\, a_i^2 f_u}{t_i} C_{wi}\mu_{ni} \\
&\frac{EJ_{vi}}{L_i} \Rightarrow k_{Mvi} \Rightarrow \frac{0.298 \cdot A_i\, l_i^2 f_u}{t_i} C_{wi}\mu_{ni} \\
&\frac{GJ_{ti}}{L_i} \Rightarrow k_{Ti} \Rightarrow 0.298 \cdot A_i \frac{l_i^2 f_u}{t_i} C_{vi}\mu_{ti}\mu_{li}
\end{aligned}
\tag{6.51}
$$

If the initial stiffness is needed, it can be obtained by using these different values:

$$
\begin{aligned}
&\frac{EA_i}{L_i} \Rightarrow k_{Ni} \Rightarrow 546\frac{A_i f_u}{t_i} C_{wi}\mu_{ni} \\
&\frac{GA_{ui}}{L_i} \Rightarrow k_{Vui} \Rightarrow \frac{546 \cdot A_i f_u}{t_i} C_{ui}\mu_{ti} \\
&\frac{GA_{vi}}{L_i} \Rightarrow k_{Vvi} \Rightarrow \frac{546 \cdot A_i f_u}{t_i} C_{vi}\mu_{ti}\mu_{li} \\
&\frac{EJ_{ui}}{L_i} \Rightarrow k_{Mui} \Rightarrow \frac{45.3 \cdot A_i\, a_i^2 f_u}{t_i} C_{wi}\mu_{ni} \\
&\frac{EJ_{vi}}{L_i} \Rightarrow k_{Mvi} \Rightarrow \frac{45.3 A_i\, l_i^2 f_u}{t_i} C_{wi}\mu_{ni} \\
&\frac{GJ_{ti}}{L_i} \Rightarrow k_{Ti} \Rightarrow 45.3 \cdot A_i \frac{l_i^2 f_u}{t_i} C_{vi}\mu_{ti}\mu_{li}
\end{aligned}
\tag{6.52}
$$

6.7.4 Instantaneous Center of Rotation Method in 3D

6.7.4.1 Normal Fillet Welds

The linear method of evaluating the shear forces in the welds has been proved by experiments to be too conservative, and not well related to the experimental results. In particular, the isotropic model does not take into account the difference of stiffness related to the relative direction of loading of the welds (longitudinal or transverse).

So the instantaneous center of rotation method (ICRM) has been proposed for fillet welds, which correctly takes into account both the difference in the transverse and longitudinal behavior and the non-linear stress–strain relationship.

The ICRM was initially used only for in-plane eccentricity: for $\boldsymbol{\sigma}_{\mathbf{ip}} = \{0, V_u, V_v, M_w, 0, 0\}^T$. Later, after the work of Gomez et al. 2008, the method was extended to out-of-plane eccentricity, that is, to stress states $\boldsymbol{\sigma}_{\mathbf{op}} = \{N, 0, 0, 0, M_u, M_v\}^T$.

Considering first the in-plane eccentricity, the method assumes that the displacement of each point of each weld is obtained by multiplying the radius connecting that point to a unique point named the instantaneous center of rotation (IC) times a suitable rotation r_w (Figure 6.27).

This is merely a different way to consider the kinematics of the rigid rotation assumed for the weld group effective area. If the displacements of a generic point i are expressed as a function of the displacement of the centroid, d_u and d_v, plus the effect of the rotation r_w, a subset of the Equations 6.26 is found:

$$d_{ui} = d_u - r_w v_i$$

$$d_{vi} = d_v + r_w u_i$$

By imposing null displacements for the point i, the coordinates of the IC are readily found:

$$u_{IC} = -\frac{d_v}{r_w}$$

$$v_{IC} = \frac{d_u}{r_w}$$

This allows the replacement of the centroid displacements d_u and d_v by the IC point coordinates:

$$d_{ui} = -r_w(v_i - v_{IC})$$

$$d_{vi} = r_w(u_i - u_{IC})$$

So, the two descriptions are equivalent. The first one uses the displacements of the centroid and the rotation, while the second one uses the coordinates of the IC and the rotation.

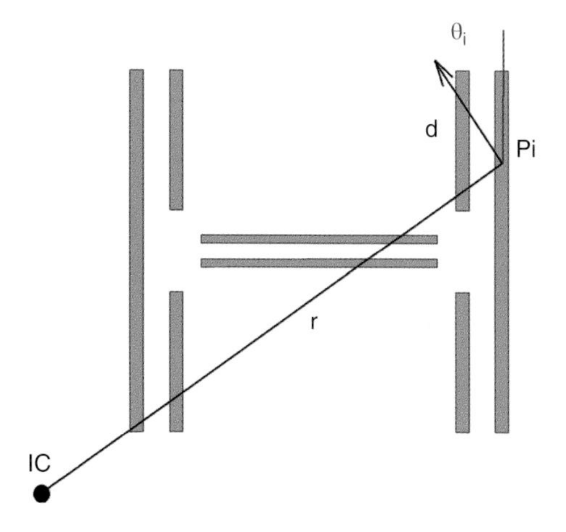

Figure 6.27 The ICRM method.

The radius connecting each point i to the IC has a length r equal to

$$r = \sqrt{(u_i - u_{IC})^2 + (v_i - v_{IC})^2}$$

and the amount of displacement of a point i is

$$d = r \cdot r_w$$

The displacement vector is perpendicular to the radius ($\mathbf{P_i}$-$\mathbf{IC}$) and forms an angle θ_i with the weld axis to which the point $\mathbf{P_i}$ belongs (Figure 6.27). The angle θ_i can be found (in radians) by

$$\theta_i = \frac{\pi}{2} - \arccos\left[\frac{(\mathbf{P_i} - \mathbf{IC}) \cdot \mathbf{ui}}{r}\right]$$

The problem is that depending on the angle θ_i, a different limit for the displacement $d_{max,i}$ will be found (see the third Equation 6.50), so the measure of the strain

$$p_i = \frac{d_i(u_{IC}, v_{IC}, r_w)}{d_{\max}(\theta_i)}$$

that is needed to compute the stress related to d, is a complex function of the IC position and of r_w.

To solve the problem, an iterative solution is needed. For in-plane eccentricity, Brandt (1982) proposed an algorithm with a computer program written in Fortran, which solves the problem finding a suitable position for the IC.

When considering the analysis of a weld layout in a general 3D context, it is not possible to rely on the assumption that the eccentricity is only in plane or only out of plane. In order to use the ICRM method, a general solution needs to be found, and an iterative procedure must be adapted to a more general framework.

In general, the displacement components of single weld points will be three, d_{wi}, d_{ui}, d_{vi}, and the rotation of the whole section will not be only r_w, but also r_u and r_v. The kinematic description of the act of motion can still use the IC concept, but the act of motion is now a rotation with IC center plus a translation parallel to the rotation axis $\{r_w, r_u, r_v\}^T$; each infinitesimal rigid body motion can be reduced to a helical motion. So the IC point will not have null displacements, but will move in the direction of the rotation axis, by a displacement $\{gr_w, gr_u, gr_v\}^T$ where g is a suitable constant.

The displacements of a generic point $\mathbf{P_i}$ can be expressed in the usual way as a function of the centroid displacements d_w, d_u, d_v and the rotations r_w, r_u, r_v about the principal axes of the weld layout, or it can be expressed as a function of g, the coordinates of the IC u_{IC} and v_{IC}, and the rotations r_w, r_u, r_v.

This can be shown as follows. The displacements of a generic point $\mathbf{P_i}$ are

$$d_{wi} = d_w + v_i r_u - u_i r_v$$

$$d_{ui} = d_u - v_i r_w$$

$$d_{vi} = d_v + u_i r_w$$

At the IC, which is the intersection of the rotation axis to the weld plane $w = 0$, the displacements are directed as the vector $\mathbf{r}$

$$gr_w = d_w + v_{IC} r_u - u_{IC} r_v$$

$$gr_u = d_u - v_{IC} r_w$$

$$gr_v = d_v + u_{IC} r_w$$

so that

$$u_{IC} = \frac{gr_v - d_v}{r_w}$$

$$v_{IC} = \frac{-gr_u + d_u}{r_w}$$

$$g = \frac{r_v d_v + r_u d_u + r_w d_w}{r_w^2 + r_u^2 + r_v^2} = \frac{\mathbf{r} \cdot \mathbf{d}}{\mathbf{r} \cdot \mathbf{r}}$$

Once more, the two descriptions of the kinematics of the weld group effective area are equivalent. Finding the coordinates of the IC, however, is not enough and is needed, in general, to determine the limit condition for the weld group. This is because there is a part of the displacement governed by g, which is not dependent on the IC position. Moreover, if $r_w = 0$, that is, for bending with no torsion, the IC position is undetermined. The problem is usually solved by a change of reference system, and keeping the problem 2D.

However, what is needed in general is not the IC position, but a suitable set of displacements of the centroid $\mathbf{d} = \{d_w, d_u, d_v, r_w, r_u, r_v\}^T$ which imply internal forces balanced with the externally applied loads. So, the name of the method "instantaneous center of rotation", related to the geometrical interpretation for 2D problems, may perhaps be changed into "non-linear method for welds": we can see that, at the very base of both methods, is a kinematic assumption and a stress–strain relationship, not the instantaneous center concept.

The incremental-iterative procedure to find a configuration for the displacements balanced with the externally applied load is not easy, as the non-linearity is very strong (Figures 6.28 and 6.29). The curve $h(p)$ has infinite slope for $p = 0$, and has a wide range where the derivative is almost null, around $p = 1$ (Figure 6.28).

To solve this problem, the general tools available for non-linear analysis should be used (e.g. Crisfield 1991). Here a simplified procedure will be outlined, but more advanced and refined procedures can also be set up, using automatic stepping, and more.

The automatic procedure outlined here has been set up and tested for a good number of general loading conditions (i.e. loading vector $\boldsymbol{\sigma} = \{N, V_u, V_v, M_w, M_u, M_v\}^T$ having both in-plane and out-of-plane eccentricity) and weld geometries, with success.

First of all, as the function $h(p)$ of Equation 6.50 has vertical slope for $p = 0$, and a horizontal one for $p = 1$, the following replacement is used:

$$h(p) = 152.6p \qquad p \le 0.001$$

$$h(p) = [p \cdot (1.9 - 0.9p)]^{0.3} \quad 0.001 < p \le 0.95$$

$$h(p) = 1.0503357p \qquad p > 0.95$$

This modification is useful as it introduces a linear branch in p for very small strains, and avoids the decrease of loads at increasing strain p. The problem is, however, non-linear

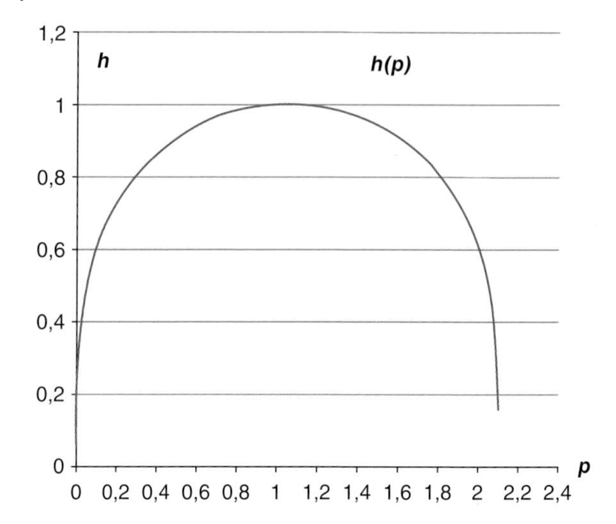

Figure 6.28 unmodified curve $h(p)$.

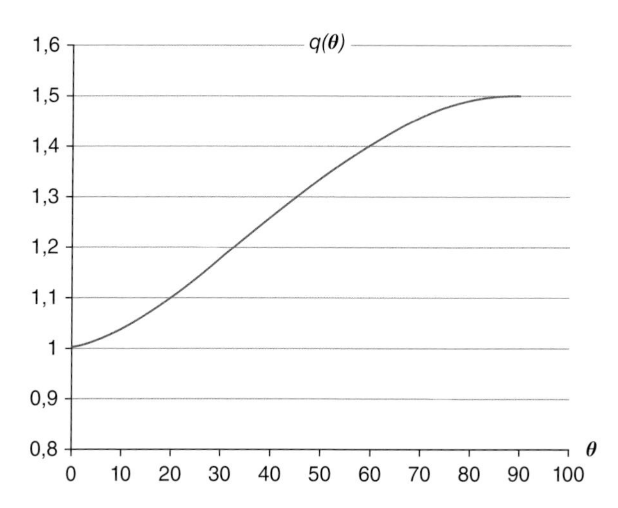

Figure 6.29 The curve $q(\theta)$.

also in this initial branch, as the dependency on θ is also non-linear ($q = q(\theta)$ – see Figure 6.29). The linear branch is abandoned when the strain reaches 0.001 and the stress is at 0.1526 of its limiting value, at equal angle θ.

For the moment, it will be assumed that all the fillets have a 90° fusion faces angle, and that they are all fully resistant FR with no modifications (*normal* fillet welds).

If a displacement vector $\mathbf{d_0} = \{d_{w0}, d_{u0}, d_{v0}, r_{w0}, r_{u0}, r_{w0}\}^T$ of the centroid of a weld layout is known, referred to the weld layout principal axes reference system (**w**, **u**, **v**), then the internal forces due to the displacements, $\boldsymbol{\sigma}_0$, in the principal reference system, can be computed as follows.

A loop on the welds will add a single contribution to a total sum:

$$\boldsymbol{\sigma}_0 = \sum_{i=0}^{nw} \boldsymbol{\sigma}_{0i}$$

The problem is now to compute the contribution of each single weld. To do that, the displacements of the single weld centroid $\boldsymbol{d}_{0i}$, in the local system of the weld seam (**wi**, **ui**, **vi**), are computed as

$$\mathbf{d}_{0i} = \mathbf{Q}_i \mathbf{d}_0 = \begin{vmatrix} \mathbf{T}_i & \mathbf{S}_i \\ \mathbf{0} & \mathbf{T}_i \end{vmatrix} \cdot \mathbf{d}_0$$

The forces and moments in the reference system of the weld seam are then computed by means of Equation 6.50, as follows. Dividing the weld seam i into n_{si} segments, or elements, the displacements of the centroid of each segment at a distance u_{ij} from the weld center (u_{ij} is comprised between $-l_i/2$ and $+ l_i/2$), is (Figure 6.30)

$$\mathbf{d}_{0ij} = \begin{vmatrix} d_{0wij} \\ d_{0uij} \\ d_{0vij} \end{vmatrix} = \begin{vmatrix} d_{0wi} - u_{ij} \cdot r_{0vi} \\ d_{0ui} \\ d_{0vi} + r_{0wi} \cdot u_{ij} \end{vmatrix} = \begin{vmatrix} 1 & 0 & 0 & 0 & 0 & -u_{ij} \\ 0 & 1 & 0 & 0 & 0 & 0 \\ 0 & 0 & 1 & u_{ij} & 0 & 0 \end{vmatrix} \cdot \mathbf{d}_{0i} = \mathbf{S}_{ij} \cdot \mathbf{d}_{0i}$$

The dot product of $\mathbf{d}_{0ij}$ with the unit vector of axis **ui** is merely d_{0ui}, and so

$$\theta_{0ij} = \arccos\left(\frac{d_{0ui}}{\left\|\mathbf{d}_{0ij}\right\|}\right)$$

and is measured in degrees. If $\theta > 90°$, θ is set equal to $(180° - \theta)$. The force F_{0ij} exerted by the weld segment j is

$$F_{0ij} = 0.6 \cdot f_{uw} \cdot h(p_{0ij}) \cdot q(\theta_{0ij}) \cdot \frac{l_i a_i}{n_{si}} = 0.6 \cdot f_{uw} \cdot \left[p_{0ij} \cdot (1.9 - 0.9 p_{0ij})\right]^{0.3} \cdot \left[1 + 0.5 \sin^{1.5}(\theta_{0ij})\right] \cdot \frac{l_i a_i}{n_{si}}$$

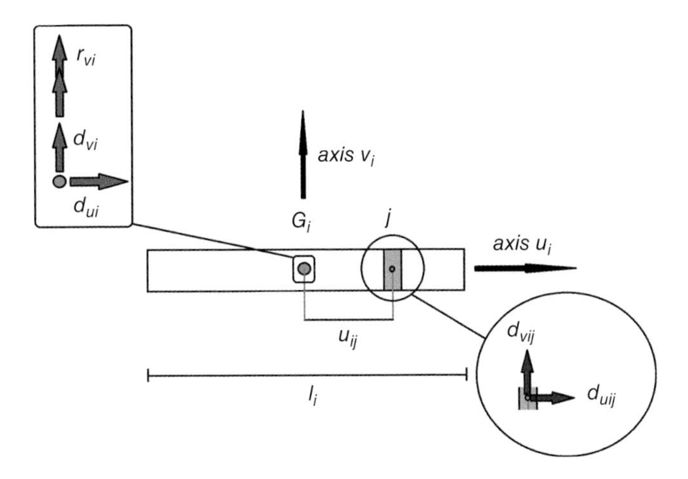

Figure 6.30 The displacements of a weld segment, j.

where

$$p_{0ij} = \frac{\|\mathbf{d_{0ij}}\|}{d_{\max}(\theta_{0ij})} = \frac{\sqrt{d_{0wij}^2 + d_{0uij}^2 + d_{0vij}^2}}{0.209\cdot(\theta_{0ij}+2)^{-0.32}\cdot t_i}$$

The forces and moments acting over the single weld seam i, in its local reference system (**wi, ui, vi**), $\boldsymbol{\sigma}_{\mathbf{0i,loc}}$, are now the sum of the contributions of all the segments:

$$\boldsymbol{\sigma}_{\mathbf{0i,\,loc}} = \sum_{j=1}^{nsi} \begin{vmatrix} F_{0wij} \\ F_{0uij} \\ F_{0vij} \\ u_{ij}F_{0vij} \\ 0 \\ -u_{ij}F_{0wij} \end{vmatrix} = \sum_{j=1}^{nsi} \begin{vmatrix} 1 & 0 & 0 \\ 0 & 1 & 0 \\ 0 & 0 & 1 \\ 0 & 0 & u_{ij} \\ 0 & 0 & 0 \\ -u_{ij} & 0 & 0 \end{vmatrix} \cdot \begin{vmatrix} F_{0wij} \\ F_{0uij} \\ F_{0vij} \end{vmatrix} = \sum_{j=1}^{nsi} \frac{F_{0ij}}{\|\mathbf{d_{0ij}}\|} \mathbf{S_{ij}^T} \cdot \mathbf{d_{0ij}} = \sum_{j=1}^{nsi} \frac{F_{0ij}}{\|\mathbf{d_{0ij}}\|} \mathbf{S_{ij}^T S_{ij} d_{0i}}$$

The number of segments should be a good compromise between speed of execution and precision. The procedure outlined has used a number of segments computed considering a reference length of each segment equal to 1 mm.

If these forces are moved to the layout center and expressed in the global layout reference system (**w**, **u**, **v**) the contribution $\boldsymbol{\sigma}_{\mathbf{0i}}$ is found:

$$\boldsymbol{\sigma}_{\mathbf{0i}} = \mathbf{Q_i^T} \boldsymbol{\sigma}_{\mathbf{0i,\,loc}}$$

and finally, as mentioned

$$\boldsymbol{\sigma}_{\mathbf{0}} = \sum_{i=0}^{nw} \boldsymbol{\sigma}_{\mathbf{0i}} = \sum_{i=1}^{nw} \left[\mathbf{Q_i^T} \cdot \left(\sum_{j=1}^{nsj} \frac{F_{0ij}}{|\mathbf{d_{0ij}}|} \mathbf{S_{ij}^T S_{ij}} \right) \cdot \mathbf{Q_i} \right] \cdot \mathbf{d_0}$$

The activity described allows the computation of an internally balanced force vector $\boldsymbol{\sigma}_{\mathbf{0}}$, related to a displacement vector of the centroid of the weld layout, $\mathbf{d_0}$, and can be symbolically described as

$$\boldsymbol{\sigma}_{\mathbf{0}} = \mathbf{K_{ls}}(\mathbf{d_o}) \cdot \mathbf{d_0} = \Theta(\mathbf{d_0})$$

where $\mathbf{K_{ls}}(\mathbf{d})$ is the secant stiffness matrix of the weld layout, a function of the displacement vector, $\mathbf{d}$. Given a displacement vector $\mathbf{d}$, the forces $\boldsymbol{\sigma}$ balanced to it are evaluated by the $\Theta()$ operator, which is a shorter way to express the operations needed to evaluate the layout forces and moments $\boldsymbol{\sigma}$ balanced to a given layout displacement vector $\mathbf{d}$.

This is one of the two workhorses of the non-linear analysis of fillet weld layouts. The second is the computation of the (6×6) tangent stiffness matrix $\mathbf{K_{ltk}}$, as a function of an existing generic displacement vector $\mathbf{d_k}$.

If it is assumed that at the displacement level $\mathbf{d_k}$ the balanced forces are $\boldsymbol{\sigma}_{\mathbf{k}}$, that is,

$$\boldsymbol{\sigma}_{\mathbf{k}} = \Theta(\mathbf{d_k})$$

If the internal force vector $\boldsymbol{\sigma}_{\mathbf{kl}}$ is found by a displacement vector $\mathbf{d_k} + \boldsymbol{\delta}\mathbf{d_{kl}}$, that is if,

$$\boldsymbol{\sigma}_{\mathbf{kl}} = \Theta(\mathbf{d_k} + \boldsymbol{\delta}\mathbf{d_{kl}})$$

then the increase in the internal force vector is

$$\delta\boldsymbol{\sigma}_{\mathbf{kl}} = \boldsymbol{\sigma}_{\mathbf{kl}} - \boldsymbol{\sigma}_{\mathbf{k}} = \Theta(\mathbf{d}_{\mathbf{k}} + \boldsymbol{\delta}\mathbf{d}_{\mathbf{kl}}) - \Theta(\mathbf{d}_{\mathbf{k}})$$

This can be written shortly as

$$\boldsymbol{\delta\sigma}_{\mathbf{kl}} = \mathbf{K}_{\mathbf{ltk}}\boldsymbol{\delta}\mathbf{d}_{\mathbf{kl}}$$

if it is used the tangent stiffness matrix $\mathbf{K}_{\mathbf{ltk}}$. In Appendix 2 we derive the closed form of this stiffness matrix, which requires many analytical manipulations (Equation A2.2). In the following, this matrix is obtained by a different, easier approach, derived by its definition; the increase of the internal forces $\boldsymbol{\delta\sigma}_{\mathbf{kl}}$ can be computed increasing one displacement component at a time, as it is now explained.

If a generic component l of the displacement vector $\mathbf{d}_{\mathbf{k}}$ is null (or a number so small to be compared to null), its increment should be a suitable small number, such as 10^{-6}. If it is not null its increment is a small fraction of its value, for instance, 1/1000. Assuming that all six the displacements are not null, the following increment vectors $\boldsymbol{\delta}\mathbf{d}_{\mathbf{kl}}$ would be used:

$$\boldsymbol{\delta}\mathbf{d}_{\mathbf{k1}} = \{0.001{\cdot}d_{wk},\ 0,\ 0,\ 0,\ 0,\ 0\}^T$$

$$\boldsymbol{\delta}\mathbf{d}_{\mathbf{k2}} = \{0,\ 0.001{\cdot}d_{uk},\ 0,\ 0,\ 0,\ 0\}^T$$

$$\dots$$

$$\boldsymbol{\delta}\mathbf{d}_{\mathbf{k6}} = \{0,\ 0,\ 0,\ 0,\ 0,\ 0.001{\cdot}r_{vk}\}^T$$

and

$$\boldsymbol{\delta\sigma}_{\mathbf{kl}} = \Theta(\mathbf{d}_{\mathbf{k}} + \boldsymbol{\delta}\mathbf{d}_{\mathbf{kl}}) - \Theta(\mathbf{d}_{\mathbf{k}})$$

Each internal force increment vector $\boldsymbol{\delta\sigma}_{\mathbf{kj}}$, divided by the pertinent displacement increment, i.e. for instance 0.001 d_{wk} for $\boldsymbol{\delta\sigma}_{\mathbf{k1}}$, is a column of the stiffness matrix needed. So, if all the $\mathbf{d}_{\mathbf{k}}$ components are not null, the matrix $\mathbf{K}_{\mathbf{lk}}$ is formed by adding the columns from 1 to 6 to it:

$$\mathbf{K}_{\mathbf{lk}} = \left| \frac{1000\boldsymbol{\delta\sigma}_{\mathbf{k1}}}{d_{kw}} \quad \frac{1000\boldsymbol{\delta\sigma}_{\mathbf{k2}}}{d_{ku}} \quad \frac{1000\boldsymbol{\delta\sigma}_{\mathbf{k3}}}{d_{kv}} \quad \frac{1000\boldsymbol{\delta\sigma}_{\mathbf{k4}}}{r_{kw}} \quad \frac{1000\boldsymbol{\delta\sigma}_{\mathbf{k5}}}{r_{ku}} \quad \frac{1000\boldsymbol{\delta\sigma}_{\mathbf{k6}}}{r_{kv}} \right|$$

The activity carried on is needed in order to evaluate the tangent stiffness matrix of the weld layout at a given displacement vector $\mathbf{d}_{\mathbf{k}}$, and can be symbolically described as

$$\mathbf{K}_{\mathbf{ltk}} = \mathbf{K}_{\mathbf{ltk}}(\mathbf{d}_{\mathbf{k}})$$

Now that these two fundamental steps have been described, we can outline the incremental-iterative procedure needed to compute in non-linear range a weld layout under the effect of an externally applied load $\boldsymbol{\sigma}$, expressed with reference to the principal axes of the weld layout itself.

Two results are interesting:

- The weld stresses balanced with the applied load $\boldsymbol{\sigma}$, if an equilibrium condition is possible.
- The load multiplier α_L, which takes the weld layout to its limit $\alpha_L\boldsymbol{\sigma}$.

The utilization factor U of the weld layout will then be

$$U = \frac{1}{\alpha_L}$$

In order to get the multiplier α_L, the loads applied must be scaled so as to reach the limit for the weld layout. This limit is considered numerically reached when one segment j of one weld i reaches a p value higher than 0.9. As the $h(p)$ curve is very flat around $p = 1$, this assumption does not imply a significant loss in the load carrying capacity estimate. For $p = 0.9$ $\text{h}(p) = 0.994$, which is practically 1.0.

A first rough estimate of the utilization related to the applied load $\boldsymbol{\sigma}$ is needed. To do that, and considering the principal reference system (**w**, **u**, **v**) for the whole weld layout, the maximum distance of the weld seams extremities from the center of the weld layout r_{max}, and the maximum u and v, u_{max} and v_{max} are needed. These are all positive numbers. A rough upper estimate of the utilization can be obtained assuming a reference maximum force F_{max} equal to

$$F_{\max} = 1.5 \cdot 0.6 \cdot A f_{uw}$$

where A is the weld layout area. So the rough estimate of U can be

$$U_{rough} = \frac{|N|}{F_{\max}} + \frac{|V_u|}{F_{\max}} + \frac{|V_u|}{F_{\max}} + \frac{|M_w|}{F_{\max} r_{\max}} + \frac{|M_u|}{F_{\max} v_{\max}} + \frac{|M_v|}{F_{\max} u_{\max}}$$

The next step is to assume a reasonable precision ε for the needed utilization U. A good value can be $\varepsilon = 0.025$, i.e. 2.5%. This is also the chosen step size, not too large due to the very strong non-linearity. The ε value can be chosen appropriately, depending on the needs.

Now, the number of steps n_{sP}, related to the applied load $\boldsymbol{\sigma}$ can be evaluated as

$$n_{sP} = rnd\left(\frac{U_{rough}}{\varepsilon}\right)$$

If n_{sP} is not 0, then the fraction of the applied load related to one single step, α_{ss} can be evaluated as

$$\alpha_{ss} = \frac{1}{n_{sP}}$$

and the maximum number of steps required to surely reach the limit can be guessed as

$$n_{sMax} = rnd\left(\frac{1.5 n_{sP}}{U_{rough}}\right)$$

Usually the incremental process will be stopped well before having applied all these steps. At the step n_{sP}, the load will be equal to the applied one ($n_{sP} \cdot \alpha_{ss} = 1$).

If n_{sP} is zero, then

$$n_{sP} = 1$$

$$\alpha_{ss} = \frac{\varepsilon}{U_{rough}}$$

$$n_{sMax} = \text{rnd}\left(\frac{1.5}{\varepsilon}\right)$$

This condition means that the applied load is very low (its U_{rough} is estimated lower than ε), and the minimum step considered is higher than the applied loads.

Now, a factor per single step α_{ss} and a maximum number of steps n_{sMax} are known. An incremental-iterative procedure is then started. Initially the load applied is null. At each step k, a fraction $\alpha_{ss}\,\boldsymbol{\sigma}$ of the load is added, and the displacement $\mathbf{d_k}$ related to that level of applied forces is searched for; searching for the correct $\mathbf{d_k}$ is the aim of the iterative procedure, within a step. The index of the step is k, and ideally goes from 1 to n_{sMax}. However, if at the end of a step $k = n_{Lim}$ one weld segment has reached a p value equal to or higher than 0.9, the procedure is stopped and the utilization is evaluated as

$$U = \frac{1}{(n_{Lim}-1)\cdot\alpha_{ss}} = \frac{n_{sP}}{(n_{Lim}-1)}$$

The procedure is also stopped if during the iterative procedure inside one step k no convergence has been reached, that is, if no $\mathbf{d_k}$ has been found so that

$$k\alpha_{ss}\boldsymbol{\sigma} \approx \Theta(\mathbf{d_k})$$

after a suitable number of iterations. However, if the convergence has not been reached and the maximum p_{ijk} of weld segments is higher than 0.9, then the procedure is stopped considering that is has reached the limit load, and computing U as before.

If the convergence is reached, remaining below the limit strain, then at the end of a step k a new tangential stiffness matrix of the weld layout is computed,

$$\mathbf{K_{ltk}} = \mathbf{K_{ltk}}(\mathbf{d_k})$$

and this will be kept constant during the iterations of the next step.

Each step k begins with an initial guess for the displacement increment

$$\delta\mathbf{d_{k1}} = \mathbf{K_{ltk-1}^{-1}}\Delta\boldsymbol{\sigma}_{\mathbf{k1}}$$

At the very beginning of a step, the unbalanced force vector is just

$$\Delta\boldsymbol{\sigma}_{\mathbf{k1}} = \alpha_{ss}\boldsymbol{\sigma}$$

Then the new displacement estimate is computed as

$$\mathbf{d_{k1}} = \mathbf{d_{k-1}} + \boldsymbol{\delta}\mathbf{d_{k1}}$$

By using this displacement estimate, a new internal force vector is computed as

$$\boldsymbol{\sigma}_{\mathbf{k1}} = \Theta(\mathbf{d_{k1}})$$

and the unbalanced force vector is formed, checking for the tolerance:

$$\Delta\boldsymbol{\sigma}_{\mathbf{k2}} = k\alpha_{ss}\boldsymbol{\sigma} - \boldsymbol{\sigma}_{\mathbf{k1}}$$

$$\frac{\|\Delta\boldsymbol{\sigma}_{\mathbf{k2}}\|}{\alpha_{ss}\|\boldsymbol{\sigma}\|} < tol$$

where tol is a suitable tolerance, such as 0.005.

If the tolerance is not reached, then a new iteration is made:

$$\delta\mathbf{d_{k2}} = \mathbf{K_{ltk-1}^{-1}}\Delta\boldsymbol{\sigma}_{\mathbf{k2}}$$

$$\mathbf{d_{k2}} = \mathbf{d_{k1}} + \boldsymbol{\delta}\mathbf{d_{k2}}$$

$$\boldsymbol{\sigma}_{\mathbf{k2}} = \Theta(\mathbf{d_{k2}})$$

$$\Delta \boldsymbol{\sigma}_{\mathbf{k3}} = k\alpha_{ss}\boldsymbol{\sigma} - \boldsymbol{\sigma}_{\mathbf{k2}}$$

$$\frac{\|\Delta \boldsymbol{\sigma}_{\mathbf{k3}}\|}{\alpha_{ss}\|\boldsymbol{\sigma}\|} < tol$$

and so on with (m is an iteration index)

$$\delta \mathbf{d}_{\mathbf{km}} = \mathbf{K}_{\mathbf{ltk-1}}^{-1} \Delta \boldsymbol{\sigma}_{\mathbf{km}}$$

$$\mathbf{d}_{\mathbf{km}} = \mathbf{d}_{\mathbf{km-1}} + \boldsymbol{\delta}\mathbf{d}_{\mathbf{km}}$$

$$\boldsymbol{\sigma}_{\mathbf{km}} = \Theta(\mathbf{d}_{\mathbf{km}})$$

$$\Delta \boldsymbol{\sigma}_{\mathbf{km+1}} = k\alpha_{ss}\boldsymbol{\sigma} - \boldsymbol{\sigma}_{\mathbf{km}}$$

$$\frac{\|\Delta \boldsymbol{\sigma}_{\mathbf{km+1}}\|}{\alpha_{ss}\|\boldsymbol{\sigma}\|} < tol$$

At the very beginning of the whole process, the (6 × 6) stiffness matrix of the weld layout $\mathbf{K}_{\mathbf{11,mod}}$ has to be used, not the secant stiffnesses of the weld seams, but their initial stiffnesses. However, it would not be efficient to use that stiffness matrix, tangent to the null displacement condition as the stiffness matrix for the whole first step. Instead, at the end of the first iteration of the first step, the tangent stiffness matrix of step 1 can be formed as

$$\mathbf{K}_{\mathbf{1}} = \mathbf{K}_{\mathbf{1}}(\delta \mathbf{d}_{\mathbf{11}})$$

that is, using the first displacement estimate computed by the very first iteration. This is useful when a complex mix of shears and moments is applied; otherwise a lack of convergence can be detected on the very first step.

During the incremental process, when the step considered is $k = n_{sP}$, the stresses in the welds balanced with the applied loads are stored, because they are the response of the weld layout to the applied loads, $\boldsymbol{\sigma}$. Usually this is not the limit condition for the weld

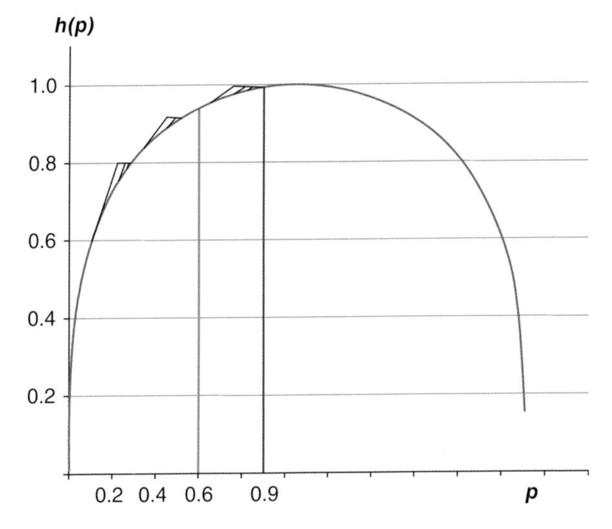

Figure 6.31 Schematic representation of the incremental-iterative process. It is implicitly assumed that all the welds are equal and equally loaded. The true curve for generic configurations does not follow $h(p)$ or $q(\theta)$.

layout. In the very special case $n_{sL} = 1$, after having found $n_{sL} = 0$ at the initial guess, the stresses stored are not in general the ones related to the loads applied, but are slightly different.

In order to avoid possible problems related to the lowering of the stiffness to zero near to the limit load, if at the end of a step the maximum value of p is higher than a given limit, say 0.6, the stiffness matrix is not updated anymore. This avoids the problems related to null determinant that would otherwise be obtained, but this is at the cost of an increased number of iterations needed to bring convergence.

When applying the standards, the reference ultimate stress f_u used in this section can be obtained by applying proper reduction factors, such as for LRFD of AISC:

$$f_{uw} = \phi \cdot f_{uweld} = 0.75 \cdot f_{uweld}$$

6.7.4.2 Modified Fillet Welds

Up to now it has been assumed that the welds all have a 90° angle between the fusion faces, and that no modification is applied to them, that is, a flag like shear only (SO) or longitudinal only (LO). Indeed, as the non-linear method already takes into account the difference in stiffness, ductility, and resistance depending on the loading, the need for such modification factors seems less important. However, a generalization able to include such effects can be outlined as follows.

If the modification factors μ_{ni}, μ_{ti}, μ_{li} are used, then the local displacements can be obtained by the following modified rules:

$$\mathbf{d_{0ij}} = \begin{vmatrix} d_{0ijw} \\ d_{0iju} \\ d_{0ijv} \end{vmatrix} = \begin{vmatrix} \mu_{ni} d_{0wi} - \mu_{ni} u_{ij} \cdot r_{0vi} \\ \mu_{ti} d_{0ui} \\ \mu_{ti} \mu_{li} d_{0vi} + \mu_{ti} \mu_{li} r_{0wi} \cdot u_{ij} \end{vmatrix} = \begin{vmatrix} \mu_{ni} & 0 & 0 & 0 & 0 & -\mu_{ni} u_{ij} \\ 0 & \mu_{ti} & 0 & 0 & 0 & 0 \\ 0 & 0 & \mu_{ti}\mu_{li} & \mu_{ti}\mu_{li} u_{ij} & 0 & 0 \end{vmatrix} \cdot \mathbf{d_{0i}}$$

This implies a different way of computing the strain and so the stresses. As the rule is used to generate the forces $\boldsymbol{\sigma} = \Theta(\mathbf{d})$, and then the tangent stiffness, the only part which needs to be modified is that one.

Due to the high first derivative of the curve $h = h(p)$ (Figure 6.28), we may need to decrease the μ value used to set the modification factors, to a value of the order of 10^{-6}, when the ICRM is used. Otherwise, very small values of p will lead to non-negligible unwanted forces in the welds.

Considering now the welds having angle between the fusion faces not equal to 90°, it has been found by elastic finite element analyses that the elastic stiffness is strongly related to that angle. The variation of the throat size with the angle γ between the fusion faces is not enough to take this effect into consideration, at least in the linear range. No clear mention of the issue has been found in the literature. So, it is questionable how the non-linear method outlined in this section should be modified to consider this effect. The problem is not academic: in 3D analyses inclined welds with angles γ different from 90° can easily be found (Figure 6.32).

If the difference in stiffness due to the angle γ is not considered in weld stress computation, the forces flowing in the welds having angle γ greater than 90°, usually weaker and less stiff, will be overestimated, while the forces flowing in the welds having γ lower than 90° will be underestimated. This is on the safe side, as the throat size of the former is lower than that of the latter, at equal thickness.

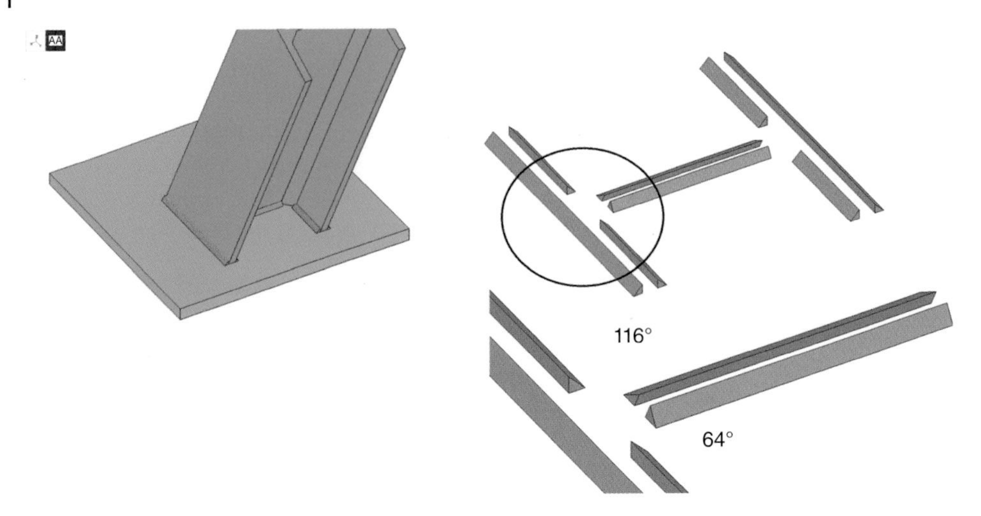

Figure 6.32 Weld layout with fillets having different γ. An inclined element is considered.

One way to solve the issue would be to factor the strain by the factors C_w, C_u, and $C_v = C_w$, as already done with the factors μ, leaving the maximum stress only as a function of p and θ with no modification. This would imply altering both p and θ. However, the maximum stresses would not be changed, and the same functions $h(p)$ and $q(\theta)$ would be used, that is

$$\mathbf{d_{0ij}} = \begin{vmatrix} d_{0ijw} \\ d_{0iju} \\ d_{0ijv} \end{vmatrix} = \begin{vmatrix} C_w\left(\mu_{ni}d_{0wi} - \mu_{ni}u_{ij}\cdot r_{0vi}\right) \\ C_u\mu_{ti}d_{0ui} \\ C_v\left(\mu_{ti}\mu_{li}d_{0vi} + \mu_{ti}\mu_{li}r_{0wi}\cdot u_{ij}\right) \end{vmatrix} = \begin{vmatrix} C_w\mu_{ni} & 0 & 0 & 0 & 0 & -C_w\mu_{ni}u_{ij} \\ 0 & C_u\mu_{ti} & 0 & 0 & 0 & 0 \\ 0 & 0 & C_v\mu_{ti}\mu_{li} & C_v\mu_{ti}\mu_{li}u_{ij} & 0 & 0 \end{vmatrix} \cdot \mathbf{d_{0i}}$$

Tests carried out using this approach, and considering the non-linear method outlined in this section, have not shown convergence problems, and they lead to reasonable results, not far from those obtained when not considering the γ angle effect.

For instance, a column HEB 200 clamped at the bottom and skewed in both directions has been tested (Figure 6.33). The member axis vector is $\mathbf{v_1}$ = (2000 mm, 2000 mm, 4000 mm), which means an angle of about 30° in both directions. The fillet weld layout used to fix the member to the base plate has eight weld seams, with γ angles ranging from 63.4° to 116.6°. Different thicknesses have been used for the web welds and the flange welds. The ICRM has been applied both with and without the correction related to the C_w, C_u, and C_v coefficients, getting the utilization factors summarized in Table 6.5. Twelve loading conditions have been tested, using the six elementary internal forces of the member alone, at 0.3 of the elastic cross-section limits, and a mix of axial force and bending moments, so as to test the weld layout under a generic stress state.

As can be seen, the effect of the introduction of the C coefficients is to increase the utilization ratios, in this case of about 5–6%. As the angles γ are near the limit allowed by the standards, this example would suggest that the effect related to the angle γ is low.

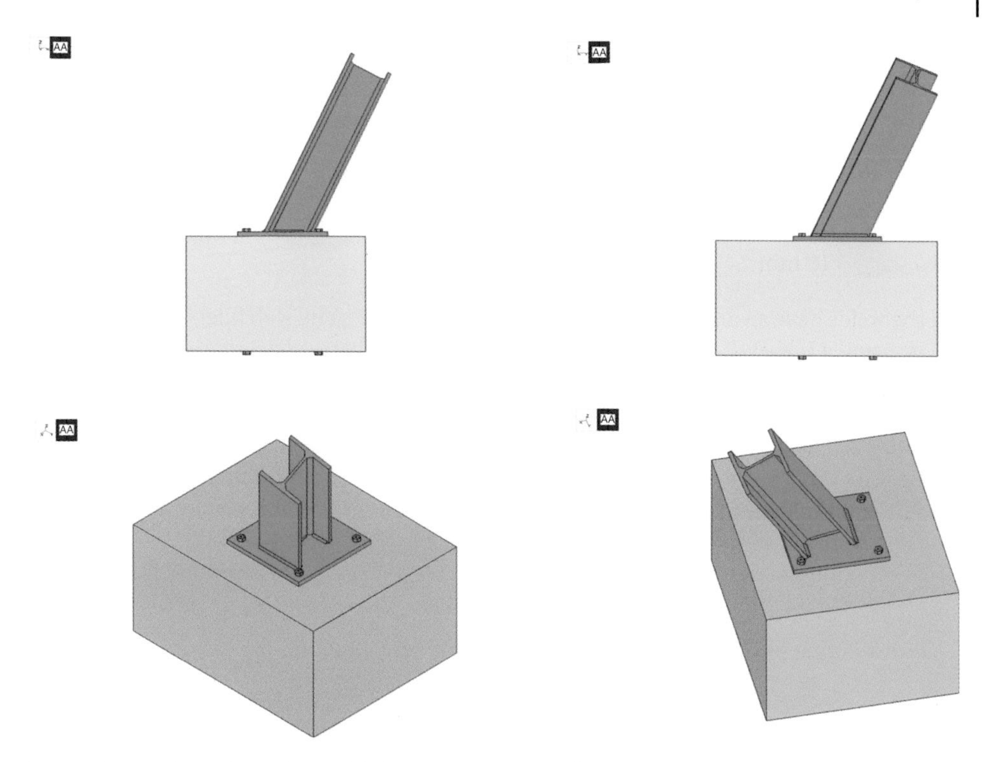

Figure 6.33 A skewed member welded to a base plate. The fillet welds have γ angle ranging from 63.4° to 116.6°.

Table 6.5 Comparison between the utilization factors computed by ICRM in 3D, with and without the effect due to the angle between the fusion faces (C_w, C_u, and C_v coefficients).

Load	Utilization factors not using C_w, C_u, C_v	Utilization factors using C_w, C_u, C_v	Difference %
1	0.407	0.429	5.4
2	0.474	0.500	5.5
3	0.143	0.148	3.5
4	0.017	0.018	5.9
5	0.303	0.312	3.0
6	0.200	0.211	5.5
7	0.309	0.321	3.9
8	0.286	0.291	1.7
9	0.246	0.259	5.3
10	0.246	0.259	5.3
11	0.286	0.291	1.7
12	0.309	0.321	3.9

6.7.4.3 Examples

A weld layout applied to a HEB 200 cross-section is considered (Figure 6.34). The HEB cross-section has the following sizes:

- $h_{HEB\ 200} = b_{HEB\ 200} = 200$ mm
- $t_{f.HEB\ 200} = 15$ mm
- $t_{w.HEB\ 200} = 9.0$ mm
- $r_{HEB\ 200} = 18$ mm

All the welds have an angle between the fusion faces of 90°. The web fillets have a leg t_w = 4.5 mm and a throat a_w = 3.2 mm, while the flange fillets have a leg size t_f = 7.5 mm and a throat a_f = 5.3 mm. An ultimate stress f_u = 500 MPa is assumed, which is around 70 ksi. The weld layout is referred to its principal axes (**w**, **u**, **v**) as in Figure 6.34. The computations are according to AISC LRFD, so a factor $\phi = 0.75$ is used.

The limit force in the **w** direction is found as follows, considering that the first fillets to reach the ultimate strain are the thinner ones. When they reach their limit, $p = 1$, their displacement is

$$0.049 \cdot 4.5 = 0.2205 \text{ mm}$$

The same value of displacement is related to a lower value of p for the thicker welds:

$$p_{flange} = \frac{0.2205}{0.049 \cdot 7.5} = 0.6$$

So while at limit the web fillets are loaded by their maximum stress, the flange fillets are at a fraction equal to

$$h(0.6) = [0.6 \cdot (1.9 - 0.9 \cdot 0.6)]^{0.3} = 0.94$$

It is now possible to compute the limit force as

$$N_{\max} = 1.5 \cdot 0.6 \cdot 0.75 \cdot 500 \cdot (355 \cdot 2 \cdot 5.3 \cdot 0.94 + 134 \cdot 2 \cdot 3.2) = 1{,}483{,}251 \text{ N}$$

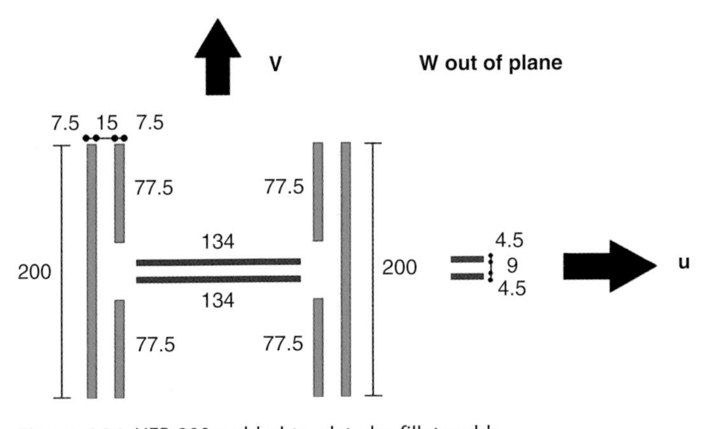

Figure 6.34 HEB 200 welded to plate by fillet welds.

Considering now the force parallel to the web, **u** direction, the fillets loaded transversely will be the flange ones. At their limit the displacement will be

$$0.049 \cdot 7.5 = 0.3675\,\text{mm}$$

Considering now the fillets loaded longitudinally, their p value is

$$p = \frac{0.3675}{0.1674 \cdot 4.5} = 0.4878$$

so that the fraction of limit stress is

$$h(0.4878) = [0.4878 \cdot (1.9 - 0.9 \cdot 0.4878)]^{0.3} = 0.903$$

and the force is now computed as

$$V_{u,\max} = 0.6 \cdot 0.75 \cdot 500 \cdot (355 \cdot 2 \cdot 5.3 \cdot 1.5 + 134 \cdot 2 \cdot 3.2 \cdot 0.903) = 1,444,255\,\text{N}$$

Finally, considering a force perpendicular to the web, direction **v**, the web fillets are loaded transversely, while the flange fillets are loaded longitudinally, so

$$0.049 \cdot 4.5 = 0.2205\,\text{mm}$$

$$p = \frac{0.2205}{0.1674 \cdot 7.5} = 0.1756$$

$$h(0.1756) = [0.1756 \cdot (1.9 - 0.9 \cdot 0.1756)]^{0.3} = 0.700$$

$$V_{v,\max} = 0.6 \cdot 0.75 \cdot 500 \cdot (134 \cdot 2 \cdot 3.2 \cdot 1.5 + 355 \cdot 2 \cdot 5.3 \cdot 0.700) = 882,112\,\text{N}$$

The incremental-iterative method described in this section has been tested by loading the weld layout concentrically and applying a unique force equal to 500,000 N in all directions, **w**, **u**, **v**, in three different loading conditions. The following settings are used for this run and for the following:

- $\alpha_{ss} = 0.025$
- $n_{si} \approx \text{rnd}(l_i/1\,\text{mm}) + 1$, which means 200, 78, and 135 elements depending on welds
- $p_{max} = 0.9$
- $tol = 0.05$
- $maximum\ number\ of\ iterations = 100$

The results are in Table 6.6.

Now, a set of more complex loading conditions is considered, as listed in Table 6.7.

Table 6.6 Concentrically loaded weld layout: loading, related utilizations, and difference between utilization factors. Comparison between expected and computed results. Here a precision of about 2.5% on utilization is expected.

Load	U_{Exact}	$U_{Computed}$	Δ
$N_w = 500$ kN	0.337	0.342	0.014
$V_u = 500$ kN	0.346	0.342	0.005
$V_v = 500$ kN	0.566	0.565	0.001

Table 6.7 A set of generic loading conditions tested.

Load	*N* (kN)	V_u (kN)	V_v (kN)	M_w (kNm)	M_u (kNm)	M_v (kNm)
1	0	200	300	10	0	0
2	100	0	0	0	20	30
3	100	200	300	10	20	30

Table 6.8 Utilization and displacement at limit related to the loading conditions of Table 6.7.

Load	$U_{Computed}$	p_{max}	d_w (mm)	d_u (mm)	d_v (mm)	r_w (mrad)	r_u (mrad)	r_v (mrad)
1	0.306	1.01	≈0	−0.175	0.136	0.970	≈0	≈0
2	0.536	0.976	0.0219	≈0	≈0	≈0	−2.51	0.862
3	0.588	0.942	0.0228	−0.0479	0.0330	0.167	−2.45	0.799

These loading conditions are too complex to be tested by hand. The first is an in-plane eccentric loading condition. The second is an out-of-plane loading condition, and the third is the sum of the previous two.

In Table 6.8, the utilization calculated and the computed displacements at limit are listed.

Considering the second loading condition, the maximum displacement in the flange welds is

$$d_{w,\max,flange} = 0.0219 + 0.00251 \cdot 100 + 0.000862 \cdot (100 + 0.5 \cdot 5.3) = 0.361\,\text{mm}$$

and the related p is

$$p_{\max,flange} = \frac{0.361}{0.049 \cdot 7.5} = 0.982$$

For the web weld, instead, the maximum displacement is

$$d_{w,\max,web} = 0.0219 + 0.00251 \cdot (4.5 + 0.5 \cdot 3.2) + 0.000862 \cdot 0.5 \cdot 134 = 0.0949\,\text{mm}$$

and the related p is

$$p_{\max,web} = \frac{0.0949}{0.049 \cdot 4.5} = 0.430$$

The maximum value of p, considering the rounding errors, is equal to that found by the process, and is higher than 0.9, as expected.

6.7.5 Computing the Stresses in Fillet Welds from the Forces Applied to the Layout

The method to be used in order to compute the stresses of the single weld seams starting from the forces applied to the whole weld layout follows the path already explained for penetration weld layouts.

Clearly, if the ICRM is used, the stresses are computed using the method itself, so what will be explained in this section is to be used when the ICRM is not used.

The general path is to compute the modified stiffness matrix of the weld layout, and from this the single weld seams displacements and then the stresses (Equations 6.46–6.48).

The methods available differ as to how the stiffness matrix of the single weld seam is defined, and it will be here summarized.

6.7.5.1 Conventional, or Isotropic Approach

Using the conventional approach, the stiffness of each weld is basically related to its effective area, and the differences in stiffness related to the direction of loading and to the thickness of the welds are neglected: for this reason it could also be called an *isotropic approach*. This is what is implicitly done using traditional approaches like the one of Section 6.6.2. Here, however, the working flags μ_n, μ_t, μ_l, and the weld angle coefficients C_w, C_u, and C_v are also possibly used. If not, they must be set equal to 1.

The stiffness considered as reference is the one of the longitudinally loaded fillet. If this approach is used, then the terms of the stiffness matrix of a single weld seam must be replaced as follows (t_{ave} is the average thickness of the weld seams of the weld layout).

Using the (unmodified) secant stiffness (0.298 = 3.58/12):

$$
\begin{aligned}
&\frac{EA_i}{L_i} \Rightarrow k_{Ni} \Rightarrow 3.58\frac{A_i f_u}{t_{ave}} C_{wi}\mu_{ni} \\
&\frac{GA_{ui}}{L_i} \Rightarrow k_{Vui} \Rightarrow \frac{3.58 \cdot A_i f_u}{t_{ave}} C_{ui}\mu_{ti} \\
&\frac{GA_{vi}}{L_i} \Rightarrow k_{Vvi} \Rightarrow \frac{3.58 \cdot A_i f_u}{t_{ave}} C_{vi}\mu_{ti}\mu_{li} \\
&\frac{EJ_{ui}}{L_i} \Rightarrow k_{Mui} \Rightarrow \frac{0.298 \cdot A_i\, a_i^2 f_u}{t_{ave}} C_{wi}\mu_{ni} \\
&\frac{EJ_{vi}}{L_i} \Rightarrow k_{Mvi} \Rightarrow \frac{0.298 \cdot A_i\, l_i^2 f_u}{t_{ave}} C_{wi}\mu_{ni} \\
&\frac{GJ_{ti}}{L_i} \Rightarrow k_{Ti} \Rightarrow \frac{0.298 \cdot A_i\, l_i^2 f_u}{t_{ave}} C_{vi}\mu_{ti}\mu_{li}
\end{aligned}
\tag{6.53}
$$

Using the initial stiffness (45.3 = 546/12):

$$\begin{aligned}
&\frac{EA_i}{L_i} \Rightarrow k_{Ni} \Rightarrow 546\frac{A_i f_u}{t_{ave}} C_{wi}\mu_{ni} \\
&\frac{GA_{ui}}{L_i} \Rightarrow k_{Vui} \Rightarrow \frac{546 \cdot A_i f_u}{t_{ave}} C_{ui}\mu_{ti} \\
&\frac{GA_{vi}}{L_i} \Rightarrow k_{Vvi} \Rightarrow \frac{546 \cdot A_i f_u}{t_{ave}} C_{vi}\mu_{ti}\mu_{li} \\
&\frac{EJ_{ui}}{L_i} \Rightarrow k_{Mui} \Rightarrow \frac{45.3 \cdot A_i\, a_i^2 f_u}{t_{ave}} C_{wi}\mu_{ni} \\
&\frac{EJ_{vi}}{L_i} \Rightarrow k_{Mvi} \Rightarrow \frac{45.3 \cdot A_i\, l_i^2 f_u}{t_{ave}} C_{wi}\mu_{ni} \\
&\frac{GJ_{ti}}{L_i} \Rightarrow k_{Ti} \Rightarrow \frac{45.3 \cdot A_i\, l_i^2 f_u}{t_{ave}} C_{vi}\mu_{ti}\mu_{li}
\end{aligned} \tag{6.54}$$

6.7.5.2 Orthotropic Approach

This approach considers the difference of stiffness related to the direction of loading and to the different thicknesses of the weld seams. Using this approach, when the fillet is loaded transversely it is stiffer than when loaded longitudinally, and the weld seams with lower thickness are stiffer than those with higher stiffness, at equal effective areas. In elastic range this approach tends to overload the transversely loaded fillets, and the ones having small thickness.

If this approach is used, then the terms of the stiffness matrix of a single weld seam must be replaced as follows (t_i is the thickness of the single weld seam i).

Using the secant stiffness (1.52 = 18.3/12):

$$\begin{aligned}
&\frac{EA_i}{L_i} \Rightarrow k_{Ni} \Rightarrow 18.3\frac{A_i f_u}{t_i} C_{wi}\mu_{ni} \\
&\frac{GA_{ui}}{L_i} \Rightarrow k_{Vui} \Rightarrow \frac{3.58 \cdot A_i f_u}{t_i} C_{ui}\mu_{ti} \\
&\frac{GA_{vi}}{L_i} \Rightarrow k_{Vvi} \Rightarrow \frac{18.3 \cdot A_i f_u}{t_i} C_{vi}\mu_{ti}\mu_{li} \\
&\frac{EJ_{ui}}{L_i} \Rightarrow k_{Mui} \Rightarrow \frac{1.52 \cdot A_i\, a_i^2 f_u}{t_i} C_{wi}\mu_{ni} \\
&\frac{EJ_{vi}}{L_i} \Rightarrow k_{Mvi} \Rightarrow \frac{1.52 \cdot A_i\, l_i^2 f_u}{t_i} C_{wi}\mu_{ni} \\
&\frac{GJ_{ti}}{L_i} \Rightarrow k_{Ti} \Rightarrow \frac{1.52 \cdot A_i\, l_i^2 f_u}{t_i} C_{vi}\mu_{ti}\mu_{li}
\end{aligned} \tag{6.55}$$

Using the modified secant stiffness:

$$\begin{aligned}
&\frac{EA_i}{L_i} \Rightarrow 18.3\frac{A_i f_u}{t_i} C_{wi}\mu_{ni}\\
&\frac{GA_{ui}}{L_i} \Rightarrow \frac{10.14 \cdot A_i f_u}{t_i} C_{ui}\mu_{ti}\\
&\frac{GA_{vi}}{L_i} \Rightarrow \frac{18.3 \cdot A_i f_u}{t_i} C_{vi}\mu_{ti}\mu_{li}\\
&\frac{EJ_{ui}}{L_i} \Rightarrow \frac{1.52 \cdot A_i\, a_i^2 f_u}{t_i} C_{wi}\mu_{ni}\\
&\frac{EJ_{vi}}{L_i} \Rightarrow \frac{1.52 \cdot A_i\, l_i^2 f_u}{t_i} C_{wi}\mu_{ni}\\
&\frac{GJ_{ti}}{L_i} \Rightarrow \frac{1.52 \cdot A_i\, l_i^2 f_u}{t_i} C_{vi}\mu_{ti}\mu_{li}
\end{aligned} \tag{6.56}$$

Using the initial stiffness (232 = 2793/12):

$$\begin{aligned}
&\frac{EA_i}{L_i} \Rightarrow 2793\frac{A_i f_u}{t_i} C_{wi}\mu_{ni}\\
&\frac{GA_{ui}}{L_i} \Rightarrow \frac{546 \cdot A_i f_u}{t_i} C_{ui}\mu_{ti}\\
&\frac{GA_{vi}}{L_i} \Rightarrow \frac{2793 \cdot A_i f_u}{t_i} C_{vi}\mu_{ti}\mu_{li}\\
&\frac{EJ_{ui}}{L_i} \Rightarrow \frac{232 \cdot A_i\, a_i^2 f_u}{t_i} C_{wi}\mu_{ni}\\
&\frac{EJ_{vi}}{L_i} \Rightarrow \frac{232 \cdot A_i\, l_i^2 f_u}{t_i} C_{wi}\mu_{ni}\\
&\frac{GJ_{ti}}{L_i} \Rightarrow 232 \cdot A_i\, \frac{l_i^2 f_u}{t_i} C_{vi}\mu_{ti}\mu_{li}
\end{aligned} \tag{6.57}$$

6.7.6 Fillet Welds Using Contact and Friction

A mechanical model taking into account contact and friction for willet welds can be outlined as follows.

Three states can be defined for the weld segment:

1) Unlocked, when the displacement $(d_{w2} - d_{w1}) > 0$. These are pulled weld segments. Unlocked fillet weld segments behave like weld segments not using contact and friction. The safe-side model assumes unlocked weld segments also if a compression is applied.
2) Locked and no sliding. This is when the segment is compressed and the shear applied is lower than the normal compressive force times a friction coefficient, μ. Friction blocks the sliding, and so the fillet is totally unstressed. The mechanical model will

imply forces in the weld segments, but these forces will not be used to check the weld segment resistance.

3) Locked and sliding. This is when the compressive forces plus the friction are not sufficient to avoid sliding movement. In this case the fillet segment behaves like a fillet weld segment loaded by a force in the plane of the faying surface. So, only d_u and d_v will generate stresses in the weld segment, to be computed with the methods outlined in the previous sections.

In the locked and sliding state, the tangent stiffness matrix of the weld segment can be obtained as explained in Appendix 2.

In the locked not sliding state, the stiffness of the weld segment is high, both in the axial direction (k_N) and in the shear direction (k_{Vu}, k_{Vv}). The normal Timoshenko beam element stiffness may be assumed.

In the locked sliding state, the axial stiffness k_N is high, while the shear stiffnesses k_{Vu} and k_{Vv} are those of the unlocked fillet weld segments.

If the whole layout is to be modeled in the elastic range, the stiffness of the single welds can be set high, equal to that of normal Timoshenko beam elements in both the axial and shear directions. Once the force packets flowing into the layout are known, coherently with this stiffness assumption, a simple check can be done to validate the assumption.

If all the welds are compressed, because the axial compressive stresses are everywhere higher than the bending stresses (i.e. if $n < 0$ always), then the initial assumption of "contact" weld layout is validated. The shear stresses t will have to be locally compared to the normal compressive stresses factored by a friction coefficient μ, in this way:

$$\sqrt{t_{par}^2 + t_{per}^2} < \mu \cdot |n|$$

If this check is passed everywhere, the fillet weld layout is actually unloaded and all the stresses are transferred by contact and friction. A relaxed kind of test would instead check the sliding at the layout level using the following rule:

$$\sqrt{V_u^2 + V_v^2} < \mu \cdot |N|$$

The welds connecting members highly compressed by permanent forces, like columns, and in direct contact, are expected to behave in this way. However, the author believes that the use of contact and friction as a resisting mechanism is potentially very dangerous, because if the permanent compression is not permanent as believed, of if higher bending is applied, the stress state could change dramatically. So, this must be used with extreme care.

If instead a tensile stress is applied somewhere in the welds, or if the no-sliding condition is violated, then a more refined model for evaluating the stresses in the single weld seams is needed.

One possible way to solve the issue is to extend the bearing surface approach, which will be explained for the bolt layouts, replacing the no-tension constitutive law of the bearing surface with a constitutive law using different stiffnesses in tension and compression (that of the fillets, low, in tension, that of the contact, high, in compression), and considering the weld layout effective surface projected onto the faying surface, as *bearing surface* (see Section 7.5.2.2.). This approach has not been tested yet, and so is not described in detail.

This refers to the weld layout treated as a whole.

As will be explained more fully in Chapter 10, in the context of a finite element analysis explicitly modeling the welds, the single weld seams are divided into weld segments, and are independent of one another. If a non-linear analysis is run, these weld segments can use a formulation that also embeds contact and friction, simply modifying the tangential stiffness matrix according to the state of the weld segment.

In the unlocked state, the tangent stiffness matrix of the weld segment is that described in Appendix 2, that is, the one used for the instantaneous center of rotation method.

In the locked no-sliding state, the tangent stiffness matrix of the weld segment uses the normal Timoshenko k_N, k_{Vu}, and k_{Vv} (see Appendix 2).

In the locked sliding state, the tangent stiffness matrix of the weld segment can be modified as explained in Appendix 2.

If linear finite element analyses are run, the assumption of "weld using contact and friction" can be checked in the post-processing stage, simply verifying that the weld segment is compressed, and that the shear resultant is lower than the compression multiplied by the friction coefficient. For the welds of highly compressed members in contact this can be a good modeling approach.

6.8 Mixed Penetration and Fillet Weld Layouts

In ideally good design, penetration welding and fillet welding should not be used in parallel to weld the same two constituents. This is because of the very different stiffnesses of the two welding procedures, and on the uncertainties related to the exact computation of the single contributions.

If it is always possible to exclude one of the two from the checks, so as to keep only the other, then, in the real world, if both are present it is not really predictable how the connection will behave. In particular, what seems dangerous is the onset of stress concentration, which may imply fractures and the sudden change of load paths which, due to the dynamic nature of the transition, may trigger stresses higher than those computed.

It is possible to analyze connections where more than one type of welding is applied to carry in parallel the loads applied. However, this may result in an unsafe design.

References

AISC 360-10, 2010, *Specifications for Structural Steel Buildings*, American Institute for Steel Construction.

Ballio, G. and Mazzolani, F.M. 1983, *Theory and Design of Steel Structures*, Taylor and Francis.

Brandt, G.D. 1982, *A General Solution for Eccentric Loads on Weld Groups*, Engineering Journal, American Institute for Steel Construction, **19**, N° 2, 1982.

Bruneau, M., Uang, C.M., and Sabelli, S. 2011, *Ductile Design of Steel Structures*, McGraw Hill.

Chen, W.F. 2000 (ed.), *Practical Analysis for Semi-Rigid Frame Design*, World Scientific Press.

Crisfield, M.A. 1991, *Non-linear Finite Element Analysis of Solids and Structures*, Vol. 1 and 2, John Wiley & Sons.

EN 1993-1.3 2006, *Design of Steel Structures – General rules – Supplementary rules for cold formed thin gauge members and sheeting*, CEN.

EN 1993-1.8 2006, *Design of Steel Structures – Design of Joints*, CEN.

Gomez, I., Grondin, G., Kanvinde, A., and Kwan, Y.K. 2008, *Strength and Ductility of Welded Joints Subjected to Out-of-Plane Bending*, American Institute for Steel Construction.

Heyman, J. 1995, *The Stone Skeleton*, Cambridge University Press, Cambridge.

Jaspart, J.P. 1991, *Etude de la semi-rigidité des nœuds poutre-colonne et son influence sur la résistance et la stabilité des ossatures en acier*, PhD Thesis, Department MSM, University of Liège.

Jaspart, J.P. and Weynand, K. 2016, *Design of Joints in Steel and Composite Structures*, ECCS Eurocode Design Manuals, Wiley Ernst & Sohn.

Lesik, D.F. and Kennedy, D.J.L. 1990, *Ultimate Strength of Fillet Welded Connections Loaded in Plane, Canadian Journal of Civil Engineering*, **17**, 1, pp 57–67.

Rugarli, P. 2010, *Structural Analysis with Finite Elements*, Thomas Telford.

Rugarli, P. 2014, *Validazione Strutturale Volume 1: Aspetti Generali*, EPC Libri.

Tamboli, A.R. 2010, *Handbook of Structural Steel Connection Design and Details*, McGraw Hill.

7

Connectors: Bolt Layouts and Contact

7.1 Introduction to Bolt Layouts

Bolts are very important in steel connections, as they provide a quick and simple way to connect parts, in a standard manner. Bolts have gradually replaced rivets, which were still widely used some decades ago.

Computing bolt layouts is not an easy task because their behavior is strictly related to that of the neighboring parts, especially when normal force (i.e. axial for the bolts) and bending moments are applied. The behavior of bolt layouts can be considered decoupled:

- One part is the response of the bolt layout to the shear and torque applied. This part of the response induces shears in the bolt shafts or friction in the connected plates when no-slip preloaded bolts are used.
- Another part is the response to axial force and bending moments applied to the layout. This part causes axial forces in the bolt shafts and an exchange of pressures at the interface between the bolted constituents.

Moreover, the behavior of slip resistant bolt layouts is different from that of bearing-type bolt layouts, so this important difference also has to be carefully considered.

As will be seen, the forces loading each single bolt can be found by assuming some organization of the displacement field, and the most frequently used assumption is that of a linear variation of displacements. This in turn means that it is possible to predict the displacements and forces of each bolt for a given bolt layout knowing how its center has moved, for each thickness connected. So, three displacements and three rotations of n points, or better nodes, the extremities of the bolt layout, are enough to assess the level of the forces taken by the layout and the individual forces flowing into each bolt.

Later in the chapter, it will be seen how the stiffness matrices of single bolts can be prepared, and in strict analogy to what has been done for welds, how the stiffness matrix of the whole layout can be seen as an organized assembly of the stiffness matrix of the single bolts.

There are, however, some important differences.

1) The deformability at the interface bolt-plate must be taken into account, at least for bolts in bearing.
2) The element describing the layout has not two nodes only, but n nodes, where n is the multiplicity of the bolt layout.

Steel Connection Analysis, First Edition. Paolo Rugarli.
© 2018 John Wiley & Sons Ltd. Published 2018 by John Wiley & Sons Ltd.

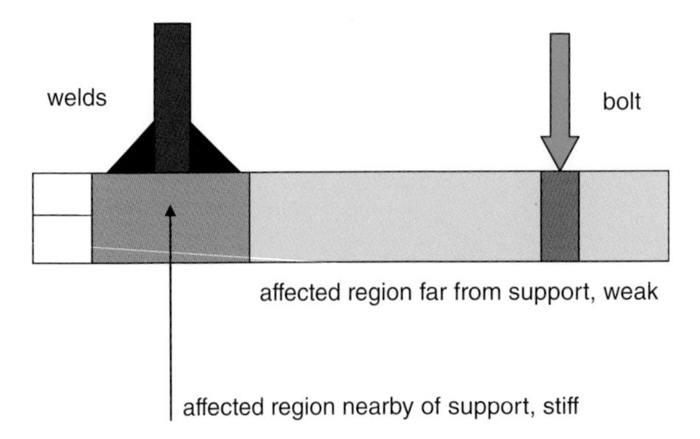

Figure 7.1 As the bolts can be applied anywhere, they usually imply bending stresses and severe bending of plates.

3) The pressures exchanged at the interface do help to carry the loads, and this cannot usually be neglected, both for the statics of the bolt and for the statics of the part connected.

As the bolts that connect two plates may be quite far from the supports of the plates themselves, bolt layouts in bending, and in tension and compression, deliver forces and pressures normal to the plane of the plates connected; in turn, these forces are responsible of the plate bending, which is a fundamental feature to be considered (Figure 7.1). In welds, this is much less important, as for instance in a T-joint the fillet welds deliver normal stresses n which are very near or at the support.

Before getting into the detail of the stiffness matrix of the bolts (Section 7.9) and of the procedures needed to compute individual bolt forces (Section 7.10), as well as the pressures exchanged at the interface, the different typical problems of bolt layouts will be discussed (Sections 7.2–7.8). This discussion is a fundamental preliminary step in order to evaluate properly the behavior of bolt layouts, and to make the correct design choices.

7.2 Bolt Sizes and Classes

In European practice, the bolts most frequently used are those in Table 7.1.

In this book, to avoid confusion with the symbol d used for displacements, the symbol D will be used for bolt diameter.

It can be seen that the following relations between the diameter of the bolt D and the height of bolt-head k, and the nut height m, can be applied (mm):

$$k = 0.613D + 0.61$$

$$m = 0.793D + 0.26$$

The available material grades are those of Table 7.2, and the American practice uses most frequently the bolts described in Table 7.3.

Table 7.1 Data for European bolts. D is the diameter of the bolt, D_0 is the normal hole diameter, A is the bolt-shank area, A_{res} is the threaded area, s is the distance between two opposite faces of the hexagon, e is the diameter of the circle circumscribed to the hexagon, k is the height of the bolt-head, and m is the height of the nut.

Bolt	D (mm)	D_0 (mm)	A (mm^2)	A_{res} (mm^2)	s (mm)	e (mm)	k (mm)	m (mm)
M8	8	9	50.3	38.6	13	14.38	5.5	6.5
M10	10	11	78.5	58.0	16	18.90	7	8
M12	12	13	113.1	84.3	18	21.1	8	10
M14	14	15	153.9	115	21	24.49	9	11
M16	16	18	201.1	157	24	26.75	10	13
M18	18	20	254.5	192	27	30.14	12	15
M20	20	22	314.2	245	30	33.53	13	16
M24	22	24	380.1	303	34	35.72	14	18
M27	24	26	452.4	353	36	39.98	15	19
M30	27	30	572.5	459	41	45.63	17	22
M33	30	33	706.8	561	46	51.28	19	24
M36	33	36	855.3	694	50	55.80	21	26
M39	36	39	1017.8	817	55	61.31	23	29
M42	39	42	1194.6	976	60	66.96	25	31
M45	42	45	1385.4	1120	65	72.61	26	34
M48	45	48	1590.4	1310	70	78.26	28	36
M52	48	51	1809.5	1470	75	83.91	30	38

Table 7.2 Material grades for European bolts. f_{yb} is the yield strength and f_{ub} is the ultimate strength.

Grade	f_{yb} (MPa)	f_{ub} (MPa)
4.6	240	400
4.8	320	400
5.6	300	500
5.8	400	500
6.6	360	600
6.8	480	600
8.8	640	800
10.9	900	1000
12.9	1080	1175

Table 7.3 US most frequently used bolt sizes.

Bolt	D (mm)	D_0 (mm)	A (mm^2)	A_{res} (mm^2)	s (mm)	e (mm)	k (mm)	m (mm)
5/8	15.87	17.46	197,8	145.81	26.99	31.16	9.92	15.48
3/4	19.05	20.64	285,0	215.48	31.75	36.66	11.91	18.65
7/8	22.22	23.81	387,8	298.06	36.51	42.16	13.89	21.82
1	25.4	26.99	506,7	390.97	41.28	47.66	15.48	25.00
1 1/8	28.57	30.16	641,1	492.26	46.04	53.16	17.46	28.17
1 1/4	31.75	33.34	791,7	625.16	50.8	58.66	19.84	30.96
1 3/8	34.92	36.51	957,7	748.39	55.56	64.16	21.43	34.13
1 1/2	38.10	39.69	1140,1	909.68	60.32	69.66	23.81	37.30

Table 7.4 Material grades for US bolts. f_{yb} is the yield strength and f_{ub} is the ultimate strength.

Grade	f_{yb} (MPa)	f_{ub} (MPa)
A307	248	413
A325-1,2,3	634	827
A490	986	1034

The following relationships can be used to evaluate k and m (using millimeters):

$$k = 0.617D + 0.058$$

$$m = 0.977D + 0.06$$

The grades are those of Table 7.4.

7.3 Reference System and Stresses for Bolt Layouts

Bolt layouts are referred to an initial reference system (1, 2, 3), using abscissas (x_1, x_2, x_3) whose origin lies over the entry face of the first object drilled (see Chapter 4). Then, for each thickness drilled k it is possible to define a bolt layout center, $\mathbf{E_k}$. In this book it is assumed that all the bolts of a bolt layout drill the very same objects, and are identical, that is, they have the same diameter and class.

So, it is easily possible to find the bolt layout center $\mathbf{E_k}$ at each extremity k, and the bolt layout principal axes, $(\mathbf{u}, \mathbf{v}, \mathbf{w})$, where axis $\mathbf{w}$ is parallel to axis 3, by using the same methods already seen for the weld layouts.

From a computational point of view, if **G** is the geometrical center of the bolt sections placed over the plane of object O_1, that is, over the entry face, the extremities of the bolt layout $\mathbf{E_1} \ldots \mathbf{E_k}.\mathbf{E_{ne}}$. are obtained as

$$\begin{aligned}
\mathbf{E}_1 &= \mathbf{G} - 0.5t_1 \cdot \mathbf{v}_3 \\
\mathbf{E}_2 &= \mathbf{E}_1 - 0.5(t_1 + t_2) \cdot \mathbf{v}_3 \\
&\ldots\ldots \\
\mathbf{E}_{\mathbf{ne}} &= \mathbf{G} + \left(0.5t_{ne} - \sum_{k=1}^{ne} t_k\right) \cdot \mathbf{v}_3
\end{aligned} \tag{7.1}$$

where ne is the number of extremities of the bolt layout, t_k is the thickness of the k-th object drilled, and $\mathbf{v}_3$ is the unity vector directed as the axis 3 of the bolt layout and axis **w** (i.e. as the normal **n** of the chosen face defined in Chapter 4). These extremities $\mathbf{E_k}$ are also the nodes of the overall elements modeling a weld layout.

The angle between the principal axis **u** and axis 1 is named β (Figure 7.2), as has been done for weld layouts. However, for bolts, the following simplifications apply:

1) There is no need to introduce axes local to each single bolt, $(\mathbf{u_i}, \mathbf{v_i}, \mathbf{w_i})$, as was done for welds.
2) There is no difference in the behavior of a single bolt, displacing it in the **u** or **v** direction, so the shear areas are equal: $A_u = A_v$.

If the bolt layout is a regular grid, made of rows and columns, possibly translated and rotated over the entry face, it is also possible to define axes (**x**, **y**), that is, the axes of the grid. As some bolt of the grid may miss, the axes (**x**, **y**) are not generally speaking parallel to the principal axes (**u**, **v**). The present formulation does not rely on regular grids. Bolts can be freely placed over the entry face and no specific pattern is assumed.

The force packets taken by a bolt layout will be three forces and three moments for each extremity of the bolt layout $\mathbf{E_k}$, and will be considered referred to the principal system of the bolt layout. They will be applied at point $\mathbf{E_k}$, for the thickness k. Each force packet will be referred to as $\mathbf{s_k}$, or, if it is not important to underline the extremity k to

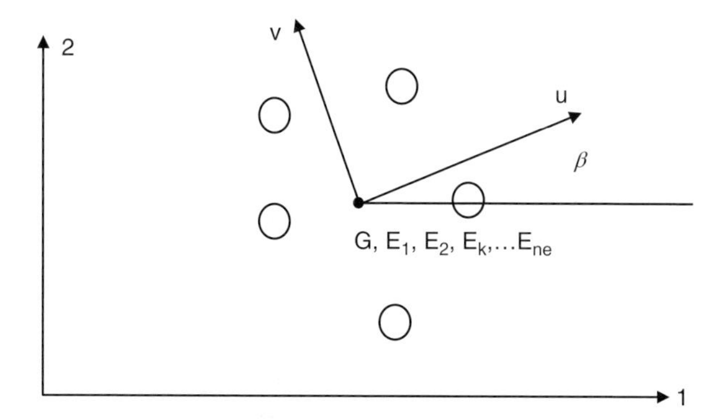

Figure 7.2 Reference system for a generic bolt layout.

which it is applied, merely as **s**, where **s** is a vector of six real numbers, expressed in the principal reference system of the bolt layout:

$$\mathbf{s} = \{F_w, F_u, F_v, C_w, C_u, C_v\}^T$$

where F are forces and C are moments.[1]

To get the internal forces $\boldsymbol{\sigma}$ applied at points **Z** – that is, the internal force vector $\boldsymbol{\sigma}$ applied at points $\mathbf{Z_k}$ that is, over the planes where checks are done (Figure 7.3) – some transformation of the force packets **s** applied at $\mathbf{E_k}$ is needed.

Along the bolt layout center line, in the drilling direction ($-\mathbf{w} = -\mathbf{v_3}$), the following points are found in sequence:

$$\mathbf{E_1}, \mathbf{Z_1}, \mathbf{E_2}, \mathbf{Z_2}, \mathbf{E_3}, \ldots .\mathbf{Z_{ne-2}}, \mathbf{E_{ne-1}}, \mathbf{Z_{ne-1}}, \mathbf{E_{ne}}$$

and, if t_k is the kth bolted thickness, the following rule applies:

$$\mathbf{Z_k} = \mathbf{E_k} - 0.5 t_k \cdot \mathbf{v_3}$$

It should be noted that the internal forces in the bolt shaft depends on the extremity of the bolt, exactly as the shear and moments of a beam depend of the point along the axis line. Now, as seen in Chapter 5, a force packet applied at a 3D point **P** can be transferred to **Q** by the rule

$$\mathbf{s_Q} = \mathbf{T_{PQ}} \mathbf{s_P}$$

So, using as point **P** the extremity $\mathbf{E_j}$, and as point **Q** the check point $\mathbf{Z_k}$, th results is that the internal forces $\boldsymbol{\sigma_k}$ at $\mathbf{Z_k}$ are the sum of a number of k terms:

$$\boldsymbol{\sigma_k} = \sum_{j=1}^{k} \mathbf{T_{EjZk}} \mathbf{s_j} \tag{7.2}$$

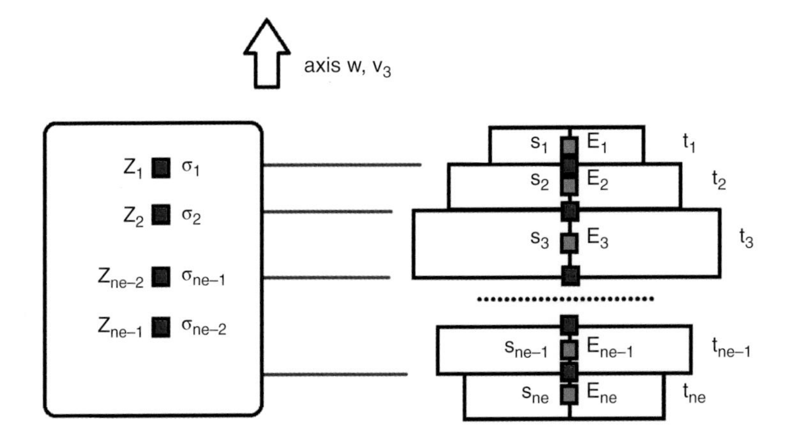

Figure 7.3 Extremities $\mathbf{E_k}$, check sections **Z** and points $\mathbf{Z_k}$, force packets $\mathbf{s_k}$ and internal forces $\boldsymbol{\sigma_k}$ at check sections $\mathbf{Z_k}$, for a generic bolt layout.

1 The symbol C is used to avoid confusion with the moments M, used for the checks.

The vector $\boldsymbol{\sigma}_{\mathbf{k}}$ is the vector of the overall internal forces in the layout at the check point $\mathbf{Z}_{\mathbf{k}}$. It can be written as

$$\boldsymbol{\sigma}_{\mathbf{k}} = \{N_k, V_{uk}, V_{vk}, M_{wk}, M_{uk}, M_{vk}\}^T$$

where:

- N_k is the axial force in the layout; it applies traction or compression to the layout, and so possibly a compression field over a bearing surface.
- V_{uk} and V_{vk} are the shears applied to the layout. They cause shears in the bolt shafts.
- M_{wk} is the torque applied to the layout; it causes shears in the bolt shafts.
- M_{uk} and M_{vk} are bending moments applied to the layout; they cause axial force in the bolt shafts, bending moments in the bolt shafts, and possibly a compression field over a bearing surface.

In bolt layouts there can be several extremities, so in theory two indices are needed: the index k, first, to specify the check plane or the extremity, and the index i, to specify the bolt. For this reason, the forces flowing into each subconnector i, that is, each bolt, are marked also by the index i. Using two indices the forces in single bolts are

- N_{ki}, the axial force in the bolt shaft of the i-th bolt, at check section k
- V_{uki}, V_{vki}, the shears in the bolt shaft of the i-th bolt, at check section k
- M_{uki}, M_{vki}, the bending moments applied to the bolt shaft of the i-th bolt, at check section k.

If it is not important to stress the specific check section k to which the internal forces are referred, the index k can be omitted. In this case, it is agreed that a specific (generic) check section is considered, and a single index i can be used to distinguish between single bolts, that is, between subconnectors.

So, in order to save an index, the following notation will be used in the remaining part of this chapter:

- The check section index k will be applied to layout internal force *vectors*, $\boldsymbol{\sigma}_{\mathbf{k}}$.
- The bolt index i will be applied to single bolt internal force components, N_i, V_{ui}, etc., implicitly considering a generic check section k.
- The overall internal forces acting at a generic check section k will be identified as N, V_u, V_v, with no i or k index.
- Where needed, both the indices will be used, for example N_{ki}, V_{uki}, etc.

7.4 Geometrical Limitations

Geometrical limitations on the spacing and edge distances of bolts are provided by both AISC 360-10 and Eurocode 3. These limitations are very important and must be respected in order to allow the application of the methods used to check against the possible failure modes (see Chapter 8).

7.4.1 Eurocode 3

The minimum distance between the bolts, in the direction of the force applied, is $2.2D_0$, where D_0 is the diameter of the hole. In the direction perpendicular to the force applied, the minimum distance is $2.4D_0$. If the force changes direction, and regular bolt rows and columns are used, it is clearly the value 2.4 that must be used, in all directions.

The maximum distance between bolts is the smaller of $14\,t$ or 200 mm, where t is the thickness of thinner bolted part.

The minimum edge distance is $1.2D_0$.

The maximum edge distance is $4\,t + 40$ mm.

Special provisions apply to steels conforming to EN 10025-5 (steel with improved resistance to atmospheric corrosion).

7.4.2 AISC 360-10

The minimum distance between the bolts is set at 2.66 times the nominal diameter, and a distance of $3D$ is preferred.

The maximum distance between bolts depends on the surface treatment. If there is no risk of corrosion (e.g. painted plates), the maximum distance is 24 times the thickness of the thinner part, or 305 mm (12″). If there is a risk of corrosion, the maximum distance is 14 times the thickness of the thinner part, or 180 mm (7″).

The minimum edge distance is a function of the nominal diameter and varies from 1.5 to 1.25 times the nominal diameter (see Table J3.4 and J3.4 M of AISC 360-10). The higher values are for thinner bolts (e.g. if $D = 0.5''$, it's $1.5D$, i.e. 0.75″).

The maximum edge distance must be 12 times the thickness of the connected part, with a maximum of 150 mm (6″).

7.5 Not Preloaded Bolt Layouts (Bearing Bolt Layouts)

7.5.1 Shear and Torque

If a bolt layout at a given check section is under the effect of a shear applied in its center, V, and of a torque M_w, the bolt layout will react only with shears V_i in the bolts, having direction and intensity different from bolt to bolt, in general. The overall shear can be decomposed into two components, aligned with the principal axes of the bolt layout, V_u and V_v. Also, the shear of bolt i, V_i, can be decomposed in V_{ui} and V_{vi}.

The shear stiffness k_V related to each bolt, depends mainly on two effects:

1) the shear stiffness of the bolt, k_s
2) the bearing stiffness due to the deformability of the plates in the region near the contact area between bolt shaft and plate connected, k_{be}

There are several available formulations for k_s and k_{be}, very different from one another. The differences are relevant not only in the results (especially for low thicknesses), but also in the formulations, that is, in the parameters taken as relevant. Early attempts used Et for the bearing stiffness k_{be} (Tate and Rosenfeld's approach was as early as 1946, and yet it treated the system as an assembly of springs in series), and GA/L for k_s, where E is

Young's modulus, G the shear modulus, A the area of the bolt, and L its length. If a linear assumption is adopted, the choice of the secant vs the initial stiffness is also relevant. The first choice is optimal for high loads, the second one for low loads. If the bolt layout is isoconnected, these differences do not imply a difference in the evaluation of the force packet taken by the layout. The difference is mainly related to the total amount of displacement and then to the stiffness. However, if more than one bolt layout acts in parallel, as in hyperconnected joints, then the relative stiffness is related to the proportion of the loads flowing into each bolt layout.

According to Eurocode 3, Part 1.8, the first stiffness, k_s, is to be evaluated as

$$k_s = \frac{8D^2 f_{ub}}{D_{M16}}$$

where D_{M16} is 16 mm, f_{ub} is the ultimate stress of the bolt, and D its nominal diameter. This stiffness is to be considered an initial stiffness.

According to Eurocode 3 the second stiffness, related to the bearing of the bolt, is a function of plate thickness, bolt spacing, and edge distance.

$$k_{be} = 12 \cdot \alpha_t \cdot \alpha_b \cdot D \cdot f_u$$

The term α_t is

$$\alpha_t = \frac{1.5 \cdot t}{D_{M16}} \le 2.5$$

The term α_b depends on the edge distance and bolt pitch in the direction of the load, its maximum value 1.25 occurs when the distance from the edge in the direction of the force is

$$e \ge 3D$$

and when the pitch is

$$p \ge 3.5D$$

These values are not so far from the minimum set by Eurocode 3, and so these are values quite frequently reached and crossed in real world connections. It seems then reasonable to set a constant value for $\alpha_b = 1.25$, in order to evaluate the stiffness k_{be}.

So finally

$$k_{be} = \frac{22.5 \cdot t}{D_{M16}} \cdot D \cdot f_u \le 37.5 \cdot D \cdot f_u$$

The final local stiffness can be obtained by assuming two springs in series, which gives

$$k_{V,EC3} = \frac{k_s \cdot k_{be}}{k_s + k_{be}} \tag{7.3}$$

Eurocode 3 Part 1.8 does not provide a formulation for the shear vs displacement relation. Only the stiffnesses already mentioned are provided. However, the work by Rex and Easterling (2003) provides a valuable alternative. In any case, this work is more recent, and the experimental results presented indicate that the new formulation proposed is better than that of Eurocode 3.

In this very useful work, the stiffness and the force–displacement relation for a bolt in bearing is analyzed. The force–displacement curve is in the form of the so-called Richard equation (Richard and Abbott 1975):

$$R = R_{be} \cdot \left\{ \frac{1.74 \dfrac{K_{ini}\Delta}{R_{be}}}{\left[1 + \left(\dfrac{K_{ini}\Delta}{R_{be}}\right)^{0.5}\right]^2} - 0.009 \cdot \frac{K_{ini}\Delta}{R_{be}} \right\} \quad (7.4)$$

and takes into account only the bolt bearing (k_{be}). The meaning of the symbols is as follows:

- R_{be} is the ultimate load due to bearing, usually equal to $\alpha f_u Dt$, where D is the bolt diameter, t the plate thickness, and f_u the ultimate stress of plate. The factor α is not greater than 2.4–2.5.
- Δ is the displacement.
- K_{ini} is the initial elastic stiffness of the bearing, and will be discussed below.

Setting

$$r = \frac{R}{R_{be}}$$

the secant stiffness K_{sec} at $r = 0.8$ is

$$K_{\text{sec}} = 0.149 K_{ini}$$

In Rex and Easterling (2003), the initial stiffness K_{ini} is evaluated by three springs in series. The first is the bearing properly expressed; the second and third are related to the bending and shear deformation of the plate seen as a double clamp beam. The cross-section of this beam is a rectangle of width equal to t and height equal to the clear distance of the bolt hole from the free edge. The span is the bolt diameter. As the minimum distance from the edge is enforced by the standards, the two stiffnesses related to these resisting mechanisms are much higher than the first, and so negligible in springs in series. So with good approximation, we can write (using millimeters for the lengths)

$$K_{ini} = 120 \cdot t \cdot f_y \cdot \left(\frac{D}{25.4}\right)^{0.8}$$

and this provides a better approximation than the one by Tate and Rosenfeld (1946), $K_{ini} = Et$, where E is Young's modulus.

The curve proposed by Rex and Easterling is shown in Figure 7.4. The derivative to displacement, that is, the tangential stiffness, is

$$\frac{dR}{d\Delta} = \left\{ \frac{1.74}{\left[1 + \left(\dfrac{K_{ini}\Delta}{R_{be}}\right)^{0.5}\right]^3} - 0.009 \right\} \cdot K_{ini}$$

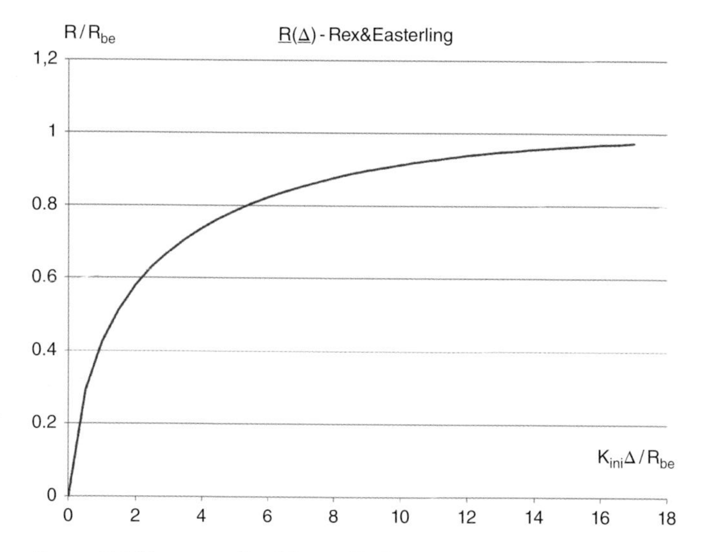

Figure 7.4 The normalized force–displacement curve according to Rex and Easterling (2003). In the abscissa, the normalized displacement $\Delta = K_{ini}\Delta/R_{be}$. In the ordinate, the ratio is R/R_{be}.

This curve can be expressed as a function of r, leading to the following interpolating curves (all having coefficient of determination $R^2 > 0.99$):

$$\begin{aligned}
&0 \le r \le 0.005 && \frac{dR}{d\Delta} = 1.731\cdot(1-17.31\cdot r)\cdot K_{ini} \\
&0.005 < r \le 0.1 && \frac{dR}{d\Delta} = [-0.267\ \ln(r) + 0.1665]\cdot K_{ini} \\
&0.1 < r \le 0.6 && \frac{dR}{d\Delta} = [-0.36\ \ln(r) - 0.0913]\cdot K_{ini} \\
&0.6 < r \le 0.98 && \frac{dR}{d\Delta} = [-0.194\ \ln(r) - 0.0021]\cdot K_{ini}
\end{aligned} \tag{7.5}$$

So, once the force is known, the tangential stiffness can also be easily evaluated. Note that for $r = 0$, the initial stiffness is $1.731\ K_{ini}$, and not K_{ini}.

The curve can be used in non-linear analyses.

The formulation used by AISC 360-10 is different. Reference is made to the pioneering work by Crawford and Kulak (1968), and the load–displacement curve relating the shear force V to the lateral displacement Δ of a single bolt is set as

$$R = R_{ult}\left(1 - e^{-10\Delta}\right)^{0.55}$$

where the displacement Δ is in inches. Using millimeters

$$R = R_{ult}\left(1 - e^{-0.3937\Delta}\right)^{0.55} \tag{7.6}$$

The maximum displacement Δ_{max}, is set equal to 0.34″, that is, 8.636 mm. Naming $q(\Delta)$ the function

$$q = \left(1 - e^{-0.3937\Delta}\right)^{0.55}$$

and plotting it as in Figure 7.5, it can be seen that the function is strongly non-linear and that the initial slope is high.

Taking the derivative of the curve to Δ, we can find the tangent stiffness as a function of Δ:

$$k_{V,AISC,\tan} = \frac{dR}{d\Delta} = 0.55\frac{R_{ult}}{\left(1-e^{-0.3937\Delta}\right)^{0.45}}0.3937\cdot e^{-0.3937\Delta} = 0.216\frac{R_{ult}\cdot e^{-0.3937\Delta}}{\left(1-e^{-0.3937\Delta}\right)^{0.45}}$$

and it tends to infinity when Δ tends to 0. This expression can be used for a non-linear analysis considering the non-linear shear displacement relationship of a single bolt, between two bolted thicknesses. Once the displacement Δ is known, it is also known the tangent stiffness, which can be updated from step to step.

In this load–displacement curve, Δ takes into account the deformation of the bolt, the deformation of the bearings (both bolt and plate), and the bending of the bolt. The value R_{ult} is the minimum resistance considering bolt shear *and* bolt bearing limits. No mention is made about the possible reduction of bolt shear limit due to the contemporary presence of axial force in the bolt shaft, as required by interaction formulae for bolt resistance. The issue will be treated in Section 7.9.3.

The secant stiffness according to AISC 360-10, can be obtained by dividing the force R_{ult} by the maximum allowed displacement, $\Delta_{max} = 0.34''$, and so

$$k_{V,AISC} = \frac{R_{ult}}{\Delta_{\max}} \tag{7.7}$$

More possible formulations are available in the literature, and recent advances propose a more complex formulation in order to also take into account hysteresis loops and pinching (Weigand 2017), also for preloaded bolts.

No torsional stiffness is related to a bolt.

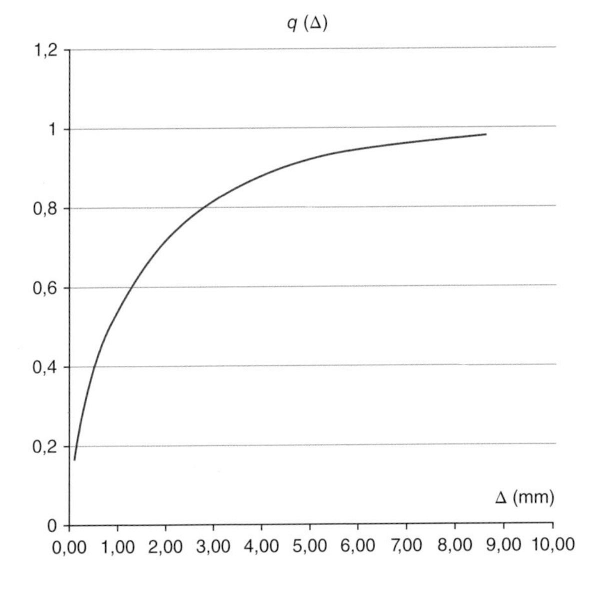

Figure 7.5 The curve $q(\Delta)$ defining the shear vs displacement law of a bolt.

7.5.2 Axial Force and Bending

7.5.2.1 Not Using a Bearing Surface

If a bolt layout is under the effect of an axial force applied in its center **G**, N, and of two bending moments referred to its principal system, M_u and M_v, a first resisting mechanism can be considered that related to the bolt shafts only.

It is assumed, then, that the tensile forces are exerted by the bolt shafts (Figure 7.6), while the compressive forces are exerted by the region immediately surrounding the bolt shaft. When the constituent has to be checked, both tensile and compressive forces can be assumed concentrated at the bolt center, which is on the safe side.

So, if N_i is the axial force (positive or negative) taken by the ith bolt and an elastic distribution is assumed, the following rule does apply (it is assumed that all bolts have the same area and that all do react):

$$N_i = \frac{N}{n_b} + \frac{M_u}{\sum_{i=1}^{nb} v_i^2} v_i - \frac{M_v}{\sum_{i=1}^{nb} u_i^2} u_i \tag{7.8}$$

where n_b is the number of bolts in the layout and u_i and v_i are the coordinates of the ith bolt in the principal reference system.

In this model, the force exerted in bending by a bolt is proportional to its distance from the bending principal axis, **u** or **v** respectively for M_u and M_v, and no bending flows in the bolt shafts.

A modified model assumes a plastic distribution, that is, the forces exerted are all equal. If this is the rule adopted, in case of pure bending M_u

$$N_i = \frac{M_u}{\sum_{i=1}^{nb} |v_i|} \cdot \frac{v_i}{|v_i|}$$

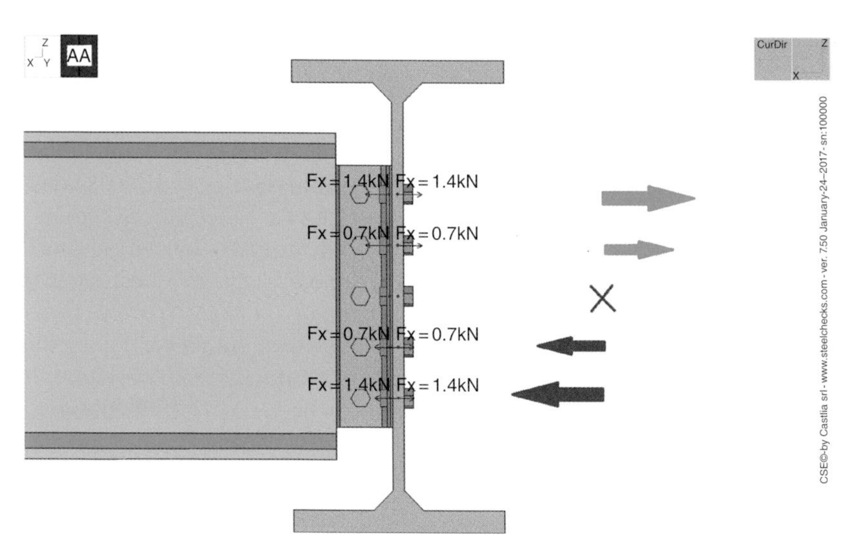

Figure 7.6 A bolt layout under a slight bending carried only by the axial forces in the shaft.

and in case of pure bending M_v

$$N_i = -\frac{M_v}{\sum_{i=1}^{nb} |u_i|} \cdot \frac{u_i}{|u_i|}$$

If a mixed bending plus axial force is applied, we have to find a plastic neutral axis (PNA) with equation

$$au + bv + c = 0$$

that respects the following three conditions:

$$N = F\sum_{i=1}^{nb} \frac{au_i + bv_i + c}{|au_i + bv_i + c|}$$

$$M_u = F\sum_{i=1}^{nb} \frac{au_i + bv_i + c}{|au_i + bv_i + c|} \cdot v_i$$

$$M_v = -F\sum_{i=1}^{nb} \frac{au_i + bv_i + c}{|au_i + bv_i + c|} \cdot u_i$$

where F is the axial force in the bolts. All the bolts will take the same force F, positive or negative depending on the position relative to the PNA, that is, below or above it.

According to AISC 360-10, the plastic distribution is to be preferred, because the method is rather conservative. However, if an axial force plus two bending moments are applied, the method is not as straightforward as the elastic one and in general requires an iterative procedure.

Independently, by the use of plastic or elastic distribution, the method will only imply concentrated forces in the bolt shafts. The shafts will not be checked for the compressive forces, which will be taken by the surrounding region via pressures exchanged in a limited area nearby.

By the Newton's third law, i.e. the action and reaction principle, the constituents bolted will be loaded by a set of concentrated forces that will generate bending stresses in them. Following the static theorem, the constituents will have to be checked against those forces.

This model is clearly a very simplified one, as it completely neglects the contact problem related to bending, confining what it is called *bearing surface* to the ideal circular regions around the bolt shafts.

However, if the bending and axial force are low, as the primary forces carried by the bolts are shears (i.e. if the bolt layout tends to be *shear only*), the model is very useful, as it can take the possible slight extra bending moments and axial forces computed due to equilibrium considerations, in a very robust and easy way.

If instead, as in what are usually called moment connections, the bending moments carried by the bolt layout are not negligible, it is then necessary to adopt a much more refined model that uses the concept of bearing surface. This will be described in the next section.

7.5.2.2 Using a Bearing Surface

7.5.2.2.1 What is a Bearing Surface?

When a bolt layout is under compression or bending, the resisting mechanism does not rely on the forces exerted by the bolt shafts only: at the interface between the bolted constituents a field of compressive contact stresses is exchanged, and these, together with the tensile forces exerted by the bolt shafts, carry the load applied. The surface that is involved in this exchange of pressures is named the *bearing surface,*[2] and its extent strongly depends on the loads applied, the geometry of the connection, including the position of the stiffened part, and the thickness of the parts connected.

Considering the typical example of Figure 7.7, the pressures are exchanged at the interface between the steel plate and the foundation concrete block. Several regions can be identified, with different mechanical roles.

- The bolt (here anchors), drill the plate and connect it to the constraint block, that is, the reinforced concrete foundation. The holes of the bolts are the white circles in the figure. Here are applied, if any, the tensile forces of the bolts.
- The plate is connected by welding to other parts, which "come from above"; these are the *constraints* of the plate and are marked black in the top view on the right in Figure 7.7. This means that the out-of-plane displacement of the plate in those regions will be negligible. The constrained part is related to the footprint of the column, and to the vertical plates acting as force transferrers from the horizontal plate to the column (the so-called "stiffeners"). The faces in contact with the entry face of the first constituent

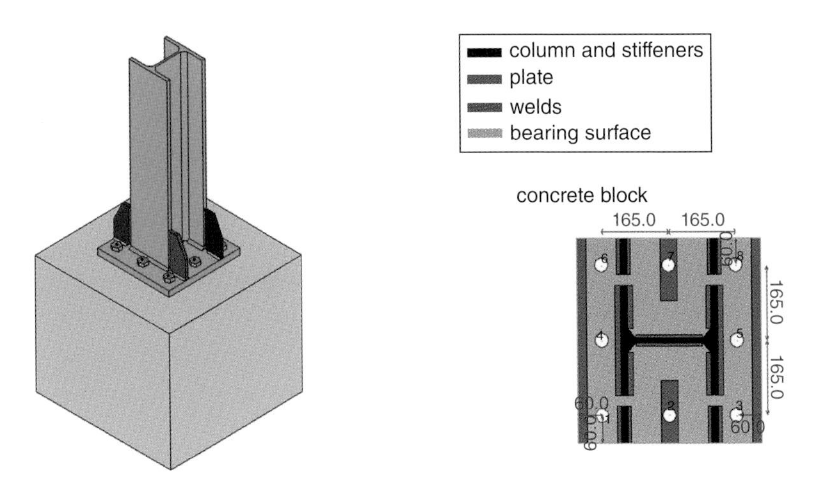

Figure 7.7 A typical example of bolt layout in bending using a bearing surface.

2 A possible alternative would be to name it the *contact surface*, also to avoid possible confusion with the bolt bearing, i.e. the exchange of pressures at the bolt holes, due to shear in the shafts. However, for this latter mechanism, there is no surface to be determined, while the physical mechanism, an exchange of pressures, is indeed the same. Moreover, as will be better seen when dealing with preloaded bolts, a mere contact does not necessarily imply the exchange of pressures, as happens, by definition, along the *bearing surface*.

drilled by bolts, have all their normal opposite to it, and are related to the parts connected. So, they are easily found in the scene, considering the faces:
  - coplanar and contained in the entry face of the bolt layout
  - having normal opposite to the entry face.
- The footprints of the (fillet) welds mark a part immediately near the constraints, but not exactly equal. Also, they are part of the drilled plane inaccessible to bolts, if clashes are to be avoided.
- A complex area marked gray is the (candidate) *bearing surface*, that is, the surface where it is expected and allowed that the plate and the concrete block can exchange pressures. In general, it extends in the vicinity of the constraints, and as it will be seen in the next section is the result of Boolean operations between polygons. The bearing surface may contain the bolts, only some of the bolts, or no bolts at all. Only a part of the candidate bearing surface will react. The reacting part will be found by the non-linear methods, as will be explained in the next sections.

The extent of the bearing surface and the amount of the pressures exchanged at the interface are unknowns of the problem. As will be better seen in Chapter 9, the refined solution of this problem would involve a complex 3D analysis using contact non-linearity.

Simplifying, a *candidate* bearing surface is chosen, independent of the loads applied. This is done thanks to some simplifying hypotheses and using specific geometrical tools. In each single combination, the part of the candidate bearing surface effectively reacting will be found by selecting an appropriate neutral axis. In what follows the candidate bearing surface will also be referred to as the bearing surface: it will be implicitly assumed that only a part of it will react, and that this part will change from load combination to load combination. So, only two unknowns will be kept: the neutral axis and the amount of the pressures at a given distribution of strain. The distribution of the pressures will depend, as it will be shown in the coming sections, on the *constitutive law* of the bearing, that is, by its stiffness. The amount of pressure is chosen so as to guarantee equilibrium to the externally applied loads.

Also, due to the contra-flexure of the plate caused by the tensile forces of the bolts applied to it as forces normal to its middle plane, prying forces may in general develop. These prying forces are one of the most complex issues of connection analysis, because their appropriate evaluation requires complex and time-consuming non-linear analyses. A simple and useful model due to Thornton (1985), and referring to T-stub is available but at the moment it has not been generalized, to the best of the author's knowledge.

Generally speaking, the effect of these prying forces is to

- increase the tensile forces in some of the bolts, which are then loaded by forces higher than those predicted by the methods not considering prying
- decrease the bending effects on the plate, which finds an "unexpected bearing" usually at the corners, far from the constraints.

Up to now an example referring to a base plate has been shown. However, the concept and application of bearing surface is much more general, and it can be applied also to bolt layouts connecting flanges or any other part. The model can be described as follows:

- Two or more plates are bolted, say two: A and B.
- Coming from other constituents usually welded to plate A, normal force and bending moments are applied to the plate A and from it to the bolt layout. These generalized forces will be transferred to plate B and the constituents connected to it.

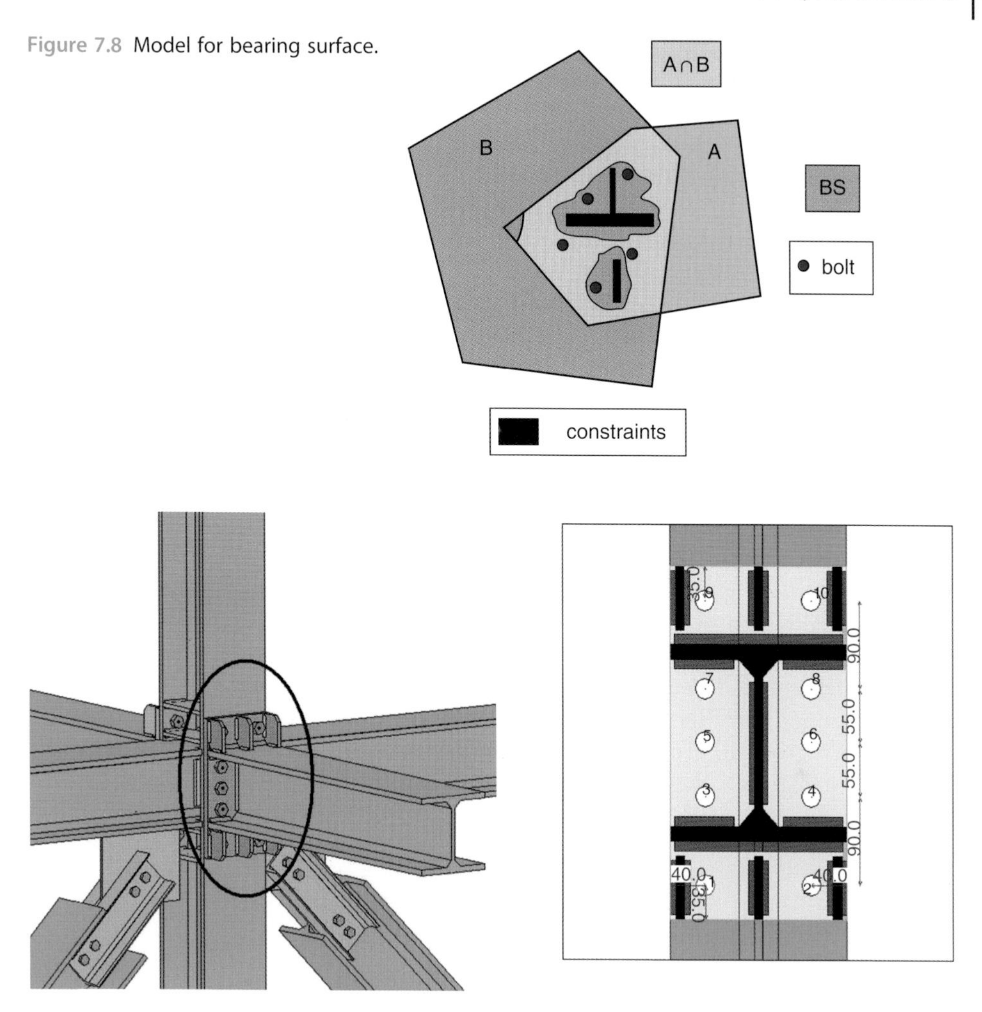

Figure 7.8 Model for bearing surface.

Figure 7.9 A plate bolted to a flange and one possibly related bearing surface – the whole plate extent. Bearing surface can also be limited to a subpart of the bolted plate. However, no part of the bearing surface may be external to the bolted plate.

- In order to balance these loads, tensile forces develop in the bolts, and pressures are exchanged at a bearing surface BS contained in, but not necessarily coincident with, the intersection of the faces in contact of A and B (Figure 7.8). In no way can the bearing surface be more extended than the intersection of the two faces in contact (A ∩ B – see Figure 7.8).

As has been mentioned, the bearing surface model is general and may be applied also to flanges in contact as in the example of Figure 7.9.

7.5.2.2.2 Extent of the Bearing Surface

When two or more constituents are bolted, the force packets flowing in the bolts come from the constituents themselves. In the vicinity of the footprint of the part which is imposing the bending or the axial force, it is expected that the displacements are low, while the main part of the displacements is expected far from these loading footprints.

In Figure 7.10 a plate is loaded from above by a centered compression acting through an HEB column and four "stiffeners" (Figure 7.7). The force is balanced by a field of pressures exchanged at the interface with a concrete block, coming from below. It is here assumed that the plate is clamped in the nodes related to the welding of the column and of the stiffeners to the plate itself.

The displacement field clearly shows that there is a region of the plate around the footprint of the loading HEB plus "stiffeners", which is stiffer, and another region, far from these regions, which is more flexible. The pressures exchanged at the interface will tend to be confined in the stiffer region, while the more flexible ones will also accept a very low pressure field, or null, because they tend to be displaced far from the contact.

It is intuitive to understand that if the plate is very thin, the region in the vicinity of the footprint will be very limited, while if it is thick, and more stiffeners are added, then the reacting region will be wider.

A simple model (Figure 7.11) can be used to enable us to evaluate the amount of plate region affected by the exchanged pressure.

It is assumed that the bolted part cannot be displaced normal to its plane in the regions of the loading footprint. A unit-width strip spanning c from one point of this region, normal to the boundary of the region, is considered to be loaded by a constant pressure p, which is the pressure level designed at the interface where pressures are exchanged. If the plate has a thickness t, then the bending moment at the clamp is

$$M = \frac{pc^2}{2}$$

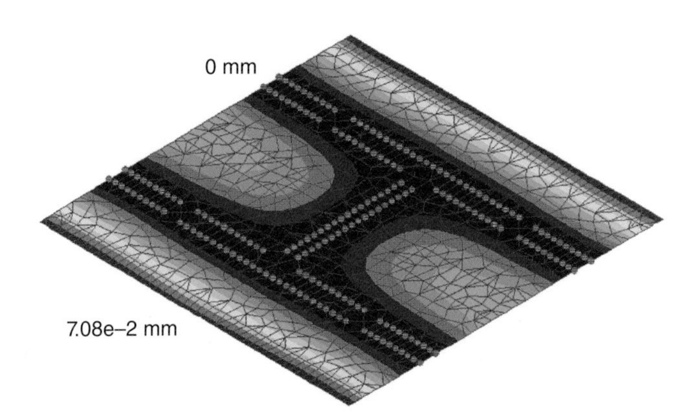

Figure 7.10 Finite element model of a plate loaded from below. The magenta dots are nodal constraints simulating the welds to the column and to the stiffeners. The dark blue and blue area is where displacements are null or very low: this is where the main part of the pressures are exchanged, as the plate is flexible.

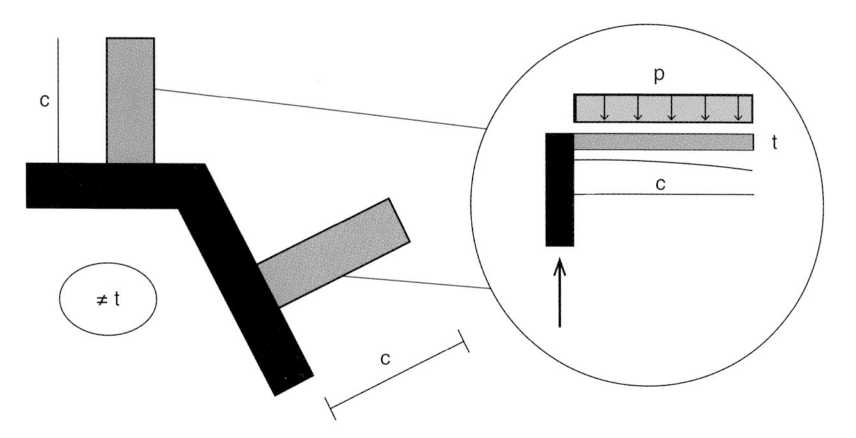

Figure 7.11 Cantilever model for bearing surface near to the footprint of loads applied. The span of the cantilever c is a function of the desired load p, the thickness of the plate t, and the yield stress of the plate, f_y.

Equating this moment to the elastic maximum moment of the unit width, the following relation is obtained (see also Eurocode 3 Part 1.8 §6.2.5, where this simple model is named "equivalent T-stub in compression"):

$$f_y \cdot \frac{t^2}{6} = \frac{pc^2}{2}$$

Once the pressure has been set, the maximum span of the cantilever is found as

$$c = t\sqrt{\frac{f_y}{3p}}$$

a linear function of the thickness t, and inverse function of the square root of the desired pressure, p.

There are two typical situations: base node and flanged connections.

In base nodes, the interface where the pressures are exchanged is between a steel plate and some foundation material, such as concrete or grout. Here the pressure p may be set equal to the crushing stress of the concrete, if the base plate is stiff and stiffened enough to resist such pressure.

In flanged connections, there is no "crushing" stress, but the pressures exchanged assume the only meaning of pressure applied to the plates in contact, that is, they assume the meaning of external loads for the plates. As the plate is usually thin, the bending stresses in the plate will usually govern. In turn, they depend on the geometry of the connection: the position of the stiffeners, the thickness of the plate, and the loading applied. Note that if the aim is – as indeed it is – to assess the stresses in the plate, then the loads applied to the bolt layout (N, M_u, M_v) should be considered to act at the same time, so that the proper neutral axis can be found.

Setting a low value for p, the extension of the bearing surface will increase. However, it will also increase the span of the elementary cantilevers, and the bending stresses in the plate. A simple way to control the stresses is to consider the *crushing stress of the bearing surface* limited by the pressure p, set to determine the bearing surface amount. This may help to give a quick idea of the ability of the plate to carry the applied loads.

If for example a flange of thickness 30 mm is used, and a reference value for the pressure is considered as 50 MPa, with a yielding $f_y = 235$ MPa, the value of c is

$$c = 30\sqrt{\frac{235}{3 \cdot 50}} = 37.55\,\text{mm} \Rightarrow 35\,\text{mm}$$

If after solving, the maximum pressure exchanged at the interface between the plates in contact turns out to be 89 MPa, it is likely that the plate will not be able to carry the load. If, instead, the value computed is 35 MPa, for instance, it will be probable that the plate will be able to carry the loads.

The final proof that the plate is able to resist the applied pressure field can be obtained quickly and easily by an automated finite element model of the plate, loaded by appropriate pressures and bolt tensile forces, both balanced with the externally applied loads (N, M_u, M_v).

It is now possible to set a reasonable amount for the bearing surface:

1) If the bolt layout is a base plate, use the crushing stress as p.
2) If the bolt layout connects two steel parts, use a reasonable amount of limiting pressure, clearly lower than, or at least equal to, the yield stress of the material of the plate (which possibly can be set if the bearing surface is equal to the footprint of the beam and the "stiffeners", with no borders applied). If $c = 2\,t$, $p = f_y/12$.

The cantilevers are clamped at the constraints, and these constraints are diffused over the plate, depending on the footprint of the elements transferring the loads and of the additional force transferrers ("stiffeners") provided by the analyst, so in general the problem of finding an appropriate bearing surface is not trivial.

An example will explain better than words. Consider the column with four force transferrers of Figure 7.12.

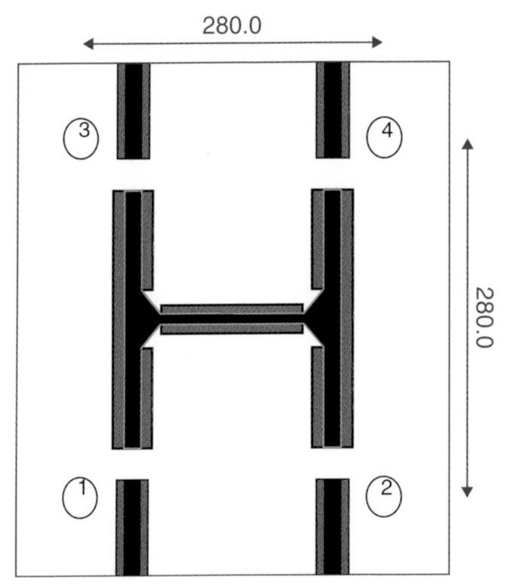

Figure 7.12 A column and four force transferrers welded to a base plate. All the faces are coplanar. The face with red border is the plate face, and the entry face is drilled for bolts.

The drawing is typical of what an analyst has to deal with. There are a number of coplanar faces:

1) the face of the plate that will receive the bolts; its normal points to the viewer
2) the faces, black, of the column and of the force transferrers; they all have normal pointing into the paper, normal to the paper sheet
3) the faces, gray, of the fillet welds; they all have normal pointing into the paper, normal to the paper sheet.

The crushing stress for the concrete is set to $f_{jd} = p = 14.1$ MPa, the thickness of the plate is $t = 20$ mm, and its yield stress is $f_y = 235$ MPa. So $c = 47$ mm.

The first step will be to consider as bearing surface Σ the face of the cross-section of the column S_c, bordered by c. To do that, the face of the column S_c, which is a closed polyline, will have to be transformed into another polyline, S_{cc}, obtained by bordering S_c by c. The result is in Figure 7.13.

The next step will be to select one by one the faces of the four force transferrers, and to add to the existing bearing surface a new surface, obtained by bordering each face by c. The result is in Figure 7.14.

The operations are done between polygons and are *unions*. Up to now the following operations have been done:

$$\Sigma_1 = S_{cc}$$
$$\Sigma_2 = \Sigma_1 \cup S_{s1,c}$$
$$\Sigma_3 = \Sigma_2 \cup S_{s2,c}$$
$$\Sigma_4 = \Sigma_3 \cup S_{s3,c}$$
$$\Sigma_5 = \Sigma_4 \cup S_{s4,c}$$

where $S_{s1,c}$ is the face of the force transferrer 1 (stiffener 1), bordered by c. Now it is noticed that the bearing surface is partly outside of the face S_P of the plate bolted, which

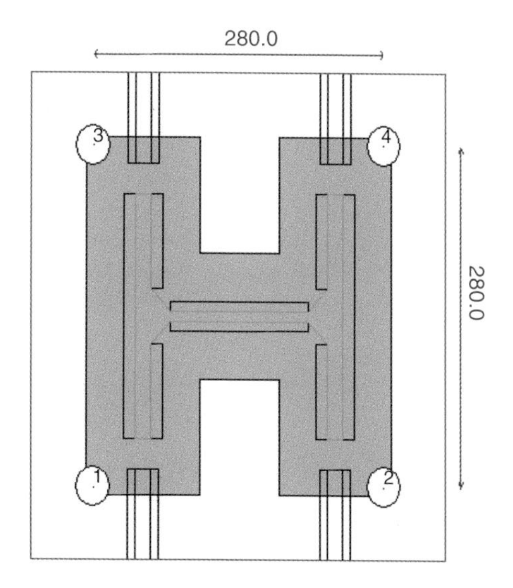

Figure 7.13 First step: the bearing surface is now the face of the column (green) bordered by c.

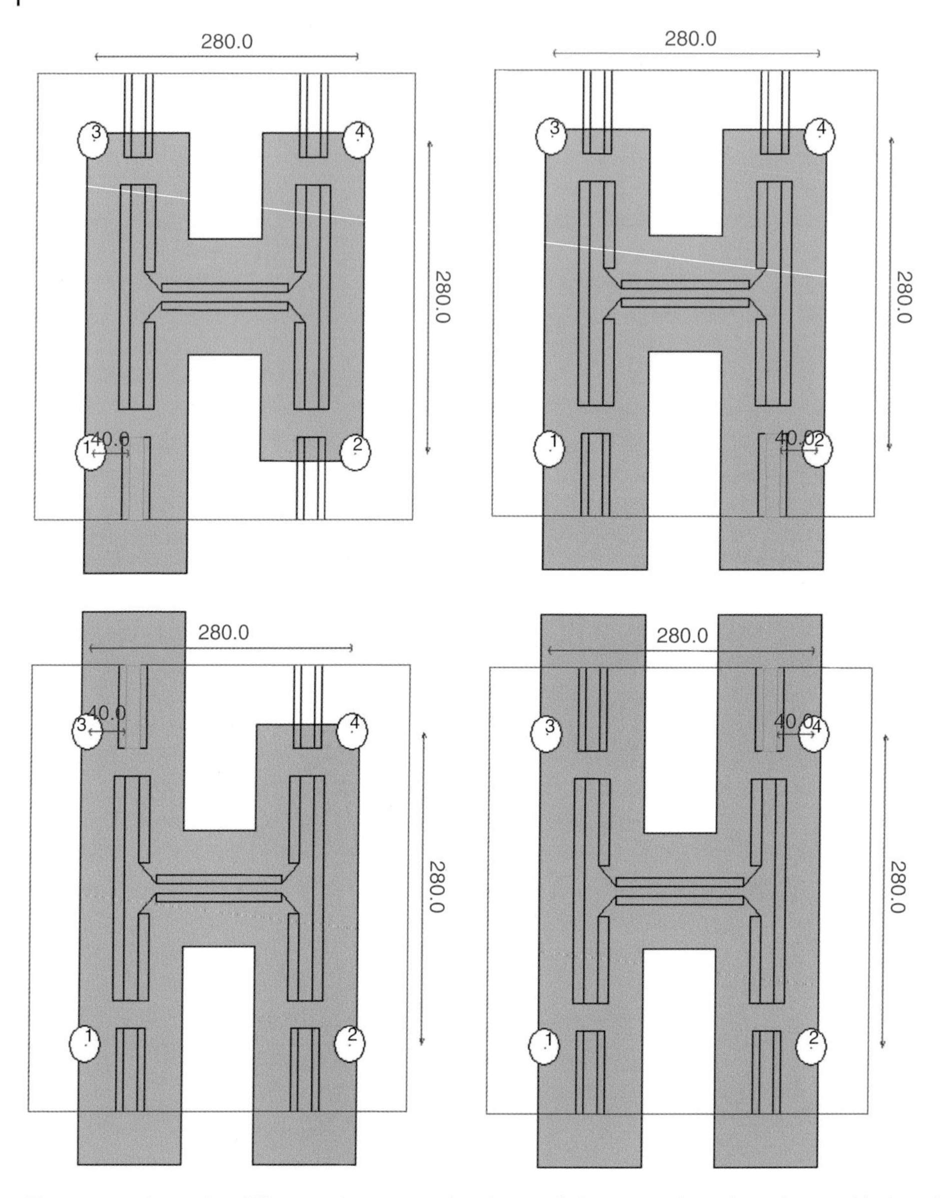

Figure 7.14 Second to fifth step: the terms related to each force transferrer have been added to the bearing surface. The resulting bearing surface becomes outside of the face of the bolted plate, an inadmissible condition.

is not acceptable. So, an *intersection* with this face must be applied, getting to the final bearing surface (see Figure 7.15, on the left).

$$\Sigma_{final} = \Sigma_5 \cap S_P$$

If the span c is different, a different bearing surface is obtained: for instance, in Figure 7.15 on the right, the bearing surface has been obtained by assuming $c = 10$ mm. Sometimes,

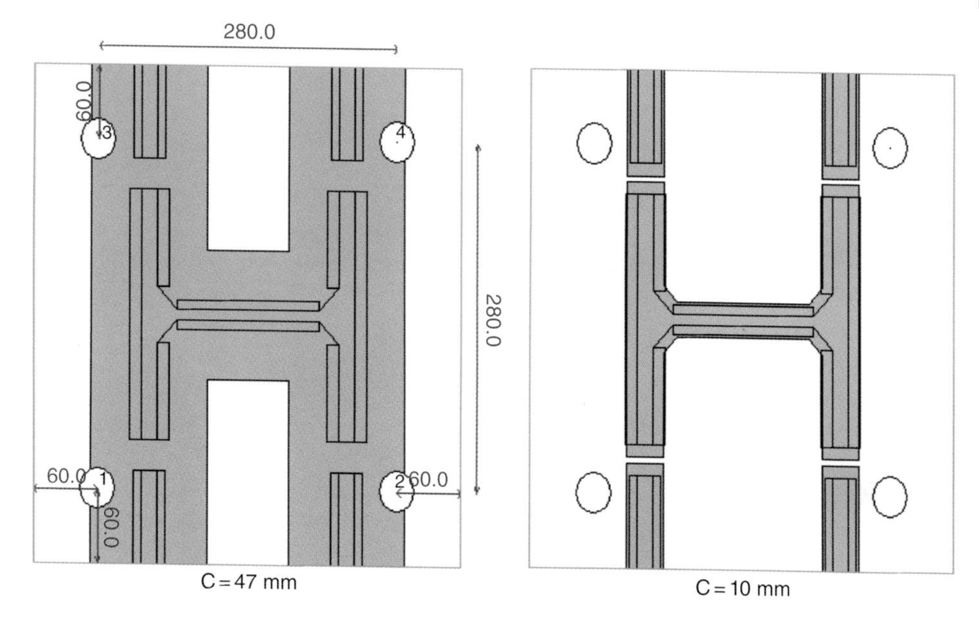

Figure 7.15 Final bearing surface: on the left, $c = 47$ mm, on the right $c = 10$ mm.

the extent of the bearing surface is such that it almost entirely covers the plate entry face. If this happens, the decision to assume the whole entry face as bearing surface can be made with good reason.

Depending on the geometries, very complex bearing surfaces can be defined. The method here outlined has no limitations, it can be applied to every cross-section face, and to every set of force transferrers. Also, hollow sections can be considered (see some examples in Figure 7.16).

In the author's experience, the following basic operations should be available in order to define the bearing surface (see also Figure 7.17). It is presumed that all the faces coplanar to the bolt-layout entry face have been collected and can be selected, one by one.

1) Set the current face as bearing surface (assign).
2) Set the current face bordered by c as bearing surface (assign).
3) Add the current face bordered by c to the bearing surface (union).
4) Add the border c (plus or minus) of the current face to the bearing surface (union).
5) Set as bearing surface the intersection of the current bearing surface to the current face (intersection).
6) Subtract from the current bearing surface the current face (subtraction).

Boolean operations between planar polygons, and polygon clipping, are not as easy a task as it might seem. The issue has been extensively studied in computer graphics and the interested reader may find algorithms and methods in the pertinent literature (e.g. Schneider and Eberly, 2003).

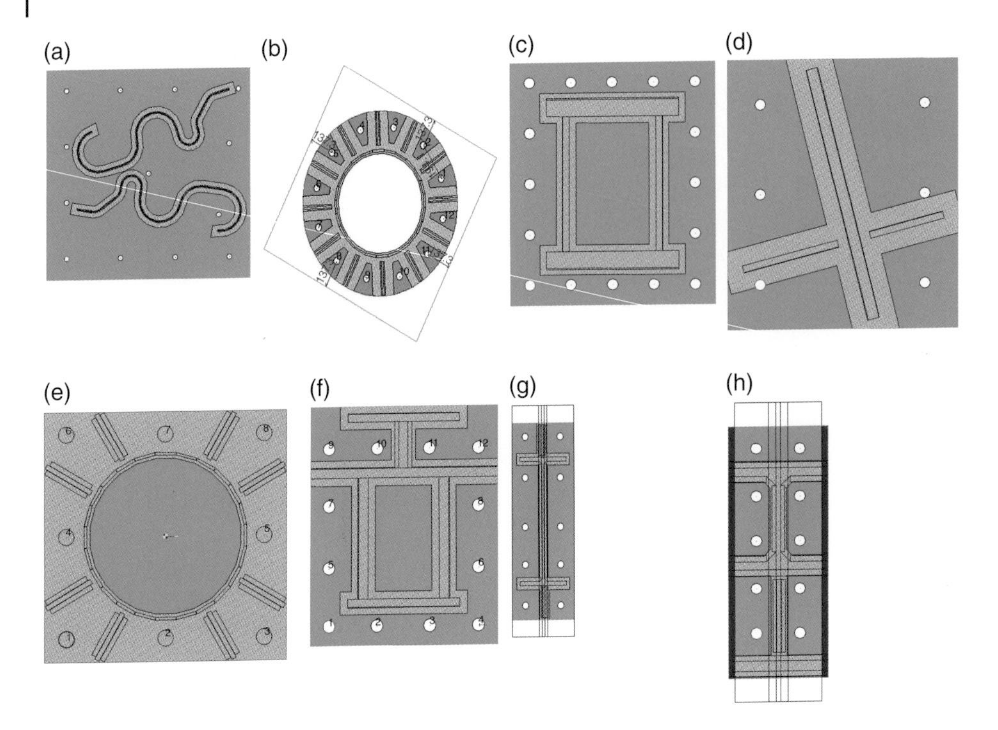

Figure 7.16 Bearing surface examples. With the exception of case A, a cold-formed "scribble" connected to a plate for test, all cases are taken from real engineering analyses. Case D is a bracket (courtesy B&M, France). Cases B and E use polygon subtraction due to the hole. Case H is a case where the plate bolted to the column flange is wider than the column flange: the red strips are not part of the bearing surface.(Courtesy Boema SpA, Neive, Italy, CE-N Civil Engineering Network, Bochum, Germany, Studio Capè Ingegneria srl, Milan, Italy.)

The final goal of the superposition of the pertinent Boolean operations is the candidate bearing surface, described as a collection of closed polylines, circulating anti-clockwise for full regions and clockwise for empty regions (holes).

This is where the exchange of contact pressures between the constituents bolted is expected and allowed. So, referring to all the principal axes of the bolt layout (**w**, **u**, **v**), the data will be

1) a bearing surface, i.e. a collection of closed polylines
2) a set of points where the bolts are positioned, their diameter and their hole diameter
3) the applied loads (N, M_u, M_v) for each loading combination.

The unknowns of the problem are:

1) the plastic neutral axis, dividing the bearing surface in a part reacting and in a part not reacting
2) the distribution of pressures over the reacting part of the bearing surface, and their intensity
3) the tensile forces in the bolts reacting
4) the bolts not reacting, as positioned in the compressed part of the plane.

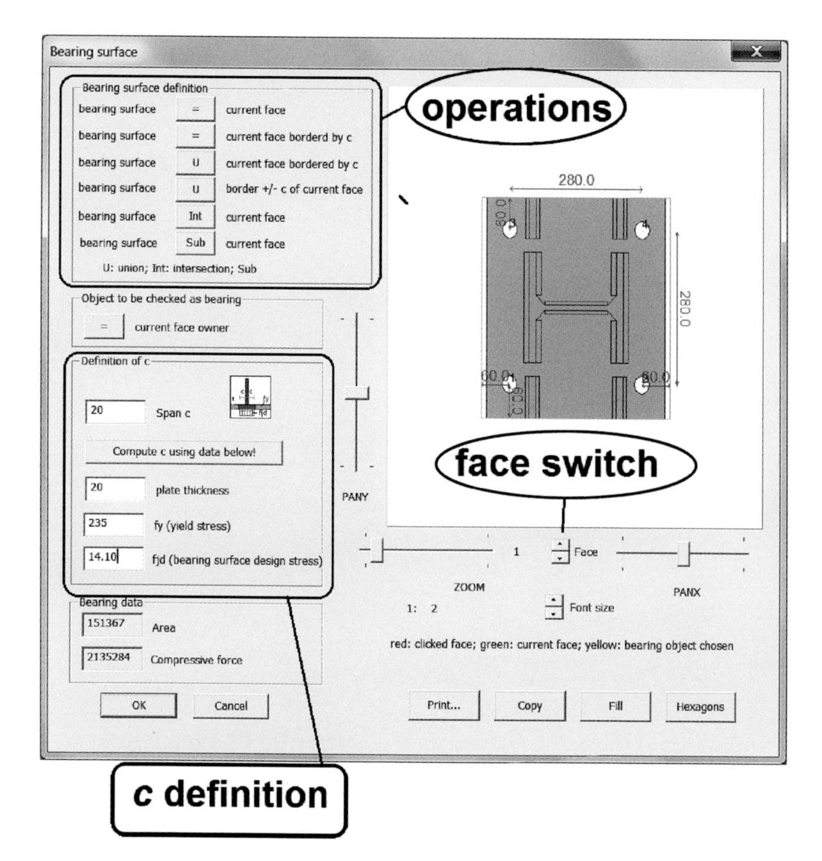

Figure 7.17 A set of controls designed in such a way as to allow the general definition of the bearing surface (*Connection Study Environment*, by the author).

7.5.2.2.3 Linear Strain Model

To solve the problem, a linear strain field may be assumed, so that if ε is the strain

$$\varepsilon = au + bv + c$$

and a, b, c, are three constants to be determined. The linear strain field assumes that the bearing surface rigidly translates and rotates with reference to an axis, named the plastic neutral axis, whose equation is immediately found once the strain is equated to zero:

$$au + bv + c = 0$$

Assuming that a well defined no-tension constitutive law $\sigma = \sigma(\varepsilon)$ has been assigned to the bearing (see next section), in each point of it, we can assume that

$$\sigma = \sigma(\varepsilon) \quad \text{if } \varepsilon < 0$$

$$\sigma = 0 \quad \text{if } \varepsilon \geq 0$$

On the other hand, for each bolt i, having principal coordinates (u_i, v_i), the force is

$$N_i = EA \cdot \varepsilon_i \quad \text{if } \varepsilon > 0$$

$$N_i = 0 \quad \text{if } \varepsilon \leq 0$$

with

$$\varepsilon_i = au_i + bv_i + c$$

This can be written as

$$N_i = E_i A \cdot \varepsilon_i$$

where

$$E_i = E \quad \text{if } \varepsilon > 0$$
$$E_i = 0 \quad \text{if } \varepsilon \le 0$$

The three conditions that must be satisfied are

$$N = \int_{\Sigma} \sigma dA + A\sum_{i=1}^{nb} E_i \varepsilon_i$$

$$M_u = \int_{\Sigma} \sigma v dA + A\sum_{i=1}^{nb} E_i \varepsilon_i v_i + n_b \cdot \frac{\pi D^4 E}{64} \cdot b$$

$$M_v = -\int_{\Sigma} \sigma u dA - A\sum_{i=1}^{nb} E_i \varepsilon_i u_i - n_b \cdot \frac{\pi D^4 E}{64} \cdot a$$

E being the Young's modulus of the bolts, D their shaft diameter, and A their tensile area (gross or threaded, but usually threaded). The unknowns are a, b, and c, that is, the position of the neutral axis is unknown. The integrals are extended over the bearing surface Σ. The solution of the problem is iterative, and is similar to those currently used in the analysis of reinforced concrete cross-sections. However, there are some differences:

- The constitutive law of the bearing is not in general that of concrete, although for base connections to concrete this may be a good choice.
- The bolts can be positioned out of the bearing surface, and not necessarily inside.
- The compressive stresses in the bearing surface are not usually enough to check the constituent acting as bearing. Instead, they must be considered as loads, normal to the mid-plane, applied to plated constituents.
- The relative efficiency of the bolts will not be the same, from bolt to bolt, because bolts far from the plate supports will be less efficient than bolts near it, it being the distance from the constraining part the element that has to be considered.

The last remark could suggest giving to the area of the bolt i an "efficiency factor" μ_{ni}, ranging from 0 to 1. If a bolt is declared shear only, for instance, as has been seen for welds, then taking μ as a very small number ($\approx 10^{-4}$) would result in

$$\mu_{ni} = \mu$$

With this correction, the previous equations are now

$$N = \int_{\Sigma} \sigma dA + A\sum_{i=1}^{nb} E_i \mu_{ni} \varepsilon_i$$

$$M_u = \int_{\Sigma} \sigma v dA + A \sum_{i=1}^{nb} E_i \mu_{ni} \varepsilon_i v_i + \frac{\pi D^4 Eb}{64} \cdot \sum_{i=1}^{nb} \mu_{ni}$$

$$M_v = -\int_{\Sigma} \sigma u dA - A \sum_{i=1}^{nb} E_i \mu_{ni} \varepsilon_i u_i - \frac{\pi D^4 E}{64} \cdot a \cdot \sum_{i=1}^{nb} \mu_{ni}$$

which are identical to the previous ones if

$$\mu_{ni} = 1 \quad \forall i$$

7.5.2.2.4 Stiffness of the Bearing Surface

The bearing constitutive law has a fundamental importance in deciding the distribution of the pressures.

If the bearing is stiff, the amount of the bearing surface reacting is limited, and the internal lever is maximum. This usually implies high concentrated pressures, and lower tensile forces in the bolts.

If, instead, the bearing is flexible, then a large amount of it will react, with lower pressures, and with a lower internal lever, which will result in higher tensile forces in the bolts.

All the constitutive laws assigned to the bearing must be compression-only, that is, it is assumed that the bearing surface does not react with tensile stresses.

There are several typical choices for the bearing constitutive law. This should not be considered a realistic choice, as the strain distribution can only occasionally be approximated linear (and this happens for thick plates or plates that are greatly stiffened). This should be considered as a way to drive the distribution of pressures and forces, under the shelter of the static theorem of limit analysis; irrespective of the choice of the constitutive law, all the distributions will be balanced with the externally applied loads. However, some are preferable as they are nearer to the ones experienced by the true connection.

The first natural choice is that of linear no-tension constitutive law ((a) in Figure 7.18). If the ratio m of the Young's modulus of bolts E_b to the notional Young's modulus of the bearing surface E_s is assigned, so that

$$m = \frac{E_b}{E_s}$$

very low values of m implying a very stiff bearing surface. Usually this implies that only a small surface in the vicinity of an edge in simple bending, or nearby a corner in biaxial bending, are going to react.

The second typical choice, at least if using Eurocode, is the parabola–rectangle constitutive law of concrete ((b) in Figure 7.18). This is typically used for the steel-to-concrete interface, as in base joints. The first strain ε_1 is usually −0.002, while the ultimate strain ε_u is usually −0.0035. The typical maximum compressive stress is the characteristic concrete crushing stress divided by a safety factor γ_c usually equal to 1.5. A drawback of this choice is that if the compression is excessive, no convergence can be reached in the iterative procedure: this means that the applied loads are higher than the limit for the bearing material, which is unable to exert the needed reacting pressures.

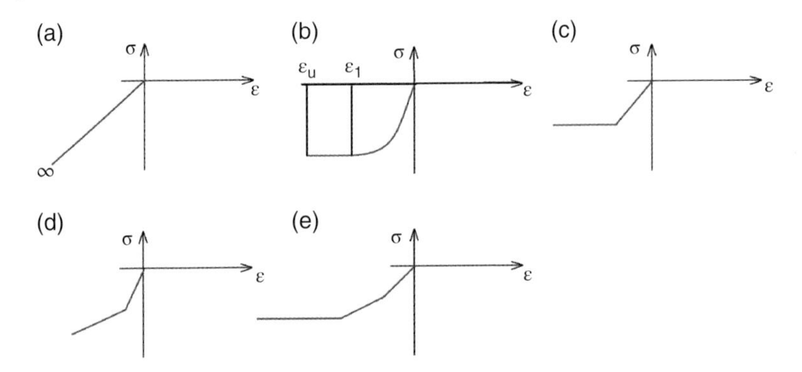

Figure 7.18 Typical constitutive laws for bearing surface.

Other possible laws are elastic perfectly plastic (c), bilinear (d), or trilinear ((e) in Figure 7.18). If a maximum compressive stress exists, then the convergence issue already mentioned for parabola–rectangle will possibly appear.

The "true" distribution of pressures at the interface can only be computed by means of advanced contact non-linear analysis. What it is happening here is that a balanced configuration of the internal pressures and tensile forces is searched.

If the constituents are stiff, and the candidate bearing surface has been set properly, the linear assumption is valid. But if the constituents are not stiffened, and the amount of bearing surface has not been properly set, then the distribution will not be very realistic. This will be made clear by the stress analysis of the plates that are used to exchange the forces and the pressures: under the computed pressures they will be found unable to resist. So, also if "some result" has been found, balanced with the externally applied loads, it is only when the stress analysis coherent with the results found confirms the capacity of the constituents to carry those loads, that the distribution will eventually be accepted. It must in particular be underlined that a mere check of the bolts is not enough.

7.5.2.2.5 *Examples*

Consider a base plate, and in order to have an idea of the role played by the bearing surface amount and constitutive law, a simple bending condition. The applied strong axis bending moment is 0.25 times the elastic limit for the column cross-section (HE220B, S235, $M_y = 43.21$ kNm, plate size 450 mm, thickness 50 mm). Several constitutive laws, at equal bearing surface are tested, at increasing stiffness:

1) parabola rectangle, with max. stress equal to −21.16 MPa, divided by $\gamma_c = 1.5$, $\varepsilon_1 = -0.002$ $\varepsilon_u = -0.0035$
2) no tension, linear elastic, with $m = 1$
3) no tension, linear elastic, with $m = 0.1$
4) no tension, linear elastic with $m = 0.01$
5) no tension linear elastic with $m = 0.001$

The results in Figure 7.19 clearly show that:

- the amount of the compressed part of the bearing surface decreases with increasing bearing surface stiffness
- the amount of the maximum compressive stress increases with increasing bearing surface stiffness

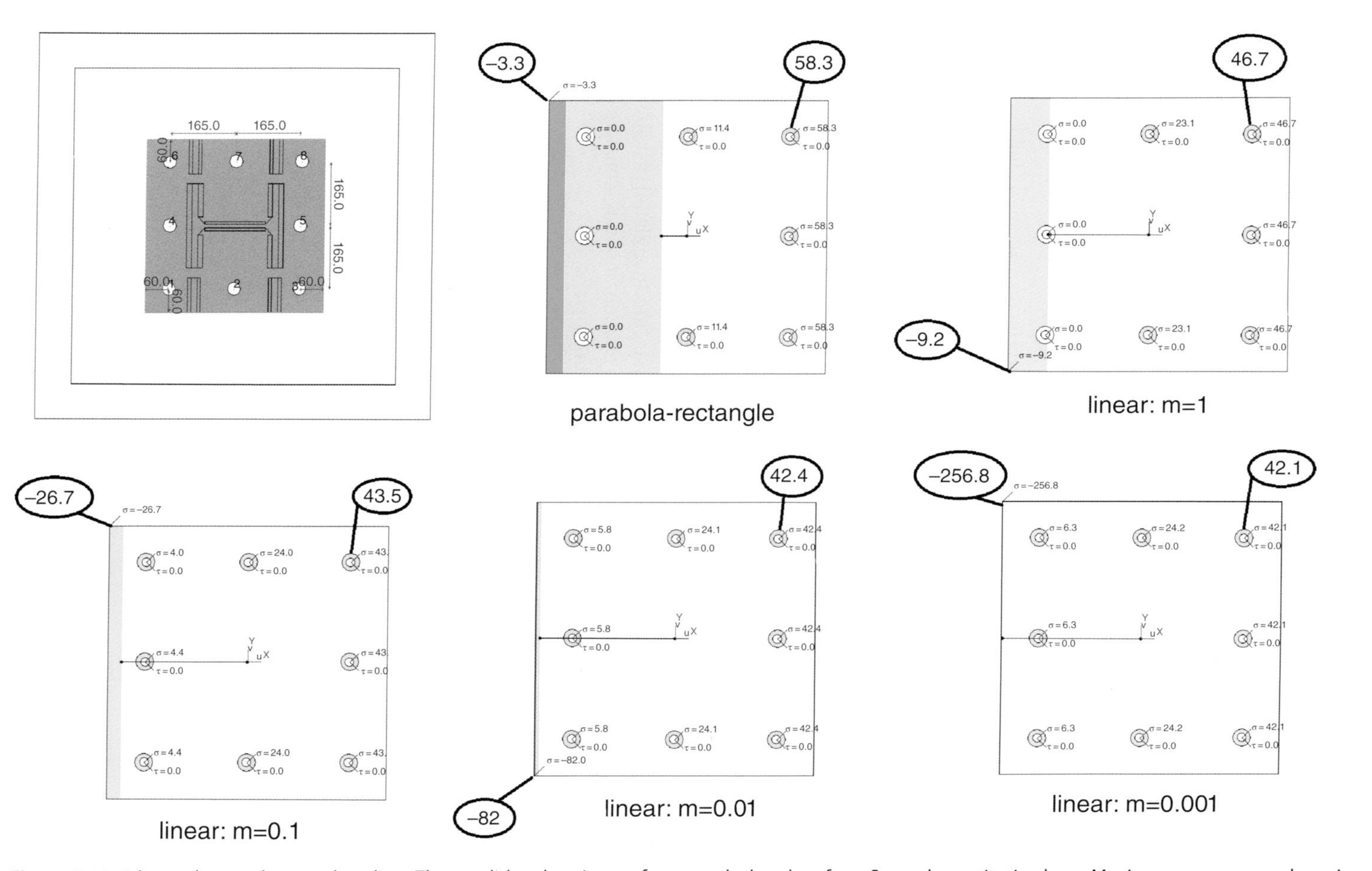

Figure 7.19 A base plate under pure bending. The candidate bearing surface equals the plate face. Several constitutive laws. Maximum pressure and maximum tensile stress in bolts are highlighted, MPa (results obtained from the CSE software, Connection Study Environment, written by the author).

- the amount of the maximum tensile stress in bolts decreases with increasing bearing surface stiffness
- rigid bearing surfaces tend to simulate a rotation along an edge of the plate, as it is often assumed in traditional approaches; increasing the stiffness of the bearing surface, the tensile stress in the bolts tends to assume an asymptotic value.

Now, a different bearing surface is considered, using the cantilever model explained in the previous sections, assuming $c = 47$ mm. The results are plotted in Figure 7.20.

As the amount of the bearing surface is lower, the maximum compressive stresses are higher, and so is the tensile stresses in the bolts. This latter effect is due to a very much lower internal lever arm. The remarks already made for the previous case still apply.

7.5.2.2.6 FEM Approach by Contact Non-Linearity

The linear strain assumption is not usually verified, and the displacement field of a stiffened plate in contact is not linear. If a finite element analysis is run, using contact non-linearity, the field of the pressures exchanged is usually complex and has a highly non-linear distribution. As contact non-linearity problems are usually more complex and time-consuming to solve, the linear hypothesis is useful and is still very often adopted, explicitly, or sometimes also implicitly.

In Chapter 9, dealing with finite element analysis, the issue of non-linear contact analysis will be covered in greater depth. For the moment, note that there are two alternative ways of computing the forces in the bolts and the pressures exchanged:

1) Assume a linear strain field and compute the forces and pressures accordingly, solving a non-linear problem at the section level. This does not require finite elements. It is the quicker way.
2) Use non-linear contact techniques and let the analysis find the related distribution of forces and stresses. This requires a non-linear finite element analysis, considering the contact between the parts. This is more precise, but slower.

7.6 Preloaded Bolt Layouts (Slip Resistant Bolt Layouts)

7.6.1 Preloading Effects

Preloading of bolts has a good effect on the stiffness of the joints, and the fatigue behavior of bolts, as it reduces the variation of tensile stress in bolts, depending on the externally applied loads. The bolts are preloaded by applying an initial tensile force N_i usually evaluated as

$$N_i = 0.7A_{res}f_{ub}$$

and divided by a suitable safety factor.

The plates bolted are then pushed one against the other, leading to high local compressive pressures. This mean that:

- the bolts are preloaded to a high tensile force
- the plates are locally compressed by the washers and by the bolt head and nut, and so their thickness is reduced, locally; the plate area involved in the compressive stresses depends on the geometry and it is roughly in the range of 2–3 times the bolt diameter (see Figure 7.21)

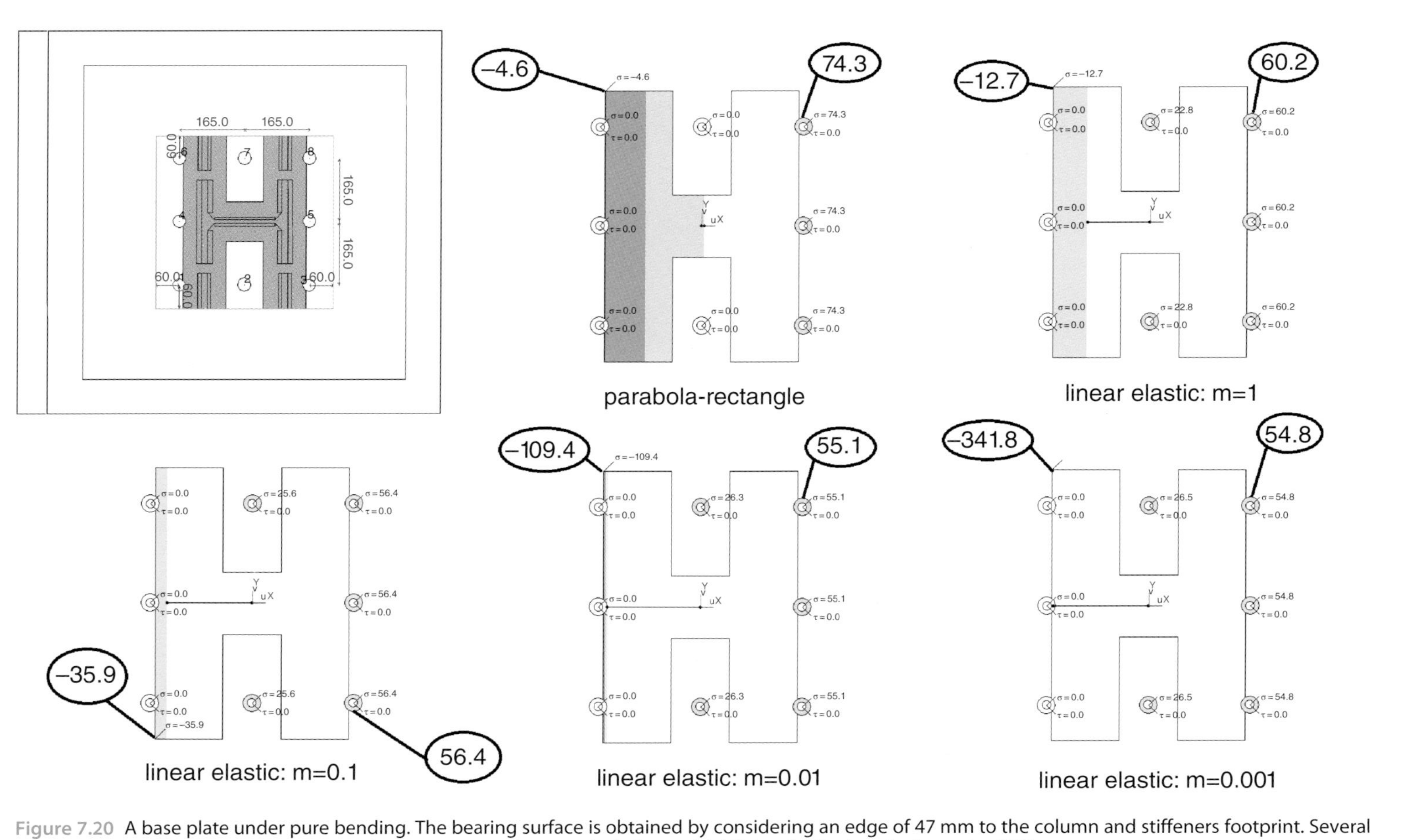

Figure 7.20 A base plate under pure bending. The bearing surface is obtained by considering an edge of 47 mm to the column and stiffeners footprint. Several constitutive laws. Maximum pressure and maximum tensile stress in bolts are highlighted, MPa (results from the CSE).

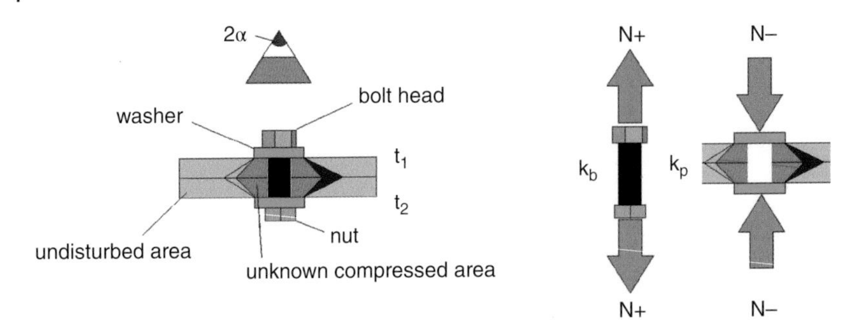

Figure 7.21 The effect of preload of bolts on the plates connected.

- the plates that are pushed one against the other cannot slide, as long as the tangential force applied is lower than the friction force available; this depends linearly on N_i and the friction coefficient μ, usually between 0.2 and 0.5, and the value of the friction coefficient depends on the surface treatment and on the standard used (see Chapter 8).

The amount of the area compressed has been extensively studied and there are several available methods for computing it. However, what must be first understood is the behavior of such a connection when a tensile or compressive force is applied to the bolt layout and then to the single preloaded bolt.

The compressed part of the plates and the bolt are usually considered as two springs in parallel, having different stiffness: the stiffness of the bolt, k_b, and the stiffness of the plate packet, k_p. Moreover, the stiffness of the whole packet can be considered, or the stiffness of each thickness separately.

Assuming this model, an external tensile force ΔN applied to the preloaded bolt will cause an increase of tensile force in the bolt, ΔN_b, and a *decrease in the compression* of the plate, ΔN_p. By equilibrium:

$$\Delta N_b = \frac{k_b}{k_b + k_p} \Delta N$$

$$\Delta N_p = \frac{k_p}{k_b + k_p} \Delta N$$

It turns out that the stiffness related to the plates, k_p, is often much higher than the stiffness of the bolt, k_b. So, the most part of the applied force ΔN will imply a decrease or an increase in the compressive forces carried by the plates. The increment of length of the bolt will exactly match the increment of thickness of the plates, as the model assumes springs in parallel. However, as the plates were compressed, their thickness after preload in the bolt region was lower than the nominal thickness before preload. So an increase of thickness of the plates will lead to a new thickness lower than the nominal one. When the detachment of the plates is reached, the stiffness k_p drops to 0, and only the bolt reacts as for normal bolts not preloaded.

The combined stiffness of the coupled system, bolt plus plates, is

$$k = k_p + k_b$$

The stiffness of the bolt, can be assumed equal to

$$k_b = \frac{EA_{res}}{L}$$

where A_{res} is the threaded area of the bolt, E the Young's modulus, and L the total thickness of the plates bolted. Some more refined models take into account the difference in length of the threaded part having area A_{res}, and the washer, bolt-head, and nut length. However, they are probably trying to get too much from what is a simple model, also considering that the plate stiffness k_p is much higher than k_b, and what is relevant is the ratio between the stiffnesses.

The stiffness related to the plates k_p depends on a number of factors:

- the ratio between the plate packet thickness, L and the bolt diameter, D
- the amount of material available in the vicinity of the bolt
- the confinement effect due to the ineffective material in the undisturbed region of the plates
- the friction forces exchanged at the interface between the plates and the related confinement effect
- the relative thickness of the different plates

As is well explained in the work by Brown et al., 2008 (which also has a good list of references), there are several possible methods for evaluating the stiffness of the plates. It is assumed here that all the plates are made of steel.

A first model considers the compressed part as a hollow cylinder, having external diameter QD and internal hole diameter D_0, that is the bolt hole diameter. The factor Q depends on the geometry and is often taken as 3. The stiffness then, is that of a compressed bar having length L, and area A_p equal to

$$A_p = \frac{\pi\left(Q^2D^2 - D_0^2\right)}{4}$$

and the stiffness is predicted as

$$k_p = \frac{EA_p}{L}$$

The ratio of the stiffnesses R would then be simply

$$R = \frac{A_p}{A_{res}} \approx \frac{\left(Q^2D^2 - D_0^2\right)}{D^2} \approx Q^2 - 1$$

using $Q = 3$, $R = 8$.

The model considers only a circular hollow limited region of the plates, under compression, not the whole of the plates connected. This is confirmed by finite element analyses, which show that at some distance from the bolt there is a detachment of the two surfaces (Figure 7.22). This is an effect that only 3D models can capture. So the plates are in contact only in limited regions around the bolts. There, in those limited regions, high pressures are exchanged.

This model of simple bar in compression does not take into account the confinement of the undisturbed material and the confinement at the friction interface. So, the true stiffness k_p would be higher. A more refined approach, due to Shigley, considers a

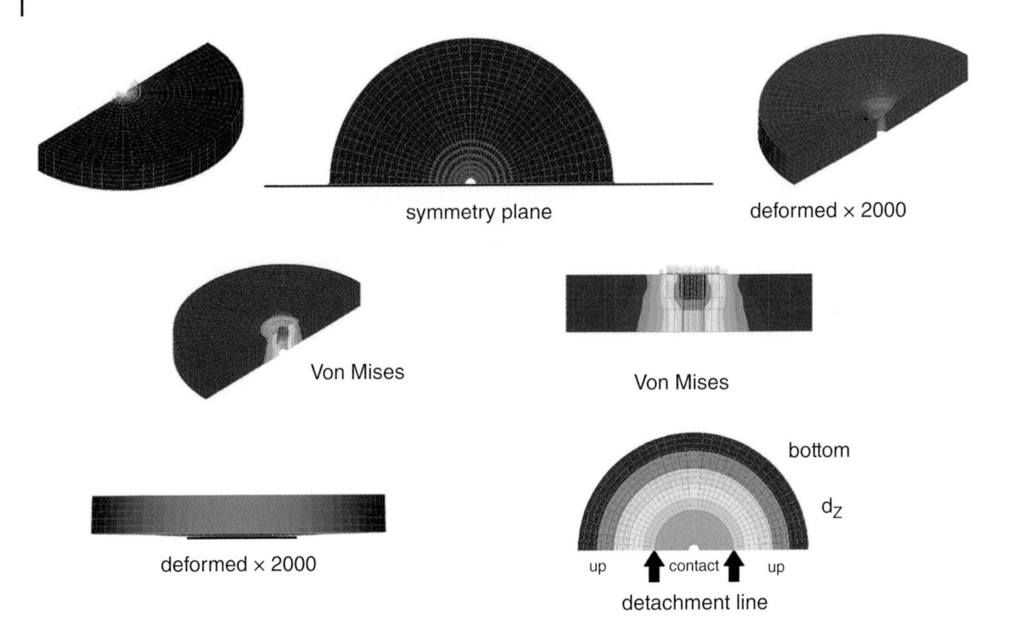

Figure 7.22 A test model for preload effects. Thickness of plate: 75 mm. Brick elements used. Number of elements 3780. Number of degrees of freedom 13625. The load is applied at the top surface as a constant pressure over a ring. The model rests on a set of frictionless non-linear no-tension springs. The green area (bottom right) is in contact. The remaining part lifts up.

resisting part having the shape of a hollow cone. The angle of the cone is taken as 2α (Figure 7.21), and the best results are usually obtained by assuming $\alpha = 30°$. The formula for the stiffness k_{pi} of one layer of thickness t_i, is

$$k_{pi} = \frac{\pi E D \tan\alpha}{\ln\left[\frac{(2t_i \tan\alpha + e - D)\cdot(e + D)}{(2t_i \tan\alpha + e + D)\cdot(e - D)}\right]}$$

where e is the diameter of the bolt head or nut. The stiffness of a packet of plates can be computed considering the springs in series. So if for instance there are two plates, the final plate stiffness would be

$$k_p = \frac{k_{p1} k_{p2}}{k_{p1} + k_{p2}}$$

As the thicknesses can be different, the hollow cone must match one another, so as to avoid discontinuities in the stress field. So the model is more complex and tricky to adopt than the previous one.

Other approaches use finite element techniques, and find best fit curves considering the finite element model results as target. A particularly simple one is that of Wileman et al. (1991). The method is applicable to a two thickness bolting, and if materials are equal.[3] There must be sufficient space around the bolt, so that edge effects can be

3 If the materials are different, then the method is still applicable with some modification.

neglected. If the thickness related to the packet is L, and the Young's modulus of steel is E, then

$$k_p = 0.78715 \cdot E \cdot D \cdot e^{0.62873\frac{D}{L}} \tag{7.9}$$

The confining effect of the surrounding plate metal is correctly taken into account, but the model used by Wileman et al. does not consider friction, and the nodes are free to move laterally, at the bottom nodes. So, a certain amount of increase in stiffness is to be expected.

Equating the stiffness predicted by Equation 7.9 to that of a hollow cylinder having external diameter D_{eq} and internal diameter D_0, and length L, the following relation for D_{eq} is found:

$$\frac{\pi E\left(D_{eq}^2 - D_0^2\right)}{4L} = 0.78715 \cdot E \cdot D \cdot e^{0.62873\frac{D}{L}}$$

$$D_{eq} = \sqrt{1.0022 \cdot L \cdot D \cdot e^{0.62873\frac{D}{L}} + D_0^2} \approx D\sqrt{\frac{L}{D} e^{0.62873\frac{D}{L}} + 1}$$

Naming the square root and its content Q, so that

$$Q = \sqrt{\frac{L}{D} e^{0.62873\frac{D}{L}} + 1} \tag{7.10}$$

and plotting it as a function of (L/D) as in Figure 7.23, we can see that the equivalent external diameter varies from 1.65 to 3.25 times D, when (L/D) varies between 0.5 and 10.[4] Moreover, in a wide range of (L/D), say from 0.5 to 1.5, Q is around 1.75. The curve may be interpolated by using

$$Q = 0.192 \cdot \frac{L}{D} + 1.58 \tag{7.11}$$

Using a resisting plate area circular, with no hole, and neglecting the overlapping that would result, the equivalent diameter $D_{eq}{}^*$ would be

$$D_{eq}^* = Q^* D$$

and

$$Q^* = \sqrt{\frac{L}{D} e^{0.62873\frac{D}{L}}} \tag{7.12}$$

This may be interpolated by

$$Q^* = 0.210 \cdot \frac{L}{D} + 1.26 \tag{7.13}$$

In civil engineering, in order to avoid a brittle shear fracture of the bolt, and to ensure ductility, the diameter of the bolt in bearing is usually higher than the thickness of the

4 The work by Wileman et al. considered (L/D) between 0.5 and 10.

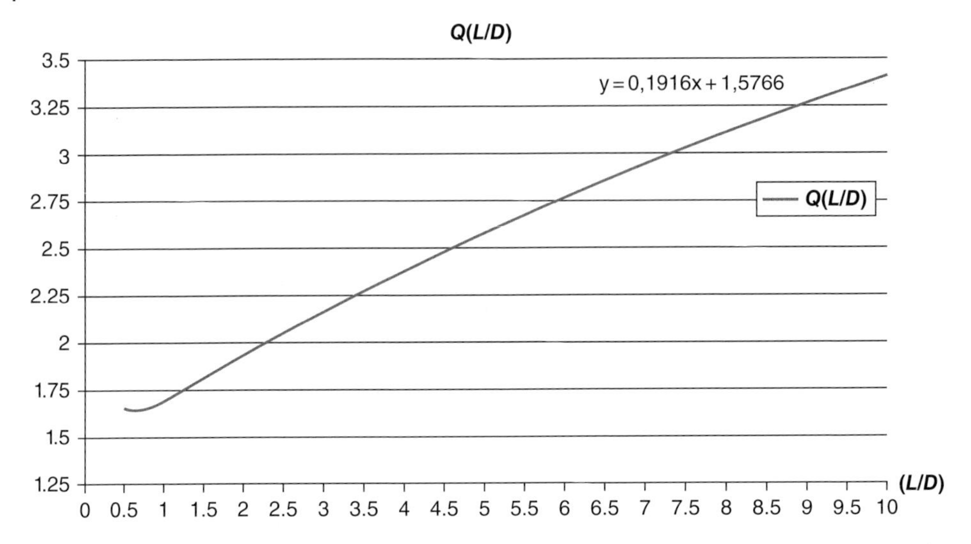

Figure 7.23 The curve $Q = Q(L/D)$. Q is the factor to be applied to diameter D, to get the external diameter D_{eq} of the equivalent resisting hollow cylinder, so as to give the same results of Wileman et al., 1991.

single plate bolted. Using Eurocode 3, and imposing that an average bolt bearing resistance is lower than the bolt shear resistance

$$2tDf_u < 0.6\frac{\pi D^2}{4}f_{ub}$$

meaning

$$D > 4.2t\frac{f_u}{f_{ub}} \approx 2t$$

The factor 2 is found assuming that

$$\frac{f_u}{f_{ub}} = 0.48$$

More precise evaluation is obtained considering Table 7.5 where several bolt classes and materials are considered.

So, if two thicknesses are bolted, (L/D) may be around 1, and Q around 1.75 (Figure 7.24 gives an idea of the proportions). For values of (L/D) higher than 1, as happens when the bolt diameter is a fraction of the total thickness bolted, the value of Q increases to 3 or more. If the bolts are closely spaced, and so thinner, the ratio (L/D) may become high, and Q reaches values of something like 3 or 4. Note that according to European standards, preloaded bolts resisting at ultimate limit state need not be checked for bolt bearing, while according to AISC standards they do have to be checked for bolt bearing.

The preload force has no effect on the plates bolted, with the exception of the local through-thickness compression shown in Figure 7.22. No bending of the connected

Table 7.5 Estimate of the minimum multiplier to be applied to single plate thickness, to get a bolt diameter such that the bolt bearing resistance is lower than the bolt shear resistance. It is assumed that the bolt resisting shear area is the gross one. The first column lists bolt classes, the first row material grades. The shaded cells are the most frequent couplings.

	S235	S275	S355
4.6–4.8	3.8	4.5	5.4
5.6–5.8	3.0	3.6	4.3
6.6–6.8	2.5	3.0	3.6
8.8	1.9	2.3	2.7
10.9	1.5	1.8	2.1
12.9	1.3	1.5	1.8

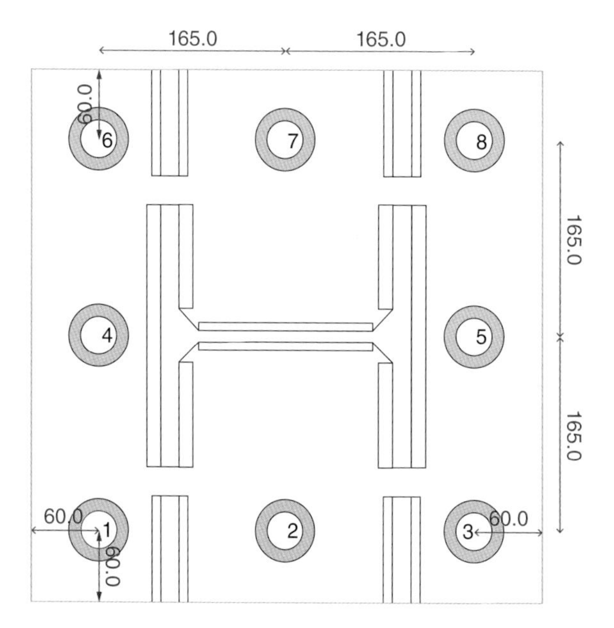

Figure 7.24 An example of bearing surface of a preloaded bolt layout. This is obtained by considering for each bolt a circular annulus of external diameter 1.75D and internal diameter D_0.

plates, in particular, is related to preload, unless imperfections and lack of planarity are taken into account.

However, if an external load is applied, the net force change flowing into the plates, normal to their mid-planes, will be that related to each single bolt.

Part of this net change in the force applied will be related to a variation of the compressive stresses in the circular region at plate interface (Figure 7.24). The other part will be related to a variation of the preload force in the bolt.

7.6.2 Shear and Torque

As long as no sliding occurs, the shear stiffness related to preloaded bolt layouts is very high, and so is the torsional stiffness. A possible lower bound for the shear stiffness k_s of no-slip bolts is to consider the sum of the stiffnesses related to the bolt shaft and to the equivalent compressed hollow section. With the symbols already introduced, instead of using

$$k_s = \frac{GA}{L} = \frac{G\pi D^2}{4L}$$

it is possible to use

$$k_s = \frac{G\pi D^2}{4L} + \frac{G\pi\left(Q^2D^2 - D_0^2\right)}{4L} \approx \frac{\pi GQ^2D^2}{4L} = Q^2\frac{GA}{L}$$

which means an increment of the factor Q^2 relative to the normal stiffness of a bolt alone.

If a complete circle is considered, then

$$k_s = \frac{G\pi D^2}{4L} + \frac{G\pi Q^{*2}D^2}{4L} = \frac{GA}{L}\left(1 + Q^{*2}\right)$$

which means an increment of a factor $(1 + Q*^2)$ relative to the normal stiffness of a bolt alone.

This model does not consider the confinement of the material around the cylinder, and so it is an underestimate of the stiffness. A reasonable rough estimate can be considered as

$$k_s = \frac{10GA}{L} \Leftrightarrow \frac{50GA}{L} \tag{7.14}$$

As slip-critical bolt layouts are usually isoconnected, the assumptions about individual bolt stiffness, and then group stiffness, do not imply a variation in the force packets flowing into the bolt group. If instead the bolt layouts connecting the two or more constituents are more than one, they will probably *all* be slip resistant, and so the relative stiffness will not be violated.

The stiffness related to the membrane deformation of the bolted plate is usually very high. However, it is also possible that a bolted constituent under the effect of the bolt shears undergoes non-negligible deformation. This additional flexibility, and the related stiffness, can only be evaluated, in general, by means of global finite element models.

The methods applied to bolt layouts in bearing, in order to evaluate the bolt forces starting from the layout force packets, are used also for preloaded bolts: namely, both the elastic method and the instantaneous center of rotation method.

The latter may trigger some doubts, as the typical force–displacement relation used by it is referred to bolts in bearing, and not to no-slip bolts. However, experimental tests and analytical studies (Kulak 1975) indicate that it is possible to conservatively extend the methods valid for bolt layouts in bearing and also to slip-critical connections. One important difference is that the forces exerted by the bolts at limit do not depend on the distance from the instantaneous center of rotation, but are all equal. While for bolts in bearing a single bolt is at maximum displacement, and the forces of the others are scaled according to their different displacements, for no-slip bolts it can be assumed that all bolts are at the slip-critical force (Kulak 1975).

7.6.3 Axial Force and Bending

When considering the behavior of a no-slip bolt layout under the effect of bending moments and axial force, two effects must be considered.

The first refers to the local stiffness of the bolt group, considering only the individual bolts. This stiffness can be obtained by considering that, due to preload, each bolt has an axial stiffness equal to $(k_p + k_b)$. The bending stiffness of the group is obtained by applying a unit rotation to each principal axis of the bolt layout. If for instance a rotation along u is considered, then the displacement of each bolt i is

$$d_{w,i} = v_i \cdot r_u$$

The elementary moment exerted by the bolt is

$$M_{ui} = \left(k_p + k_b\right) \cdot v_i^2 \cdot r_u$$

and the bending stiffness of the group k_{uu} is

$$k_{uu} = \frac{M_u}{r_u} = \frac{\sum_{i=1}^{nb} M_{ui}}{r_u} = \left(k_p + k_b\right) \cdot \sum_{i=1}^{nb} v_i^2$$

The translational axial stiffness of the group k_{ww} would be (assuming that all bolts are equal)

$$k_{ww} = nb \cdot \left(k_p + k_b\right)$$

The second effect (i.e. the second part of the stiffness) is the one "in series" with the previous one, and it depends on the flexibility of the plates connected, in the points where the bolts are placed. This is an out-of-plane flexibility, much higher than the in-plane one related to shear forces. As already seen for the shear forces, there is no general rule to assess this part of the stiffness; this depends on the number of bolts, the distance from welds acting as constraints, the thickness of the plates, and so on. Indeed, this problem is like the one already seen for the bolts in bearing under tensile forces: using a finite element model of the connection this stiffness would be correctly considered, otherwise it is very difficult to assess.

Once more, if the connector is isoconnected, then the force packet flowing into it is directly set by equilibrium. Then, the effect of the stiffness of the plates connected is managed by setting a proper distribution of forces between the bolts. The efficiency factor μ_{ni} already mentioned is a possible tool, otherwise pure fem must be used (see Chapter 10).

Two models can be used for preloaded bolt layouts, in order to compute the tensile and compressive forces flowing into each bolt, from the layout overall applied forces (i.e. from the force packet).

The first is the simple model using only the bolt shafts as reacting areas. Paradoxically, this model is good, as it considers the same stiffness for all bolts, that is, k_b. Even if the true stiffness is higher, as has been shown, since the stiffness of the bolts is always the same ($k_p + k_b$, assuming that all bolts have the same diameter and that there are no edge effects), the resulting forces are the same. The only difference is that this model assumes the stiffnesses concentrated in the bolt centers, with no extension, and so possibly the internal lever is underestimated, resulting in slightly higher bolt forces (Figure 7.25).

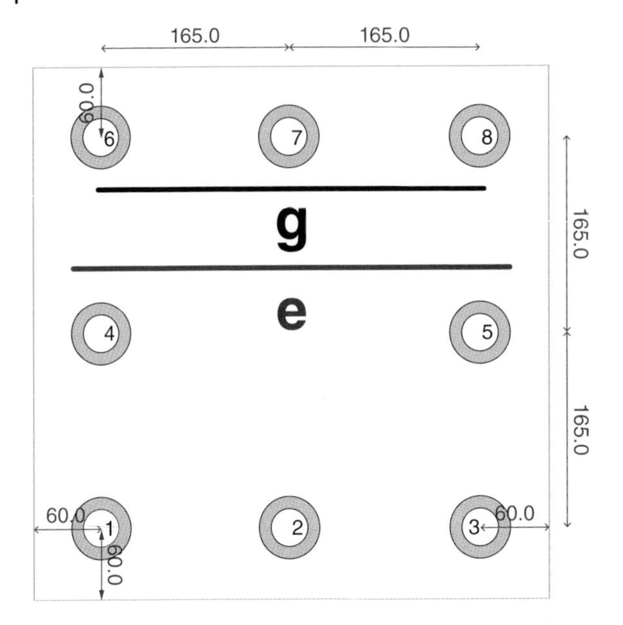

Figure 7.25 Variation of internal lever when bearing surface is assumed, bending around vertical axis.

The forces computed by this method will be the sum of two effects: the variation of compression of the plates, and the variation of the preload in the bolts. For the reasons explained, the first term is usually higher, and can be evaluated as

$$\Delta N_{ip} = \frac{k_p}{k_p + k_b} N_i$$

where N_i is the ith computed force, assumed acting in the bolt shaft of bolt i. The true variation of force flowing in the bolt is

$$\Delta N_{ib} = \frac{k_b}{k_p + k_b} N_i$$

and it can be used to check the bolt shaft and to check fatigue.

The second model that can be used computes the distribution of forces in the bolts and of compressive stresses in the reacting parts of the plates, assuming a bearing surface consisting of circles of proper radii, one for each bolt (Figure 7.25). This bearing surface, however, is considered to be reacting *both in tension and compression*, and the bolts also will react both in tension and compression. So the total reacting section is obtained by:

- the bearing surface, that is, all the circles of diameter Q^*D, if Equation 7.11 is used, or all the annuli of external diameter QD and internal hole diameter D_0, if Equation 7.10 is used, plus
- the areas of the bolts.

Moreover, the tensile and compressive stresses will vary over the resisting areas (bearing surface), increasing the internal lever as part of the bending is absorbed as bending in the bolt shaft and in the reacting circles.

The same constitutive laws of the bearing already seen for the bolts with no preload can still be used, but now *symmetrical in tension and compression*. If all the elements bolted are made of steel, and if a linear elastic constitutive law for the bearing is assumed, then the correct ratio between the Young's modulus of the bolts, and that of the bearing surface, should be 1. That is

$$m = \frac{E_b}{E_s} = 1$$

This model correctly computes the forces flowing into the bearing surface in the vicinity of each bolt, a force related to k_p, and the variation of force flowing into the bolt, related to k_b. If $m = 1$, then it is the ratio of the areas to drive the forces properly.

If this model is assumed, then the forces in the bolts are directly the variation of forces in the bolt shafts, that is

$$\Delta N_{ib} = N_i$$

while the forces flowing at the compressed interface between plates can be evaluated as

$$\Delta N_{ip} = \int_{Circle,\, i} \sigma_{com} dA = \frac{k_p}{k_b} N_i$$

As will be better seen in Chapter 9, the availability of these compressive stresses σ_{com} over the reacting circles (or annuli) may enable the correct creation of finite element model using plate–shell elements, loaded out of plane with forces balanced to these pressures.

7.7 Anchors

Anchors are typically used at the interface between the steel structure and other structural constituents, such as walls, plinths, and slabs.

The typical anchor is connected to concrete.

The evaluation of the stiffness of the bolted connection depends on a number of parameters, and on the way the anchor is realized. There are indeed very many types of anchors, with specific failure modes, and specific technical features. An in-depth analysis of anchors, also considering the reinforced concrete details, and using the component method, can be found in Kuhlmann, Wald et al., 2014.

One important remark is that, depending on the axial force applied to the bolt layout, the bearing surface can be always in compression or, for some load combinations, may be inactive due to tensile forces. Sometimes, when a bending is applied, a part of the candidate bearing surface reacts, on the compression side, while only bolts react on the tension side.

This means that the stiffness of the connection in general depends on the loads applied.

If it is possible to ensure that the bearing surface will always react, as when the permanent compressive forces are high, then the stiffness of the connection should take into account the concrete behavior, and this must be assumed with a symmetrical constitutive law.

If this is not the case, as when low permanent compressive forces are applied, it is possible to evaluate the *permanent* stiffness only considering the bolt shafts. In turn, the stiffness related to them depends on a number of parameters, the main one being the length of the anchors. Long lengths imply low stiffness.

In order to ensure that the compression in the bolt layout will always be present, a preload can be applied to bolts. This would ensure that the stiffness, both axial and rotational, is related to the concrete core reaction, in the region of the bearing surface.

The existence of compression between steel and grout ensures that friction resistance is available. However, if no preload is applied, the final compression on the concrete is a function that is very sensitive to the loads applied, and so, in order to rely on this friction mechanism to resist to horizontal forces, we must be sure that no tension will ever be applied, and that the compression force will always be sufficient. It is the author's opinion that at least in seismic areas, considering the high uncertainness of the evaluation of the loads, and of the load combinations, the friction resisting mechanism in the absence of preload should not be relied upon. A tragic example was seen in Italy after the Emilia 2012 earthquake, where many industrial pre-stressed reinforced concrete buildings failed because the portal transverse was simply supported by the columns, with no lateral force resisting mechanism except the friction.

The proper evaluation of the stiffness of the anchor connection may have high importance in order to properly model the behavior of the structure to lateral loads (and especially in evaluating P-Δ effects). However, the bolt layout used in a base plate or similar connections is rarely hyperconnected. That means that the force packet flowing into the bolt layout is known merely from equilibrium conditions, and no stiffness information is necessary for that aim if it is assumed that the member forces at extremities are already known, as has been done in this book.

The issue then becomes to correctly compute the forces flowing into the bolts, and the pressures stressing the concrete. This can be done by using the same models already seen for bolts in bearing and for preloaded bolt layouts.

If preloaded bolts are used, then it must be considered a problem related to the existence of two materials so as to arrive at a proper evaluation of the stiffness k_p. Usually the depth of concrete is enough to reach (L/D) values out of the range studied by Wileman et al. (i.e. $L/D > 10$).

Two springs in series may be considered equivalent to k_p.

The first spring, k_{ps}, is related to the steel plate compressed. Its stiffness may be evaluated by using the Wileman et al. formulation. So if the thickness of the base plate is t, and the diameter of the anchor is D, and E_p is the Young's modulus of the steel plate

$$k_{ps} = 0.78715 \cdot E_p \cdot D \cdot e^{0.62873\frac{D}{t}}$$

The second term is related to the stiffness of a concrete half space, loaded by constant pressure acting over a circular annulus having internal diameter D_0 and external diameter D_a, evaluated as

$$D_a = 1.5D + 2t\tan 30^\circ = 1.5D + 1.15t$$

The displacement δ of a half space under a pressure q applied to a circle of radius R is

$$\delta = \frac{2qR(1-\nu^2)}{E_c}$$

where ν is Poisson's ratio of concrete.

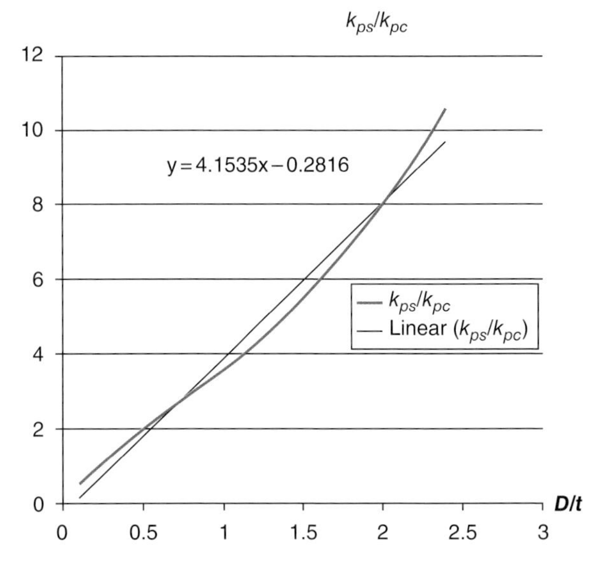

Figure 7.26 Ratio (k_{ps}/k_{pc}) as a function of (D/t).

The stiffness k_{pc} related to the circular annulus can be evaluated as[5]

$$k_{pc} = \frac{\pi(D_a + D_0)}{4} \cdot (1 - \nu^2) \cdot E_c$$

where E_c the Young's modulus of concrete. The resulting stiffness of the compressed part due to preloading can now be evaluated as

$$k_p = \frac{k_{ps} \cdot k_{pc}}{k_{ps} + k_{pc}}$$

Now if the ratio between the two stiffnesses is written as

$$\frac{k_{ps}}{k_{pc}} = \frac{0.78715 \cdot E_p \cdot D \cdot e^{0.62873\frac{D}{t}}}{\frac{\pi(1.5D + 1.15t + D_0)}{4} \cdot (1 - \nu^2) \cdot E_c} \cong \frac{1}{\left(2.5\frac{D}{t} + 1.15\right)} \cdot \frac{E_p}{E_c} \cdot \frac{D}{t} \cdot e^{0.62873\frac{D}{t}}$$

Assuming $E_p/E_c = 7$ ($E_p = 210$ GPa, $E_c = 30$ GPa) it can be seen (Figure 7.26) that

$$\frac{k_{ps}}{k_{pc}} \cong 4.15\frac{D}{t} - 0.281$$

As usual, to ensure proper repartition of preload stresses over concrete, $(D/t) < 0.5$, the two stiffnesses are comparable. The final stiffness can then be expressed as ($\nu = 0.18$)

5 The final displacement is computed by subtracting from this displacement, the displacement due to a pressure q, acting on a circle having radius r equal to the hole radius. The force applied is $F = \pi(R^2 - r^2)q$.

$$k_p = \frac{\left(4.15\frac{D}{t} - 0.281\right)k_{pc}^{\;2}}{\left(4.15\frac{D}{t} + 0.719\right)k_{pc}} = \frac{\left(4.15\frac{D}{t} - 0.281\right)}{\left(4.15\frac{D}{t} + 0.719\right)} \frac{\pi(2.5D + 1.15t)}{4} \cdot \left(1 - \nu^2\right) \cdot E_c =$$

$$= \frac{0.76\left(2.5\frac{D}{t} + 1.15\right) \cdot \left(4.15\frac{D}{t} - 0.281\right)}{\left(4.15\frac{D}{t} + 0.719\right)} \cdot t \cdot E_c = G\left(\frac{D}{t}\right) \cdot t \cdot E_c$$

and the curve $G(D/t)$ is plotted in Figure 7.27.

A linear regression shows that the curve $G(D/t)$ in the range $0.1 < D/t < 1$ may be approximated by

$$G \cong 2.24\frac{D}{t}$$

so

$$k_p \cong 2.24 \cdot D \cdot E_c$$

The stiffness of the preloaded anchor of length L, and area A, k_b, can be estimated as

$$k_b = \frac{E_b A}{L} = \frac{E_p A}{L}$$

In order to evaluate the forces in the bolts, both methods already seen for non-anchor preloaded bolts can be used.

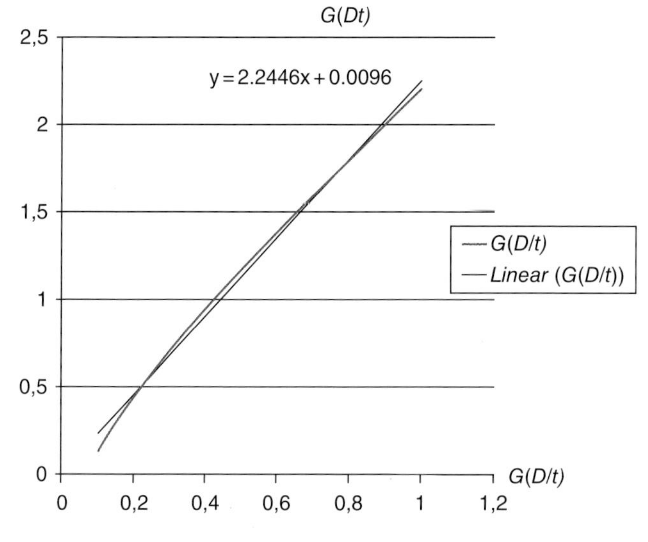

Figure 7.27 Curve $G = G(D/t)$. D is the anchor diameter, t is the steel plate thickness.

If the bearing surface method is used, and the assumed anchor length is L, then, equating the stiffnesses,

$$k_p = \frac{E_{bearing} A_{bearing}}{L}$$

and assuming

$$E_{bearing} = E_p = E_b$$

i.e. $m = 1$, the bearing surface area can be found as

$$A_{bearing} = \frac{k_p L}{E_p} = \frac{2.24 \cdot D \cdot L \cdot E_c}{E_p} = 0.32 \cdot D \cdot L$$

If a circular annulus of external diameter Qd and internal diameter D_0 is assumed,

$$A_{bearing} = \pi \frac{\left(Q^2 D^2 - D_0^2\right)}{4} = 0.32 \cdot D \cdot L$$

and $(D_0 \approx D)$, so

$$Q \cong \sqrt{0.32 \frac{L}{D} + 1} \quad (7.15)$$

If instead, a circle of diameter $Q^* D$ is assumed, then

$$Q^* \cong \sqrt{0.32 \frac{L}{D}} \quad (7.16)$$

The values of Q and Q^* are listed in Table 7.6 for several possible L/D values. The Q values are usually higher than the values previously seen for preloaded bolts steel to steel, as the concrete half space is stiffer than a cone of compressed material.

Summing up, if preloaded bolts are used in an anchor connection using a steel plate, the bearing surface may be obtained by considering for each anchor an annulus of diameter QD (or a circle of diameter $Q^* D$), and assuming $m = 1$. The value of Q (or Q^*) is a function or the ratio between the length of the anchor and its diameter, L/D (Table 7.6).

As the stiffness k_p does not depend on L, while the stiffness k_b is inversely proportional to L, to keep the ratio of the stiffnesses correct, the area of the bearing must be increased linearly with L.

Table 7.6 Values of Q and $Q*$ as a function of L/D.

L/D	25	50	75	100	125	150
Q	3	4.1	5	5.7	6.4	7
Q^*	2.8	4	4.9	5.6	6.3	6.9

7.8 Stiffness Matrix of Bolt Layouts and of Single Bolts

7.8.1 Generality

In this section the stiffness matrix of single bolts and the stiffness matrix of the whole bolt layout will be described.

The basic hypothesis, when considering the stiffness matrix of the whole layout, will be that of organized displacement, that is, of a linear displacement field. This means that the displacement of the extremity k $\mathbf{E_{ki}}$ of a bolt i can be described once the displacements of the whole bolt layout at that extremity, $\mathbf{E_k}$, are known.

The displacement vector of the whole bolt layout at extremity k will be referred to as $\mathbf{d_k}$, while the displacement vector of the bolt i at extremity k, will be referred to as $\mathbf{d_{ki}}$. Each vector has six components, three translations d_w, d_u, d_v, and three rotations r_w, r_u, r_v, so

$$\mathbf{d_k} = \{d_{wk} \;\; d_{uk} \;\; d_{vk} \;\; r_{wk} \;\; r_{uk} \;\; r_{vk}\}^T$$

and

$$\mathbf{d_{ki}} = \{d_{wki} \;\; d_{uki} \;\; d_{vki} \;\; r_{wki} \;\; r_{uki} \;\; r_{vki}\}^T$$

The displacements and rotations are referred to the principal axes of the bolt layout, (**w**, **u**, **v**). The position of a bolt in this reference system, for each extremity k, is identified by two coordinates (u_i, v_i). The position of a bolt with reference to the initial reference system is identified by the two coordinates (x_1, x_2).

If an efficiency factor μ_{ni} is assigned to the bolt i tensile stiffness (like the one introduced for welds but ranging with continuity in the interval $\mu - 1$), the coordinates (x_{1G}, x_{2G}) of the center **G** of the bolt layout are

$$x_{1G} = \frac{\sum_{i=1}^{nb} x_{1i}\mu_{ni}}{\sum_{i=1}^{nb} \mu_{ni}}$$

$$x_{2G} = \frac{\sum_{i=1}^{nb} x_{2i}\mu_{ni}}{\sum_{i=1}^{nb} \mu_{ni}}$$

It is now possible to compute the second area moment, relative to axes parallel to (1, 2) and having center G (D is the bolt nominal diameter, A_{res} the threaded area).

$$J_{1G} = \frac{\pi D^4}{64} \cdot \sum_{i=1}^{nb} \mu_{ni} + \sum_{i=1}^{nb} (x_{2i} - x_{2G})^2 \mu_{ni} A_{res}$$

$$J_{2G} = \frac{\pi D^4}{64} \cdot \sum_{i=1}^{nb} \mu_{ni} + \sum_{i=1}^{nb} (x_{1i} - x_{1G})^2 \mu_{ni} A_{res}$$

$$J_{12G} = \sum_{i=1}^{nb} (x_{1i} - x_{1G}) \cdot (x_{2i} - x_{2G}) \mu_{ni} A_{res}$$

and from this, the angle β between principal axis **u** and reference axis 1 (Figure 7.2) is

$$\beta = \frac{1}{2}\tan^{-1}\frac{2J_{12G}}{J_{2G} - J_{1G}}$$

The displacement vector of the extremity k of a bolt i can be computed once the displacement vector of the extremity k of the whole layout is known:

$$\mathbf{d_{ki}} = \begin{vmatrix} \mathbf{I} & \mathbf{S_i} \\ \mathbf{0} & \mathbf{I} \end{vmatrix} \cdot \mathbf{d_k} \equiv \mathbf{Q_i} \cdot \mathbf{d_k}$$

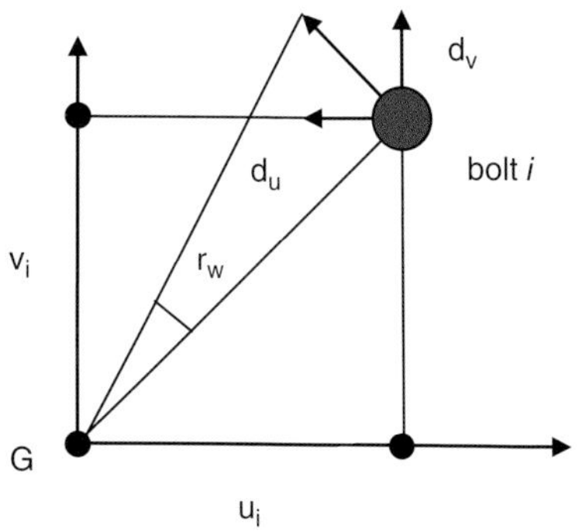

Figure 7.28 Displacements d_{ui} and d_{vi} of a bolt i, depending or r_w.

where $\mathbf{I}$ is a unit matrix (3 × 3), and $\mathbf{S_i}$ (Figure 7.28 to understand the first column of this matrix)

$$\mathbf{S_i} = \begin{vmatrix} 0 & v_i & -u_i \\ -v_i & 0 & 0 \\ u_i & 0 & 0 \end{vmatrix}$$

From the extremity k to the extremity $(k + 1)$, the stiffness matrix of a bolt can be expressed as a two-node matrix having 12 rows and 12 columns (12, 12), following the path already described for welds.

The stiffness matrix of a single bolt having n_e extremities, will be the assembly of the stiffness matrix of $(n_e - 1)$ two-nodes matrices (and therefore 12 × 12), referring to extremities k to $(k + 1)$, with k ranging from 1 to $(n_e - 1)$.

The stiffness matrix of the whole bolt layout can be obtained by assembling the n_b stiffness matrices of all the bolts having the extremities k to $(k + 1)$, and then assembling all the $(n_e - 1)$ stiffness matrices, each (12 × 12), obtained in this way.

So, for instance, if a three extremity bolt layout having nine bolts is analyzed ($n_e = 3, n_b = 9$), each stiffness matrix of each bolt can be seen as the assembly of the stiffness matrix 1-2 and the stiffness matrix 2-3, where 1, 2, 3 are the extremities. The stiffness matrix of the whole layout can be seen as the assembly of the whole layout stiffness matrix 1-2 plus the whole layout stiffness matrix 2-3. In turn, each of these matrices is the assembly of nine matrices $(1\text{-}2)_i$ for extremities 1-2, and nine matrices $(2\text{-}3)_i$ for extremities 2-3. The next sections will give formal rules.

7.8.2 Not Preloaded Bolts

7.8.2.1 General Form of the Stiffness Matrix of a Single Bolt and Its Assembly

The stiffness matrix of a single bolt i, between extremities k and $(k + 1)$, will be referred to as $\mathbf{K_{ik}}$, and it is a (12 × 12) square matrix.

Following the path already described for weld layouts, the stiffness matrix of a single bolt i, between extremity k and $(k + 1)$ will be expressed in the form of Equation 6.33, i.e. using 16 blocks (3 × 3), in the local system of the bolt.

$$\mathbf{K}_{ik} = \begin{vmatrix} \mathbf{A}_{ik} & \mathbf{B}_{ik} & -\mathbf{A}_{ik} & \mathbf{B}_{ik} \\ \mathbf{B}_{ik}^{T} & \mathbf{D}_{ik} & -\mathbf{B}_{ik}^{T} & \mathbf{E}_{ik} \\ -\mathbf{A}_{ik} & -\mathbf{B}_{ik} & \mathbf{A}_{ik} & -\mathbf{B}_{ik} \\ \mathbf{B}_{ik}^{T} & \mathbf{E}_{ik} & -\mathbf{B}_{ik}^{T} & \mathbf{D}_{ik} \end{vmatrix}$$

The local system of the bolt uses axes (**wi, ui, vi**) that are parallel to global layout axes (**w, u, v**), the only difference being the origin: the origin of the local reference system of a single bolt is merely (u_i, v_i).

The matrix $\mathbf{K}_{ik}$ valid in the reference system with center at the center of the bolt can be transformed to the principal system of the bolt layout, so getting $\mathbf{K}_{ikl}$, by the following rule (i is the bolt index, k is the extremity index, l stands for "layout"):

$$\mathbf{K}_{ikl} = \begin{vmatrix} \mathbf{Q}_i^T & \mathbf{0} \\ \mathbf{0} & \mathbf{Q}_i^T \end{vmatrix} \cdot \mathbf{K}_{ik} \cdot \begin{vmatrix} \mathbf{Q}_i & \mathbf{0} \\ \mathbf{0} & \mathbf{Q}_i \end{vmatrix}$$

Now, the assembled stiffness matrix of the bolt layout between extremities k and $(k+1)$ is just

$$\mathbf{K}_{kl} = \sum_{i=1}^{nb} \mathbf{K}_{ikl}$$

This means that the following blocks are obtained in the principal reference system of the bolt layout:

$$\mathbf{K}_{kl} = \sum_{i=1}^{nb} \begin{vmatrix} \mathbf{A}_{ik} & \mathbf{A}_{ik}\mathbf{S}_i + \mathbf{B}_{ik} & -\mathbf{A}_{ik} & -\mathbf{A}_{ik}\mathbf{S}_i + \mathbf{B}_{ik} \\ \mathbf{S}_i^T\mathbf{A}_{ik} + \mathbf{B}_{ik}^T & \mathbf{S}_i^T\mathbf{A}_{ik}\mathbf{S}_i + \mathbf{S}_i^T\mathbf{B}_{ik} + \mathbf{B}_i^T\mathbf{S}_i + \mathbf{D}_i & -\mathbf{S}_i^T\mathbf{A}_{ik} - \mathbf{B}_{ik}^T & -\mathbf{B}_{ik}^T\mathbf{S}_i - \mathbf{S}_i^T\mathbf{A}_{ik}\mathbf{S}_i + \mathbf{E}_{ik} + \mathbf{S}_i^T\mathbf{B}_{ik} \\ -\mathbf{A}_{ik} & -\mathbf{A}_{ik}\mathbf{S}_i - \mathbf{B}_{ik} & \mathbf{A}_{ik} & \mathbf{A}_{ik}\mathbf{S}_i - \mathbf{B}_{ik} \\ -\mathbf{S}_i^T\mathbf{A}_{ik} + \mathbf{B}_{ik}^T & -\mathbf{S}_i^T\mathbf{A}_{ik}\mathbf{S}_i - \mathbf{S}_i^T\mathbf{B}_{ik} + \mathbf{E}_{ik} + \mathbf{B}_{ik}^T\mathbf{S}_i & \mathbf{S}_i^T\mathbf{A}_{ik} - \mathbf{B}_{ik}^T & \mathbf{S}_i^T\mathbf{A}_{ik}\mathbf{S}_i - \mathbf{S}_i^T\mathbf{B}_{ik} + \mathbf{D}_{ik} - \mathbf{B}_{ik}^T\mathbf{S}_i \end{vmatrix}$$

The following different blocks are found:

1) block number 1: $\mathbf{A}_{ik}$
2) block number 2: $\mathbf{A}_{ik}\mathbf{S}_i$
3) block number 3: $\mathbf{B}_{ik}$
4) block number 4: $\mathbf{S}_i^T\mathbf{A}_{ik}\mathbf{S}_i$
5) block number 5: $\mathbf{S}_i^T\mathbf{B}_{ik}$
6) block number 6: $\mathbf{D}_{ik}$
7) block number 7: $\mathbf{E}_{ik}$

The matrix $\mathbf{K}_{kl}$ can be partitioned into four blocks (6 × 6) as follows

$$\mathbf{K}_{kl} = \begin{vmatrix} \mathbf{K}_{k,l,11} & \mathbf{K}_{k,l,12} \\ \mathbf{K}_{k,l,21} & \mathbf{K}_{k,l,22} \end{vmatrix}$$

The assembled matrix of the whole layout, $\mathbf{K}_l$, takes the form

$$\mathbf{K}_l = \begin{vmatrix} \mathbf{K}_{1,l,11} & \mathbf{K}_{1,l,12} & \mathbf{0} & \dots & \mathbf{0} \\ \mathbf{K}_{1,l,21} & \mathbf{K}_{1,l,22} + \mathbf{K}_{2,l,11} & \mathbf{K}_{2,l,12} & \dots & \\ \mathbf{0} & \mathbf{K}_{2,l,21} & \mathbf{K}_{2,l,22} + \mathbf{K}_{3,l,11} & \dots & \\ \dots & \dots & \dots & \dots & \dots \\ \mathbf{0} & \mathbf{0} & \mathbf{0} & \dots & \mathbf{K}_{(ne-1),l,22} \end{vmatrix}$$

The force packets acting at each extremity can be computed as

$$\begin{vmatrix} \mathbf{s}_1 \\ \mathbf{s}_2 \\ \dots \\ \mathbf{s}_{ne} \end{vmatrix} = \mathbf{K}_l \cdot \begin{vmatrix} \mathbf{d}_1 \\ \mathbf{d}_2 \\ \dots \\ \mathbf{d}_{ne} \end{vmatrix}$$

once the displacement vectors at each extremity k, $\mathbf{d}_k$, are known by solving.

The problem is now to express the blocks of the stiffness matrix of a single bolt, $\mathbf{A}_{ik}$, $\mathbf{B}_{ik}$, $\mathbf{D}_{ik}$ and $\mathbf{E}_{ik}$.

The block $\mathbf{A}_{ik}$ is related to the direct translational stiffnesses of the bolt. If k_{Nik} is the axial stiffness and k_{Vik} the shear stiffness, then $\mathbf{A}_{ik}$ can be written as

$$\mathbf{A}_{ik} = \begin{vmatrix} k_{Nik} & 0 & 0 \\ 0 & k_{Vik} & 0 \\ 0 & 0 & k_{Vik} \end{vmatrix}$$

The block $\mathbf{B}_{ik}$, collects the indirect rotational stiffness due to translations. It can be written as

$$\mathbf{B}_{ik} = \begin{vmatrix} 0 & 0 & 0 \\ 0 & 0 & \frac{k_{Vik} L_k}{2} \\ 0 & -\frac{k_{Vik} L_k}{2} & 0 \end{vmatrix}$$

and

$$L_k = \frac{t_k + t_{k+1}}{2}$$

The block $\mathbf{D}_{ik}$ collects the direct rotational stiffnesses. The torsional stiffness of a single bolt is assumed null, so

$$\mathbf{D}_{ik} = \begin{vmatrix} 0 & 0 & 0 \\ 0 & R_{ik} & 0 \\ 0 & 0 & R_{ik} \end{vmatrix}$$

The block $\mathbf{E_{ik}}$ collects the indirect translational stiffnesses due to rotations:

$$\mathbf{E_{ik}} = \begin{vmatrix} 0 & 0 & 0 \\ 0 & K_{ik} & 0 \\ 0 & 0 & K_{ik} \end{vmatrix}$$

In a normal Timoshenko beam element, considering only the bolt, the terms R_{ik} and K_{ik} can be expressed as

$$R_{ik} = k_{Mik} + k_{Vik}\,\frac{L_k^2}{3}$$

$$K_{ik} = -k_{Mik} + k_{Vik}\,\frac{L_k^2}{6}$$

where

$$k_{Mik} = \frac{\pi D^4 E}{64 L_k}$$

So, the whole matrix $\mathbf{K_{ik}}$ of a single bolt from extremity k to $(k+1)$ can be written in the general form:

$$\mathbf{K}_{\mathbf{ik}} = \left|\begin{array}{ccc|ccc|ccc|ccc}
k_{Nik} & 0 & 0 & 0 & 0 & 0 & -k_{Nik} & 0 & 0 & 0 & 0 & 0 \\
0 & k_{Vik} & 0 & 0 & 0 & \frac{k_{Vik}L_k}{2} & 0 & -k_{Vik} & 0 & 0 & 0 & \frac{k_{Vik}L_k}{2} \\
0 & 0 & k_{Vik} & 0 & -\frac{k_{Vik}L_k}{2} & 0 & 0 & 0 & -k_{Vik} & 0 & -\frac{k_{Vik}L_k}{2} & 0 \\
\hline
0 & 0 & 0 & 0 & 0 & 0 & 0 & 0 & 0 & 0 & 0 & 0 \\
0 & 0 & -\frac{k_{Vik}L_k}{2} & 0 & R_{ik} & 0 & 0 & 0 & \frac{k_{Vik}L_k}{2} & 0 & K_{ik} & 0 \\
0 & \frac{k_{Vik}L_k}{2} & 0 & 0 & 0 & R_{ik} & 0 & -\frac{k_{Vik}L_k}{2} & 0 & 0 & 0 & K_{ik} \\
\hline
-k_{Nik} & 0 & 0 & 0 & 0 & 0 & k_{Nik} & 0 & 0 & 0 & 0 & 0 \\
0 & -k_{Vik} & 0 & 0 & 0 & -\frac{k_{Vik}L_k}{2} & 0 & k_{Vik} & 0 & 0 & 0 & -\frac{k_{Vik}L_k}{2} \\
0 & 0 & -k_{Vik} & 0 & \frac{k_{Vik}L_k}{2} & 0 & 0 & 0 & k_{Vik} & 0 & \frac{k_{Vik}L_k}{2} & 0 \\
\hline
0 & 0 & 0 & 0 & 0 & 0 & 0 & 0 & 0 & 0 & 0 & 0 \\
0 & 0 & -\frac{k_{Vik}L_k}{2} & 0 & K_{ik} & 0 & 0 & 0 & \frac{k_{Vik}L_k}{2} & 0 & R_{ik} & 0 \\
0 & \frac{k_{Vik}L_k}{2} & 0 & 0 & 0 & K_{ik} & 0 & -\frac{k_{Vik}L_k}{2} & 0 & 0 & 0 & R_{ik}
\end{array}\right| \tag{7.17}$$

Up to now it has been assumed that the bolts react also with bending. It is also possible to use a model that totally neglects the bending in the bolt shafts, assuming null for all the rotational terms, and using a (6×6) stiffness matrix instead of a (12×12) one. This matrix would then be as follows (using the vector $\mathbf{d} = \{d_{w1}, d_{u1}, d_{v1}, d_{w2}, d_{u2}, d_{v2}\}^{\mathrm{T}}$):

$$\mathbf{K}_{\mathbf{ik}} = \left|\begin{array}{ccc|ccc}
k_{Nik} & 0 & 0 & -k_{Nik} & 0 & 0 \\
0 & k_{Vik} & 0 & 0 & -k_{Vik} & 0 \\
0 & 0 & k_{Vik} & 0 & 0 & -k_{Vik} \\
\hline
-k_{Nik} & 0 & 0 & k_{Nik} & 0 & 0 \\
0 & -k_{Vik} & 0 & 0 & k_{Vik} & 0 \\
0 & 0 & -k_{Vik} & 0 & 0 & k_{Vik}
\end{array}\right| \tag{7.18}$$

In the following, the formulation will be carried out using the matrix (12 × 12). However, it is straightforward to adopt the simpler (6 × 6) version, when needed.

Considering now the summed blocks of the layout matrix $\mathbf{K_{kl}}$, the following expressions are found for the blocks obtained by matrix multiplications:

$$\mathbf{A_{ik}S_i} = \begin{vmatrix} 0 & k_{Nik}v_i & -k_{Nik}u_i \\ -k_{Vik}v_i & 0 & 0 \\ k_{Vik}u_i & 0 & 0 \end{vmatrix}$$

$$\mathbf{S_i^T A_{ik} S_i} = \begin{vmatrix} k_{Vik}\left(u_i^2 + v_i^2\right) & 0 & 0 \\ 0 & k_{Nik}v_i^2 & -k_{Nik}u_iv_i \\ 0 & -k_{Nik}u_iv_i & k_{Nik}u_i^2 \end{vmatrix} = \begin{vmatrix} k_{Vik}r_i^2 & 0 & 0 \\ 0 & k_{Nik}v_i^2 & -k_{Nik}u_iv_i \\ 0 & -k_{Nik}u_iv_i & k_{Nik}u_i^2 \end{vmatrix}$$

$$\mathbf{S_i^T B_{ik}} = \begin{vmatrix} 0 & -\dfrac{k_{Vik}L_k u_i}{2} & -\dfrac{k_{Vik}L_k v_i}{2} \\ 0 & 0 & 0 \\ 0 & 0 & 0 \end{vmatrix}$$

The center of the bolt layout and its principal axes are computed in such a way that for each k, if C_k is a suitable value depending on the extremities connected, but not on the bolt, then

$$\sum_{i=1}^{nb} k_{Nik}u_i = C_k \sum_{i=1}^{nb} \mu_{ni}u_i = 0$$

$$\sum_{i=1}^{nb} k_{Nik}v_i = C_k \sum_{i=1}^{nb} \mu_{ni}v_i = 0$$

$$\sum_{i=1}^{nb} k_{Nik}u_iv_i = C_k \sum_{i=1}^{nb} \mu_{ni}u_iv_i = 0$$

If all μ_{ni} are equal, the familiar relationships are found:

$$\sum_{i=1}^{nb} u_i = 0$$

$$\sum_{i=1}^{nb} v_i = 0$$

$$\sum_{i=1}^{nb} u_iv_i = 0$$

In the formulation of the blocks, no difference has been assumed between the shear behavior in u and v direction. This is coherent with the force displacement law of AISC, but Eurocode 3 makes the shear stiffness depend also on the distance from the edges, and the bolt pitch, in the direction of the force. A more refined model would then have to consider that each bolt might have different shear stiffness from the siblings, due to differences in bolt bearing stiffnesses. It is the author's opinion that this is an unnecessary complexity; moreover the experimental work by Rex and Easterling (Rex and Easterling 2003) has shown that the stiffness does not change much with width and edge distance. The main variation of the shear stiffness is related to the thickness t of the bolted plate, not to the edge distance and pitch. So, for these reason this difference is here neglected.

Due to the existence of bolt bearing, the terms of the stiffness matrix related to shear and bending must be considered.

7.8.2.2 The Effect of Bolt Bearing

The problem of the coupling of bolt and bearing stiffness is tackled considering a stiffness matrix limited to translation d_u and rotation r_v, at first and second node, where two more translational degrees of freedom are added, one related to the bearing at first node, d_{be1}, and one related to the bearing at second node, d_{be2}. In the other direction, the behavior is identical.

The stiffnesses related to bearing are named k_{be1} and k_{be2}, the stiffness related to the bolt, k_s. The rows and columns are then reordered, so as to keep d_{u1} and d_{u2} as the last two terms, so that $\mathbf{d} = \{d_{be1}, r_{v1}, d_{be2}, r_{v2}, d_{u1}, d_{u2}\}^T$.

This stiffness matrix is (the indices i and k are here omitted for brevity)

$$\mathbf{K} = \begin{vmatrix} k_{be1} & 0 & 0 & 0 & -k_{be1} & 0 \\ 0 & R & 0 & K & \frac{k_s L}{2} & -\frac{k_s L}{2} \\ 0 & 0 & k_{be2} & 0 & 0 & -k_{be2} \\ 0 & K & 0 & R & \frac{k_s L}{2} & -\frac{k_s L}{2} \\ -k_{be2} & \frac{k_s L}{2} & 0 & \frac{k_s L}{2} & k_s + k_{be1} & -k_s \\ 0 & -\frac{k_s L}{2} & -k_{be2} & -\frac{k_s L}{2} & -k_s & k_s + k_{be2} \end{vmatrix}$$

Now the degrees of freedom d_{u1} and d_{u2} are condensed out using static condensation (see Figure 7.29 for a basic explanation of static condensation, e.g. see Rugarli 2010), getting back to a (4×4) matrix where the translational stiffness considers both the bolt stiffness and the bearing stiffness. The result can be written as

$$\mathbf{K} = \begin{vmatrix} T_V & \frac{T_V L}{2} & -T_V & \frac{T_V L}{2} \\ \frac{T_V L}{2} & R_{\text{mod}} & -\frac{T_V L}{2} & K_{\text{mod}} \\ -T_V & -\frac{T_V L}{2} & T_V & -\frac{T_V L}{2} \\ \frac{T_V L}{2} & K_{\text{mod}} & -\frac{T_V L}{2} & R_{\text{mod}} \end{vmatrix}$$

where

$$T_V = \frac{k_s k_{be1} k_{be2}}{k_s k_{be1} + k_s k_{be2} + k_{be1} k_{be2}} \tag{7.19}$$

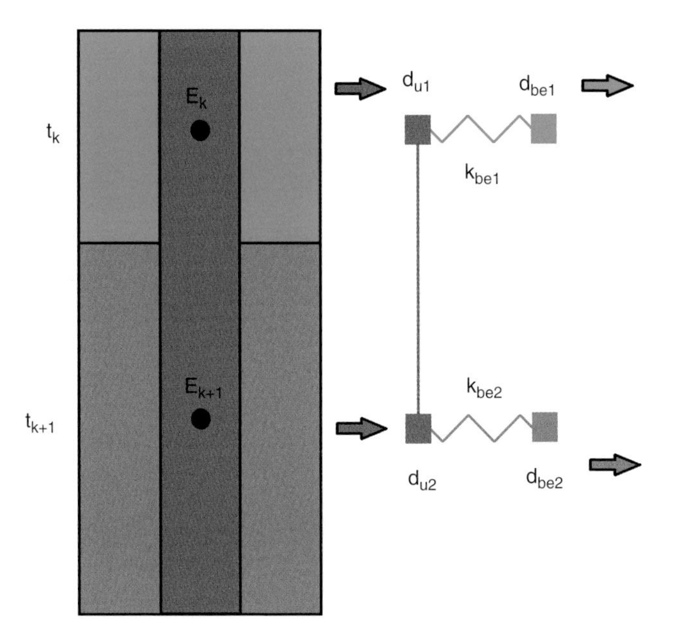

Figure 7.29 The model for bolt bearing and bolt shear. Degrees of freedom d_{u1} and d_{u2} are condensed out.

$$R_{\text{mod}} = R - \frac{k_s^2 L^2 (k_{be1} + k_{be2})}{4(k_s k_{be1} + k_s k_{be2} + k_{be1} k_{be2})} = k_M + \frac{k_s L^2}{3} - \frac{k_s^2 L^2 (k_{be1} + k_{be2})}{4(k_s k_{be1} + k_s k_{be2} + k_{be1} k_{be2})} \tag{7.20}$$

$$K_{\text{mod}} = K - \frac{k_s^2 L^2 (k_{be1} + k_{be2})}{4(k_s k_{be1} + k_s k_{be2} + k_{be1} k_{be2})} = -k_M + \frac{k_s L^2}{6} - \frac{k_s^2 L^2 (k_{be1} + k_{be2})}{4(k_s k_{be1} + k_s k_{be2} + k_{be1} k_{be2})} \tag{7.21}$$

It can easily be proved that, as expected

$$\frac{T_V L^2}{2} = K_{\text{mod}} + R_{\text{mod}}$$

The final matrix can be obtained using the following replacements in matrices $\mathbf{A_{ik}}$, $\mathbf{B_{ik}}$, $\mathbf{D_{ik}}$, and $\mathbf{E_{ik}}$, or equivalently in the complete bolt matrix $\mathbf{K_{ik}}$:

$$\begin{aligned} k_{Vik} &\Rightarrow T_{Vik} \\ R_{ik} &\Rightarrow R_{ik,\text{mod}} \\ K_{ik} &\Rightarrow K_{ik,\text{mod}} \end{aligned} \tag{7.22}$$

7.8.2.3 Final Form of the Stiffness Matrix

Adopting the Eurocode 3 choices, the following settings apply:

$$k_{Nik} = \frac{EA_{res}}{L_k} \cdot \frac{\mu_{ni}}{f} \tag{7.23}$$

where

- E is the Young's modulus of steel
- A_{res} is the threaded area of the bolt
- L_k is the average thickness bolted between extremity k and $(k + 1)$
- μ_{ni} is the efficiency factor of bolt in tension, in the range from μ to 1
- f is the flexibility index of the bolt layout.

According to Eurocode 3, a higher length should be taken, considering the washer thickness, the head height, and the nut height, where applicable (i.e. at first and last extremity). Besides a factor 0.8 should be applied, valid for a T-stub in the presence of prying forces. So if on one hand the length should be measured with greater accuracy, on the other hand a factor 0.8 is applied which has no general validity (see also Jaspart and Weynand, 2016). The author's view is that there are not many reasons to push apparent precision beyond a reasonable limit; there are other errors, unavoidable, that are much worst. Besides, if the aim is to compute the forces in the bolts, only relative errors are important, and so the role played by μ_{ni} is much higher. This is, as has been shown, an efficiency factor of bolts in tension, and takes into account the difference in flexibility of the connected plates, from point to point. If the stiffness matrix of the bolt has to be used in a global pure fem model (all constituents modeled by plate–shell elements), it should possibly be set equal to 1.

Referring to the shear part, the following term should be used:

$$T_{Vik} = \frac{k_{s,ik} \cdot k_{be,ik1} \cdot k_{be,ik2}}{k_{s,ik} k_{be,ik1} + k_{s,ik} k_{be,ik2} + k_{be,ik1} k_{be,ik2}} \cdot \frac{\mu_{ti}}{f} \tag{7.24}$$

where

- $k_{s,ik}$ is the shear stiffness of the bolt, between extremities k and $(k + 1)$
- $k_{be,ik1}$ is the bearing stiffness of the element at extremity k
- $k_{be,ik2}$ is the bearing stiffness of the element at extremity $k + 1$
- μ_{ti} is the shear factor of bolt i, equal to a small number μ if the bolt is no-shear, and equal to 1 otherwise
- f is the flexibility index of the bolt layout.

If Eurocode 3 is used, the term $k_{s,ik}$ can be expressed as

$$k_{s,ik} = \frac{8D^2 f_{ub}}{D_{M16}}$$

and so it does not depend on i or k.

The term $k_{be,ik1}$, can be expressed as (f_{uk} is the ultimate stress of the constituent at extremity k)

$$k_{be,ik1} = \frac{22.5 \cdot h_k \cdot t_k}{D_{M16}} \cdot D \cdot f_{uk} \leq 37.5 \cdot D \cdot h_k \cdot f_{uk}$$

where the term h_k is 1 at first and last extremity, and 0.5 otherwise:

$$h_k = 1 \quad \text{if}(k = 1) \text{ OR } (k = n_e)$$

$$h_k = 0.5 \quad (k \neq 1) \text{ AND } (k \neq n_e)$$

and the term $k_{be,ik2}$ is

$$k_{be,ik2} = \frac{22.5 \cdot h_{k+1} \cdot t_{k+1}}{D_{M16}} \cdot D \cdot f_{uk} \leq 37.5 \cdot D \cdot h_{k+1} \cdot f_{uk}$$

If the Rex and Easterling formulation is used, with the secant stiffness, then the three stiffnesses can be expressed as (D is expressed in mm)

$$k_{s,ik} = \frac{8D^2 f_{ub}}{D_{M16}} \tag{7.25}$$

$$k_{be,ik1} = 0.149 \cdot 120 \cdot h_k t_k \cdot f_{yk} \cdot \left(\frac{D}{25.4}\right)^{0.8} \tag{7.26}$$

$$k_{be,ik2} = 0.149 \cdot 120 \cdot h_{k+1} t_{k+1} \cdot f_{yk+1} \cdot \left(\frac{D}{25.4}\right)^{0.8} \tag{7.27}$$

If the initial stiffness is used instead of the secant (which would be a good choice for low shear loads), then

$$k_{be,ik1} = 120 \cdot h_k t_k \cdot f_{yk} \cdot \left(\frac{D}{25.4}\right)^{0.8} \tag{7.28}$$

$$k_{be,ik2} = 120 \cdot h_{k+1} t_{k+1} \cdot f_{yk+1} \cdot \left(\frac{D}{25.4}\right)^{0.8} \tag{7.29}$$

In non-linear analysis these last stiffnesses $k_{be,ik1}$ and $k_{be,ik2}$, can be made dependent on displacement, using the tangential stiffness instead of the secant one (see Equations 7.5). The stiffness of the bolt $k_{s,ik}$ can drop to zero if the plasticization of the shaft is reached.

The direct and indirect rotational stiffness related to the bolt only, that is, not considering the bearing, can be set as

$$R_{ik} = k_{Mik} + \frac{k_{sik} L_k^2}{3}$$

$$K_{ik} = -k_{Mik} + \frac{k_{sik} L_k^2}{6}$$

The modified direct and indirect stiffness are

$$R_{ik,\text{mod}} = \frac{\pi D^4 E}{64 L_k} \cdot \frac{\mu_{ni}}{f} + \frac{k_{sik} L_k^2}{3} \cdot \frac{\mu_{ti}}{f} - \frac{k_{sik}^2 L_k^2 (k_{bek1} + k_{bek2})}{4(k_{sik} k_{be1k} + k_{sik} k_{be2k} + k_{be1k} k_{be2k})} \cdot \frac{\mu_{ti}}{f}$$

$$K_{ik,\text{mod}} = -\frac{\pi D^4 E}{64 L_k} \cdot \frac{\mu_{ni}}{f} + \frac{k_{sik} L_k^2}{6} \cdot \frac{\mu_{ti}}{f} - \frac{k_{sik}^2 L_k^2 (k_{bek1} + k_{bek2})}{4(k_{sik} k_{be1k} + k_{sik} k_{be2k} + k_{be1k} k_{be2k})} \cdot \frac{\mu_{ti}}{f}$$

having set

$$k_{Mik} = \frac{\pi D^4 E}{64 L_k} \cdot \frac{\mu_{ni}}{f}$$

The absence of the term μ_{ni} from k_{Mik} would imply that the bending stiffness of a shear-only bolt layout is not negligible, because also if the bolts are not axially loaded, they are able to absorb a bending moment with the bending of their shaft. Adding μ_{ni} to k_{Mik}, then, the bending stiffness of a shear-only bolt layout is negligible. However, if the bolts act in tension or compression, the bending stiffness related to their elongation is much higher than the stiffness related to the bending of the shafts, which will be considered in the summations.

Using AISC, the same choices already seen for k_{Nik} and k_{Mik} can be made, while the following replacement must be used, because AISC formulation (see Crawford and Kulak 1968) takes into account both the stiffness of the bolt and the stiffness of the bearing:

$$T_{Vik} = \frac{R_{ult}}{\Delta_{\max}} \cdot \frac{\mu_{ti}}{f}$$

$$R_{ik,\text{mod}} = \frac{\pi D^4 E}{64 L_k} \cdot \frac{\mu_{ni}}{f} + \frac{T_{Vik} \cdot L_k^2}{3} \cdot \frac{\mu_{ti}}{f}$$

$$K_{ik,\text{mod}} = -\frac{\pi D^4 E}{64 L_k} \cdot \frac{\mu_{ni}}{f} + \frac{T_{Vik} \cdot L_k^2}{6} \cdot \frac{\mu_{ti}}{f}$$

At the current state of research it appears that the model by Rex and Easterling is the best available (see also Rex and Easterling, 2003).

7.8.3 Preloaded Bolts

The stiffness matrix of preloaded bolt layouts can be set by following the same path already seen for bearing bolt layouts, but the terms k_{Nik}, k_{Vik}, and k_{Mik} should be changed keeping in mind what was shown in Section 7.6.

Considering the model that uses a circular annuluses of external diameter QD, and internal diameter D_0, the following expressions are found.

The axial stiffness is

$$k_{Nik} = \frac{EA_{res}}{L_k} \cdot \frac{\mu_{ni}}{f} + \frac{\pi E\left(Q^2 D^2 - D_0^2\right)}{4 L_k} \cdot \frac{\mu_{ni}}{f}$$

and Q is evaluated using Equation 7.10, where L is the total thickness bolted. That is

$$Q = 0.192 \cdot \frac{L}{D} + 1.58$$

If the complete circle model for bearing is used, instead

$$k_{Nik} = \frac{EA_{res}}{L_k} \cdot \frac{\mu_{ni}}{f} + \frac{\pi E Q^{*2} D^2}{4 L_k} \cdot \frac{\mu_{ni}}{f}$$

and

$$Q^* = 0.210 \cdot \frac{L}{D} + 1.26$$

The shear stiffness is (Equation 7.13)

$$k_{Vik} = \frac{10 GA}{L_k} \cdot \frac{\mu_{ti}}{f}$$

The bending stiffness is, if the bearing is an annulus,

$$k_{Mik} = \frac{\pi E D^4}{64 L_k} \cdot \frac{\mu_{ni}}{f} + \frac{\pi E\left(Q^4 D^4 - D_0^4\right)}{64 L_k} \cdot \frac{\mu_{ni}}{f}$$

and, if the bearing is a full circle,

$$k_{Mik} = \frac{\pi E D^4}{64 L_k} \cdot \frac{\mu_{ni}}{f} + \frac{\pi E Q^{*4} D^4}{64 L_k} \cdot \frac{\mu_{ni}}{f}$$

The direct and indirect rotational stiffnesses are

$$R_{ik} = k_{Mik} + \frac{k_{Vik} \cdot L_k^2}{3}$$

$$K_{ik} = -k_{Mik} + \frac{k_{Vik}\, L_k^2}{6}$$

If the bolt layout is a preloaded anchor, the expressions to be used are the same, but different Q values must be used, specifically, Equation 7.14:

$$Q = \sqrt{0.32 \cdot \frac{L}{D} + 1}$$

or Equation 7.15:

$$Q^* = \sqrt{0.32 \cdot \frac{L}{D}}$$

7.8.4 Non-Linear Analysis of Bolts

If a non-linear analysis is used, and the single bolt has to be modeled, the stiffness matrix that must be used during the incremental iterative procedure is not the secant stiffness matrix but the tangent one. Basically, the terms of the stiffness matrix will be modified at the end of each step or iteration of the analysis, so as to take into account the displacements of the nodes of the element.

7.8.4.1 Bolts in Bearing

Two terms must be considered: the first one is related to the tangent stiffness of the bolt, the second to the tangent stiffness of the bearing.

Considering first the behavior of the bolt shaft, this experiences an axial force N, bending moment M, and shear V. The shear and bending considered are the resultant:

$$V = \sqrt{V_u^2 + V_v^2}$$

$$M = \sqrt{M_u^2 + M_v^2}$$

The tangent stiffness matrix depends on the three stiffnesses already discussed, k_{Ni}, k_{Vi}, and k_{Mi}.

The term k_{Ni} should be considered null if the bolt is compressed and the contact non-linearity is activated. Otherwise, if no contact non-linearity is activated, then the evaluation of the pressure exchanged at the bearing surface can only be done by considering the bolts reacting also in compression, or replacing the bolt elements with the forces and pressures exchanged, as evaluated by simplified means (see Chapters 9 and 10). If the first method is used, the model is that of forces only in the bolt shafts.

Assuming the following simple rules to get the normal and tangential stress in the shaft (it is supposed that the shear loads the threaded part of the bolt shaft),

$$\sigma = \frac{N}{A_{res}}$$

$$\tau = \frac{V}{A_{res}}$$

and imposing the Von Mises stress plastic criterion, but with the ultimate stress of bolt f_{ub}, the limit domain of the bolt under shear and axial force can be found:

$$\left(\frac{N}{N_{pl}}\right)^2 + \left(\frac{V}{V_{pl}}\right)^2 = 1$$

where

$$N_{pl} = A_{res} f_{ub}$$

$$V_{pl} = A_{res} \frac{f_{ub}}{\sqrt{3}}$$

This is basically the limit domain assumed by AISC and Eurocode (the issue will be discussed in Chapter 8).

Considering now the situation where a bending moment and an axial force are applied, the plastic condition can be well approximated by[6]

$$\left(\frac{M}{M_{pl}}\right) + \left(\frac{N}{N_{pl}}\right)^2 = 1$$

This can be shown as follows.

If a generic intermediate plastic condition is considered, the plastic axial force and plastic bending moment can be expressed as (Figure 7.30)

$$N = \left(\pi r^2 - 2\alpha r^2 + 2r^2 \sin\alpha \cos\alpha\right) \cdot f_{ub}$$

$$M = 2 f_{ub} \int_{\alpha}^{0} (2r\sin\theta) \cdot (r\cos\theta) \cdot (-r\sin\theta) d\theta = \frac{4r^3}{3} \cdot (\sin\alpha)^3 f_{ub}$$

so that

$$\frac{N}{N_{pl}} = \frac{\pi - 2\alpha + 2\sin\alpha\cos\alpha}{\pi}$$

$$\frac{M}{M_{pl}} = (\sin\alpha)^3$$

6 It has been found to be in very good agreement with the more complex rule $m + 1.084n^2 - 0.0935n = 0.9927$, where $(N/N_{pl}) = n$, and $(M/M_{pl}) = m$.

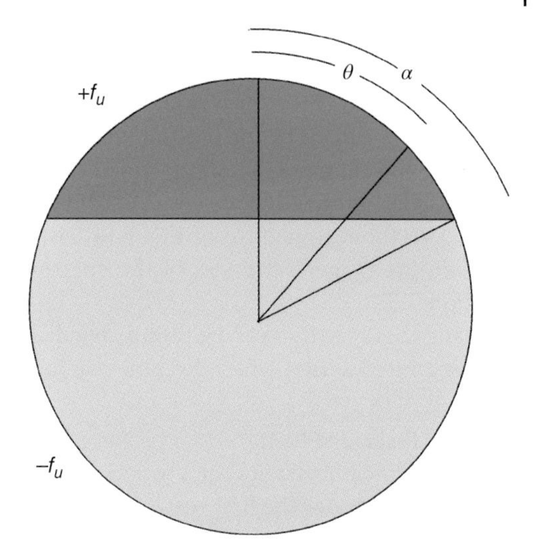

Figure 7.30 Generic plastic condition for (N, M) limit domain.

If the angle α is considered a parameter, the curve relating (N/N_{pl}) to (M/M_{pl}) can be drawn, and a suitable interpolation found.

The application of a shear reduces the available yield stress by a factor ρ, which can be evaluated as

$$\rho = \sqrt{1 - \left(\frac{T}{T_{pl}}\right)^2}$$

So, the limit domain referring to the three actions could be written as

$$\left(\frac{M}{\rho M_{pl}}\right) + \left(\frac{N}{\rho N_{pl}}\right)^2 = 1 \tag{7.30}$$

If the normalized internal forces are introduced as

$$n = \frac{N}{N_{pl}}$$

$$m = \frac{M}{M_{pl}}$$

$$v = \frac{V}{V_{pl}}$$

The limit condition takes the form

$$\frac{m}{\sqrt{1-v^2}} + \frac{n^2}{1-v^2} = 1 \tag{7.31}$$

and solving for v gives

$$v = \sqrt{1 - \frac{\left(m + \sqrt{m^2 + 4n^2}\right)^2}{4}}$$

This is the maximum applicable normalized shear, once an axial force and a bending moment are applied.

A simple non-linear model for the bolt behavior can then be: use the normal stiffness k_{Vi}, k_{Ni}, and k_{Mi} up to the limit condition of Equation 7.31, which is found in the check section, and then set the stiffnesses to null considering the bolt "broken".

Considering now the behavior of the bearing springs k_{be1} and k_{be2}, it can be modeled thanks to the formulation of Rex and Easterling, Equation 7.4. Once the shear force is known, the tangent stiffness of the spring can be evaluated directly, as was shown in Section 7.5.1.

Static condensation can be then applied and the final tangent matrix assembled. As the springs are in series, when the bolt reaches the limit, it becomes inactive.

7.8.4.2 Preloaded Bolts

The non-linear behavior of a preloaded bolt can be modeled in two different ways: assuming that once the bolt reaches the limit slip condition the tangential stiffness is null, or replacing the tangential stiffness with the bearing stiffness.

If the first approach is used, the axial and bending stiffnesses k_N and k_M are the ones of the bolt taken alone, increased by the application of a factor depending on Q, or Q^*, as has been shown in the previous sections. The high tangential stiffness k_V is kept until the slip condition is reached, and then it is replaced by 0. The axial and bending stiffnesses related to preloading are kept as long as there has been no detachment. The detachment condition is signaled by the limit conditions provided by the standards. When the detachment is reached, the axial and bending stiffnesses become equal to that of the normal bolt, that is, $k_N = EA_{res}/L$, and $k_M = [\pi ED^4/(64\,L)]$. If, considering the true forces flowing in the bolt shaft – that is, the applied axial forces multiplied by a factor $1/(1 + Q^{*2})$ plus the initial preload, and the applied bending moments multiplied by a factor $1/(1 + Q^{*4})$ – the bolt reaches the condition of Equation 7.31, is considered broken, and all the stiffnesses are set to 0.

The second approach can be modeled using two superimposed elements.

The first has the axial and shear stiffness of the bolt in bearing (i.e. with the bearing stiffness, which can be considered much lower than the friction stiffness and so negligible during the no-slip phase).

The second superimposed element has the axial and shear stiffness of a hollow cylinder of diameter QD, or of a full cylinder of diameter Q^*D, as discussed in the previous sections. The shear stiffness is much higher. The two elements act in parallel.

When this second element reaches its axial "yield", that is when the initial preload has been canceled by an applied tensile force, its axial stiffness is set to null, as is its shear stiffness. If the slip condition is crossed, but preload is still applied, the friction shear stiffness drops to zero, while the axial stiffness is kept. In this way, the two elements can be kept independent, and behave globally as expected.

7.9 Internal Force Distribution

7.9.1 General Method

The general method for computing the forces in bolts, not using the bearing surface, and not using the instantaneous center of rotation method, is conceptually similar to that already described for weld layouts. The first step is to compute the internal forces at the check sections starting from the force packets $\mathbf{s_{lk}}$ at the extremities. These force packets are the result of the analysis and can be computed as

$$\mathbf{s_l} = \begin{vmatrix} \mathbf{s_{l,1}} \\ \mathbf{s_{l,2}} \\ \dots \\ \mathbf{s_{l,k}} \\ \dots \\ \mathbf{s_{l,ne}} \end{vmatrix} = \mathbf{K_l d_l}$$

where $\mathbf{K_l}$ is the stiffness matrix of the layout in the principal reference system of the bolt layout (**w**, **u**, **v**), and $\mathbf{d_l}$ are the nodal (i.e. at the extremities) displacements in the same reference system.

The internal forces at the check sections k $\boldsymbol{\sigma}_\mathbf{k}$ (with $1 \le k \le n_e - 1$) are computed by means of Equation 7.2, resulting in the six generalized forces applied to the layout at check section k:

$$\boldsymbol{\sigma}_\mathbf{k} = \{N_k \quad V_{uk} \quad V_{vk} \quad M_{wk} \quad M_{uk} \quad M_{vk}\}^\mathrm{T}$$

The next step is to compute the modified stiffness matrix $\mathbf{K_{lk,mod}}$ that will be used to compute the notional displacement $\mathbf{d_{k,mod}}$ assigned to check section k, from the applied internal forces $\boldsymbol{\sigma}_k$, that is

$$\mathbf{d}_{\mathbf{k},mod} = \{d_{wk,\mathrm{mod}} \quad d_{uk,\mathrm{mod}} \quad d_{vk,\mathrm{mod}} \quad r_{wk,\mathrm{mod}} \quad r_{uk,\mathrm{mod}} \quad r_{vk,\mathrm{mod}}\}^\mathrm{T} = \mathbf{K}_{\mathbf{l},\mathbf{k},mod}^{-1} \cdot \boldsymbol{\sigma}_\mathbf{k}$$

Considering the elementary blocks $\mathbf{A_{ik}}$, $\mathbf{B_{ik}}$, $\mathbf{D_{ik}}$, and $\mathbf{E_{ik}}$, when L_k tends to zero, the following relationships hold true:

$$\mathbf{B_{ik}} \to \mathbf{0}$$
$$R_{ik} \to k_{Mik}$$
$$K_{ik} \to -k_{Mik}$$

and so

$$\mathbf{K}_{\mathbf{l},\mathbf{k},mod} = \sum_{i=1}^{nb} \begin{vmatrix} \mathbf{A_{ik}} & \mathbf{A_{ik}S_i} \\ \mathbf{S_i^T A_{ik}} & \mathbf{S_i^T A_{ik} S_i} + \mathbf{D_{ik}} \end{vmatrix}$$
$$= \sum_{i=1}^{nb} \begin{vmatrix} k_{Nik} & 0 & 0 & 0 & k_{Nik}v_i & -k_{Nik}u_i \\ 0 & k_{Vik} & 0 & -k_{Vik}v_i & 0 & 0 \\ 0 & 0 & k_{Vik} & k_{Vik}u_i & 0 & 0 \\ 0 & -k_{Vik}v_i & k_{Vik}u_i & k_{Vik}r_i^2 & 0 & 0 \\ k_{Nik}v_i & 0 & 0 & 0 & k_{Mik} + k_{Nik}v_i^2 & -k_{Nik}u_iv_i \\ -k_{Nik}u_i & 0 & 0 & 0 & -k_{Nik}u_iv_i & k_{Mik} + k_{Nik}u_i^2 \end{vmatrix}$$

The extra diagonal terms (1, 5), (1, 6), (5, 6), and symmetrical,

$$\sum_{i=1}^{nb} k_{Nik} u_i, \ \sum_{i=1}^{nb} k_{Nik} v_i, \ \sum_{i=1}^{nb} k_{Nik} u_i v_i$$

are always null because of how the bolt layout center and principal axes have been defined. So they can be removed from the matrix $\mathbf{K_{I,k,mod}}$.

The extra diagonal terms

$$\sum_{i=1}^{nb} k_{Vik} u_i, \ \sum_{i=1}^{nb} k_{Vik} v_i$$

do not, in general, vanish, as the factors μ_{ti} are different, from bolt to bolt.

So the matrix $\mathbf{K_{I,k,mod}}$ is block-diagonal, with the first block of size 1, the second of size 3, and the third of size 2. The second block is not diagonal owing to the coupling of shear and torque. The first and last blocks are also diagonal themselves. That is

$$\mathbf{K_{I,k,}}_{mod} = \sum_{i=1}^{nb} \begin{vmatrix} k_{Nik} & 0 & 0 & 0 & 0 & 0 \\ 0 & k_{Vik} & 0 & -k_{Vik} v_i & 0 & 0 \\ 0 & 0 & k_{Vik} & k_{Vik} u_i & 0 & 0 \\ 0 & -k_{Vik} v_i & k_{Vik} u_i & k_{Vik} r_i^2 & 0 & 0 \\ 0 & 0 & 0 & 0 & k_{Mik} + k_{Nik} v_i^2 & 0 \\ 0 & 0 & 0 & 0 & 0 & k_{Mik} + k_{Nik} u_i^2 \end{vmatrix}$$

This is the modified matrix considering the assembly of normal Timoshenko beam elements. Also considering the correction due to bolt bearing (i.e. the static condensation) or the effect of preloaded bolts, the results are similar.

In fact, considering the effect of bolt and bearing stiffness, some terms depend on $(1/L_k)$, namely k_{Nik} and k_{Mik}, while other terms, coupled with these depend on L_k^2. Considering the rows referring to shear and torque, the term T_{Vik} remains, while all the terms $T_{Vik}L_k/2$ vanish. In practice, as already seen, k_{Vik} is replaced by T_{Vik} (which does not change), R_{ik} is replaced by $R_{ik,mod}$, and K_{ik} is replaced by $K_{ik,mod}$, and these last change as follows

$$R_{ik,\text{mod}} = \frac{\pi D^4 E}{64 L_k} \cdot \frac{\mu_{ni}}{f} + \frac{k_{sik}\, L_k^2}{3} \cdot \frac{\mu_{ti}}{f} - \frac{k_{sik}^2\, L_k^2 (k_{bek1} + k_{bek2})}{4(k_{sik} k_{be1k} + k_{sik} k_{be2k} + k_{be1k} k_{be2k})} \cdot \frac{\mu_{ti}}{f} \quad \rightarrow \quad \frac{\pi D^4 E}{64 L_k} \cdot \frac{\mu_{ni}}{f}$$

$$K_{ik,\text{mod}} = -\frac{\pi D^4 E}{64 L_k} \cdot \frac{\mu_{ni}}{f} + \frac{k_{sik}\, L_k^2}{6} \cdot \frac{\mu_{ti}}{f} - \frac{k_{sik}^2\, L_k^2 (k_{bek1} + k_{bek2})}{4(k_{sik} k_{be1k} + k_{sik} k_{be2k} + k_{be1k} k_{be2k})} \cdot \frac{\mu_{ti}}{f} \quad \rightarrow \quad -\frac{\pi D^4 E}{64 L_k} \cdot \frac{\mu_{ni}}{f}$$

In the case of preloaded bolts, the terms depending on $(1/L_k)$ prevail on all the others, depending on L_k and L_k^2. So the results are those already listed, replacing properly the terms k_{Nik}, k_{Vik}, and k_{Mik} with those valid for preloaded bolts.

The forces flowing into the single bolt can then be computed by multiplying the modified local stiffness matrix of a bolt times its displacement vector.

The modified stiffness matrix of a single bolt is, as $(\mathbf{B_{ik}} \rightarrow 0)$,

$$\mathbf{K_{ik,\,mod}} = \begin{vmatrix} \mathbf{A_{ik,\,mod}} & \mathbf{0} \\ \mathbf{0} & \mathbf{D_{ik,\,mod}} \end{vmatrix}$$

The local displacement of the bolt is

$$\mathbf{d}_{\mathbf{ik,\,mod}} = \begin{vmatrix} \mathbf{I} & \mathbf{S_i} \\ \mathbf{0} & \mathbf{I} \end{vmatrix} \cdot \mathbf{d}_{\mathbf{k,\,mod}} = \begin{vmatrix} \mathbf{I} & \mathbf{S_i} \\ \mathbf{0} & \mathbf{I} \end{vmatrix} \cdot \mathbf{K}_{\mathbf{l,\,k,\,mod}}^{-1} \cdot \boldsymbol{\sigma}_{\mathbf{k}}$$

and so the forces flowing in the bolt i are

$$\boldsymbol{\sigma}_{\mathbf{ik}} = \begin{vmatrix} \mathbf{A}_{\mathbf{ik,\,mod}} & \mathbf{0} \\ \mathbf{0} & \mathbf{D}_{\mathbf{ik,\,mod}} \end{vmatrix} \cdot \begin{vmatrix} \mathbf{I} & \mathbf{S_i} \\ \mathbf{0} & \mathbf{I} \end{vmatrix} \cdot \mathbf{K}_{\mathbf{l,\,k,\,mod}}^{-1} \cdot \boldsymbol{\sigma}_{\mathbf{k}}$$

If the terms k_{Nik}, k_{Vik}, and k_{Mik} are expressed, splitting the part which depends only on k, by the part depending on the bolt i (because not considering the working-mode flags, all the axial, shear and bending stiffnesses of the single bolts are equal for a given k),

$$k_{Nik} = k_{Nk} \cdot \frac{\mu_{ni}}{f}$$

$$k_{Vik} = k_{Vk} \cdot \frac{\mu_{ti}}{f}$$

$$k_{Mik} = k_{Mk} \cdot \frac{\mu_{ni}}{f}$$

then the forces flowing into each single bolt can be expressed directly. Besides, it can be noted that if not preloaded bolts are used, then

$$\frac{k_{Mk}}{k_{Nk}} = \frac{\frac{\pi D^4 E}{64 L_k}}{\frac{E A_{res}}{L_k}} = \frac{\pi D^4}{64 A_{res}} = \frac{D^2}{16} \cdot \frac{A}{A_{res}}$$

while if preloaded bolts are used then

$$\frac{k_{Mk}}{k_{Nk}} = \frac{\frac{\pi E D^4}{64 L_k} + \frac{\pi E \left(Q^4 D^4 - D_0^4\right)}{64 L_k}}{\frac{E A_{res}}{L_k} + \frac{\pi E \left(Q^2 D^2 - D_0^2\right)}{L_k}} = \frac{1}{64} \cdot \left[\frac{\pi D^4 + \pi \left(Q^4 D^4 - D_0^4\right)}{\left[A_{res} + \pi \left(Q^2 D^2 - D_0^2\right)\right]}\right]$$

and both do not depend on k, as L_k disappears.

The equivalent properties of the bolt layout (and therefore marked by $*$), not depending on k, can then be defined as

$$A_n^* = \sum_{i=1}^{nb} \mu_{ni}$$

$$J_u^* = \sum_{i=1}^{nb} \left(\frac{k_M}{k_N} \mu_{ni} + v_i^2 \mu_{ni} \right)$$

$$J_v^* = \sum_{i=1}^{nb} \left(\frac{k_M}{k_N} \mu_{ni} + u_i^2 \mu_{ni} \right)$$

$$J_p^* = \sum_{i=1}^{nb} \mu_{ti} r_i^2$$

$$A_s^* = \sum_{i=1}^{nb} \mu_{ti}$$

$$S_u^* = \sum_{i=1}^{nb} \mu_{ti} v_i$$

$$S_v^* = \sum_{i=1}^{nb} \mu_{ti} u_i$$

The coordinates of the shear center can be defined as

$$u_{SC} = \frac{S_v^*}{A_s^*}$$

$$v_{SC} = \frac{S_u^*}{A_s^*}$$

and the torsional radius as

$$r_p^* = \sqrt{\frac{J_p^*}{A_s^*}}$$

Using these definitions, the axial force, shears, and bending moments flowing into each bolt i can be expressed as

$$N_{ik} = \frac{N_k \cdot \mu_{ni}}{A_n^*} + \frac{M_{uk}}{J_u^*} \cdot \mu_{ni} \cdot v_i - \frac{M_{vk}}{J_v^*} \cdot \mu_{ni} \cdot u_i$$

$$V_{uik} = \frac{V_{uk}}{A_s^*} \mu_{ti} - \left[\frac{M_{wk} + V_{uk} v_{SC} - V_{vk} u_{SC}}{A_s^* \left(r_p^{*2} - u_{SC}^2 - v_{SC}^2 \right)} \right] \cdot (v_i - v_{SC}) \cdot \mu_{ti}$$

$$V_{vik} = \frac{V_{vk}}{A_s^*} \mu_{ti} + \left[\frac{M_{wk} + V_{uk} v_{SC} - V_{vk} u_{SC}}{A_s^* \left(r_p^{*2} - u_{SC}^2 - v_{SC}^2 \right)} \right] \cdot (u_i - u_{SC}) \cdot \mu_{ti}$$

$$M_{uik} = \frac{M_{uk}}{J_u *} \cdot \frac{k_M}{k_N} \mu_{ni}$$

$$M_{vik} = \frac{M_{vk}}{J_v^*} \cdot \frac{k_M}{k_N} \mu_{ni}$$

It is now possible to define the resistance modulus of each bolt, so that the computation of the forces flowing into each bolt gets easier:

$$A_{ni}^* = \frac{A_n^*}{\mu_{ni}}$$

$$W_{ui}^* = \frac{J_u^*}{v_i \mu_{ni}}$$

$$W_{vi}^* = -\frac{J_v^*}{u_i \mu_{ni}}$$

$$A_{uui}^* = \frac{A_s^* \left(r_p^{*2} - u_{SC}^{*2} - v_{SC}^{*2} \right)}{\mu_{ti} \cdot \left(r_p^{*2} - u_{SC}^{*2} - v_{SC}^* v_i \right)}$$

$$A^*_{uvi} = \frac{A^*_s\left(r^{*2}_p - u^{*2}_{SC} - v^{*2}_{SC}\right)}{\mu_{ti}\cdot(u_{SC}v_i - u_{SC}v_{SC})}$$

$$A^*_{vvi} = \frac{A^*_s\left(r^{*2}_p - u^{*2}_{SC} - v^{*2}_{SC}\right)}{\mu_{ti}\cdot\left(r^{*2}_p - v^{*2}_{SC} - u_{SC}u_i\right)}$$

$$A^*_{vui} = \frac{A^*_s\left(r^{*2}_p - u^{*2}_{SC} - v^{*2}_{SC}\right)}{\mu_{ti}\cdot(v_{SC}u_i - u_{SC}v_{SC})}$$

$$W^*_{Vui} = -\frac{A^*_s\left(r^{*2}_p - u^{*2}_{SC} - v^{*2}_{SC}\right)}{(v_i - v_{SC})\cdot\mu_{ti}}$$

$$W^*_{Vvi} = \frac{A^*_s\left(r^{*2}_p - u^{*2}_{SC} - v^{*2}_{SC}\right)}{(u_i - u_{SC})\cdot\mu_{ti}}$$

$$W^*_{Mui} = \frac{J^*_u k_N}{k_M \mu_{ni}}$$

$$W^*_{Mvi} = \frac{J^*_v k_N}{k_M \mu_{ni}}$$

Having defined these resistance moduli, the forces and moments in each bolt i are computed as

$$N_{ik} = \frac{N_k}{A^*_{ni}} + \frac{M_{uk}}{W_{ui}} + \frac{M_{vk}}{W_{vi}}$$

$$V_{uik} = \frac{V_{uk}}{A^*_{uui}} + \frac{V_{vk}}{A^*_{uvi}} + \frac{M_{wk}}{W^*_{Vui}}$$

$$V_{vik} = \frac{V_{uk}}{A^*_{vui}} + \frac{V_{vk}}{A_{vvi}*} + \frac{M_{wk}}{W^*_{Vvi}}$$

$$M_{uik} = \frac{M_{uk}}{W^*_{Mui}}$$

$$M_{vik} = \frac{M_{vk}}{W^*_{Mvi}}$$

This formulation takes into account that each single bolt may receive "no shear" or "shear only" flags, and that the normal-stress efficiency of bolts may be varied from 0 (or better a suitable small constant μ) to 1. If all the bolts have the same behavior, the formulation defaults to the normally used one. That is

$$N_i = \frac{N_k}{n_b} + \frac{M_{uk}}{\left(\frac{n_b D^2}{16} + \sum_{i=1}^{nb} v_i^2\right)}\cdot v_i - \frac{M_{vk}}{\left(\frac{n_b D^2}{16} + \sum_{i=1}^{nb} u_i^2\right)}\cdot u_i$$

$$V_{uki} = \frac{V_{uk}}{n_b} - \frac{M_{wk}}{\sum_{i=1}^{nb} r_i^2}\cdot v_i$$

$$V_{vki} = \frac{V_{vk}}{n_b} + \frac{M_{wk}}{\sum_{i=1}^{nb} r_i^2} \cdot u_i$$

$$M_{uki} = \frac{M_{uk}}{\left(\frac{n_b D^2}{16} + \sum_{i=1}^{nb} v_i^2\right)} \cdot \frac{D^2}{16} \cdot \frac{A}{A_{res}}$$

$$M_{vki} = \frac{M_{vk}}{\left(\frac{n_b D^2}{16} + \sum_{i=1}^{nb} u_i^2\right)} \cdot \frac{D^2}{16} \cdot \frac{A}{A_{res}}$$

This way to compute the forces in bolts is fit for:

1) bolt layouts not using a bearing surface and not preloaded; the forces in the bolts balance the actions applied to the layout
2) preloaded bolt layouts not using a bearing surface, i.e. when the total stiffness is concentrated in bolts. It must be noted that for preloaded bolt layouts the flags SO (shear only, $\mu_{ni} = 0.0001$), and NS (no shear, $\mu_{ti} = 0.0001$) have by definition in principle no meaning. However, the efficiency of the preloaded bolt in transferring the axial forces and shears, might be related to the flexibility of the plate and so the flags SO and also possibly NS, and the efficiency factor μ_{ni} preserve their utility.

If a bearing surface is used, the axial force N_{ki} and bending moments M_{uki}, M_{vki} in single bolts are computed in a different way.

Also, if the instantaneous center of rotation method is used, the way shears V_{uki} and V_{vki} are computed changes.

These two topics are addressed in the next sections.

7.9.2 Bearing Surface Method to Compute Forces in Bolts

It is assumed that the internal forces $\boldsymbol{\sigma} = \{M_v, M_u, N\}^T$ acting at a given check section k of bolt layout are known. In this section the index k is omitted for brevity.

Also, it has been assigned a bearing surface, which can be seen as the union of a number of closed polygons S_j, obtained by the Boolean operations described in Section 7.5.2.2. If the polygons sides are ordered counterclockwise, they are full regions, but if the sides are ordered clockwise, the polygon is a hole. Each polygon S_j is defined by the vector of points $\mathbf{P_{js}} = (u_{js}, v_{js})$, defined in the principal system of the bolt layout.

Moreover, a constitutive law $\sigma = \sigma(\varepsilon)$ has been assigned to the bearing. This is a no-tension law for bolt layout in bearing (i.e. not preloaded), or a tension-compression law for preloaded bolts.

It is assumed that the strain distribution over the plane referred to (u, v) is linear, that is

$$\varepsilon = au + bv + c$$

where a, b, c are three constant values, whose physical meaning is:

- a is the v “curvature”, χ_v
- b is the u “curvature”, χ_u
- c is a constant strain, ε_0.

The problem is to find a plastic neutral axis (PNA) of equation

$$au + bv + c = 0$$

so that the following relationships are satisfied:

$$N = \sum_{j=1}^{npolygons} \int_{S_j} \sigma dA + \sum_{i=1}^{nb} \sigma_i A_i \cdot \mu_{ni}$$

$$M_u = \sum_{j=1}^{npolygons} \int_{S_j} \sigma v dA + \sum_{i=1}^{nb} \sigma_i A_i \cdot v_i \cdot \mu_{ni} + \frac{E\pi D^4 b}{64} \cdot \sum_{i=1}^{nb} \mu_{ni}$$

$$M_v = -\sum_{j=1}^{npolygons} \int_{S_j} \sigma u dA - \sum_{i=1}^{nb} \sigma_i A_i \cdot u_i \cdot \mu_{ni} - \frac{E\pi D^4 a}{64} \cdot \sum_{i=1}^{nb} \mu_{ni}$$

where A_i is the area of bolt i. This can be set as the gross area of the bolt or as the threaded area of the bolt or as an intermediate value.

The final terms in the second and third equilibrium equations are related to the bending moments in the bolt shafts, and have the form $EJ\chi$, where E is the Young's modulus of steel, J is the second area moment of the cross-section of the bolt shaft, and a and b are curvatures χ. If these terms are omitted, no bending moments in the bolt shafts are computed.

Depending on the PNA position (i.e. on a, b, c) the bearing surface will react in different ways and the forces flowing into the bolts will be different. The effect of the constitutive law of the bearing surface has already been discussed in Section 7.5.2.2.4.

It turns out that by Green's formula (see Rugarli 1998 for the application to cross-sections of Green's formula[7]) the following relation holds true:

$$\int_{S_j} u^p v^q dA = \int_{\Gamma_j} \frac{u^{p+1} v^q}{p+1} dv$$

where Γ_j is the boundary of S_j. As the boundary of S_j is defined using segments, the integral over the boundary can be seen as the sum of a number of line integrals,

$$\int_{\Gamma_j} \frac{u^{p+1} v^q}{p+1} dv = \sum_{s=1}^{nside,j} \int_{\mathbf{P_s}}^{\mathbf{P_{s+1}}} \frac{u^{p+1} v^q}{p+1} dv$$

where $\mathbf{P_s}$ and $\mathbf{P_{s+1}}$ are the two extremity points of the sth segment of the boundary Γ_j. Now since along the segment $\mathbf{P_s} - \mathbf{P_{s+1}}$

$$u = u_s + \lambda \cdot (u_{s+1} - u_s) = u_s + \lambda \cdot \Delta u$$

7 The author acknowledges that the algorithm using Green's formula was originally developed internally at Castalia srl in 1994, by Ing. Giorgio Borré, and was intended for use with reinforced concrete (unpublished). Four years later, in 1998, the author adapted it to the automatic computation of plastic moduli of generic cross-sections (Rugarli 1998), and in 2008 to the problem of bolt layouts using a bearing surface.

$$v = v_s + \lambda \cdot (v_{s+1} - v_s) = v_s + \lambda \cdot \Delta v$$
$$dv = d\lambda \cdot \Delta v$$

As λ is a non-dimensional abscissa between 0 and 1, the previous integral becomes easy to solve (p and q are 0, 1, or 2):

$$\int_{\Gamma_j} \frac{u^{p+1}v^q}{p+1} dv = \sum_{s=1}^{nside,j} \int_0^1 \frac{(u_s + \lambda \Delta u)^{p+1}(v_s + \lambda \Delta v)^q}{p+1} \Delta v d\lambda$$

Let vector **x** be defined as

$$\mathbf{x} = \{u \;\; v \;\; 1\}^{\mathrm{T}}$$

It can be assumed, with no loss of generality, that the normal stress at a point **P** of the bearing surface can be expressed as

$$\sigma = E_s(P) \cdot \varepsilon = E_s(P) \cdot (au + bv + c) = E_s(P) \cdot \mathbf{x}^{\mathrm{T}}\mathbf{a}$$

and that the normal stress of a bolt i, placed in point $\mathbf{Q_i}$ can be expressed as

$$\sigma_i = E_b(Q_i) \cdot \varepsilon = E_b(Q_i) \cdot (au_i + bv_i + c) = E_b(Q_i) \cdot \mathbf{x_i^{\mathrm{T}}}\mathbf{a}$$

where E_s and E_b are suitable stiffness values and the vector **a** is defined as

$$\mathbf{a} = \{a \;\; b \;\; c\}^{\mathrm{T}}$$

In order to ensure equilibrium, it must be verified that

$$\boldsymbol{\sigma} = \begin{vmatrix} M_v \\ M_u \\ N \end{vmatrix} = \sum_{j=1}^{npolygons} \int_{S_j} E_s \begin{vmatrix} -1 & 0 & 0 \\ 0 & 1 & 0 \\ 0 & 0 & 1 \end{vmatrix} \cdot \mathbf{x}\mathbf{x}^{\mathrm{T}}\mathbf{a} dA + \sum_{i=1}^{nb} E_{bi}A_i \cdot \mu_{ni} \cdot \begin{vmatrix} -1 & 0 & 0 \\ 0 & 1 & 0 \\ 0 & 0 & 1 \end{vmatrix} \cdot \mathbf{x_i}\mathbf{x_i^{\mathrm{T}}}\mathbf{a}$$
$$+ \sum_{i=1}^{nb} \mu_{ni} \cdot \frac{E_{bi}\pi D^4}{64} \cdot \begin{vmatrix} -1 & 0 & 0 \\ 0 & 1 & 0 \\ 0 & 0 & 0 \end{vmatrix} \cdot \mathbf{a} = \mathbf{Ka}$$

where

$$\mathbf{K} = \sum_{j=1}^{npolygons} \int_{S_j} E_s \begin{vmatrix} -1 & 0 & 0 \\ 0 & 1 & 0 \\ 0 & 0 & 0 \end{vmatrix} \cdot \mathbf{x}\mathbf{x}^{\mathrm{T}} dA + \sum_{i=1}^{nb} E_{bi}A_i \cdot \mu_{ni} \cdot \begin{vmatrix} -1 & 0 & 0 \\ 0 & 1 & 0 \\ 0 & 0 & 0 \end{vmatrix} \cdot \mathbf{x_i}\mathbf{x_i^{\mathrm{T}}}$$
$$+ \sum_{i=1}^{nb} \mu_{ni} \cdot \frac{E_{bi}\pi D^4}{64} \cdot \begin{vmatrix} -1 & 0 & 0 \\ 0 & 1 & 0 \\ 0 & 0 & 0 \end{vmatrix}$$

Given a generic vector **a**, the related matrix **K** can then be computed using the boundary integrals already discussed. The PNA will divide the plane into two regions, one with tensile strain and one with compressive strain, and the strains will linearly increase with the distance from the PNA. So, in short,

$$\mathbf{K} = \Omega(\mathbf{a})$$

The problem of finding the exact **a** can be solved by an iterative procedure. Initially the vector $\mathbf{a_0}$ can be evaluated as

$$\begin{vmatrix} a_0 \\ b_0 \\ c_0 \end{vmatrix} = \begin{vmatrix} \dfrac{M_v}{E_{si}J_{sv} + \sum_{i=1}^{nb} E_{bi}A_i \cdot \mu_{ni} \cdot u_i^2 + \sum_{i=1}^{nb} E_{bi}\mu_{ni} \cdot \dfrac{\pi D^4}{64}} \\ \dfrac{M_u}{E_{si}J_{su} + \sum_{i=1}^{nb} E_{bi}A_i \cdot \mu_{ni} \cdot v_i^2 \cdot + \sum_{i=1}^{nb} E_{bi}\mu_{ni} \cdot \dfrac{\pi D^4}{64}} \\ \dfrac{N}{E_{si}A_s + n_b E_{bi}A_i \cdot \mu_{ni}} \end{vmatrix}$$

where E_{si} and E_{bi} are the initial (elastic) stiffnesses of the bearing surface and of the bolts, and A_s, J_{us}, and J_{vs} are the total area and the total second area moment of the bearing surface, to principal axes (u, v).

Then we can start a numerical procedure

$$\mathbf{K_1} = \Omega(\mathbf{a_0})$$

$$\mathbf{a_1} = \mathbf{K}_1^{-1} \cdot \boldsymbol{\sigma}$$

$$\mathbf{K_2} = \Omega(\mathbf{a_1})$$

$$\mathbf{a_2} = \mathbf{K}_2^{-1} \cdot \boldsymbol{\sigma}$$

which is stopped when

$$\frac{\|\mathbf{a}_{iter} - \mathbf{a}_{iter-1}\|}{\|\mathbf{a}_{iter}\|} < tol$$

In this way, a balanced distribution is found. As already discussed, it may also happen that a convergence is impossible, for instance because the constitutive law of the bearing surface or of the bolts has a plateau, a maximum value that cannot be crossed.

Once the vector **a** is known, the forces in the bolts and the pressures exchanged at the bearing surface are known (including the part of the bearing surface reacting, and the part ineffective). This analysis must be done for every check section *k* of the bolt layout, and for every load combination.

This method can be applied

1) to bolt layouts in bearing using a bearing surface
2) to preloaded bolt layouts using a bearing surface (sum of annuli, or sum of circles, as discussed in the previous sections), but the constitutive law of the bearing has both the tensile and compressive branches.

7.9.3 Instantaneous Center of Rotation Method

The instantaneous center of rotation method is suggested by AISC 360-10, and uses the curve developed by Crawford and Kulak in 1968 (Equation 7.5).

The problem can be considered equivalent to finding a correct position for the instantaneous center of rotation, $\mathbf{IC}(u_{IC}, v_{IC})$, and a proper rotation r_w, so that at least one bolt reaches its maximum displacement Δ_{max}. As such, it has been investigated by Brandt who also provided a numerical procedure (Brandt 1982).

In this book a general method valid for welds in 3D has been proposed, and this method will now be adapted to bolt groups. In this section, the index s will be used for the load step; it is implicitly assumed that the computation is to be repeated for every check section k, but the index k is omitted for simplicity.

The numerical procedure by Brandt is not directly applicable here, because due to the efficiency factors μ_{ni} and the activation shear factors μ_{ti}, the center and principal axes of the bolt layout are not in general the ones that are obtained by considering all the bolts to be equal. So, by coupling, extra diagonal terms appear.

When considering bolt layouts also under axial force and bending (and not just shear force and torque), the force in the bolt shafts is not only a shear force, but also possibly a tensile force. Also, a bending moment may appear. This reduces the shear available to the bolt, and makes it dependent on the bolt position and on (N, M_u, M_v). So, a utilization factor computed by only taking into account the shears in the bolts, i.e. (V_u, V_v, M_w), would not be correct. Ideally, what should be done is to gradually increase the applied forces to the bolt layout, and to gradually increase the shear, tensile force, and bending moment, tracing the first level of the applied loads implying a failure of one bolt or of one bearing. In the following, the instantaneous center of rotation method will be described not considering the tensile forces and the bending moments in the bolt shafts.

The initial guess for the utilization ratio can be computed as

$$U_{rough} = \frac{|V_u|}{R_{ult}\sum_{i=1}^{nb}\mu_{ti}} + \frac{|V_v|}{R_{ult}\sum_{i=1}^{nb}\mu_{ti}} + \frac{|M_w|}{R_{ult}\cdot\sum_{i=1}^{nb}\mu_{ti}\cdot\sqrt{\left(u_i^2 + v_i^2\right)}}$$

The displacement vector of bolt layout center at step s is $\mathbf{d_s} = \{d_{u,s}, d_{v,s}, r_{w,s}\}^T$. The displacement vector $\mathbf{d_{si}} = \{d_{usi}, d_{vsi}\}^T$ of bolt i can be obtained by

$$\mathbf{d_{si}} = \begin{vmatrix} d_{usi} \\ d_{vsi} \end{vmatrix} = \begin{vmatrix} 1 & 0 & -v_i \\ 0 & 1 & u_i \end{vmatrix} \cdot \begin{vmatrix} d_{us} \\ d_{vs} \\ r_{ws} \end{vmatrix} = \mathbf{Q_i d_s}$$

The force exerted at step s by the bolt i, R_{si} can be computed as

$$R_{si} = R_{ult,i}\cdot\left(1 - e^{-0.3937\cdot d_{si}}\right)^{0.55}$$

where $R_{ult,i}$ is the maximum shear force that can be applied considering both the bolt bearing and the bolt shear, and

$$d_{si} = \sqrt{\mathbf{d_{si}^T}\cdot\mathbf{d_{si}}} = \sqrt{d_{usi}^2 + d_{vsi}^2}$$

If the displacement d_i is greater than the maximum possible displacement d_{max}, 8.63 mm, the bolt is declared to be in a failure condition.

The vector $\boldsymbol{\sigma}_{\mathbf{si}}$ of the forces V_{usi}, V_{vsi} in the bolt is

$$\boldsymbol{\sigma}_{\mathbf{si}} = \begin{vmatrix} V_{usi} \\ V_{vsi} \end{vmatrix} = \frac{R_{si}}{d_{si}} \cdot \begin{vmatrix} d_{usi} \\ d_{vsi} \end{vmatrix} = \frac{R_{si}}{d_{si}} \cdot \mathbf{Q_i d_s}$$

The global forces can be obtained by summing up all the terms related to each bolt:

$$\boldsymbol{\sigma}_{\mathbf{s}} = \begin{vmatrix} V_{us} \\ V_{vs} \\ M_{ws} \end{vmatrix} = \sum_{i=1}^{nb} \begin{vmatrix} 1 & 0 \\ 0 & 1 \\ -v_i & u_i \end{vmatrix} \cdot \begin{vmatrix} V_{usi} \\ V_{vsi} \end{vmatrix} = \sum_{i=1}^{nb} \left(\frac{R_{si}}{d_{si}} \right) \cdot \mathbf{Q_i^T Q_i d_s}$$

So, introducing the secant stiffness matrix at step s, $\mathbf{K_s}$,

$$\boldsymbol{\sigma}_{\mathbf{s}} = \begin{vmatrix} V_{us} \\ V_{vs} \\ M_{ws} \end{vmatrix} = \sum_{i=}^{nb} \left(\frac{R_{is}}{d_{is}} \right) \cdot \begin{vmatrix} 1 & 0 & -v_i \\ 0 & 1 & u_i \\ -v_i & u_i & r_i^2 \end{vmatrix} \cdot \mathbf{d_s} = \mathbf{K_s d_s} = \Theta(\mathbf{d_s})$$

and

$$r_i^2 = u_i^2 + v_i^2$$

The tangent stiffness matrix of the bolt layout $\mathbf{K_t}$ is such that

$$\boldsymbol{\delta\sigma} = \mathbf{K_t \delta d}$$

Its derivation in closed form can be found in Appendix 3 (Equation A3.3). The numerical procedure then follows what has already been discussed for weld layouts.

If the bolt layout is also subject to a bending moment and an axial force, the utilization factor is not meaningful. An utilization index (see Chapter 8) can be computed by evaluating the shears balanced with the externally applied shears and torque, and then checking the bolt under the effect of the shear computed by ICRM, and of the tensile force and bending moment computed by other means. This could be useful for checking bolts, but in no way would it be able to assess how far from the boundary the checks are.

7.9.4 Examples

In order to see the effects of the flags shear only (SO), no shear (NS), and of the efficiency factors μ_{ni}, a simple and regular bolt layout with eight bolts (Figure 7.31) has been loaded by six elementary generalized internal forces, aligned with the principal axes of the column.

The layout is kept simple so as to allow an easy understanding of the changes related to the working flags. Clearly, the method set up is valid for each bolt layout, with no limitation in the bolt positions and working flags distribution. In this example, the bending moments in the bolt shafts are maintained.

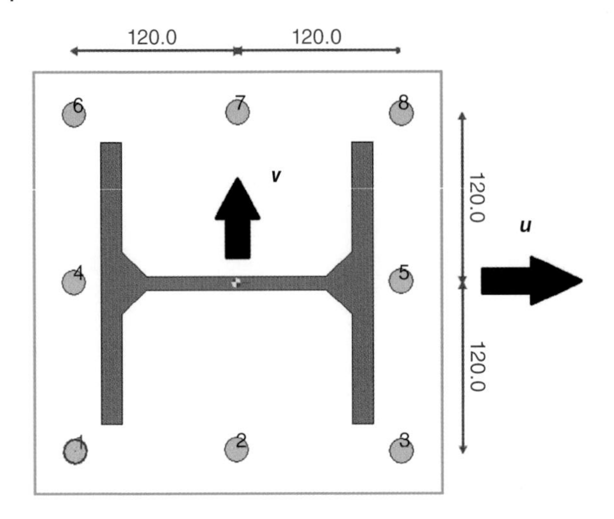

Figure 7.31 Bolt layout to be examined.

The bolts are eight M16 class 8.8 bolts at 120 mm spacing. The loads applied are

1. Load 1 $N = 183{,}488$ N
2. Load 2 $V_v = 72{,}248.2$ N, $M_u = -1{,}444{,}963.6$ Nmm
3. Load 3 $V_u = -33{,}688.7$ N, $M_v = -673{,}773.6$ Nmm
4. Load 4 $M_w = 536{,}196.7$ Nmm
5. Load 5 $M_v = 13{,}386{,}070.0$ Nmm
6. Load 6 $M_u = -4{,}707{,}990.0$ Nmm

The bending moments in loads 2 and 3 are due to the transport of the member applied shear to the proper elevation for the bolt check (e.g. 1,444,963.6/72,248.2 = 20 mm).

The first computation assumes that all the bolts are fully reactive. The forces in the bolts are given in Table 7.7.

Considering the loads 5 and 6, where a pure bending is applied, it's easy to see that the major part of the bending is carried by axial forces in the bolt shafts. Low bending moments carry a limited part of the bending (1/675 in both loads 5 and 6).

Now, if all the bolts are declared "shear only", using $\mu_{ni} = 1 \times 10^{-4}$, the forces and moments in the bolts are those of Table 7.8.

It can be seen that the forces and moments are the same as in the previous case; the only difference is that the predicted displacements are enormous, and so this means that the connector is not able to carry the loads. As there is no other connector available, the bending moments must imply axial forces and bending moments in the bolt shafts. If other connectors having higher bending stiffness had been available, these connectors would have been loaded instead.

The third computation assumes that mid-side bolts are shear only, while corner nodes are no-shear (Figure 7.32). This is merely a test, with no specific reason but the relative simplicity. The results are listed in Table 7.9.

Table 7.7 All bolts fully resistant and with full efficiency. N_B: axial force of bolt. V_{uB}: shear along axis **u**. V_{vB}: shear along axis **v**. V_B: total shear. M_{uB}: bending moment about axis **u** of bolt shaft. M_{vB}: bending moment about axis **v**. M_B: total bending moment.

Load	Bolt	N_B (N)	V_{uB} (N)	V_{vB} (N)	V_B (N)	M_{uB} (N mm)	M_{vB} (N mm)	M_B (N mm)
1	1	22936.0	0.0	0.0	0.0	0.0	0.0	0.0
1	2	22936.0	0.0	0.0	0.0	0.0	0.0	0.0
1	3	22936.0	0.0	0.0	0.0	0.0	0.0	0.0
1	4	22936.0	0.0	0.0	0.0	0.0	0.0	0.0
1	5	22936.0	0.0	0.0	0.0	0.0	0.0	0.0
1	6	22936.0	0.0	0.0	0.0	0.0	0.0	0.0
1	7	22936.0	0.0	0.0	0.0	0.0	0.0	0.0
1	8	22936.0	0.0	0.0	0.0	0.0	0.0	0.0
2	1	2003.1	0.1	9031.0	9031.0	−342.0	0.0	342.0
2	2	2003.1	0.1	9031.0	9031.0	−342.0	0.0	342.0
2	3	2003.1	0.1	9031.0	9031.0	−342.0	0.0	342.0
2	4	−1.4	0.1	9031.0	9031.0	−342.0	0.0	342.0
2	5	−1.4	0.1	9031.0	9031.0	−342.0	0.0	342.0
2	6	−2003.1	0.1	9031.0	9031.0	−342.0	0.0	342.0
2	7	−2003.1	0.1	9031.0	9031.0	−342.0	0.0	342.0
2	8	−2003.1	0.1	9031.0	9031.0	−342.0	0.0	342.0
3	1	−934.0	−4211.1	−0.0	4211.1	0.0	−159.5	159.5
3	2	−0.7	−4211.1	−0.0	4211.1	0.0	−159.5	159.5
3	3	934.0	−4211.1	−0.0	4211.1	0.0	−159.5	159.5
3	4	−934.0	−4211.1	−0.0	4211.1	0.0	−159.5	159.5
3	5	934.0	−4211.1	−0.0	4211.1	0.0	−159.5	159.5
3	6	−934.0	−4211.1	−0.0	4211.1	0.0	−159.5	159.5
3	7	−0.7	−4211.1	−0.0	4211.1	0.0	−159.5	159.5
3	8	934.0	−4211.1	−0.0	4211.1	0.0	−159.5	159.5
4	1	0.0	372.4	−372.4	526.6	0.0	0.0	0.0
4	2	0.0	372.4	0.5	372.4	0.0	0.0	0.0
4	3	0.0	372.4	372.4	526.6	0.0	0.0	0.0
4	4	0.0	0.5	−372.4	372.4	0.0	0.0	0.0
4	5	0.0	0.5	372.4	372.4	0.0	0.0	0.0
4	6	0.0	−372.4	−372.4	526.6	0.0	0.0	0.0
4	7	0.0	−372.4	0.5	372.4	0.0	0.0	0.0
4	8	0.0	−372.4	372.4	526.6	0.0	0.0	0.0
5	1	18556.6	−0.0	−0.0	0.0	0.0	3168.6	3168.6

(Continued)

Table 7.7 (Continued)

Load	Bolt	N_B (N)	V_{uB} (N)	V_{vB} (N)	V_B (N)	M_{uB} (N mm)	M_{vB} (N mm)	M_B (N mm)
5	2	13.4	−0.0	−0.0	0.0	0.0	3168.6	3168.6
5	3	−18556.6	−0.0	−0.0	0.0	0.0	3168.6	3168.6
5	4	18556.6	−0.0	−0.0	0.0	0.0	3168.6	3168.6
5	5	−18556.6	−0.0	−0.0	0.0	0.0	3168.6	3168.6
5	6	18556.6	−0.0	−0.0	0.0	0.0	3168.6	3168.6
5	7	13.4	−0.0	−0.0	0.0	0.0	3168.6	3168.6
5	8	−18556.6	−0.0	−0.0	0.0	0.0	3168.6	3168.6
6	1	6526.5	−0.0	−0.0	0.0	−1114.4	0.0	1114.4
6	2	6526.5	−0.0	−0.0	0.0	−1114.4	0.0	1114.4
6	3	6526.5	−0.0	−0.0	0.0	−1114.4	0.0	1114.4
6	4	−4.7	−0.0	−0.0	0.0	−1114.4	0.0	1114.4
6	5	−4.7	−0.0	−0.0	0.0	−1114.4	0.0	1114.4
6	6	−6526.5	−0.0	−0.0	0.0	−1114.4	0.0	1114.4
6	7	−6526.5	−0.0	−0.0	0.0	−1114.4	0.0	1114.4
6	8	−6526.5	−0.0	−0.0	0.0	−1114.4	0.0	1114.4

Table 7.8 Forces and moments in the bolts assuming that all bolts are shear only. N_B: axial force of bolt. V_{uB}: shear along axis **u**. V_{vB}: shear along axis **v**. V_B: total shear. M_{uB}: bending moment about axis **u** of bolt shaft. M_{vB}: bending moment about axis **v**. M_B: total bending moment.

Load	Bolt	N_B (N)	V_{uB} (N)	V_{vB} (N)	V_B (N)	M_{uB} (N mm)	M_{vB} (N mm)	M_B (N mm)
1	1	22936.3	0.0	0.0	0.0	0.0	0.0	0.0
1	2	22936.3	0.0	0.0	0.0	0.0	0.0	0.0
1	3	22936.3	0.0	0.0	0.0	0.0	0.0	0.0
1	4	22936.3	0.0	0.0	0.0	0.0	0.0	0.0
1	5	22936.3	0.0	0.0	0.0	0.0	0.0	0.0
1	6	22936.3	0.0	0.0	0.0	0.0	0.0	0.0
1	7	22936.3	0.0	0.0	0.0	0.0	0.0	0.0
1	8	22936.3	0.0	0.0	0.0	0.0	0.0	0.0
2	1	2003.1	0.1	9030.9	9030.9	−342.0	0.0	342.0
2	2	2003.1	0.1	9030.9	9030.9	−342.0	0.0	342.0
2	3	2003.1	0.1	9030.9	9030.9	−342.0	0.0	342.0
2	4	−1.4	0.1	9030.9	9030.9	−342.0	0.0	342.0
2	5	−1.4	0.1	9030.9	9030.9	−342.0	0.0	342.0
2	6	−2003.1	0.1	9030.9	9030.9	−342.0	0.0	342.0

Table 7.8 (Continued)

Load	Bolt	N_B (N)	V_{uB} (N)	V_{vB} (N)	V_B (N)	M_{uB} (N mm)	M_{vB} (N mm)	M_B (N mm)
2	7	−2003.1	0.1	9030.9	9030.9	−342.0	0.0	342.0
2	8	−2003.1	0.1	9030.9	9030.9	−342.0	0.0	342.0
3	1	−934.0	−4211.0	−0.0	4211.0	0.0	−159.5	159.5
3	2	−0.7	−4211.0	−0.0	4211.0	0.0	−159.5	159.5
3	3	934.0	−4211.0	−0.0	4211.0	0.0	−159.5	159.5
3	4	−934.0	−4211.0	−0.0	4211.0	0.0	−159.5	159.5
3	5	934.0	−4211.0	−0.0	4211.0	0.0	−159.5	159.5
3	6	−934.0	−4211.0	−0.0	4211.0	0.0	−159.5	159.5
3	7	−0.7	−4211.0	−0.0	4211.0	0.0	−159.5	159.5
3	8	934.0	−4211.0	−0.0	4211.0	0.0	−159.5	159.5
4	1	0.0	372.4	−372.4	526.6	0.0	0.0	0.0
4	2	0.0	372.4	0.5	372.4	0.0	0.0	0.0
4	3	0.0	372.4	372.4	526.6	0.0	0.0	0.0
4	4	0.0	0.5	−372.4	372.4	0.0	0.0	0.0
4	5	0.0	0.5	372.4	372.4	0.0	0.0	0.0
4	6	0.0	−372.4	−372.4	526.6	0.0	0.0	0.0
4	7	0.0	−372.4	0.5	372.4	0.0	0.0	0.0
4	8	0.0	−372.4	372.4	526.6	0.0	0.0	0.0
5	1	18556.5	−0.1	−0.0	0.1	0.0	3168.6	3168.6
5	2	13.4	−0.1	−0.0	0.1	0.0	3168.6	3168.6
5	3	−18556.5	−0.1	−0.0	0.1	0.0	3168.6	3168.6
5	4	18556.5	−0.1	−0.0	0.1	0.0	3168.6	3168.6
5	5	−18556.5	−0.1	−0.0	0.1	0.0	3168.6	3168.6
5	6	18556.5	−0.1	−0.0	0.1	0.0	3168.6	3168.6
5	7	13.4	−0.1	−0.0	0.1	0.0	3168.6	3168.6
5	8	−18556.5	−0.1	−0.0	0.1	0.0	3168.6	3168.6
6	1	6526.5	−0.0	−0.0	0.0	−1114.4	0.0	1114.4
6	2	6526.5	−0.0	−0.0	0.0	−1114.4	0.0	1114.4
6	3	6526.5	−0.0	−0.0	0.0	−1114.4	0.0	1114.4
6	4	−4.7	−0.0	−0.0	0.0	−1114.4	0.0	1114.4
6	5	−4.7	−0.0	−0.0	0.0	−1114.4	0.0	1114.4
6	6	−6526.5	−0.0	−0.0	0.0	−1114.4	0.0	1114.4
6	7	−6526.5	−0.0	−0.0	0.0	−1114.4	0.0	1114.4
6	8	−6526.5	−0.0	−0.0	0.0	−1114.4	0.0	1114.4

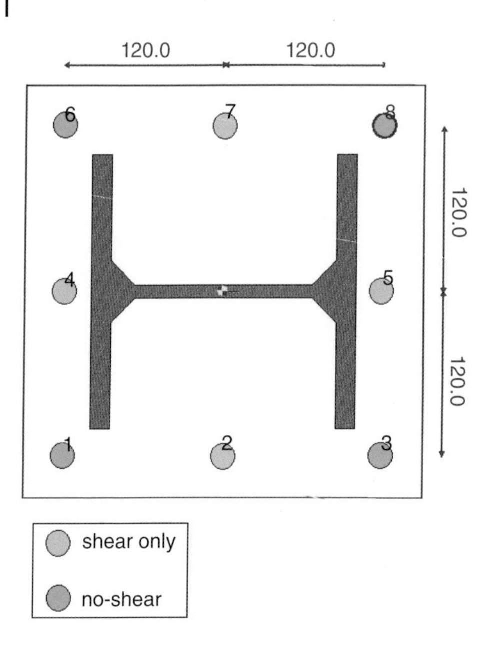

Figure 7.32 Mixed configuration of bolts: mid-side nodes are shear-only, corner-nodes are no-shear.

Table 7.9 Forces and moments in the bolts assuming mid-side nodes are shear-only, while corner nodes are no-shear. N_B: axial force of bolt. V_{uB}: shear along axis **u**. V_{vB}: shear along axis **v**. V_B: total shear. M_{uB}: bending moment about axis **u** of bolt shaft. M_{vB}: bending moment about axis **v**. M_B: total bending moment.

Load	Bolt	N_B (N)	V_{uB} (N)	V_{vB} (N)	V_B (N)	M_{uB} (N mm)	M_{vB} (N mm)	M_B (N mm)
1	1	45867.4	0.0	0.0	0.0	0.0	0.0	0.0
1	2	4.6	0.0	0.0	0.0	0.0	0.0	0.0
1	3	45867.4	0.0	0.0	0.0	0.0	0.0	0.0
1	4	4.6	0.0	0.0	0.0	0.0	0.0	0.0
1	5	4.6	0.0	0.0	0.0	0.0	0.0	0.0
1	6	45867.4	0.0	0.0	0.0	0.0	0.0	0.0
1	7	4.6	0.0	0.0	0.0	0.0	0.0	0.0
1	8	45867.4	0.0	0.0	0.0	0.0	0.0	0.0
2	1	3005.9	0.1	1.8	1.8	−513.3	0.0	513.3
2	2	0.3	0.1	18060.0	18060.0	−0.1	0.0	0.1
2	3	3005.9	0.1	1.8	1.8	−513.3	0.0	513.3
2	4	−1.4	0.1	18060.0	18060.0	−0.1	0.0	0.1
2	5	−1.4	0.1	18060.0	18060.0	−0.1	0.0	0.1
2	6	−3005.9	0.1	1.8	1.8	−513.3	0.0	513.3
2	7	−0.3	0.1	18060.0	18060.0	−0.1	0.0	0.1

Table 7.9 (Continued)

Load	Bolt	N_B (N)	V_{uB} (N)	V_{vB} (N)	V_B (N)	M_{uB} (N mm)	M_{vB} (N mm)	M_B (N mm)
2	8	−3005.9	0.1	1.8	1.8	−513.3	0.0	513.3
3	1	−1401.6	−0.8	−0.0	0.8	0.0	−239.3	239.3
3	2	−0.7	−8421.2	−0.0	8421.2	0.0	−0.0	0.0
3	3	1401.6	−0.8	−0.0	0.8	0.0	−239.3	239.3
3	4	−0.1	−8421.2	−0.0	8421.2	0.0	−0.0	0.0
3	5	0.1	−8421.2	−0.0	8421.2	0.0	−0.0	0.0
3	6	−1401.6	−0.8	−0.0	0.8	0.0	−239.3	239.3
3	7	−0.7	−8421.2	−0.0	8421.2	0.0	−0.0	0.0
3	8	1401.6	−0.8	−0.0	0.8	0.0	−239.3	239.3
4	1	0.0	0.1	−0.1	0.2	0.0	0.0	0.0
4	2	0.0	1116.9	0.5	1116.9	0.0	0.0	0.0
4	3	0.0	0.1	0.1	0.2	0.0	0.0	0.0
4	4	0.0	0.5	−1116.9	1116.9	0.0	0.0	0.0
4	5	0.0	0.5	1116.9	1116.9	0.0	0.0	0.0
4	6	0.0	−0.1	−0.1	0.2	0.0	0.0	0.0
4	7	0.0	−1116.9	0.5	1116.9	0.0	0.0	0.0
4	8	0.0	−0.1	0.1	0.2	0.0	0.0	0.0
5	1	27846.6	−0.0	−0.0	0.0	0.0	4754.9	4754.9
5	2	13.4	−0.1	−0.0	0.1	0.0	0.5	0.5
5	3	−27846.6	−0.0	−0.0	0.0	0.0	4754.9	4754.9
5	4	2.8	−0.1	−0.0	0.1	0.0	0.5	0.5
5	5	−2.8	−0.1	−0.0	0.1	0.0	0.5	0.5
5	6	27846.6	−0.0	−0.0	0.0	0.0	4754.9	4754.9
5	7	13.4	−0.1	−0.0	0.1	0.0	0.5	0.5
5	8	−27846.6	−0.0	−0.0	0.0	0.0	4754.9	4754.9
6	1	9793.9	−0.0	−0.0	0.0	−1672.3	0.0	1672.3
6	2	1.0	−0.0	−0.0	0.0	−0.2	0.0	0.2
6	3	9793.9	−0.0	−0.0	0.0	−1672.3	0.0	1672.3
6	4	−4.7	−0.0	−0.0	0.0	−0.2	0.0	0.2
6	5	−4.7	−0.0	−0.0	0.0	−0.2	0.0	0.2
6	6	−9793.9	−0.0	−0.0	0.0	−1672.3	0.0	1672.3
6	7	−1.0	−0.0	−0.0	0.0	−0.2	0.0	0.2
6	8	−9793.9	−0.0	−0.0	0.0	−1672.3	0.0	1672.3

- Load 1: as expected, the axial forces are doubled (half bolts available).
- Loads 2 and 3. The axial forces are 1.5 times higher than for the all-FR case (two bolts instead of three resisting along any side). The shears are doubled.
- Load 4. The force is obtained by dividing the torque by 4 and by 120: 536,196/ 480 = 1117.
- Loads 5 and 6. The axial forces are about 1.5 times those of all-FR, and are obtained by dividing the moment by 2 × 240. Load 5: 13,386,070/480 = 27887. Load 6: 4,707,990/ 480 = 9808. The true axial force is a bit lower because a small fraction of the applied moment is taken by bending in the shafts.

The fourth computation is like the third, but instead of flagging the mid-side nodes as "shear only", an efficiency factor of 0.5 is assigned to them, that is $\mu_{n2} = \mu_{n4} = \mu_{n5} = \mu_{n7} = 0.5$. The results are listed in Table 7.10.

Table 7.10 Forces and moments in the bolts assuming mid-side nodes have an efficiency factor $\mu_n = 0.5$, while corner nodes are no-shear. N_B: axial force of bolt. V_{uB}: shear along axis **u**. V_{vB}: shear along axis **v**. V_B: total shear. M_{uB}: bending moment about axis **u** of bolt shaft. M_{vB}: bending moment about axis **v**. M_B: total bending moment.

Load	Bolt	N_B (N)	V_{uB} (N)	V_{vB} (N)	V_B (N)	M_{uB} (Nmm)	M_{vB} (Nmm)	M_B (Nmm)
1	1	30581.3	0.0	0.0	0.0	0.0	0.0	0.0
1	2	15290.7	0.0	0.0	0.0	0.0	0.0	0.0
1	3	30581.3	0.0	0.0	0.0	0.0	0.0	0.0
1	4	15290.7	0.0	0.0	0.0	0.0	0.0	0.0
1	5	15290.7	0.0	0.0	0.0	0.0	0.0	0.0
1	6	30581.3	0.0	0.0	0.0	0.0	0.0	0.0
1	7	15290.7	0.0	0.0	0.0	0.0	0.0	0.0
1	8	30581.3	0.0	0.0	0.0	0.0	0.0	0.0
2	1	2404.1	0.1	9030.9	9030.9	−410.5	0.0	410.5
2	2	1202.1	0.1	9030.9	9030.9	−205.3	0.0	205.3
2	3	2404.1	0.1	9030.9	9030.9	−410.5	0.0	410.5
2	4	−1.4	0.1	9030.9	9030.9	−205.3	0.0	205.3
2	5	−1.4	0.1	9030.9	9030.9	−205.3	0.0	205.3
2	6	−2404.1	0.1	9030.9	9030.9	−410.5	0.0	410.5
2	7	−1202.1	0.1	9030.9	9030.9	−205.3	0.0	205.3
2	8	−2404.1	0.1	9030.9	9030.9	−410.5	0.0	410.5
3	1	−1121.0	−4211.1	−0.0	4211.1	0.0	−191.4	191.4
3	2	−0.7	−4211.1	−0.0	4211.1	0.0	−95.7	95.7
3	3	1121.0	−4211.1	−0.0	4211.1	0.0	−191.4	191.4
3	4	−560.5	−4211.1	−0.0	4211.1	0.0	−95.7	95.7

Table 7.10 (Continued)

Load	Bolt	N_B (N)	V_{uB} (N)	V_{vB} (N)	V_B (N)	M_{uB} (Nmm)	M_{vB} (Nmm)	M_B (Nmm)
3	5	560.5	−4211.1	−0.0	4211.1	0.0	−95.7	95.7
3	6	−1121.0	−4211.1	−0.0	4211.1	0.0	−191.4	191.4
3	7	−0.7	−4211.1	−0.0	4211.1	0.0	−95.7	95.7
3	8	1121.0	−4211.1	−0.0	4211.1	0.0	−191.4	191.4
4	1	0.0	372.4	−372.4	526.6	0.0	0.0	0.0
4	2	0.0	372.4	0.5	372.4	0.0	0.0	0.0
4	3	0.0	372.4	372.4	526.6	0.0	0.0	0.0
4	4	0.0	0.5	−372.4	372.4	0.0	0.0	0.0
4	5	0.0	0.5	372.4	372.4	0.0	0.0	0.0
4	6	0.0	−372.4	−372.4	526.6	0.0	0.0	0.0
4	7	0.0	−372.4	0.5	372.4	0.0	0.0	0.0
4	8	0.0	−372.4	372.4	526.6	0.0	0.0	0.0
5	1	22272.1	−0.0	−0.0	0.0	0.0	3803.0	3803.0
5	2	13.4	−0.0	−0.0	0.0	0.0	1901.5	1901.5
5	3	−22272.1	−0.0	−0.0	0.0	0.0	3803.0	3803.0
5	4	11136.0	−0.0	−0.0	0.0	0.0	1901.5	1901.5
5	5	−11136.0	−0.0	−0.0	0.0	0.0	1901.5	1901.5
5	6	22272.1	−0.0	−0.0	0.0	0.0	3803.0	3803.0
5	7	13.4	−0.0	−0.0	0.0	0.0	1901.5	1901.5
5	8	−22272.1	−0.0	−0.0	0.0	0.0	3803.0	3803.0
6	1	7833.3	−0.0	−0.0	0.0	−1337.6	0.0	1337.6
6	2	3916.6	−0.0	−0.0	0.0	−668.8	0.0	668.8
6	3	7833.3	−0.0	−0.0	0.0	−1337.6	0.0	1337.6
6	4	−4.7	−0.0	−0.0	0.0	−668.8	0.0	668.8
6	5	−4.7	−0.0	−0.0	0.0	−668.8	0.0	668.8
6	6	−7833.3	−0.0	−0.0	0.0	−1337.6	0.0	1337.6
6	7	−3916.6	−0.0	−0.0	0.0	−668.8	0.0	668.8
6	8	−7833.3	−0.0	−0.0	0.0	−1337.6	0.0	1337.6

It can be seen that systematically the bolts having efficiency 0.5 have an axial force which is half the axial force of the others. There is no difference in the shears.

As a final consideration, it should be noted that the compressive forces in the bolts, are not going to be used in bolt checks. These compressive forces depend on the assumption that no bearing surface is acting to carry pressures at the interface.

7.10 Contact

Contact connectors imply two surfaces coplanar, with or without friction; these surfaces are usually two faces of two different constituents (Figure 7.33).

In the context of general fem, this type of connector can be modeled by marking some of the plate (or solid) finite elements of the two constituents as potentially "in contact". When the distance between the elements becomes lower than a given threshold, the elements are connected, simulating the contact (the issue will be discussed in further detail in Chapter 9).

If simple models are set up (like the ones to be discussed in Chapter 9), then the connector may be simulated by a linear or non-linear beam element, able to carry three forces and three moments. Linear elements could be used if and only if it can be ensured that no detachment of the parts in contact is found, for every loading condition (tensile forces would prove that the detachment had been detected). If this is the case, then the bearing surface concept might be applied as for bolt layouts, with the unique difference that no bolt is available. In this way, a field of pressures can be found over a part of the bearing surface, and these pressures may then be applied as nodal forces to the models of single constituents, using the techniques that will be explained in Chapter 9.

The two constituents have two faces coplanar, and with opposite normals. The intersection of the two faces is where the bearing surface can be defined. The center of the bearing surface gives the position of the extremities of the simple element in plan. The two nodes of the element are obtained by moving a half thickness in both direction, that is inside constituent 1 and inside constituent 2 (Figure 7.33).

Usually, contact also uses friction. If so, the checking condition is that of no-slip, i.e. the resulting shear force V must be lower than a fraction of the applied compressive force. If there is no friction, appropriate shear end releases may be applied to the element.

The principal axes of the element can be taken as the principal axes of the bearing surface; their position, as well as the position of the center of the bearing surface, may be easily evaluated by means of the boundary integrals already discussed.

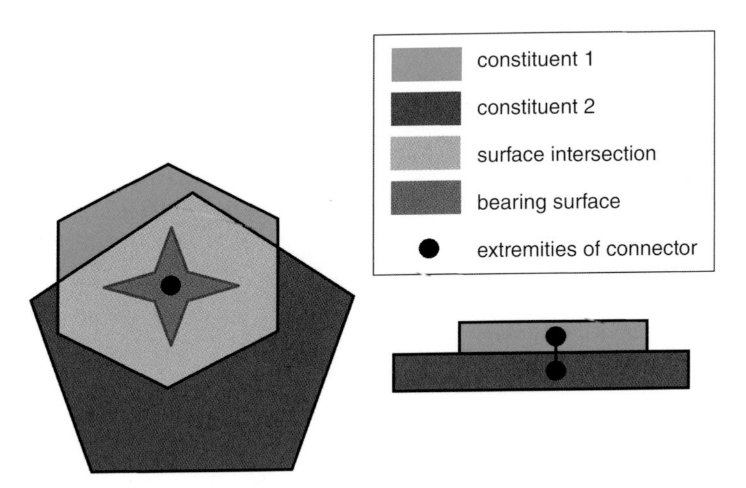

Figure 7.33 Contact element extremities in simplified models, bearing surface is inside the intersection between the faces in contact.

References

AISC 360-10, 2010, *Specifications for Structural Steel Buildings*, American Institute for Steel Construction.

Brandt, G.D. 1982, *Rapid Determination of Ultimate Strength of Eccentrically Loaded Bolt Groups*, Engineering Journal, American Institute for Steel Construction, **19**, N° 2, 1982.

Brown, K.H., Morrow, C., Durbin, S., and Baca, A. 2008, *Guideline for Bolted Joint Design and Analysis – Version* 1, Sandia Report SAND2008-0371, Sandia National Laboratories.

Bruneau, M., Uang C.M., and Sabelli, S. 2011, *Ductile Design of Steel Structures*, Mc Graw Hill.

Crawford, S.F. and Kulak, G.L. 1968, Behavior of Eccentrically Loaded Bolted Connections, *Studies in Structural Engineering, (N° 4), Department of Civil Engineering, The University of Texas at* Austin, Austin.

EN 1993-1.3 2006, *Design of Steel Structures – General rules – Supplementary rules for cold formed thin gauge members and sheeting*, CEN.

EN 1993-1.8 2006, *Design of Steel Structures – Design of Joints*, CEN.

Jaspart, J.P. and Weynand, K. 2016, *Design of Joints in Steel and Composite Structures*, ECCS Eurocode Design Manuals, Wiley Ernst & Sohn.

Kuhlmann, U., Wald, F. et al. 2014, *Design of Steel-to-Concrete Joints, Design Manual*, ECCS.

Rex, C.O. and Easterling, W.S. 2003, *Behavior and Modeling of a Bolt Bearing on a Single Plate*, Journal of Structural Engineering, ASCE, **129** (6), 792–800, June 2003.

Richard, R.M. and Abbott, B.J. 1975, *Versatile Elastic-Plastic Stress–strain Formulation*, Journal of the Engineering Mechanics, **101** (EM4):511-515, 1975.

Rugarli, P. 1998, Proprietà Flessionali Elastiche e Plastiche Calcolo Automatico di Sezioni Generiche, *Costruzioni Metalliche*, **4**.

Rugarli, P. 2010, *Structural Analysis with Finite Elements*, Thomas Telford, London.

Schneider, P.J. and Eberly, D.H. 2003, *Geometric Tools for Computer Graphics*, Morgan Kaufmann Publishers.

Tate, M.B. and Rosenfeld, S.J. 1946, *Preliminary Investigation of the Loads Carried by Individual Bolts in Bolted Joints*, Technical Note No 1051, National Advisory Committee for Aeronautics, Washington DC, 1–68.

Thornton, W.A. 1985, *Prying Action – a General Treatment*, Engineering Journal, American Institute of Steel Construction, Second Quarter 1985.

Weigand, J.M. 2017, *Component-Based Model for Single-Plate Shear Connections with Pretension and Pinched Hysteresis*, Journal of Structural Engineering, ASCE, **143** (2), February 2017.

Wileman, J., Choudhury, M. and Green, I 1991, *Computation of member Stiffness in Bolted Connections*, ASME Journal of Mechanical Design, **113**, December 1991, pp. 432–437.

8

Failure Modes

8.1 Introduction

In this chapter it will be assumed that the internal forces flowing into the connectors are known: the following two chapters will outline how this fundamental goal can be reached.

Once the forces flowing globally into the connectors, and the related pressure fields at the interface of contact surfaces, are known it is possible to compute the internal forces in the subconnectors according to the rules explained in Chapters 6 and 7.

So, in this chapter it is assumed that the following are known:

- the shear forces, the tensile force and possibly the bending moments flowing into each bolt extremity
- the pressure fields exchanged at the bearing surface interface for bolt layouts using a bearing surface, and for contact connectors
- the shear pressure fields exchanged by the contact connectors
- the stresses acting over the effective area of single welds, from point to point

Also, it will be assumed that all these forces and stresses are strictly balanced to the externally applied loads, that is, to the member forces acting at member extremities.

The methods that can be used to compute the force packets flowing into each connector will be discussed in the following two chapters.

The forces acting on the other constituents, members, force transferrers, and stiffeners are immediately known by the third law, that is, Newton's action-reaction principle. So, once the forces and stresses in the connectors are known, the forces locally loading the other constituents are also known (see Figure 8.1).

Having reached this stage, it is then possible to check the constituents against the relevant failure modes.

This chapter recalls and discusses the failure modes in connection analysis, considering both those of the connectors and those of the other parts.

Failure modes can be checked using different tools, and one of the most promising of these tools is finite element analysis.

Steel Connection Analysis, First Edition. Paolo Rugarli.
© 2018 John Wiley & Sons Ltd. Published 2018 by John Wiley & Sons Ltd.

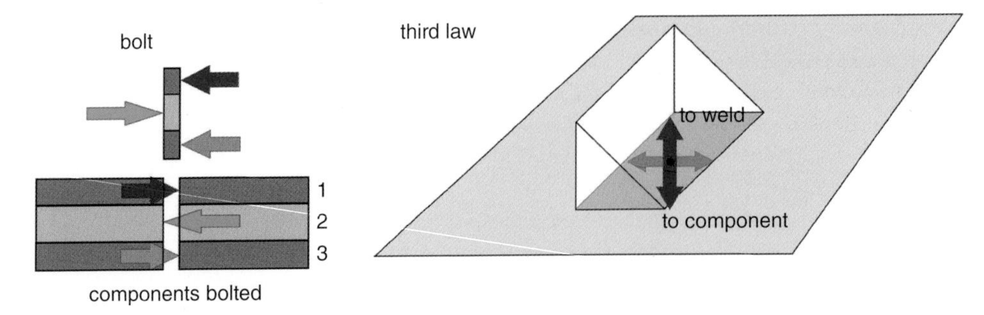

Figure 8.1 The third law applied to single connectors, bolts, and welds. Once the internal forces flowing into subconnectors are known, by the third law we also know how the constituents are loaded. Bolts also exchange bending moments and axial forces.

8.2 Utilization Factor Concept

In order to assess if a failure mode has been reached, it is necessary to compare, for each failure mode, the *demand* to the *capacity*. A useful way to compare the two is to use the *utilization index* concept.

A utilization index is a pure non-negative number which is defined in such a way that if it is less than 1 the failure mode has not been reached, while if it is greater than 1 then the failure mode has been reached (and crossed).

Depending on the failure mode and on the standard used, the utilization index can be defined in quite different ways. Very often, the utilization index, U is or can be defined as

$$U = \frac{D}{C}$$

where D is the demand and C is the capacity. These can be stresses, forces or moments, energies, displacements, or any other possible quantity. Very often, in the standards, the safe condition is written in the equivalent form

$$D \leq C$$

It must be underlined that the utilization index concept is in general notional. It strongly depends on the simplifications introduced by the standard and it is not always related to clear physical quantities. Also, computing the utilization index must be done bearing in mind that the input data of all the engineering analyses is affected by relevant uncertainness, related to loads, constraints, stiffnesses, geometries, constitutive laws, and so on. As the input quantities are themselves notional, for example a 235 MPa yield stress is a notional value, it would be meaningless to compute utilization with an excess of pretended precision. Utilization indices have meaning up to the second significant digit: no more. Sometimes also two digits, which implies an uncertainty of about 1%, are too much for the available precision of input data. In these cases it would be a good practice to round the computed index to the upper multiple of 0.05, 0.1, or 0.15, at increasing uncertainness.

So, utilization indexes of 0.894 and 0.89 are not distinguishable and it would be good engineering practice (also in the examples of the textbooks) not to push the printed digits above that limit. If the uncertainness is high, 0.894 should be written 0.9, to be coherent with the input data.

The only reason why it could be useful to keep the extra digits of a computation is for checking purposes. Moreover, in computerized analysis, extra effort is required to clean the results by these extra digits, so very often computer outputs are heavy and full of unnecessary numbers. The analyst, however, should not be fooled by this *appearance* of precision, and always keep in mind that in real-world analysis the two-digit boundary can never be crossed. It is the author's opinion that this would also improve the standards, because very often the notional methods provided seem to forget the point. Engineering is not physics, and physics is not math.

In general, a utilization index is not linearly varying with the applied loads. Therefore it is not true to say that by scaling the applied loads by $1/U$, the failure mode limit condition is met. This is very important and should be kept firmly in mind: a utilization index is not a measure of the distance from the boundary, that is, it cannot be directly used in order to evaluate how to rescale the loads so as to reach the limit for that specific failure mode. From an engineering point of view this is somewhat disturbing, as the number obtained by the check merely answers the question *Are we below or above the limit?*; it doesn't answer the other quite important question, *How far from the limit are we?*

Consider, for example, the utilization index implicitly used by AISC to check for bolt resistance.[1] This is

$$U = \left(\frac{N}{N_{lim}}\right)^2 + \left(\frac{V}{V_{lim}}\right)^2 \leq 1$$

where N is the axial force in the bolt shaft and V is the shear, while the subscript "lim" stands for *limit*.

If the analysis is linear, then a factor k can be used as load multiplier, so that if a factor k is applied to the loads, then also the forces in the shaft, N and V, will be multiplied by k. This is the luckiest situation if the aim is to measure a distance from the boundary, because in non-linear analysis the internal forces would not depend linearly on the applied load. Consequently, in the linear range, it can be written

$$U = k^2\left[\left(\frac{N}{N_{lim}}\right)^2 + \left(\frac{V}{V_{lim}}\right)^2\right]$$

If it is assumed, for example, that with $k = 1$ then $U = 0.5$, it is not true that with $k = 2$ then $U = 1$ (Figure 8.2).

Up to now it has been implicitly assumed that a single action is applied to the structure, or that all the actions applied to the structure are related, so that if one of them increases by k, then all the others will also increase by k. This is usually not true. The internal forces applied to a connection are the result of several independent loads, applied to the structure in a combination.

1 The formula provided by the standard is different, and this will be discussed later in this section. However, as explained above, and as will be discussed here, there are good reason to use the elliptic formula, instead, and this is allowed.

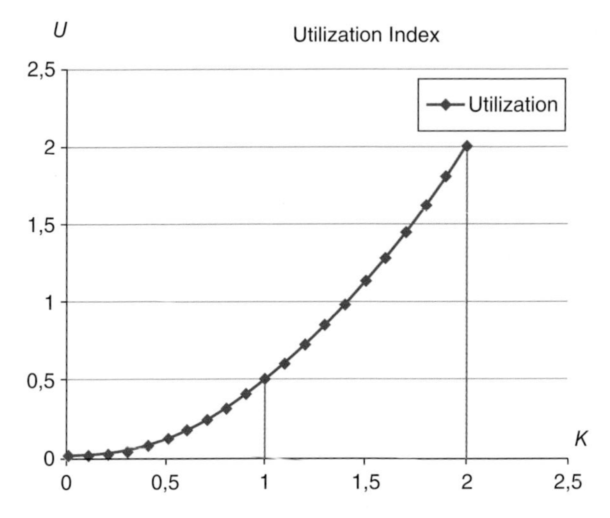

Figure 8.2 A utilization index is not a measure of the distance from the failure condition. Here, if the applied loads are related to a utilization index $U = 0.5$, the factor to be applied to get $U = 1$ is $k = 1.41$, not $k = 2$.

So, a unique measure of the "distance from the boundary" is not really possible, as probably the permanent (dead) loads will not change much,[2] while the variable, often uncertain, will possibly vary significantly.

This can be shown by considering the previous example, and assuming that the permanent load causes (N_0, V_0), while the (supposed unique) variable load causes (N_1, V_1). If it is assumed that $U = 0.5$ when ($N_0 + N_1$, $V_0 + V_1$) is applied to the bolt shaft, a correct load multiplier to get to the boundary of safety would be obtained by the equation in k

$$\left(\frac{N_0 + kN_1}{N_{lim}}\right)^2 + \left(\frac{V_0 + kV_1}{V_{lim}}\right)^2 = 1$$

If the dead loads cause the most part of the applied forces, say for simplicity

$$N_0 = 9N_1$$

$$V_0 = 9V_1$$

The equation becomes

$$(9+k)^2 \cdot \left(\frac{N_1}{N_{lim}}\right)^2 + (9+k)^2 \cdot \left(\frac{V_1}{V_{lim}}\right)^2 = 1$$

and as it has been assumed that $U = 0.5$ when $k = 1$

$$\left[\left(\frac{N_1}{N_{lim}}\right)^2 + \left(\frac{V_1}{V_{lim}}\right)^2\right] = \frac{0.5}{100}$$

2 However, it also true that frequently after structural failures dead loads are found to be much higher than the design ones, a consideration that should be kept well in mind.

this implies

$$(9+k)^2 = 200$$

and

$$k = 5.1$$

This is a multiplier much higher than the one predicted previously as $k = 1.41$. If, on the other hand, the variable load is much higher than the permanent one, so that

$$N_1 = 9N_0$$

$$V_1 = 9V_0$$

after some math, we would get

$$k = 1.46$$

which is not so far from the value 1.4 obtained by assuming that all the loads change in the very same manner.

Even if it can be assumed that it's on the safe side to evaluate the "distance from the boundary" applying the same amplification factor to all the applied loads, as in general, for the failure mode under consideration, the increment of internal forces related to the increment of single loads will not be the same, the result obtained clearly means that it is not really possible to have a rigorous "distance from the boundary". It is only possible to have an "engineering" estimate, which in turn means a "wishful thinking" measure. This is a necessary conclusion if rational reasoning is applied to the matter. It is safe to assume that all the loads will change, depending on a single k, but this is not generally true.

When a very simple renode, with two members, one plate, one weld layout, and one bolt layout is checked, for each constituent, several possible failure modes must be considered. This must be done for all the applied combinations, which, especially if using Eurocodes, can be quite numerous, each member being subjected, in general, to six internal forces. If the renode has some force transferrer, several bolt layouts and weld layouts, four or five members, and 30 or 40 combinations, the problem is much more complex. So, when the good scholars are asked, *What is the safety factor of this renode?*, or *Which is the most dangerous failure mode?*, they will probably remain silent, knowing that there's no simple answer. This is disturbing, but rational. Most of the checks will only provide utilization indices, and there will probably be at least two or three variable loads. How can we answer such questions, without abandoning rational reasoning? But the appeal of the oversimplifications is quite often too strong to resist, and meaningless statements obtained by considering only part of the issues, while forgetting the others, are too attractive to avoid. However, the truth is not the content of the answer, but the real-world behavior of the structure.

So, the aim for the experienced designer will have to be a proper (i.e. safe) evaluation of the applied loads, including the dead ones, a proper estimate of their possible combinations, while the presumed "safety factor", or "load multiplier" is substantially a myth in the most frequent and realistic situations, which are complex, and not easy. This is the uncomfortable truth emerging from a rational evaluation of the features of the problem. Using simplifications may, however, help to us discuss and to have simple numerical indices to be used for summarizing data.

To get a meaningful measure of the "distance from the boundary", specific analyses should be carried on, far beyond those required by the standards, at the current stage of development.

First of all, *utilization indices* must be replaced by *utilization factors*. A utilization factor U is a utilization index which has the property that if the loads are increased by k, then the utilization factor will also increase by k. In order to distinguish utilization indices by utilization factors, the subscripts "i" and "f" can be used, so U_i is a utilization index, and U_f is a utilization factor. Often, but not always, a utilization factor can be obtained by the methods suggested in the standards adding some modification.

For instance, considering the previous example, U_f could be defined as

$$U_f = \sqrt{\left(\frac{N}{N_{lim}}\right)^2 + \left(\frac{V}{V_{lim}}\right)^2} \leq 1$$

which, if it is assumed that N and M are both tuned by the same factor k, is indeed linear in k. This definition of U_f (Figure 8.3) has the geometrical meaning of distance of the point $\mathbf{P}(n, v)$ from the origin $\mathbf{O}$, in the plane $\mathbf{n}$-$\mathbf{v}$, being $n \equiv N/N_{lim}$ and $v = V/V_{lim}$. Indeed, as the limit distance from the origin is the radius of a circle measuring 1, **LO**, U_f is the ratio of two distances: (1) the distance of $\mathbf{P}(n, v)$ from the origin $\mathbf{O}$; (2) the distance of the limit point $\mathbf{L}$ from the origin $\mathbf{O}$. So $U_f = \mathbf{PO}/\mathbf{LO}$.

If the aim is to get a "distance from the boundary" measure, when using the elliptical rule, the additional square root, and the related increased complexity, is necessary. Otherwise, as has been shown, the result obtained is tricky.

Moreover, the checks suggested by the standards are often such that there is no continuity in the dependence of the results on the data.

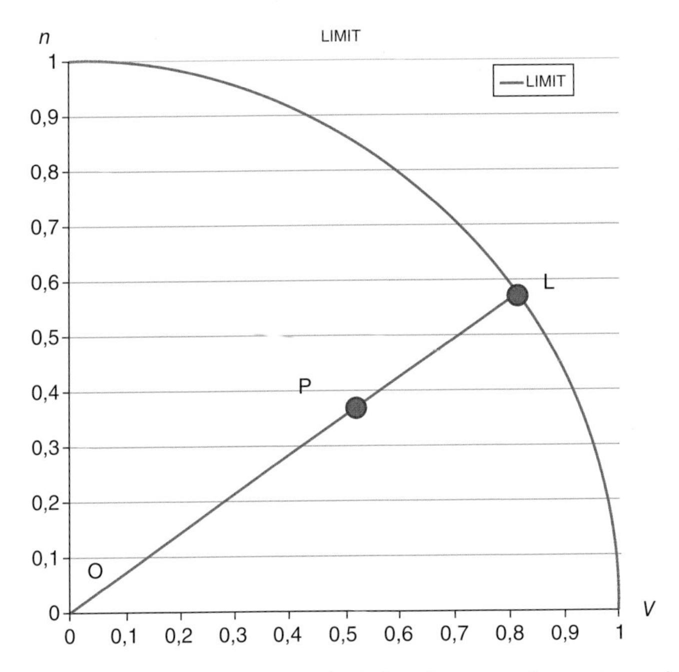

Figure 8.3 Elliptic limit domain for bolt resistance under tension and shear.

Let us once more consider the bolt shaft verification: If only an axial force is applied, the check suggested is

$$\frac{N}{N_{\text{lim}}} \le 1$$

Let us assume that $U = 0.5$. Now, if a slight shear force is *added*, say equal to $0.01\,V_{lim}$, the check suggested would be

$$0.5^2 + 0.01^2 = 0.2501 < 1$$

So now we get 0.2501 while previously we obtained 0.5. Is this wise? The modification using the square root has several advantages:

- A unique formula can manage all the three cases: axial force only, shear force only, axial force plus shear force.
- The results do not depend on slight changes in the applied loads.
- The utilization measure is a utilization factor and not just a utilization index.

Also, perhaps to avoid part of this issues, the AISC standard uses a trilinear formula, which neglects the interaction if the utilization factor related to a single component (N or V) is lower than 0.3 (Figure 8.4). Applying the same forces of the previous example, the utilization

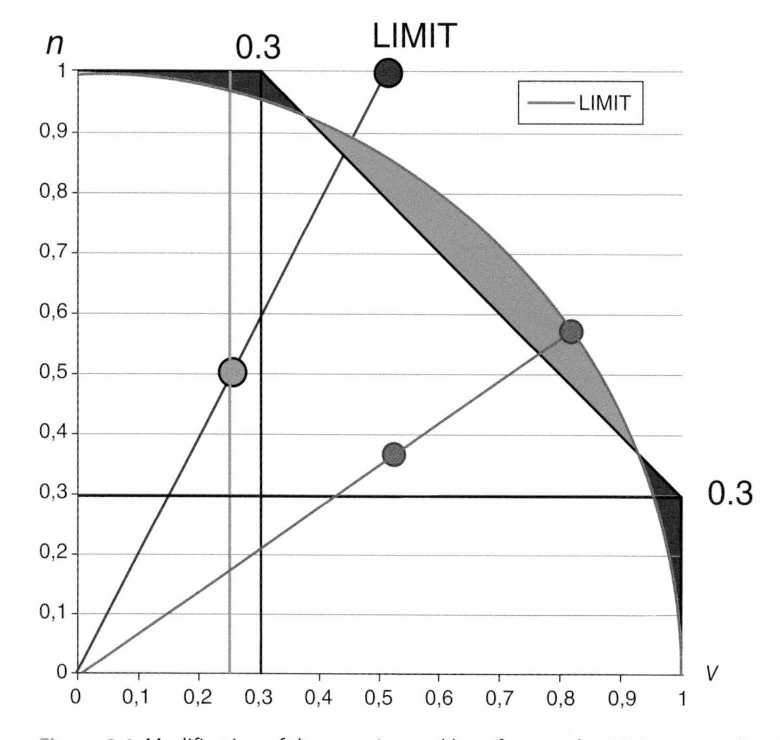

Figure 8.4 Modification of the experimental best fit curve by AISC 360-10. The aim is to make simpler computations and to avoid discontinuity. However, the approximation is made slightly worse. In the green area, the utilization would be assumed >1 instead of <1. In red area it would be assumed < 1, instead of > 1.

factor would not change at all, adding the additional V. However, in this way the computed utilization would not be a utilization factor, but a utilization index. Assuming, for instance, $n = 0.5$ and $v = 0.25$, the utilization computed would be $U_i = 0.5$ (green dot in Figure 8.4), leading to a candidate limit load multiplier $k = 1/0.5 = 2$ (green line up to $n = 1$). But, if 2 is applied to (n, v) following the red line of Figure 8.4, we find (1, 0.5), the red dot point, well outside the limit domain. So, once more, the formula suggested by the specification leads to a utilization index, not to a utilization factor. As $0.5/0.25 < 1/0.3$, even though $v < 0.3$, it is the combined formula that must be used, because *at the limit v is greater than 0.3.*

If these issues are ignored in developing the software, or in by hand computations, the results can be confusing. On the other hand, if the checks are modified by someone of good will in their own work, they are no longer compliant with the letter of the rules, and the comparisons between different software or between software and by hand calculation can become a nightmare. Assuming the elliptical formula will give different result from that of the trilinear formula, albeit they are both substantially allowed by the specification (the elliptical formula is discussed in the commentary of AISC 360-10).

So, it is apparent that specific work has to be done by whoever develops the standards. This work has not, generally speaking, been done yet, and there is apparently no attention being paid to this issue in any of the existing standards. Utilization indices are used as well as utilization factors, and usually no continuity of results from the data is ensured. This is also made easier by the "from simple to complex" approach of the standards, which (for good reasons) tends to simplify whenever possible. However, the simple design rules suggested by the standards do not exist in numerical analysis, because due to the need for machine math, almost no number is as clean as it assumed.

Sometimes, the proposed methods are such that there is no simple way to replace a formula defining a utilization index with another defining an equivalent utilization factor. Other times, computing a true "distance from the boundary" measure would imply solving a complex non-linear problem, as when the loads applied are below the limit load of a constituent but there is no exact idea of the factor that, applied to the loads, would drive to a mechanism. As the computational cost needed to get the answer may be relevant, it is legitimate to ask if such a measure is really important enough to justify the cost. It should be underlined that computing the limit load is quite a different problem from evaluating the structural response in the plastic range. The plastic range can be reached even being quite a long way from the plastic mechanism, and the related analysis can be much less expensive than the search-of-the-limit one. This is because when the loads applied are near the limit load, the convergence of numerical analysis gets very difficult to reach. Very often, elastic peaks a long way above the yield stress, instead result in very limited plastic regions, which can be computed with limited or reasonable computational effort. The final answer of this analysis would be "We are below the limit load", but would not be "This is how much we are below".

8.3 About the Specifications

In the following sections, the utilization indices and factors referring to the typical failure modes of connections are explained and briefly discussed. The specifications used are Eurocode 3 and AISC 360-10, which are presently the most commonly used specifications

in Europe and America, respectively. Eurocode 3 is also used in Asian countries (e.g. Singapore, where specific NAD, National Application Documents, have been issued), and AISC is also considered in several Middle East and Far East countries. In referring to AISC specifications, only LRFD (load and resistance factor design) will be considered here. ASD utilization can be obtained quite easily, replacing ϕ.
by

$$\frac{1}{\Omega}$$

where ϕ is lower than 1 and Ω is greater than 1. Typical values are

$$\phi = 0.75 - 0.9$$

$$\Omega = 2 - 1.67$$

These are safety factors.

In Eurocode 3, the same is obtained by the factors

$$\gamma_{Mi}$$

where $i = 0$ to 7 depending on the failure mode.

In Eurocode 3 only LRFD method is used. For resistance and stability the typical values are

$$\gamma_{M0} = \gamma_{M1} = 1.0$$

while for fracture

$$\gamma_{M2} = 1.25$$

In the Eurocode format, each single European country has to set its own "boxed values", that is, in general each country uses different gamma values. This also reflects the differences of the construction markets. The result, indeed a puzzling one, is that the utilization factor against a failure mode is a matter of citizenship, or of nationality. This is not exactly evidence of unity. Moreover, traces of sometimes bitter discussions have remained in the code, where several methods are possible (e.g. Method 1 and Method 2 for beam—column stability checks, are indeed French-Belgian and Austro-German methods, respectively).

The definition of the utilization factors or indices for the most common failure modes depending on a specification, is usually quick and simple. The methods are similar, and the definition of the utilization indices is relatively simple for many failure modes, treated in a notional way; the most complex part is the determination of the stresses and internal forces, which is unique and depends on mechanics, not on the specification used.

Both AISC 360-10 and Eurocode 3 allow the use of elastic methods for computing those forces and stresses, especially when getting from layout force packet to single subconnector forces or stresses. At the current stage of development of the specifications, there is good freedom in computing forces and checking connections. However, no explicit mention of finite element is made, as the method is still considered too complex for most engineers.

The author believes this is nowadays anachronistic.

8.4 Weld Layouts

8.4.1 Generality

Weld layouts must be checked against resistance. Once the forces flowing into a layout have been computed (the force packet), the stresses related to each single weld are computed using the methods described in Chapter 6.

Usually, weld layouts are not expected to possess relevant ductility. So, no explicit ductility check is presently required by the international standards. As their failure can be brittle, welds are usually protected and over-designed with respect to the surrounding constituents. If a connection fails first in a weld, the failure can be abrupt (especially for fillet welds transversely loaded) and cannot allow any redistribution of forces, so for this reason it is unwanted.

The failure of real-world welds is often related to weld details and to the presence of defects. From the analysis point of view, however, these considerations do not find an analytical counterpart: the weld checks are notional, and use idealized geometries that are not much related to the real geometry of real welds. For instance, fillet welds are assumed triangular, while in general they are not (Figure 8.5).

However, a sound design, regardless of the analytical results, must always carefully consider these issues and provide good detailing, so as to reduce the risk of brittle fracture and stress concentrations (e.g. see Bruneau et al. 2011 for an extended discussion of all these issues).

8.4.2 Penetration Weld Layouts

Penetration weld layouts are divided into complete and partial penetration joints (CPJ and PJP).

For full penetration, usually no special check is required by the international standards, as it is assumed that the cross-section resistance is fully restored. Thus, if the original cross-section was capable of absorbing the applied loads, so will the weld be.

Figure 8.5 Real-world welding: a stiffener fillet-welded to a circular base plate (Milan, photo by the author).

This rule often drives designers to totally neglect the checks of full penetration welds, shifting the need of the check from the analytical side to the constructional one: penetration welds must in fact be carefully inspected so as to exclude the possibility of micro defects or of stress raisers.

The simple rule "no analytical checks for full penetration weld layouts" does indeed have exceptions, which would suggest a more careful approach.

First of all, a weld will never be the same as the original cross-section, as the risk of defects and stress raisers is itself a reason for differences. At equal applied forces, a penetration weld is surely a more probable origin of failure than an undisturbed rolled or even cold-formed section. This would indeed justify a specific check, perhaps with rules more on the safe side.

Secondly, if the bevels are not really such that the fusion material entirely fills the original cross-section, as for instance around the rounded fillets of rolled sections, then the capacity cannot reach that of the original cross-section (e.g. see red zones in Figure 8.6).

Thirdly, it must be ensured that no extra bending or torque is necessary to ensure equilibrium, as these extra loads can lead to more severe checks than those of the base metal.

Finally, "full penetration" is often a work-around to avoid the tedium of making checks, as it is designed for situations where it would be physically impossible to apply. This results in tough problems when the design finally comes to construction.

A partial or full penetration weld layout can be considered as a cross-section projected onto the faying surface. Ideally, the resistance check can be done elastically or plastically. However, the checks provided by the standards, as they are referred to a point, are implicitly elastic.

Elastic checks need the evaluation of normal stress n and of the tangential stress t at any point of the cross-section. Tangential stress is expressed using the components t_{par} and t_{per} (as in Chapter 6).

In the following, the subscript "bm" stands for "base metal", "w" for weld, "u" for ultimate, and "y" for yield. The stress f_u is the ultimate stress of base metal, the stress f_y is the yield stress of base metal. The stress f_{uw} is named F_{EXX} in the AISC specification and it's the ultimate stress of the weld material.[3]

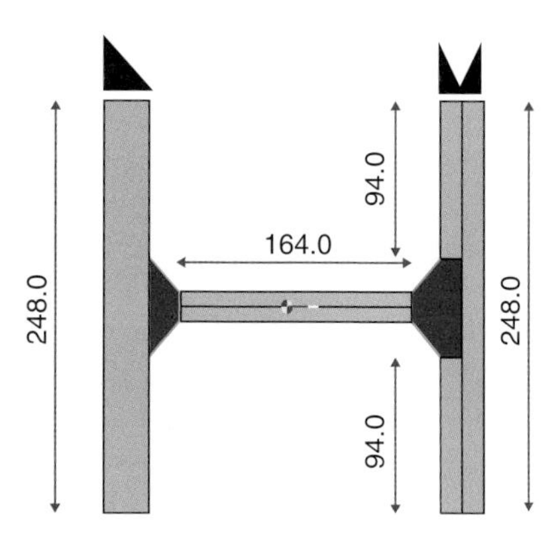

Figure 8.6 Partially welded rolled cross-section using full penetration welds.

3 In this book the upper case symbol F is usually reserved for forces.

8.4.2.1 Eurocode 3

The resistance of full penetration welds is taken to be equal to that of the weakest part connected. This means that the checks must be done relative to the connected part, using Eurocode rules for resistance, that is, the Von Mises stress for yielding, and the rupture stress for fracture under tensile stress. It is explicitly required that the weld metal is at least as strong as the base metal (both for yield and rupture stress).

The yield check is

$$U_{f,bm} = \gamma_{M0} \frac{\sqrt{n^2 + 3\,t_{par}^2 + 3\,t_{per}^2}}{f_y}$$

where γ_{M0} is usually 1.0 (but, for instance, in Italy it is 1.05).

The tensile check for rupture should also be done with (n is tensile, not compressive)

$$U_{f,bm} = \gamma_{M2} \frac{n}{f_u}$$

where $\gamma_{M2} = 1.25$, usually. So

$$U_{f,bm} = \frac{n}{0.8 f_u}$$

Partial penetration welds are computed as fillet welds, having an effective throat equal to the penetration reached (see Section 8.4.3.1).

For the special case of T-butt joints, where the weld can be considered as two couples of a partial penetration weld plus an external fillet weld, if the unwelded gap is lower than $t/5$ or 3 mm, and if the total available throat is greater than or equal to the thickness t, then the joint is assimilated to a CPJ. This is a special case of superposition of fillet weld plus partial penetration weld.

8.4.2.2 AISC 360-10

The effective area of a complete joint penetration groove weld (CPJ) is the length of the weld times the thickness of the thinner part joined.

For a partial joint penetration (PJP) groove weld the effective area is the length of the weld times an effective thickness, which is usually the depth of the groove, with the exception of some joints using 45° bevels (see Table J2.1 of AISC 360-10 specifications), where the depth of the groove is decreased by 3 mm. This can be taken into account defining a lower depth of the welds.

Both the base metal and the weld material should be checked.

CPJ – The strength of the joint is controlled by the base metal. This means that no special check is needed for the weld material. A matching filler metal is required for the weld. If only compression or tension parallel to weld axis is applied, then it is allowed to use a filler metal one class lower than the base metal.

Clearly, the base metal, that is, the constituents welded, will have to resist the stresses applied, satisfying the checks normally used for the regions far from the connections. This means an independent shear and tensile stress check, and an independent rupture and yielding check.

The general rules referring to these checks are as follows (the factor ϕ is already taken into account).

Tensile and compression yielding check:

$$U_f = \frac{n}{0.9f_y}$$

Tensile rupture check:

$$U_f = \frac{n}{0.75f_u}$$

For shear, the shear stress is

$$t = \sqrt{t_{par}^2 + t_{per}^2}$$

Shear yielding check

$$U_f = \frac{t}{0.6f_y}$$

Shear rupture check

$$U_f = \frac{t}{0.45f_u}$$

It must be underlined that for CPJ and PJP, the stresses to be used for the base metal checks are the same as those used for the weld metal checks, as the effective thicknesses of the material are the same.

PPJ – The utilization factors are computed differently depending on the nature of the stresses applied. Here, both the base metal and the weld must be checked.

For tension normal to the weld axis, the stress is n. The utilization factors are

$$U_{f,bm} = \frac{n}{0.75f_u}$$

$$U_{f,w} = \frac{n}{0.48f_{uw}}$$

For compression normal to the weld axis of connections not finished-to-bear (clearly in defining the utilization we use a positive value of n even if it's a compression)

$$U_{f,bm} = \frac{n}{0.9f_y}$$

$$U_{f,w} = \frac{n}{0.48f_{uw}}$$

Tension or compression parallel to the weld axis need not to be checked.

For shear, the utilization factors are

$$U_{f,bm} = \max\left\{\frac{t}{0.6f_y}, \frac{t}{0.45f_u}\right\}$$

$$U_{f,w} = \frac{t}{0.45f_{uw}}$$

The final utilization factor will clearly be the highest of all those computed.

8.4.3 Fillet Weld Layouts

8.4.3.1 Eurocode 3

Two methods are available: the directional method and the simplified method.

The simplified method considers the force per unit length in the fillet, $F_{w,Ed}$, evaluated point by point as

$$F_{w,Ed} = a\sqrt{n^2 + t_{par}^2 + t_{per}^2}$$

where n is the normal stress, t_{par} and t_{per} are tangential stresses, all acting over the projected effective area, and a is the throat size.

Clearly this implicitly assumes that there is no variation of stresses along the effective thickness.

The limit resistance is $F_{w,Rd}$, which is related to the effective throat size a, and to the ultimate stress of the weld material f_{uw}:

$$F_{w,Rd} = a\frac{f_{uw}}{\sqrt{3}\cdot\beta_w\cdot\gamma_{M2}}$$

The weld ultimate stress f_{uw} is defined as the stress of the weaker part connected. This means that if the constituents connected are A and B,

$$f_{uw} = \min\{f_{uA}, f_{uB}\}$$

The term

$$\frac{1}{\sqrt{3}} = 0.58$$

is not so far from the 0.6 that is met in AISC 360-10 (see below). From the engineering point of view, it is indeed the same.

Also the term

$$\frac{1}{\gamma_{M2}} = \frac{1}{1.25} = 0.8$$

is equal to the term ϕ used by AISC for the weld metal.

The term β_w, is a "correlation factor" and can be equal to 0.8, 0.85, 0.9, or 1.0 depending on the material grade. Precisely:

- $\beta_w = 0.8$ for S235, S235W, S235H
- $\beta_w = 0.85$ for S275, and all S275 X/Y grades (meaning N/NL, M/ML, ... etc.)
- $\beta_w = 0.90$ for S355, and all S355 X/Y grades
- $\beta_w = 1.0$ for S420, S420 X/Y, S460, S460 X/Y grades.

The correlation factor takes into account the relationship between the base metal and the weld metal, that is, the ratio between the limit stresses of the two materials.

The utilization factor can be defined as

$$U_f = \beta_w\cdot\gamma_{M2}\frac{\sqrt{3n^2 + 3\,t_{par}^2 + 3\,t_{per}^2}}{f_{uw}} = \frac{F_{w,Ed}}{F_{w,Rd}}$$

The resistance of the fillet considered as reference is the one of longitudinally loaded fillet welds ($0.6f_{uw}$), also if the fillet is loaded transversely. Eurocode 3 does not take into account, in this simplified method, the fact that transversely loaded fillet welds have a higher resistance.

The directional method uses the stresses read onto the effective throat plane. These are σ, τ_{par}, and τ_{per} (see Chapter 6), and can be obtained by simple transformation, from the stresses projected onto the faying plane, n, t_{par}, t_{per}. If the angle between active (fusion) faces is 90°, then the transformation becomes

$$\tau_{par} = t_{par}$$

$$\tau_{per} = \frac{\sqrt{2}}{2}\left(n + t_{per}\right)$$

$$\sigma = \frac{\sqrt{2}}{2}\left(n - t_{per}\right)$$

The directional method uses two limits that both have to be satisfied. The first one defines a first utilization factor, U_{f1} as

$$U_{f1} = \beta_w \cdot \gamma_{M2} \cdot \frac{\sqrt{\sigma^2 + 3\,\tau_{per}^2 + 3\,\tau_{par}^2}}{f_{uw}}$$

The second one defines a second utilization factor, U_{f2}, as

$$U_{f2} = \gamma_{M2} \cdot \frac{\sigma}{0.9 f_{uw}}$$

and the final utilization factor U_f is

$$U_f = \max\left\{U_{f1}, U_{f2}\right\}$$

As well explained in Ballio and Mazzolani 1983, where an in-depth discussion of the methods proposed to check fillet welds can be found, the geometrical interpretation of this two formulae in the space of the throat stresses σ, τ_{par}, and τ_{per} is that of an ellipsoid cut by a plane.

For the special case of right angle fillets, expressing the stresses referred to the inclined effective throat plane as a function of the projected stresses, the following equivalent formulae are obtained:

$$U_{f1} = \beta_w \cdot \gamma_{M2} \cdot \frac{\sqrt{2n^2 + 2\,t_{per}^2 + 2n t_{per} + 3\,t_{par}^2}}{f_u}$$

and

$$U_{f2} = \gamma_{M2} \cdot \frac{\sqrt{2}}{2 \cdot 0.9} \cdot \frac{\left|n - t_{per}\right|}{f_u} = \gamma_{M2} \cdot 0.7857 \cdot \frac{\left|n - t_{per}\right|}{f_u} = \gamma_{M2} \cdot 0.79 \frac{\left|n - t_{per}\right|}{f_u}$$

These can directly be compared to that of the simplified method, for this very special and important case of right-angle fillets.

Neglecting the effect of the safety and correlation factors, β_w and γ_{M2}, if a fillet is loaded transversely (only n or t_{per}), the limit stress is

$$f_u \cdot \min\left\{\frac{1}{0.79}, \frac{1}{\sqrt{2}}\right\} = 0.71 f_u$$

which is slightly higher than the $0.58f_u$ allowed by the simplified method, but not the $0.87f_u$ that would be obtained multiplying 0.58 by 1.5 as the experimental results would allow (see below what allowed by AISC).

If the fillet is loaded longitudinally, and once more neglecting the safety and correlation factors, then the Eurocode 3 limit stress is almost $0.6f_u$ as in AISC:

$$\frac{f_u}{\sqrt{3}} = 0.58 f_u$$

and equal to the value obtained by using the simplified method. The directional method, therefore allows a slight increase in the reference stress for transversely loaded fillet welds.

No explicit check of base metal is required by Eurocode 3.

8.4.3.2 AISC 360-10

In AISC, both the weld and the base metal have to be checked.

The base metal should be checked for tensile and shear *rupture* (this is governed by section J4).

The weld metal should be checked for rupture (section J2).

8.4.3.2.1 Base Metal Checks

The stresses should be evaluated sufficiently far from the weld fusion area (see AISC 360-10 C J2.4 and Figure 8.7). In an automated procedure, this might be difficult because the stresses are known at the throat plane of the welds or on the projected area lying over the faying surface, while the sections to be used according to CJ2.4 strongly depend on the problem at hand, and are far from there.

On the other hand, the base metal check can be seen as resistance checks applied to force transferrers and members, using appropriate "net" cross-sections. This check is one of the possible ways to do generic resistance checks.

When referring to the base metal checks for fillet welds, as explained in the AISC commentary, the stresses that must be used are not n, t_{par}, and t_{per}, but must be computed using the true thickness of the base metal, far from the welded area.

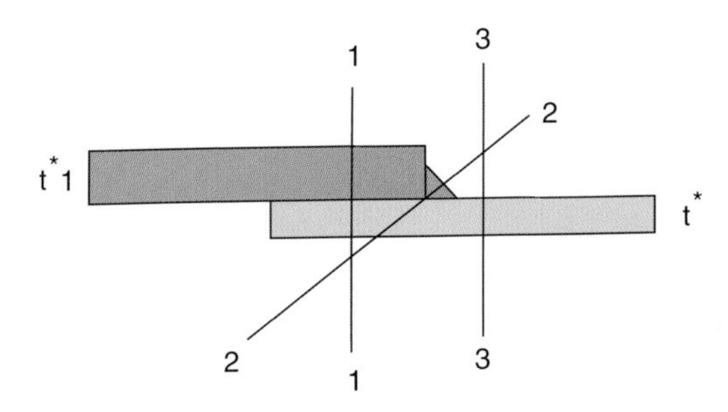

Figure 8.7 Planes where the stresses to be used for checks must be read, according to AISC 360-10. Plane 1-1 must be used for blue plate on the left. Plane 2-2 for the weld throat. Plane 3-3 for the green plate, on the right.

Let σ_{bm} and τ_{bm} be the computed stresses for the base metal, far from the welded area.

The utilization factor for rupture resistance is ("bmt" subscript stands for "base metal tensile")

$$U_{fr,\,bmt} = \frac{\sigma_{bm}}{0.75 f_u}$$

The shear check for the base metal is also for rupture. The utilization factor for rupture is ("bms" stands now for base metal shear)

$$U_{fr,\,bms} = \frac{\tau_{bm}}{0.45 f_u}$$

The final utilization factor for the base metal $U_{f,bm}$ is therefore

$$U_{f,\,bm} = \max\left\{U_{fr,\,bmt}, U_{fr,\,bms}\right\}$$

For the reasons explained, this cannot be considered a check local to welds. This check must be included in the generic resistance checks of constituents.

8.4.3.2.2 *Weld Checks*

The weld checks for fillet welds can be carried on in two ways.

The first one does not consider any difference in resistance depending on the direction of loading. This method is comparable to the simplified method of Eurocode 3.

The second method is the so-called instantaneous center of rotation method (ICRM), and is more precise albeit more complex.

The first method defines a utilization factor as

$$U_f = \frac{\sqrt{n^2 + t_{par}^2 + t_{per}^2}}{0.45 f_{uw}}$$

The term 0.45 is obtained by multiplying ϕ, 0.75, by 0.6 (f_{uw}).

The second method, the instantaneous center of rotation (ICRM) method, is based on the experimental results that have shown, as explained in Chapter 6, that the stiffness, ductility, and resistance of fillets strongly depend on the direction of loading and that the load–displacement curve for welds is strongly non-linear. Assuming a reference unfactored resistance of $0.6f_{uw}$ for longitudinally loaded fillet welds ($\theta = 0°$), the transversely loaded ones ($\theta = 90°$) have a resistance 1.5 times higher. This means that a longitudinally loaded fillet has an unfactored resistance of $0.6f_{uw}A_w$, with A_w being the effective weld area, while if loaded transversely the resistance would be $0.9f_{uw}A_w$. The variation is described by the formula

$$q(\theta) = 0.6 f_{uw}\left(1 + 0.5\sin^{1.5}\theta\right) \tag{8.1}$$

with θ in degrees between 0° and 90°. This formula takes the experimentally factor found by Lesik and Kennedy, which was equal to 0.852, and reduces it to the notional value 0.6.

Also, the stiffness of the transversely loaded fillets $\theta = 90°$ (see Chapter 6) is higher than that of the longitudinally loaded fillets, although ductility is much less. So, when the transversely loaded fillets reach their maximum resistance ($0.9f_uA_w$), at a given displacement $d_{m,90}$, for longitudinally loaded fillet welds having the same size, the displacement reached $d_{m,90}$ is related to a resistance equal only to 0.85 times the longitudinal reference resistance (i.e. $0.51f_{uw}A_w$). This means

$$d_{m,90} \ll d_{m,0}$$

The stress–strain relationship is provided by the standard:

$$f(\theta,p) = 0.6f_{uw}\left(1 + 0.5\sin^{1.5}\theta\right)\cdot\left[p(1.9 - 0.9p)\right]^{0.3}$$

where the strain p is defined as

$$p = \frac{d}{d_m}$$

that is, it is a non-dimensional ratio between the actual displacement norm d and the displacement related to the maximum resistance d_m. The displacement related to the maximum, d_m, is dependent on θ, as follows:

$$d_m = 0.209(\theta + 2)^{-0.32}t$$

where t is the thickness of the weld. For $\theta = 90°$ $d_m = 0.049\,t$, while for $\theta = 0°$, $d_m = 0.167\,t = 0.17\,t$, that is a displacement more than three times higher, at equal thickness.

For special cases, the application of the ICRM is easy. For instance, if all the fillet welds are parallel, that is for a *linear weld group* in the standard definition, and if all have a uniform leg size, and if the group is loaded concentrically in the faying plane, then the limit resistance is simply

$$R_n = A_w f(\theta) = A_w\cdot 0.6f_{uw}\left(1 + 0.5\sin^{1.5}\theta\right)$$

where θ is the angle in degrees between the applied force and the direction of the fillets. This is because all the fillets displace in the very same manner. If the force applied is R, then the utilization factor would be

$$U_f = \frac{R}{0.75R_n}$$

If the welds are not a linear group, but are transversely *or* longitudinally loaded, have a uniform leg, and the weld group is loaded concentrically in the plane, then the resistance R_n can be computed as

$$R_n = \max\left\{0.6f_{uw}A_w,\ \ 0.85\cdot 0.6f_{uw}A_{wl} + 1.5\cdot 0.6f_{uw}A_{wt}\right\}$$

where the total effective area of the welds A_w is divided into the area of the transversely loaded fillets, A_{wt}, and the area of the longitudinally loaded fillets, A_{wl}. The previous formula means that the resistance can be evaluated in two different ways:

1) assuming that all the welds have the same resistance of the longitudinally loaded welds, $0.6f_{uw}$. This assumes ductility in the transversely loaded fillet welds
2) assuming that the resistance is the sum of the ultimate resistance of the transversely loaded fillet welds, $1.5\cdot 0.6\cdot f_{uw}A_{wt}$ and the resistance of the longitudinally loaded fillet welds at the displacement $d_{m,90}$ which is related to the maximum resistance of the transversely loaded fillet welds, $0.85°0.6\cdot f_{uw}\,A_{wt}$

These simplifications may be useful in by hand calculation, for very simple loading conditions. However, they are not useful if a generic weld group is considered, under the effect of a generic loading condition for the weld layout, $\boldsymbol{\sigma}$.

In order to solve this problem, an iterative procedure has to be set up, where each weld is divided into small weld segments. The procedure outlined in Chapter 6 can be used in order to get, for each loading condition, the multiplier of the loads taking to the limit. As shown in Chapter 6, at the end of an iterative procedure the utilization *factor* U_f can directly be obtained, as well as the level of the stresses related to the applied loads $\boldsymbol{\sigma}$.

8.5 Bolt Layouts

8.5.1 Resistance of Bolt Shaft

The limit state of resistance for the bolt shaft is one important failure mode (Figure 8.8). As this failure mode is not ductile but brittle, connections must be properly protected against it. The bearing resistance and the punching shear resistance failure modes, while related to the forces in bolt shaft, are not referred to the bolt, but to the connected parts. So, they will be considered in the following sections.

For the reasons already explained, in the context of a general 3D analysis, it has little meaning to consider separately the failure mode due to shear and the failure mode related to tensile force; in general, due to rounding errors in the general context of a 3D analysis, the two components will practically always be numerically present. Also, it should be noted that compression of the bolt shaft does not have to be checked, as it is carried by the contact in the vicinity of the bolt.

A pending issue, not explicitly considered by the specifications, is the possible presence of bending in the bolt shaft. This is usually neglected in by hand computation, but it is in

Figure 8.8 One of the 10.9-class bolts connecting the engine to the frame of the author's car, bottom part, sheared and bent to failure due to a clash of the head with an uneven stone (detail). The stone rotated under the wheel and remained blocked between the other stones, acting as a wedge. The bolt fractured at a distance from the force, where the lever was enough to generate sufficient bending. Center Milan, October 2016, photo by the author.

fact usually present. As no specification about the bending contribution to the utilization is clearly set, two possibilities have to be considered:

- the bending moment in the shaft is neglected, that is, it is not taken into consideration in the checks
- the effect of the bending moment is considered, and the checking formulae adapted to include it

8.5.1.1 Eurocode 3

The limit tensile force in bolts is specified, so that if only a tensile force is applied to the shaft, then the utilization factor is

$$U_{bN} = \frac{N}{0.9 \cdot A_{res} f_{ub}} \gamma_{M2} \equiv n$$

and the value of γ_{M2} is usually 1.25.

If only a shear is applied, then the utilization factor defined by the specification is

$$U_{b,V} = \frac{V}{0.6 \cdot A \cdot f_{ub}} \gamma_{M2} \cong \frac{V}{A \cdot f_{ub}} \cdot \sqrt{3} \cdot \gamma_{M2}$$

if the shear plane passes through the shank (i.e. the unthreaded part), and

$$U_{b,V} = \frac{V}{0.6 \cdot A_{res} \cdot f_{ub}} \gamma_{M2} \cong \frac{V}{A_{res} \cdot f_{ub}} \cdot \sqrt{3} \cdot \gamma_{M2}$$

if the shear plane passes through the threaded part of the shaft.

In both formulae, the factor 0.6 is that used for bolt classes 4.6, 5.6, and 8.8. For bolt classes 4.8, 5.8–6.8, and 10.9, being less ductile, the factor 0.6 must be replaced by 0.5. So, introducing A_s as A or A_{res} and the factor β, the formula can be written as

$$U_{b,V} = \frac{V}{\beta \cdot A_s \cdot f_{ub}} \gamma_{M2}$$

In the previous expressions the shear V is obtained by vector summation of the shears along the two principal axes of the bolt layout:

$$V = \sqrt{V_u^2 + V_v^2}$$

Considering that the plastic flexural modulus of a round of diameter D is

$$W_{pl} = \pi \frac{D^3}{6}$$

a possible safe measure of the bending modulus of the threaded part of the shaft can be

$$W_{pl,res} = \pi \frac{D_{res}^3}{6}$$

where

$$D_{res} = D \cdot \sqrt{\frac{A_{res}}{A}}$$

Section 7.8.4.1 discussed the limit domain of a circular section under the effect of a shear, an axial force, and a bending moment. Defining

$$n = \frac{N}{N_{pl}} = \frac{N}{0.9 \cdot A_{res} \cdot f_{ub}} \cdot \gamma_{M2}$$

$$m = \frac{M}{M_{pl}} = \frac{\sqrt{M_u^2 + M_v^2}}{M_{pl}} = \frac{\sqrt{M_u^2 + M_v^2}}{0.9 \cdot W_{pl,res} \cdot f_{ub}} \cdot \gamma_{M2}$$

$$v = \frac{V}{V_{pl}} = \frac{\sqrt{V_u^2 + V_v^2}}{V_{pl}} = \frac{\sqrt{V_u^2 + V_v^2}}{\beta \cdot A_s \cdot f_{ub}} \cdot \gamma_{M2}$$

where A_s is the threaded area A_{res} or the gross area A, the general formulation for the utilization factor can be found as follows. It is assumed that n, m, and v are multiplied by an unknown scaling factor α, so that (see Section 7.8.4.1, Equation 7.31)

$$\frac{\alpha \cdot m}{\sqrt{1-\alpha^2 v^2}} + \frac{\alpha^2 n^2}{\sqrt{1-\alpha^2 v^2}} = 1$$

A quadratic equation in α^2 is then obtained. The utilization factor is by definition $U = 1/\alpha$. The quadratic equation degenerates to a linear equation when $n = 0$ and at the same time $v = 0$. If n, v, and m are all not null, then

$$U_{b,NVM} = \sqrt{\frac{2n^4 + 4n^2v^2 + 2v^4 + 2m^2v^2}{2n^2 + 2v^2 + m^2 - \sqrt{m^4 + 4n^2m^2}}}$$

If $n = 0$ and $v = 0$, the expression for U is merely

$$U_{bM} = m$$

If $m = 0$, then

$$U_{bNV} = \sqrt{n^2 + v^2}$$

If $n = 0$, then

$$U_{bVM} = \sqrt{v^2 + m^2}$$

If $v = 0$, then

$$U_{bNM} = \frac{2n^2}{\sqrt{m^2 + 4n^2} - m} = \frac{1}{2}\left(\sqrt{m^2 + 4n^2} + m\right)$$

In the author's experience, adding the effect due to bending moment in shafts can significantly increase the utilization. In real-world connections, the existence of bending moments in shafts can modify the stress state in bolts to a point which could be determinant to the safety of the structure.

No torque flows into the bolt shaft. In the very special case of one-bolt groups, where a very small torsional stiffness is added to avoid a singularity, the computed torque must be sufficiently small (see Section 8.6, which explains about pins).

The formula provided here (Equation 7.31) uses an elliptical interaction between V and N, and it is based on the theory of plasticity.

The formula provided by Eurocode 3 is apparently simpler:

$$U = \frac{n}{1.4} + \nu$$

but the following considerations apply.

Using the elementary utilization factors, and considering the formula provided by Eurocode 3, which is a linear one neglecting the bending moment effect, the final utilization *factor* should be computed as

$$U_b = U_{bN} = n$$

if

$$\frac{U_{bN}}{U_{bV}} = \frac{n}{\nu} > \frac{7}{2}$$

which is clearly also true if $U_{bV} = 0$ when $U_{bN} \neq 0$.

Otherwise, if

$$\frac{U_{bN}}{U_{bV}} = \frac{n}{\nu} \leq \frac{7}{2}$$

the combined utilization factor must be computed as

$$U_b = \frac{U_{bN}}{1.4} + U_{bV} = \frac{n}{1.4} + \nu$$

(Figure 8.9).

It must be noted that Eurocode 3 does not explicitly mention that for very small U_{bV} the combined formula should *not* be used. So, the letter of the standard is that if a very small or at limit negligible U_{bV} is added to a U_{bN} higher than 1, say equal to 1.2, then the result would be around 12/14 = 0.86 < 1. This problem is triggered by how the letter of the specification is written.

In the author's opinion this is an example of a dangerous specification, and it should be corrected.

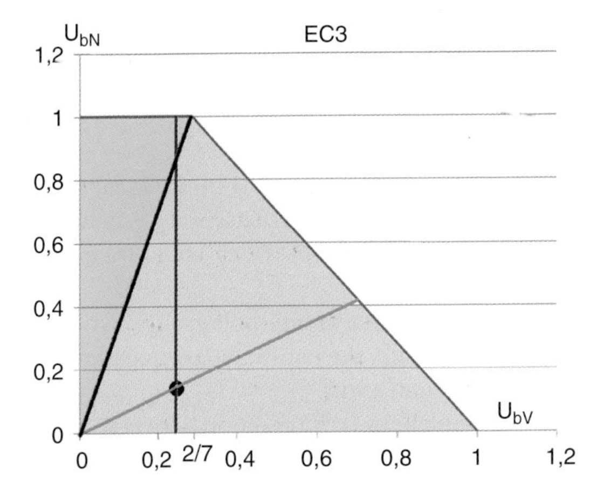

Figure 8.9 Limit domain for bolt shaft according to Eurocode 3.

Also, as Figure 8.9 clearly explains, if the aim is not to compute a utilization *index*, but to compute a utilization *factor*, the ratio between the elementary utilizations must be computed first, in order to choose the formula. If, for example the simple utilizations are

$$U_{bN} = 0.18$$

$$U_{bV} = 0.25$$

The correct final utilization would not be 0.18, implying a load multiplier to the limit of 5.55 (red line of Figure 8.9), but it would be 0.38 implying a load multiplier to the limit equal to 2.64 (green line of Figure 8.9).

Using the linear formula, the effect of bending might be taken into account replacing $U_{bN} = n$ with

$$U_{bNM} = \frac{2n^2}{\sqrt{m^2 + 4n^2} - m} = \frac{1}{2}\left(\sqrt{m^2 + 4n^2} + m\right)$$

which tends to n if m tends to zero.

In slip resistant connection according to Eurocode, the check of the bolt shaft must be done without considering the shear, which, if no slip occurs, is taken by friction (see the following section).

8.5.1.2 AISC 360-10

As explained in the excellent commentary to the specification, the limit of resistance of the bolt shaft is checked by using a linear formula just in order to have simpler by hand computation. As this book is related to the analysis of steel connection by means of software, it seems wise to suggest the use of the elliptic formula, instead of the linear one, always in compliance to the standard. In Chapter 7, we showed how the elliptic formula can be derived.

If just a tensile force is considered, the utilization factor is defined as

$$U_{bN} = \frac{N}{0.75 \cdot 0.75 \cdot A f_{ub}} = \frac{N}{0.5625 \cdot A f_{ub}} \equiv n$$

where the first 0.75 is the factor ϕ valid for LRFD design.

The utilization factor to be used for pure shear, is

$$U_{bV} = \frac{V}{0.75 \cdot 0.563 \cdot A \cdot f_{ub}} = \frac{V}{0.42225 \cdot A \cdot f_{ub}} \equiv v$$

if the shear plane passes through the shank (i.e. the unthreaded part), and

$$U_{bV} = \frac{V}{0.75 \cdot 0.45 \cdot A \cdot f_{ub}} = \frac{V}{0.3375 \cdot A \cdot f_{ub}} \equiv v$$

if the shear plane passes through the threaded part of the shaft.

The utilization due to bending moment only can be computed as

$$U_{bM} = \frac{M}{0.5625 \cdot W_{res,pl} \cdot f_{ub}} \equiv m$$

It must be emphasized that the use of four-digit precision is merely for formal compliance. The two numbers should be written 0.42, 0.34, and 0.56, respectively, in real-world considerations.

Finally, the utilization factor for combined actions, that is, the one used normally, is the same as already defined for Eurocode 3, using the pertinent n, v, and m.

8.5.2 Sliding and Resistance of No-Slip Connections

The sliding of no-slip connections can be considered a displacement limit-state: no constituent "breaks" or reaches the yield limit. The situation is unwanted because the displacement of the slip connection may induce a dramatic change on the stress field of the connected constituents, and also because the resistance mechanism immediately reverts to that of a normal bearing connection.

The slip happens when the friction forces are no longer able to keep the constituents still, and the friction forces depend on the preload applied, on the friction coefficient, and on the loads applied.

Two models are possible, the first of which – the safer – considers each bolt separately. The preload of the bolt, the tensile or compressive force applied to the bolt, and the shear applied to that very same bolt are used to check for slip.

The second model, allowed for example by Eurocode 3, considers the axial force and shear force acting on the whole layout and neglects the effect of bending, as a decrease of the compressive force in one side is balanced by an increase in the opposite side.

However, torque is not covered by this consideration, and it acts differently on different bolts. If a bolt has lost its contact, it has also lost its torque-resisting ability, and all the computations made about torque might then lose validity.

So, it seems wiser and safer to apply the no-slip criterion to each single bolt, and to consider it to have reached the limit state if in one bolt the slip condition is reached.

8.5.2.1 Eurocode 3

The initial preload force applied to a bolt is

$$F_{P,C} = \frac{0.7 \cdot f_{ub} \cdot A_{res}}{\gamma_{M7}}$$

The safety factor γ_{M7} is taken as 1.1. If either class 8.8 or 10.9 is used, and controlled tightening (see §3.91.(2) of the specification, EN 1993.1.8:2005), then $\gamma_{M7} = 1$.

If the module of the shear applied is V (i.e. the vector sum of V_u and V_v), the applied axial force is N (positive if tensile), and the friction coefficient is μ, then the utilization factor is

$$U_f = \frac{V}{(F_{P,C} - 0.8N) \cdot \mu \cdot k_s} \cdot \gamma_{M3}$$

where k_s is a hole coefficient, set according to the following rules:

- $k_s = 1.0$ for normal holes
- $k_s = 0.85$ for oversized holes, or short slotted holes with the axis of the slot perpendicular to the direction of the force transfer
- $k_s = 0.7$ for long slotted holes with the axis of the slot perpendicular to the direction of the force transfer

- $k_s = 0.76$ for short slotted hole with the axis of the slot parallel to the direction of the force transfer
- $k_s = 0.63$ for long slotted holes with the axis of the slot parallel to the direction of the force transfer

and γ_{M3} is usually equal to 1.25.

The hole coefficient in the general context of a 3D analysis – where the direction of the forces may well be inclined to the slot axis, and changes from load combination to load combination – should be applied taking the lower value. Otherwise, the angle formed between the axis of the slot and the force direction should be computed at every load combination and at every bolt, and the highest value of k_s applied, if this angle is found sufficiently near to 90°. This might be not necessary if the direction of the load applied to the bolt is always the same, as in truss-like connected elements, where only an axial force is expected.

Note that the force N has sign, and that a compressive force will increase the shear applicable. This means that the compressive force *related* to a bolt, and flowing in the near contact surface, will increase the shear available. This is not stated in the specifications, because the combined (N, V) formula is cited for "tensile forces". So the application of the concept is questionable. However, it is coherent with the assumption that bending does not affect the utilization of the bolt group, because the friction lost in the bolt under tensile force is regained in the compressed ones.

To apply the rule, it must clearly be checked that

$$F_{P,C} - 0.8N > 0$$

$$N < 1.25 \cdot F_{P,C}$$

The reason for the coefficient 0.8 is related to the existence of different levels of preload in the bolts, as explained in AISC specification for the similar coefficient 1/1.13 = 0.884. There, 1.13 is "the ratio of the mean installed bolt pretension, to the specified minimum bolt pretension" (AISC 360-10, J3.8, see below).

In the context of a general 3D analysis, as each bolt is to be considered separately from the others, the application of the coefficient 0.8 is questionable (AISC allows the modification of the factor 1.13 by the *engineer of record*, who can also set 1.0). However, if the applied tensile force reaches the preload value, i.e. when

$$N = F_{P,C} < 1.25F_{P,C}$$

it seems wise to declare the friction connection at limit.

The friction coefficient μ is between 0.2 and 0.5 depending on the surface treatment:

- surfaces blasted with shot or grit with loose rust removed, not pitted: $\mu = 0.5$
- surfaces blasted with shot or grit, pitted: $\mu = 0.4$
- surfaces cleaned by wire brushing or flame cleaning, with loose rust removed: $\mu = 0.3$
- surfaces as rolled: $\mu = 0.2$

8.5.2.2 AISC 360-10

The provisions of AISC 360-10 related to the slip failure mode are similar, but not identical, to that of Eurocode 3.

The minimum initial preload value $F_{P,C}$, a force, is directly set in Tables J3.1 and J31.M of the specification, as a function of the bolt size and of the steel grade. However, as specified in a note to the tables, this value is 0.7 times the minimum tensile strength of the bolts, possibly rounded to the integer, in kip or kN. So it's basically identical to that of Eurocode 3.

Using the same symbols already introduced for Eurocode 3, the utilization factor U when both a shear V and an axial force N are applied, is

$$U_f = \frac{V}{1.13 \cdot k_s \cdot \mu \cdot h_f \cdot F_{P,C} \cdot \left(1 - \frac{N}{1.13 \cdot F_{P,C}}\right)} = \frac{V}{1.13 \cdot k_s \cdot \mu \cdot h_f \cdot F_{P,C} \cdot \left(1 - \frac{0.885N}{F_{P,C}}\right)}$$

where:

- the hole factor k_s (named ϕ in AISC) is 1 for standard holes and for short slotted holes perpendicular to the direction of load; 0.85 for oversized holes and for short slotted holes loaded parallel to the direction of load; and 0.7 for long-slotted holes
- the friction coefficient μ is 0.3 or 0.5
- the h_f factor is 1 when there are no fillers, and might be equal to 0.85 if more than one filler has been used, and no bolt has been added to distribute the load in the fillers.

The same considerations already made for Eurocode 3 provisions are also applicable here. However, the AISC formula is more coherent than that of Eurocode 3, because the Eurocode formula is not

$$U_f = \frac{V}{1.25 \cdot k_s \cdot \mu \cdot F_{P,C} \cdot \left(1 - \frac{0.8N}{F_{P,C}}\right)} \cdot \gamma_{M3}$$

but is instead

$$U_f = \frac{V}{k_s \cdot \mu \cdot F_{P,C} \cdot \left(1 - \frac{0.8N}{F_{P,C}}\right)} \cdot \gamma_{M3}$$

This means that the statistical considerations leading to the factor 0.8 (and 1/0.8 = 1.25) are not applied when the loading is pure shear (N = 0).

A very important difference between AISC 360-10 and Eurocode 3, is that the limit state of bearing is not to be checked for slip resistant connection at ultimate loads in Eurocode 3, while it is still to be checked for the slip resistant connections at ultimate loads in AISC 360-10. This also justifies the difference in the safety factors of the two specifications, for slip (basically a factor 1.25 × 1.13 = 1.41).

8.5.2.3 Resistance of No-Slip Bolts

As was shown in Chapter 7, the external forces applied to a preloaded bolt layout, only partially flow into the bolt shafts. The preloaded bolt experiences slight variations of its initial pure tensile stress state, because the most part of the variations flow in the plates' pressure variations.

If the equivalent diameter factor is Q, or Q^*, and N and M are the force and bending moment ideally related to the bolt due to the externally applied loads, the true force and bending flowing in the bolt are

$$N_{true} = \frac{1}{Q^2} N$$

$$M_{true} = \frac{1}{Q^4} M$$

if the model with a hole is used, or

$$N_{true} = \frac{1}{(1 + Q^{*2})} N$$

$$M_{true} = \frac{1}{(1 + Q^{*4})} M$$

if the complete-circle model is used.

The resistance checks of the preloaded bolt under external loads will have to be done using the total values

$$N_{tot} = F_{P,C} + N_{true}$$

$$M_{tot} = M_{true}$$

If the variation of forces applied is important in view of fatigue checks, only the "true" values will have to be considered.

8.5.3 Pull-Out of Anchors, or Failure of the Anchor Block

Sometimes bolts are anchored to walls, foundation blocks, or other structural elements, and the pull-out condition must be verified. The resisting mechanisms related to anchors are many and varied, and strongly depend on the type of anchor. Many different systems, including chemical adhesive materials, are available. Considering each anchor bolt, in a given load combination, a value of tensile force N will have to be compared to an ultimate value N_{lim}, and the utilization will then be

$$U_f = \frac{N}{N_{lim}}$$

The limit value may depend on the position of the anchor, as anchors nearer to free edges might induce failure in the supporting block, while anchors too near one another might interact, leading to premature failure. Referring, in particular, to steel rods anchored to concrete, the observed failure modes are:

- steel failure, i.e. the failure of the steel rod, or of the anchor plate attached to the end of the rod, or of its connection to the rod
- concrete cone failure: a concrete cone is detached from the concrete block
- pull-out: the bar moves as the bond link reaches its limit
- splitting: a fracture appears below the anchor, directed as the anchor
- local blow-out: detachment of a concrete part in the vicinity of its free edges.

The structural element to which the anchor is attached is a structural subsystem that really needs great care to be taken in its design. The analyst should carefully consider these aspects when designing an anchoring system. This is, however, beyond the scope of this book. For an extended discussion of steel-to-concrete joints, see for example Kuhlmann Wald et al., 2014.

From the analytical point of view, once a limit load N_{lim} has been specified, the utilization factor of each anchor can be computed in the straightforward manner explained.

It is implicitly assumed that no interaction between the different anchors is to be expected, or that, if it is expected, it is avoided by properly reducing the limit force. This might also imply specific positioning rules, so as to avoid anchors too near to the edges or anchors too near one another.

8.6 Pins

Pins are loaded by a shear V and by a bending moment M. The axial force is limited to negligible values. Also, by definition, no torque is applied to a pin. In the frame of a 3D generic analysis this should be intended numerically.

The resistance model of pins is that of a continuous beam simply supported, loaded by constant pressure blocks, related to the thicknesses of the parts connected (Figure 8.10). This would imply null bending moment at the extremities of the pin.

8.6.1 Eurocode 3

Specific geometrical limitations are set up in order to limit the applicability of the rules (see Eurocode 3 Part 1.8 Table 3.9). If V is the shear in a section of the pin, and M is the bending moment, the uncombined utilization *factors* are defined as follows:

$$U_{pV} = \frac{V}{0.6 \cdot A \cdot f_{up}} \cdot \gamma_{M2} \equiv v$$

$$U_{pM} = \frac{M}{1.5 \cdot W \cdot f_{yp}} \cdot \gamma_{M0} \equiv m$$

where f_{up} is the ultimate stress of the pin, and f_{yp} is the yield stress of the pin. Note that one utilization refers to ultimate strength, the other to yielding.

The combined utilization factor mixes yielding and rupture and is

$$U_p = \sqrt{U_{pV}^2 + U_{pM}^2} = \sqrt{v^2 + m^2}$$

A part of the needed checks, in the context of a 3D analysis, is that the applied torque and the applied axial force should both be negligible. The related stiffnesses (see Chapter 6) are not null to avoid numerical problems in the solution due to the singularity of the stiffness matrix. So also, owing to rounding errors, in general the pin will indeed be

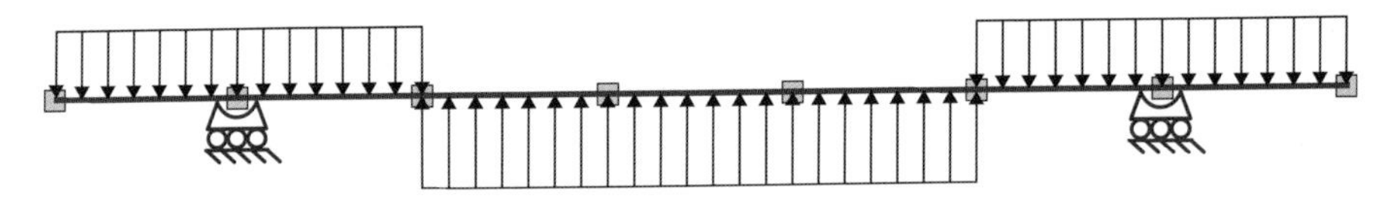

Figure 8.10 The structural simplified layout for pins.

loaded by a small axial force N and by a small torque M_t. Setting a limiting nominal stress f_{nom} equal to 1 MPa, the limit axial force would be

$$N_{lim} = \pi \cdot \frac{D^2}{4} \cdot f_{nom}$$

and for the torque

$$M_{t,lim} = \pi \cdot \frac{D^3}{16} \cdot f_{nom}$$

Axial forces and moments higher than these limits should be considered an index of numerical problems or of a wrong design of the connection.

8.6.2 AISC 360-10

No explicit rule has been found in AISC 360-10 referring to the resistance of pins (section D refers to the resistance of the plates that the pin is connected to). Using the rules valid for bolts, and considering an elliptic interaction between moment and shear like the one of Eurocode 3, the following relations for the utilization factors are found:

$$U_{pM} = \frac{M}{0.5625 \cdot W \cdot f_{up}} = \frac{M}{0.56 \cdot W \cdot f_{up}} \equiv m$$

$$U_{pV} = \frac{V}{0.75 \cdot 0.563 \cdot A \cdot f_{up}} = \frac{V}{0.42225 \cdot A \cdot f_{up}} = \frac{V}{0.42 \cdot A \cdot f_{up}} \equiv v$$

and the combined utilization is

$$U_p = \sqrt{U_{pV}^2 + U_{pM}^2} = \sqrt{v^2 + m^2}$$

8.7 Members and Force Transferrers

8.7.1 Generality

The failure modes considered up to this point have all referred to the connectors. This section instead is related to the failure modes of the connected parts, i.e. the members, the force transferrers, and the stiffeners.

Unfortunately, the geometry of the connected parts in the region surrounding the connections is in general quite complex, and no simple theory can be applied without a significant loss of meaning. In particular, the theory of beams must be abandoned for the checks of the members, because in the region surrounding the connectors the De Saint Venant principle is no longer valid. Moreover, in order to make the connection physically possible, cuts and bevels are often applied to members, so that their cross-section is no longer that of the member checks.

Some failure modes related to the connected constituents are very local, that is, they refer to a region immediately near the connectors, and these failure modes are related to the stresses induced in the constituent by the connector itself. These failure modes are those related to the bolt-bearing, to the punching shear of the bolted thicknesses, and to the crushing of the parts in contact. As they are local, they are also relatively easy to check.

All the other failure modes do strongly depend on the shape, size, loading, and constraints of the whole constituent, and for this reason they are much more complex to be considered. In particular, the *generic* resistance of the constituents and the buckling of the constituents, are failure modes that are quite difficult to study, in general. Also block shear, as will be seen, is a very complex failure mode to deal with.

Traditionally, several methods have been applied to deal with the problem of resistance and buckling. If the loading is simple, and the constituent can be assumed equivalent to simple structural elements (web panel, flange in local bending, compressed strut, T-stub, etc.), it is possible to use simple models in order to make simple checks. These are usually on the safe side, when the assumption of simple loading is valid. However, it is the assumption of simple loading condition which, in general, is not valid. This is true in particular when dealing with members acting as beam–column, that is, under the effect of six internal forces, and it's particularly true when complex 3D loading conditions like those of earthquake resistant structures, must be met.

The only possible way to properly deal with the problem, in general, is to use finite element analyses where appropriate finite element models have been set up. At the current stage of development it is indeed possible to automatically prepare finite element models that are meaningful and that automatically take into proper consideration both the geometrical complexity and the loading complexity, yet remaining under the shelter offered by the static theorem of limit analysis. These are "complex simplifications" whose complexity can be automated. The time needed to do this operation is almost always lower than that needed to compute the needed partial results, to enter into partial and simple models.

In this section, some considerations referring to failure modes and utilization factors will implicitly assume that a finite element model has been set up and run, in the linear or non-linear range. The rules, by which these finite element models can be prepared, and their limits and advantages, will be discussed in detail in Chapters 9 and 10. The finite element models that will be mentioned will be made by thick plate–shell elements, used to model in detail each constituent of the scene, i.e. each constituent to be checked.

Finite element models, however, are not the only possible method that can be used to check constituents for complex failure modes. Using the traditional methods is not wrong, as they can still have utility in the framework of a proper design activity. However, their use should be restricted to the situations where they can be applied with no loss of meaning: if the loads are more complex, or the geometry is not regular, they lose part of their effectiveness and meaning.

A software procedure should indeed be able to "learn" ad hoc rules that can be used for specific design purposes. This can be done, as mentioned in Chapter 4, adding predefined variables to each constituent, and allowing the definition of new variables as a function of the first, and the definition of inequalities that can be used as checks. These are what can be called *specific ad hoc checks*. The analyst may decide to apply such rules considering them apt for the specific design goal.

This method is usually valid for the typical connections where experience has proved the reliability of simple or simplified approaches. However, this is not a general solution.

The needed failure modes that must be checked at the constituent level can be summarized as follows:

1) constituent-local failure modes
   - bolt bearing
   - punching shear
   - crushing of the support in contact regions
2) constituent-fracture failure modes
   - block shear in bolted connections
   - shear or tensile rupture in single cross-sections
3) constituent-global failure modes
   - resistance to yielding
   - resistance to buckling
   - displacement control

Clearly, each constituent in general interacts with the others. However, when dealing with the analysis of connections, it may be useful to see the scene as the assembly of single constituents. Once the field of forces, stresses, and displacements pertinent to each constituent is known by the third law, in a given loading combinations, it is ideally possible to check the constituent against each failure mode. Sometimes, *global* models (i.e. models of the whole renode) or *partial* finite element models are set up, in order to better model the interaction between constituents. Clearly, these models can also be used for the checks of single constituents, considering their behavior inside the model. Moreover, these global models can help to have an idea of the overall global failure modes, which may affect more than one constituent at a time, typically buckling.

The local checks are related to the effects that the single connectors generate in their surrounding regions. They use simplified models, and are usually not directly covered by finite element models, which would require models using brick elements and very high levels of detail.

The fracture checks are related to the detachment of one part of the constituent under investigation, relative to the other. They do not imply any redistribution, unlike yielding, and they are usually considered with high safety factors, of the order of 1.25–1.33.

According to the standards, block shear is the fracture of a block that has been detached by its parent by two systems of fracture lines, one implying pure shear and one implying pure tensile stresses. Here it will be seen that several remarks are needed. If the fracture line is unique and straight, then a cross-section of the constituent, cut by a plane, can be considered, and the failure condition can be checked by means of simplified assumptions.

Normal (i.e. not using fracture mechanics) finite element models are not very suitable for checking against these failure modes. They can be used for this aim, limiting the Von Mises stress to values lower than ultimate stress, and if the run has been done in hardening plasticity, but, at the moment, the simplified procedures are still more effective, also due to the lack of available computer software dealing with cracks and fracture mechanics.

The constituent-global checks are those of yielding, buckling, and displacement control. Here the finite element models are very useful.

Yielding checks imply that the applied loads must be proved below the limit according to (usually perfect) plasticity theory. Buckling checks, which are quite complex, imply that no part loses its configuration and buckles, and imperfections must be considered.

Finally, also if all these failures are not possible, the constituents must not exhibit excessive displacements. This is also true when considering that the stiffness matrix of the connectors is always positive definite to avoid singularities, but can indeed imply very low stiffnesses. High displacements are then the effect of wrong assumptions about the working mode of connectors, as when a bolt layout is declared "shear only" and is instead loaded by a non-negligible bending moment.

All these checks must be done for every constituent, in every loading combinations, and considering all the components of the force packets delivered to it.

It is clearly a tremendous amount of computational work, and it's not surprising that even in mid 2010s, no one seems to check connections in this way. As in Aesop's tale *The Fox and the Grapes*, many designers have considered the goal useless, or a waste of time.

However, it is not only a tremendous amount of work, but it is also something very difficult to understand and to tackle for every designer who wishes to keep under control his or her constituents. It is difficult, in such a situation, to assess "which is the worst constituent", or "which is the relevant failure mode". The answer will probably be: it depends. It depends on loading, standards, the way to apply the checks, the checks considered, and the one avoided, and all the assumptions related to each of them.

A very interesting field of research – once the difference between utilization index and factor has been fully understood – is that of automatically finding a design such that the different failure modes are graduated and ordered in a meaningful way, taking into account the needed economical considerations.

8.7.2 Local Failure Modes

8.7.2.1 Bolt Bearing

Bolt bearing is due to the contact of the bolt shaft against the thickness of the bolted steel. It induces in the bolted plate very high local stresses, and consequent plasticization. The resistance of the plate to this kind of force depends on the "available material" that can be used to redistribute these high stresses. If another hole is too near, or a free edge in the vicinity, the available material may not be enough to redistribute as needed. So, for these reasons, the bolt bearing checks also depend on the geometry of the connection in the immediate surroundings of the bolt.

The bolt bearing check is governed by the shear V applied to a generic bolt, by the thicknesses of the plates connected, by their steel grade, by the hole size, and by the distance of the bolt considered from the free edge of the parts connected, and from the other bolts.

If a generic bolt is connecting n parts ($n \geq 2$), each part i will have in general a different thickness t_i and a different steel grade, with different ultimate stress, f_{ui} and yield stress f_{yi}. Also, the distance from the free edges of the part considered will depend on the index i. So in general, the bolt bearing check will have to be repeated n times, once for every bolt in every combination (Figure 8.11).

In the following sections it is implicitly assumed that every piece of data is referred to one of the n parts connected. The shear force V applied to the thickness t_i will have components V_u and V_v, referred to the principal axes of the bolt layout; and the principal axes of the bolt layout will not in general be aligned with the rows and columns of the bolt group, or with the free edges. It is assumed that the forces V_u and V_v of each bolt have

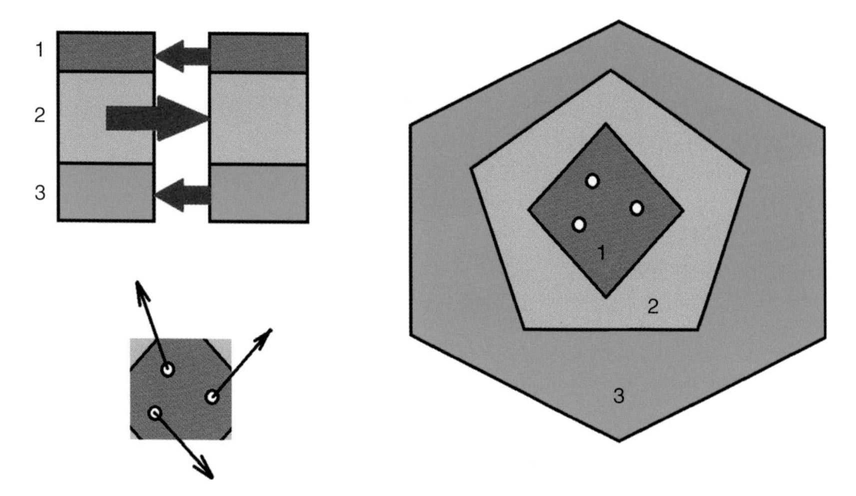

Figure 8.11 A deliberately generic configuration of the parts and of the bolts. The forces of each bolt have different direction and modulus. In the different thicknesses, all the forces are directed in the same direction, but the boundaries of the parts are different.

been computed using one of the methods described in Chapter 7, and are, by the analysis, balanced with the externally applied loads.

Moreover, in general, no "rows" or "columns" for the bolt group will always be available, if the bolt layout is, as indeed is possible, generic. This is a superset of the configurations discussed in the specifications. However, this kind of generalization is needed if the aim is to set up a method which must be able to work with generic configurations of fasteners, in true world constructions.

In the context of a generic 3D program, all the constituents of a renode are placed in *the scene* (see Chapter 4). So, when checking for bolt bearing, the provisions of the standards referring to the distance from the free edges must be applied considering the faces of the objects drilled. In general, two faces of each drilled thickness will be available, and in general these two faces will not be equal, as would happen when drilling a simple plate.

Moreover, we need to distinguish the case of free edges from the case of "stiffened" edges, because the provisions of the standards apply only to free edges.

Considering one face at a time, the faces that have an edge in common with the face considered will be different, for the top and bottom faces drilled by the bolt (Figure 8.12). By using proper vector and dot products, it is possible to distinguish the case of stiffened edge from the case of free edge.

Let:

- $\mathbf{n_f}$ is the unit normal of the drilled face considered (the green one in Figure 8.12, normal is positive if outward)
- $\mathbf{n_a}$ is the unit normal of the adjacent face
- $\mathbf{r}$ is the unit vector of the side that must be classified, when thought to belong to the drilled face[4] (it points out of the page, to the reader, in Figure 8.12)

4 When thought to belong to the face adjacent "a", the side has the opposite unit vector.

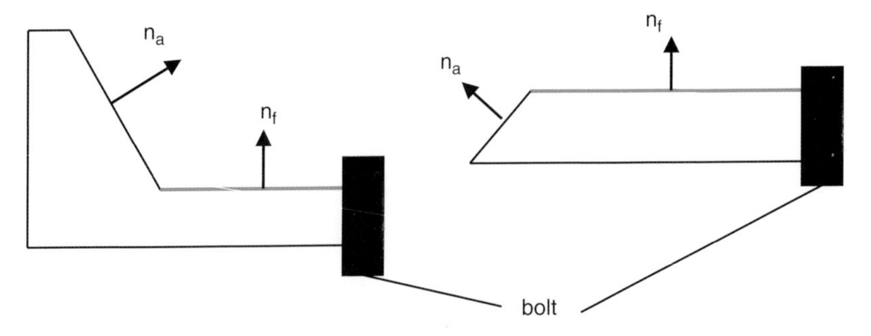

Figure 8.12 The green faces are the entry faces of a bolt. By using appropriate vector manipulation it is possible to assess if an edge is free or stiffened.

Then

$$c = \mathbf{n_f} \cdot \mathbf{n_a}$$

$$\mathbf{v} = \mathbf{n_f} \wedge \cdot \mathbf{n_a}$$

$$b = \mathbf{v} \cdot \mathbf{r}$$

$$\text{IF}\left\{ (b<0)\text{AND}\left(\text{c} < \frac{\sqrt{3}}{2} \right) \right\} \Rightarrow \text{side is stiffened}$$

Here a limit angle of 30° has been assumed.

Stiffened edges can be marked so that the distance of the bolts from them is set to a high value.

So, the distance from the edges will be computed twice for each bolt and each thickness drilled: once considering the top face, and once considering the bottom face of the thickness i considered. The results of the bolt bearing checks will be the worse of the two. In the special case of simple plates, the two operations will give the very same result.

In general, for each bolt there can be the following possible situations.

If the bolt belongs to a regular or staggered grid of bolts, that is, using rows and columns, the shear applied to each bolt can be transformed from the principal axes of the bolt layout (u, v) to the axes of the grid, (x, y). So x is now the direction of the rows, and y the direction of the columns. It is then possible to refer all the geometrical information referring to the current drilled face, the bolts, and the forces applied, to the axes (x, y). Normally the principal axes of a regular grid should be coincident with the axes of the grid (x, y), but if one or more bolts have been deleted, this is not true anymore.

Considering the forces V_x and V_y, it will be possible to define, for rows (Figure 8.13):

- if a bolt is inside a row, considering direction x, that is, if it is "innerx" or "edgex"; only two bolts will be "edgex" in a row, the first and the last
- the pitch p_x that is, the distance between the columns
- the distance e_x from the nearest edge (if the edge is stiffened, this distance can be set equal to $10D$, where D is the bolt diameter) along axis x, for end bolts of a row
- the distance e_{min} from the nearest edge considering a direction perpendicular to that edge, for the end bolts of a row.

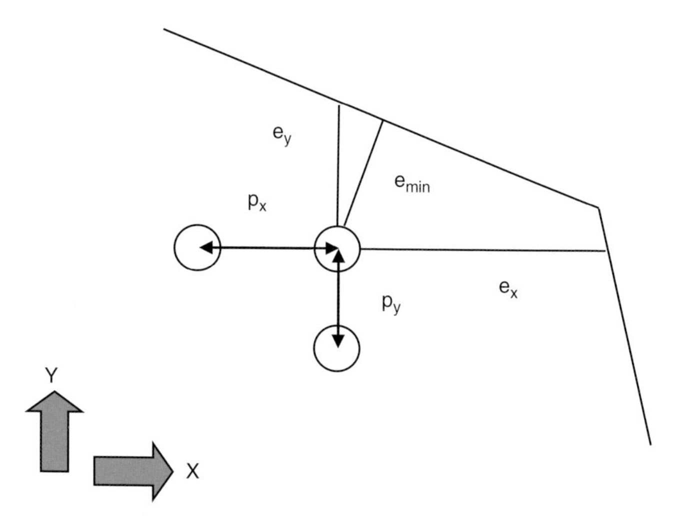

Figure 8.13 Distances from edges and pitches.

For bolt columns:

- if a bolt is inside a column, considering direction y, that is, if it is “innery” or “edgey”; only two bolts will be “edgey” in a column, the first and the last
- the pitch p_y that is, the distance between the rows
- the distance e_y from the nearest edge (as before, if the edge is stiffened, this distance can be set equal to $10D$, where D is the bolt diameter) along axis y, for the end bolts of a column
- the distance e_{min} from the nearest edge considering a direction perpendicular to that edge, for the end bolts of a column.

If the bolt does not belong to a regular grid, no “rows” or “columns” can be defined. In such cases, it seems reasonable to use as “pitch” the minimum distance from another bolt, no matter its direction. This is what is asked for by the AISC 360-10 provisions. All the bolts will then be considered “not inside a row or column” or “not inner”.

The rules to be used for the checks must pass an invariance condition, that is, when expressed using the direction x and y, or the direction of the resultant, and considering a circular plate, they must give the same result. If projecting in the x and y directions in this very special case, the result obtained is different from that obtained by considering the inclined force, the checking rule does not pass the invariance condition which is unwanted (Figure 8.14).

8.7.2.1.1 *Eurocode 3*

Given the applied shear V, it is projected into the two components V_x and V_y. (if the bolt is not in a regular grid, the principal axes u and v might be kept).

Two utilization factors are defined, $U_{bb,x}$ and $U_{bb,y}$. The final utilization is obtained by the square root of the sum of the squares, that is

$$U_{bb} = \sqrt{U_{bbx}^2 + U_{bby}^2}$$

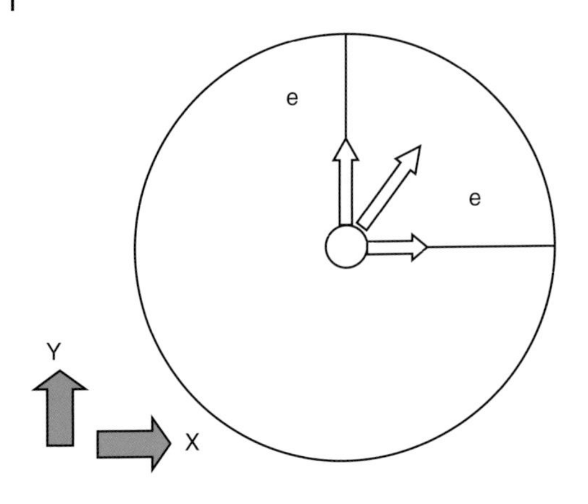

Figure 8.14 Invariance condition for bolt bearing checks.

This last rule is needed to pass the invariance check.

Considering the component x of the applied shear, the "row" utilization can be defined as

$$U_{bbx} = \frac{|V_x|}{k_{1x} \cdot \alpha_{bx} \cdot f_u \cdot D \cdot t} \cdot \gamma_{M2}$$

where:

- D is the bolt diameter
- t is the thickness of the part to be checked for bolt bearing
- f_u is the ultimate stress of that part
- V_x is the x component of the applied shear
- k_{1x} for "innery" bolt: $k_{1x} = \min\left\{1.4\frac{p_y}{D_0} - 1.7; 2.5\right\}$
- k_{1x} for an "edgey" bolt: $k_{1x} = \min\left\{2.8\frac{e_y}{D_0} - 1.7; 1.4\frac{p_y}{D_0} - 1.7; 2.5\right\}$
- D_0 is the bolt hole diameter
- α_{bx} for an "innerx" bolt: $\alpha_{bx} = \min\left\{\frac{p_x}{3D_0} - 0.25; \frac{f_{ub}}{f_u}; 1.0\right\}$
- α_{bx} for an "edgex" bolt: $\alpha_{bx} = \min\left\{\frac{e_x}{3D_0}; \frac{f_{ub}}{f_u}; 1.0\right\}$
- f_{ub} is the bolt material ultimate stress.

A possible option would be to use e_{min} for edge bolts, instead of e_x or e_y. This would be on the safe side.

The same rules with axes reversed can be used to define the other utilization:

$$U_{bby} = \frac{|V_y|}{k_{1y} \cdot \alpha_{by} \cdot f_u \cdot D \cdot t} \cdot \gamma_{M2}$$

where:

- V_y is the y component of the applied shear
- k_{1y} for "innerx" bolt: $k_{1y} = \min\left\{1.4\frac{p_x}{D_0} - 1.7; 2.5\right\}$

- k_{1y} for an "edgex" bolt: $k_{1y} = \min\left\{2.8\frac{e_x}{D_0} - 1.7; 1.4\frac{p_x}{D_0} - 1.7; 2.5\right\}$
- α_{by} for an "innery" bolt: $\alpha_{by} = \min\left\{\frac{p_y}{3D_0} - 0.25; \frac{f_{ub}}{f_u}; 1.0\right\}$
- α_{by} for an "edgey" bolt: $\alpha_{by} = \min\left\{\frac{e_y}{3D_0}; \frac{f_{ub}}{f_u}; 1.0\right\}$

The coefficient k_1 is related to the material available in a direction perpendicular to that of the load transfer, and it might reduce the available strength greatly. The coefficient α_b is related to the material available in the direction of the force and might be reduced if the distance from the edge in that direction is lower than $3D_0$, or if the bolt has a material weaker than that of the plate connected. This last is a rare condition.

According to Eurocode 3 provisions, the utilization valid for normal holes must be increased by a factor 1/0.8 = 1.25 for oversized holes. For slotted holes, if the direction of the slot is perpendicular to the direction of load transfer, the utilization valid for normal holes must be increased by a factor 1/0.6 = 1.67.

The specification, as such, seems fit for the regular situation of Cartesian grids of bolts, but it does not seem much fit to be generalized to an irregular bolt layout. Moreover, the specification adds that when the force is not parallel to the edge, two separate checks can be performed: "When the load on a bolt is not parallel to the edge, the bearing resistance may be verified separately for the bolt load components parallel and normal to the end." This is not very helpful, as it is not clear, assuming this new reference system, how the pitches would be defined, and how the distance from the edge in the direction perpendicular to load transfer would be defined. In the author's opinion these provisions should be written with more consideration for non-standard situations, indeed existing in the real world.

The maximum possible force value, if the distances are sufficiently high, is

$$F_{\lim} = \frac{2.5D \cdot t \cdot f_u}{\gamma_{M2}}$$

Projecting the force in x and y directions, we would get

$$U_{bb} = \sqrt{U_{bbx}^2 + U_{bby}^2} = \sqrt{\left(\frac{F_x}{F_{\lim}}\right)^2 + \left(\frac{F_y}{F_{\lim}}\right)^2} = \frac{F}{F_{\lim}}$$

which, indeed, is the invariance condition needed.

8.7.2.1.2 AISC 360-10

As mentioned already, AISC provisions seem easier to apply. The force is considered in its proper direction, and a length l_c is defined as *the minimum clear distance, in the direction of the force, between the edge of the hole and the edge of the adjacent hole or edge of the material.* If applied as such, this definition is also tricky, because in theory no hole can be found, "in the direction of the force" (Figure 8.15).

So, this definition also seems to rely on regular grids, where, considering the force directed as the row, or as the column, the other holes are immediately found. In the real world, however, the regular grids may have to be abandoned or modified, for specific constructional or design needs.

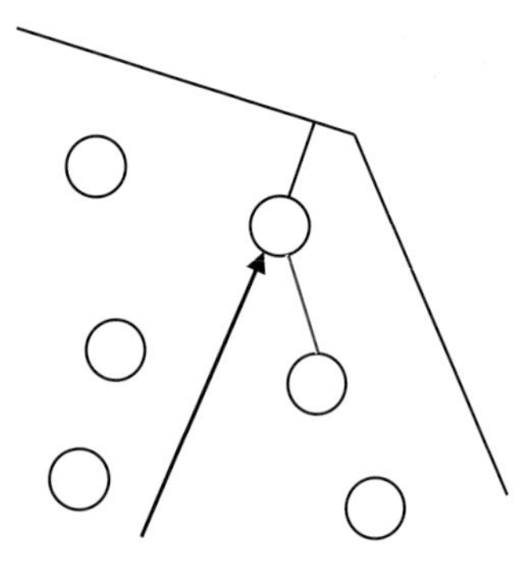

Figure 8.15 In the direction of the force, sometimes no hole can be found. The hole layout is deliberately generic.

For standard or oversized and short slotted holes, no matter the direction of the force, or for long slotted holes with the slot parallel to direction of the bearing force, the utilization factor is defined as

$$U_{bb} = \frac{V}{0.75 \cdot 1.5 \cdot l_c \cdot t \cdot f_u} \geq \frac{V}{0.75 \cdot 3.0 \cdot D \cdot t \cdot f_u}$$

The factors 1.5 and 3.0 may be applied when the deformation at the bolt hole at service load is not a design consideration. If the deformation must be limited at service loads, the two factors 1.5 and 3.0 must be replaced by 1.2 and 2.4, respectively. For long slotted holes with the slot perpendicular to the direction of the force the factors 1.5 and 3.0 must be replaced by 1.0 and 2.0, respectively.

Bolt bearing checks must be executed also for slip resistant connections.

8.7.2.2 Pin Bearing

Pin bearing checks are very similar to those of bolt bearing, but the reference limits are different, because the pin is specifically designed to carry a shear load, and is unique, no redundancy is intrinsically related to this connection. Moreover, different sizes are requested to the plates connected, in terms of distance from the edges and no other hole is expected near to the hole of a pin.

8.7.2.2.1 Eurocode 3

The utilization factor for pin bearing U_{pb} is

$$U_{pb} = \frac{V}{1.5 \cdot D \cdot t \cdot f_y} \cdot \gamma_{M0}$$

where D is the diameter of the pin, t is the thickness of the plate, f_y is the yield stress of the plate, and γ_{M0} is usually equal to 1.0.

If the pin should be replaceable, a more severe utilization is computed

$$U_{pb,\,replaceable} = \frac{V}{0.6 \cdot D \cdot t \cdot f_y} \cdot \gamma_{M6,\,ser}$$

8.7.2.2.2 AISC 360-10

The utilization factor for pin bearing is

$$U_{pb} = \frac{V}{0.75 \cdot 1.8 \cdot D \cdot t \cdot f_y} = \frac{V}{1.35 \cdot D \cdot t \cdot f_y}$$

very near the Eurocode value.

8.7.2.3 Punching Shear

Punching shear is related to the normal pressure acting at the boundary between the bolt and the plates connected. The washer is not taken into account. These normal pressures require a vertical cylindrical surface of diameter D_m and height t, equal to $\pi D_m t$, where it is notionally assumed to have a constant tangential stress (Figure 8.16). The provisions of the standards limit this equivalent tangential stress to $0.6\,f_u$.

The check is quite local, and only needs the axial tensile force in the bolt, the thickness of the material, and its strength.

8.7.2.3.1 *Eurocode 3*

The utilization factor is defined as

$$U_{ps} = \frac{N}{0.6 \cdot \pi \cdot D_m \cdot t \cdot f_u} \cdot \gamma_{M2} = 2.08 \cdot \frac{N}{\pi \cdot D_m \cdot t \cdot f_u}$$

where D_m is defined as the mean of the across points and across flats dimensions of the bolt head or the nut, whichever is smaller.

8.7.2.3.2 *AISC 360-10*

The utilization factor is defined as

$$U_{ps} = \frac{N}{0.75 \cdot 0.6 \cdot \pi \cdot D_m \cdot t \cdot f_u} = 2.22 \cdot \frac{N}{\pi \cdot D_m \cdot t \cdot f_u}$$

8.7.2.4 Crushing

Crushing is a limit condition of supports. The normal compressive stress delivered to the support is a function of the point of the support. If the support is a concrete block, a slab, or a wall, the crushing failure mode is governed by the maximum compressive stress σ_{max} read on the contact surface. Normally, if tensile stresses are assumed positive, this is a negative number, and so for this reason the modulus of this stress must be considered. Assuming that a limiting value for the compressive stress has been set, σ_{lim}, such as a proper design compressive stress for the concrete, the utilization factor is merely

$$U_{cr} = \frac{|\sigma_{max}|}{\sigma_{lim}}$$

The model is that of one constituent exerting compressive stresses at the interface with another constituent, and one of the two may fail due to an excess of compression.

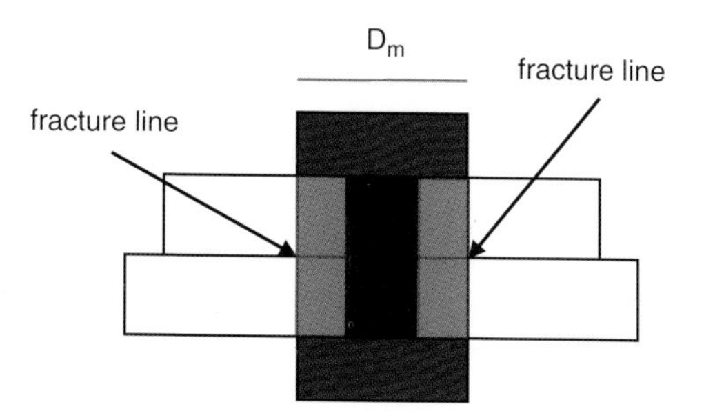

Figure 8.16 Model for punching shear.

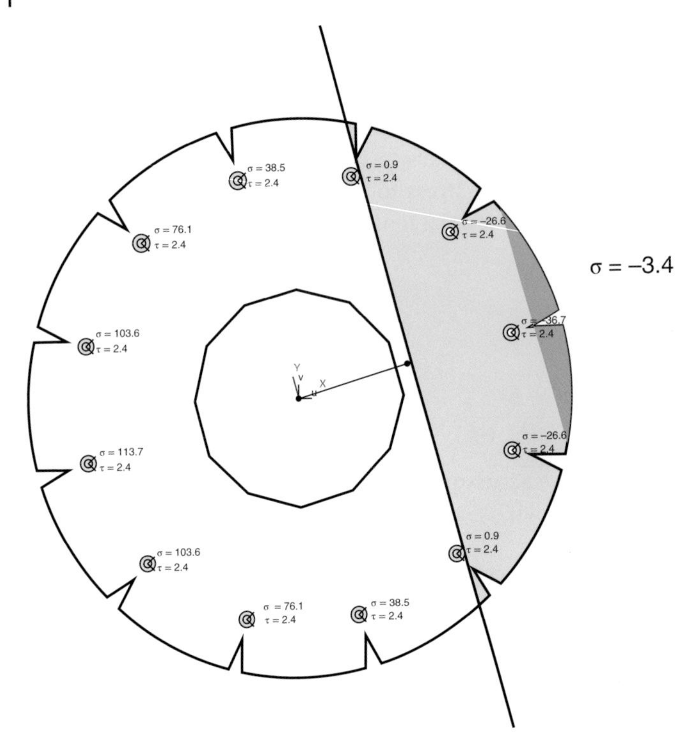

Figure 8.17 An example of compressive pressure field over a bearing surface (base plate in bending, neutral axis also drawn). The maximum compressive stress (−3.4 MPa) can be used for crushing checks.

This model might be useful also for limiting the applied compressive stresses to a constituent in bending, for instance a plate. The typical example is that of two flanges in contact, which exchange a field of compressive stresses at the interface, along the bearing surface. The "crushing stress" (Figure 8.17) might then become a limit pressure value to be applied to the flange, in order to prevent its bending failure. This limitation, in turn, may be helpful for avoiding the need of more refined analyses.

When using a bolt layout with a bearing surface defined, or a contact connector, the "crushing" stress for the constituent identified as "bearing" (usually the weakest), can be defined when defining the constitutive law of the bearing.

Crushing could ideally also be defined for the plates under the fillet welds. The compressive stresses in the welds projected effective areas are available, and these stresses might be used directly for crushing failure checks, if needed.

8.7.3 Fracture Failure Modes

8.7.3.1 Block Shear

8.7.3.1.1 Introduction

Block shear is a very subtle and important failure mode. It may trigger catastrophic failures and for this reason it is a very important failure mode to be checked. Although block shear might be found also in welded connections, the holes existing in a bolted connection make block shear failure more probable. The present discussion is limited to bolted connections, but it could be generalized also to fillet welded connections.

According to the definitions currently in use by the standards, block shear (also called block tear) is the detachment of a block of material from the parent constituent, generated by two sets of lines of fracture: one involving a pure tensile stress rupture, and another involving a shear yielding (Figure 8.18).

AISC 360-10 also considers a dual condition: shear rupture instead of shear yielding, and tensile rupture, taking the lowest resistance as reference. (For a discussion about this issue and also general considerations about block shear, see Kulak and Grondin, 2001, Driver et al. 2004, and Teh and Clements 2012). The difference is not only due to the different limit stress taken as reference, ultimate stress $0.6f_u$ for shear rupture, and yield stress $0.6f_y$, for yield, but also to the different resisting area considered, net for rupture A_{nv}, and gross for yielding, A_{gv}.

The shear rupture condition is not considered by Eurocode 3. However, Eurocode 3 uses net area and not gross area for shear yielding, which is safer. Recent works (Jönsson 2014 and Jönsson 2015), suggest that the gross area for the shearing part can be used, instead of the net area.

With reference to Figure 8.18, which summarizes the model typically used, assuming that D_0 is the bolt hole, and t is the plate thickness, the needed areas would be

$$A_{nt} = t \cdot (h - 2D_0)$$
$$A_{nv} = 2t \cdot (b - 1.5D_0)$$
$$A_{gv} = 2t \cdot b$$

With the symbols already introduced, if F if the force trying to detach the part analyzed, the utilization factor U_{bs} for block shear according to Eurocode 3 is

$$U_{bs} = \frac{F}{\dfrac{A_{nt} \cdot f_u}{\gamma_{M2}} + \dfrac{A_{nv} \cdot f_y}{\sqrt{3} \cdot \gamma_{M0}}}$$

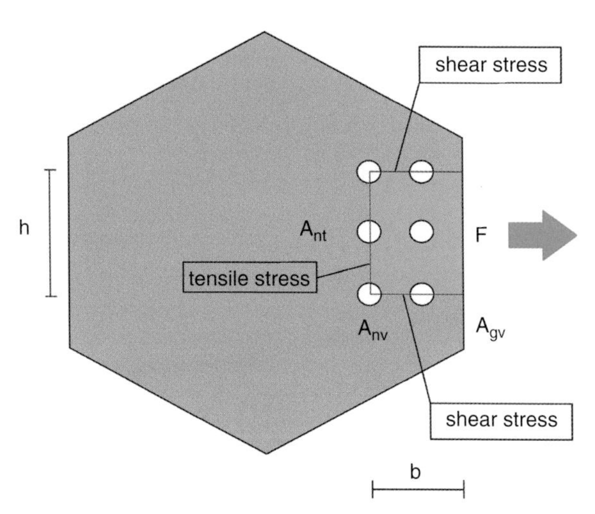

Figure 8.18 Typical model for block shear.

According to AISC 360-10, instead, the utilization factor is

$$U_{bs} = \frac{1}{0.75} \cdot \max\left\{ \frac{F}{A_{nt} \cdot f_u + 0.6 \cdot A_{nv} \cdot f_u}; \frac{F}{A_{nt} \cdot f_u + 0.6 \cdot A_{gv} \cdot f_y} \right\}$$

Here, a first difference is met: Eurocode 3 Part 1.8 mentions the failure mode specially referring to bolt groups. AISC is more general.

According to AISC 360-10, if the tensile stresses are non-uniform, a factor 0.5 must be applied to the tensile part of the resisting force. The same factor is applied to the same term by Eurocode 3, "if the bolt group is subject to eccentric loading". Both refer to an eccentric loading condition, that is, a situation where a torque is also applied to the detached part, leading to a (balancing) non-uniform tensile stress distribution.

In the general context of a 3D analysis of connections, the model briefly outlined is not general enough. Relatively few tests of block shear rupture are available, and they all refer to (very important) typical situations, where the failure paths are relatively clear. If a general numerical procedure must be set up, general rules should be found.

The tests carried on refer almost always to a regular grid of bolts, loaded parallel to the rows or to the columns of the bolt grid. Sometimes, a load eccentricity has been considered, i.e. the force plus a torque. The value of this eccentricity has usually been set very low (see Jönsson, 2014 whose work removes this limitation), so that its effect was not so important.

The standard model, as it may be called, is very important because it typically refers to pulled trusses typically acting as braces, or to coped beams in simple support, with close support. These are very frequent structural elements, but not the only possible ones. In particular, in the context of a seismic analysis, the beam might well experience an axial force, to be added to the shear and the torque, leading to a force with an inclined direction to that of the columns of the bolt grid. But, more generally, it cannot be assumed that such a dangerous failure mode could happen only when there is a bolt grid, when the applied force is parallel to the grid alignments, and when the additional moment due to force eccentricity is low.

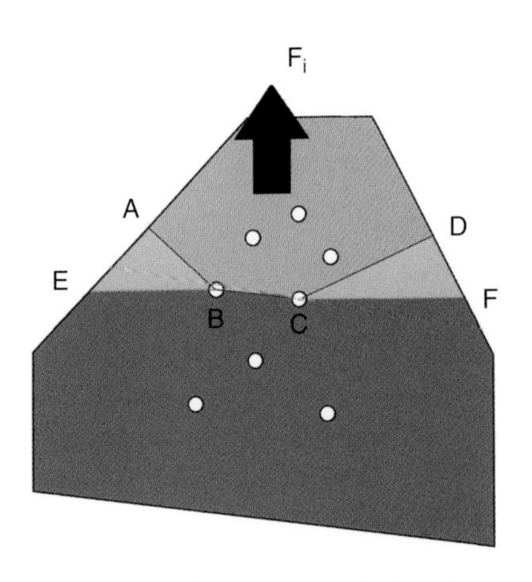

Figure 8.19 Present model for block shear: (1) totally generic bolt disposition, (2) a subset of bolts may detach, (3) failure paths may be mixed shear and tensile force.

First, totally generic bolt groups – that is, not necessarily using regular grids – are needed. (To get an idea of the generality needed, see Figure 8.19.) This is because the geometry of real-world connections may not always be as regular as would be preferred. (To break regularity, it is enough that a single bolt is displaced or removed for a number of possible reasons.) Moreover, the phenomenon is general, and not limited to gusset plates or to bracings. In this discussion, totally generic bolt layouts are considered.

Second, the block shear may include only a subset of the existing bolts, that is, a sublayout or subgroup (see Figure 8.19). This is because, in general, the direction and intensity of the forces exerted by each bolt will be different; only for very simple loading and geometrical configurations are the bolts all loaded by the same force, with equal modulus and equal direction. One subset of bolts may detach, in principle, from the others, if local failure paths, related to the force resultant of the bolt subset, are more severe. Indeed, if a bolt subset i has a resultant F_i, and would break for a failure path Π_i, what is important is to find, for all possible bolts subsets

$$U_{bs} = \max_i \left\{ U_{bs,i}\left(F_i, A_{nt,i}, A_{gv,i}, A_{nv,i}, \Pi_i\right) \right\}$$

However, considering simple grids of bolts loaded by shear in identical manner, it can be easily proved that there are configurations where the worst condition, that is, rupture, will include only some of the bolts, and not all (e.g. only the row nearest to the free edge). But this implies a search of a huge set of possible failure paths.

Third, depending on the geometrical configuration, it should be possible for shorter failure paths than the ones related to pure shear or pure tensile stresses, to be the one preferred by fracture; in Figure 8.19 failure path ABCD is much shorter than EBCF. However, as they would not be parallel or perpendicular to the applied force, along them both tensile and shear stresses would be read. So, for this reason, some kind of coupled tensile plus shear limit condition is needed. In the author's opinion, to see this kind of result, specific experimental tests should be performed with suitable geometries (extending the work initiated by Prof. Jönsson, 2014). In block shear failure, geometry is very important, because by changing geometry, failure paths also change.

If a general approach to steel connection analysis is to be found, these issues are problems that must be solved.

A reasonable failure condition on an inclined face having both normal stress σ and tangential stress τ, is

$$\sqrt{\left(\frac{\sigma}{\sigma_{\text{lim}}}\right)^2 + \left(\frac{\tau}{\tau_{\text{lim}}}\right)^2} = 1 \tag{8.2}$$

where for Eurocode 3

$$\sigma_{lim} = \frac{f_u}{\gamma_{M2}}$$

$$\tau_{lim} = \frac{\sqrt{3}}{3}\frac{f_y}{\gamma_{M0}}$$

and only *net* area is used, while for AISC, using only net area

$$\sigma_{lim} = 0.75 f_u$$

and

$$\tau_{lim} = \min\left\{ 0.75 \cdot 0.6 f_u,\ 0.75 \cdot 0.6 \cdot \mathrm{f_y} \cdot \frac{A_{gv}}{A_{nv}} \right\}$$

If the failure path is parallel to the force F_i, then only shear will be read over it, while if the failure path is perpendicular to the force, only tensile stress will be read over it. In these specific cases, the failure condition of Equation 8.2 defaults to that of the standards.

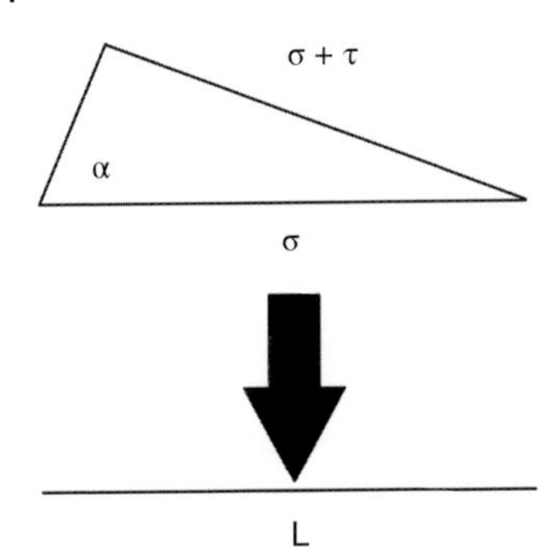

Figure 8.20 Longer tensile vs shorter mixed failure paths. Unit thickness assumed.

We will explain why a shorter path loaded with a mix of shear and tensile stress may be worse for a longer path loaded only by shear or tensile force. Let us for instance consider tensile stress (Figure 8.20).

If a unit thickness is assumed, the force exerted on the longest side at limit, would be

$$F_{tensile} = L\sigma_{lim}$$

The shortest side is only long $L\sin(\alpha)$ and the forces acting on it are $F\sin(\alpha)$ tensile, and $F\cos(\alpha)$ tangential. The related stresses are

$$\sigma = \frac{F\sin\alpha}{L\sin\alpha} = \frac{F}{L} = \sigma_{\text{lim}}$$

$$\tau = \frac{F\cos\alpha}{L\sin\alpha} = \frac{F}{L\cdot\tan\alpha} = \sigma_{lim}\cdot\frac{1}{\tan\alpha}$$

The limit condition on the inclined side is

$$\sqrt{\left(\frac{\sigma_{lim}}{\sigma_{lim}}\right)^2 + \left(\frac{\sigma_{lim}\cos\alpha}{\tau_{lim}\sin\alpha}\right)^2} \leq 1$$

which is clearly violated, meaning that the inclined face, shorter, and under mixed stresses, is in a worse condition than that of the horizontal face, loaded only by a tensile stress, which is exactly at limit.

The maximum vertical force that can be applied to the inclined face is such that

$$\sqrt{\left(\frac{F}{L\sigma_{lim}}\right)^2 + \left(\frac{F}{L\tau_{lim}\tan\alpha}\right)^2} = 1$$

so

$$F = L\sin\alpha\frac{\sigma_{lim}\tau_{lim}}{\sqrt{\sigma_{lim}^2\cos^2\alpha + \tau_{lim}^2\sin^2\alpha}}$$

which means that if a failure path is inclined of an angle α to the direction of the force applied, the limit force it can take is the net area of that face times the equivalent stress

$$\sigma_{eq} = \frac{\sigma_{lim}\tau_{lim}}{\sqrt{\sigma_{lim}^2\cos^2\alpha + \tau_{lim}^2\sin^2\alpha}}$$

If $\alpha = 90°$, the face is in pure tensile stress and it can take only $\sigma_{eq} = \sigma_{lim}$: if $\alpha = 0°$ the face is in pure shear and it can take only $\sigma_{eq} = \tau_{lim}$.

If the tangential stress limit condition is considered, similar results are found.

With these premises it is possible to propose a general method to check for block shear, as will be done in the next section.

8.7.3.1.2 General Method for Block Shear Related to Bolt Layouts

Let us consider first a simple plate, having a generic shape described by a closed polyline and possibly holes not referred to bolts inside.

If a simple plate is considered, the two faces drilled for the bolt are identical. For more complex constituents, the two faces drilled for the bolts are not identical, and the concept of a stiffened edge, already seen for bolt bearing failure mode, has to be taken into account. However, for the moment, a simple plate is considered. The generic face of a plate is referred to a reference system (x, y), that can be the principal system of the bolt layout (u, v) or any other suitable system. The position $\mathbf{P_i}$ of each bolt i will be referred to this system.

Once the forces flowing into each single bolt are known by analysis, for every thickness drilled by the bolt-layout, the problem of block tear failure check can be considered planar: a generic plate with a generic layout of bolts, each loaded by a known shear force. In general, the forces in the bolts will be different from one another, both in direction and modulus.

The system of forces applied by the bolt layout to the plate is balanced by some other sets of forces, related to other connectors. It will be assumed that the block shear failure is related to the bolt layout under consideration and does not involve other connectors. A more general model would consider the forces applied to the plate by all the connectors related to it. For welds, they would be forces per unit length of (fillet) weld.

Usually, plates are constrained by welds on their edges, and the block shear failure does not include the forces related to these constraints. This is also true for bracings such as angles of channels, which are loaded at their extremities, usually quite far from each other. In what follows it will be assumed that the forces are all related to bolts, but the model can be generalized to forces arising also from welds.

If the bolt layout has n_b bolts, there are many different subsets of bolts. The generic subset will be identified by Σ_s, and will in general be formed by n_{bs} bolts ($2 \leq n_{bs} \leq n_b$).

All the possible subsets must be checked. Let us assume that they are N.

Considering now a generic subset of bolts, the resultant of the shear forces applied to the bolts of the subset is $\mathbf{F_s}$. This force has an inclination to angle x of α_s.

A first step after having computed the resultant $\mathbf{F_s}$ merely as

$$\mathbf{F_s} = \sum_{i=1}^{nbs} \mathbf{F_i}$$

is to compute the *convex hull* of the set of points related to the n_{bs} bolt centers. The convex hull (e.g. see Schneider and Eberly, 2003), is the smallest surface S that contains the set of points, such that every straight segment connecting two generic points inside S, has all its points inside S or over its boundary. In Figure 8.21, the convex hull of the subset Σ_s has been marked with a green line. To get the convex hull a Delaunay triangulation is usually done (see Schneider and Eberly, 2003), considering the polyline defining the plate and the subset of points to be triangulated. The triangles having at least one vertex on the plate boundary are discarded, and the convex hull is found (Figure 8.22).

The convex hull boundary is made up of a set of segments, which are identified as sides of the removed triangles. The segments of the boundary may then be ordered, one after another, so as to define a closed loop.

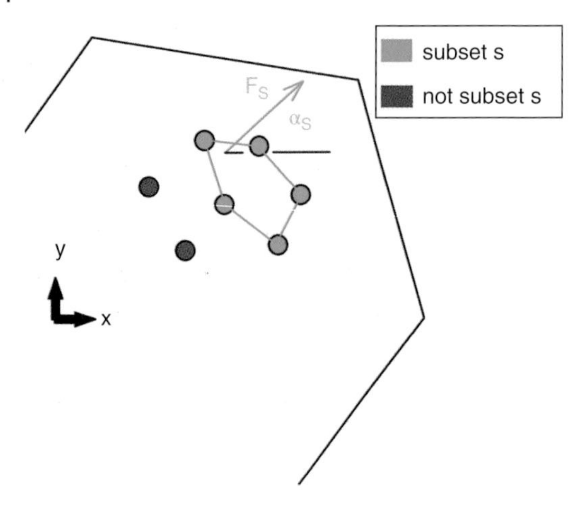

Figure 8.21 A subset Σ_s of the bolt layout (in green), and its (eccentric) resultant F_s. The violet bolts do not belong to the subset Σ_s and when the subset Σ_s is considered, they are discarded.

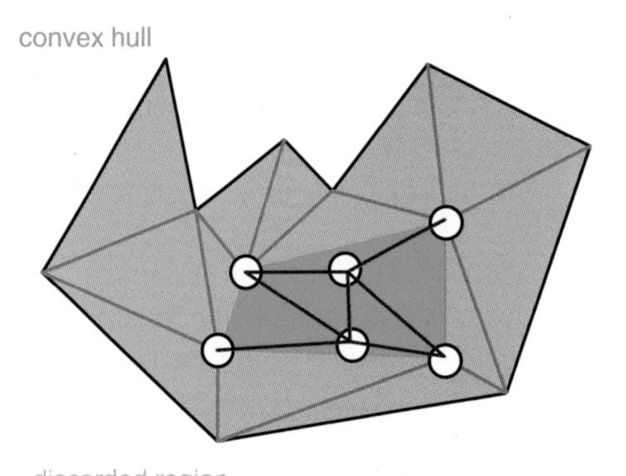

Figure 8.22 Finding the convex hull for a subset of bolts.

The convex hull is basically the block whose detachment is considered. Initially it is assumed that the detachment of the block happens by tearing out the block in the direction of the applied resultant $\mathbf{F_s}$. Later, this assumption will be discussed. Also, a possible torque, that is, a moment causing a rotation of the block, is not for the moment considered. In order to be detached in the direction of the applied force, the block must displace finding the appropriate failure paths. This means connecting the convex hull to the plate boundary. To do that, the following procedure can be considered.

Once the convex hull is found, the two farthest bolts A and B in a direction perpendicular to $\mathbf{F_s}$ are marked. If more bolts are aligned to a line parallel to $\mathbf{F_s}$, the line is oriented as the vector $\mathbf{F_s}$ and the bolt chosen is the one having the lowest abscissa (Figure 8.23). The origin over the line parallel to $\mathbf{F_s}$ is arbitrary (black dot in Figure 8.23).

At this point a first part of the assumed failure path is available. The convex hull loop can be divided in two parts: the part going from A to B and the part going from B to A (Figure 8.24). The part that must be considered is the one intersecting the segment **OZ**, where the point **O** is the center of the bolt subset Σ_s, and the point **Z** is a far point in a direction opposite to $\mathbf{F_s}$:

$$\mathbf{Z} = \mathbf{O} - a\frac{\mathbf{F_s}}{|\mathbf{F_s}|}$$

and a is a suitable very large distance (Figure 8.24).

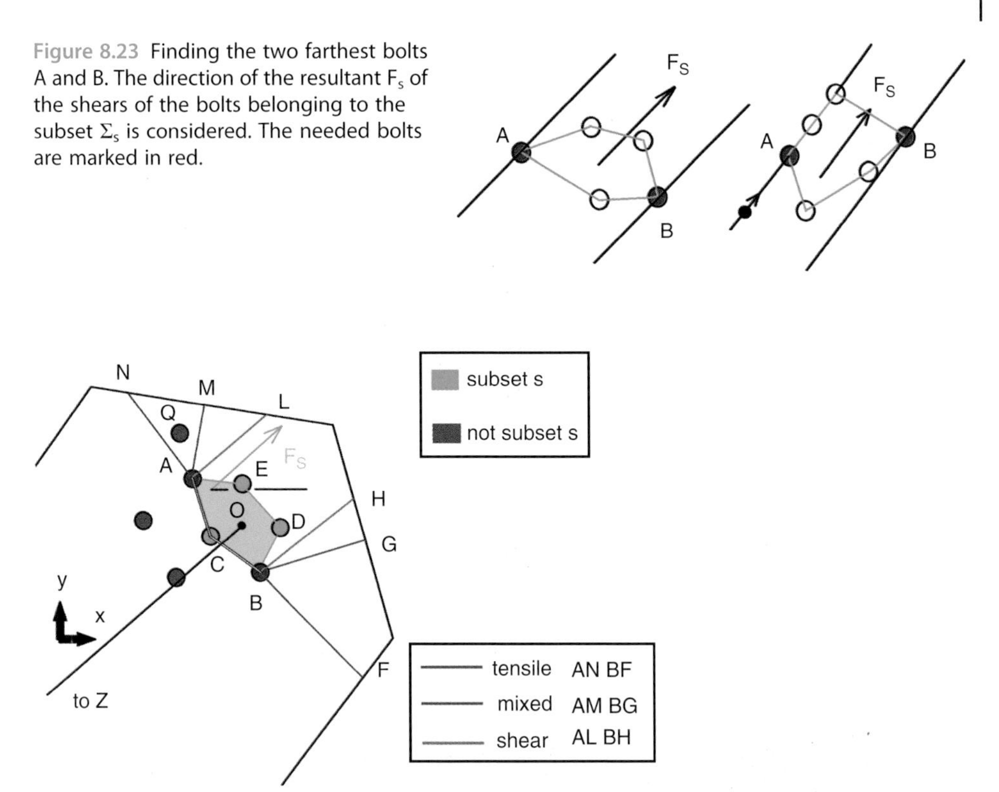

Figure 8.23 Finding the two farthest bolts A and B. The direction of the resultant F_s of the shears of the bolts belonging to the subset Σ_s is considered. The needed bolts are marked in red.

Figure 8.24 Possible failure paths from bolts A and B.

Note that, to assume the other part of the convex hull boundary as being the failure path would be contradictory to the assumption that all the bolts of Σ_s – and no others – are being considered. (Figure 8.24, where the failure path AEDB is not acceptable as the detached part, would not include bolt C.) When the candidate convex hull boundary part connecting A to B is simply AB (a special condition: as this is the shortest possible path to go from A to B), this is the right failure path also if, considering the other part, it would not leave out any bolt.

Considering Figure 8.24 the first part of the failure path, ACB is a part of the convex hull boundary.

From A there are three special failure lines to be considered:

1) line **AN**, perpendicular to $\mathbf{F_s}$ – here only tensile stresses are expected. The point **N** is the intersection of a line perpendicular to $\mathbf{F_s}$ and passing through **A**. If the part detached included some bolt not belonging to Σ_s, then this path must be discarded. In the example, failure line **AN** must be discarded because the detached part would also include bolt **Q**, which does not belong to the subset Σ_s. So, this would be contradictory. That bolt will be considered when the set $\Sigma_s \cup \mathbf{Q}$ is studied. This rule must also be applied to every possible failure line from A and B.
2) line AM, which is perpendicular to the plate boundary. – this is a minimum path line, and here, in general, both tensile stresses and tangential stresses are expected.
3) line AL, parallel to $\mathbf{F_s}$ – here only tangential stress is expected.

From B, the reasoning is the same and these possible failure lines must be considered:

1) line BF, pure tensile stress
2) line BG, minimum distance to boundary, mixed stress in general expected
3) line BH, pure tangential stress

So, there are nine possible combinations of failure paths. These generalize the C and L cutout investigated by previous researchers (Figure 8.25).

Considering bolt A, the failure path related to the minimum value of force will be considered, between the three. The same has to be done for bolt B. Each failure line k is a straight segment forming an angle α_k with the vector $\mathbf{F_s}$.

Considering bolt A, three force values will be considered, if not discarded for lack of coherence (the subscript s refers to the bolt subset).

$$F_{s,Atensile} = \mathbf{\bar{A}\bar{N}}_{s,net} \cdot \sigma_{lim}$$

$$F_{s,Ashear} = \mathbf{\bar{A}\bar{L}}_{s,net} \cdot \tau_{lim}$$

$$F_{s,Amixed} = \mathbf{\bar{A}\bar{M}}_{s,net} \cdot \frac{\sigma_{lim} \cdot \tau_{lim}}{\sqrt{\sigma_{lim}^2 \cos^2 \alpha_{s,AM} + \tau_{lim}^2 \sin^2 \alpha_{s,AM}}}$$

$$F_{s,A} = \min\{F_{s,Atensile};\ F_{s,Ashear};\ F_{s,Amixed}\}$$

Figure 8.25 Nine possible failure paths from a convex hull. It is the geometry that decides which is the worst case, depending on the distances from the plate edges.

The net lengths are computed, subtracting the diameter of the bolt hole D_0 times the number of bolts over each failure line minus 1; for example, if n_{AN} is the number of bolts over line AN,

$$\bar{\mathbf{A}}\bar{\mathbf{N}}_{s,net} = \bar{\mathbf{A}}\bar{\mathbf{N}}_s - 2\cdot\left(n_{s,AN} - 1\right)\cdot\frac{D_0}{2}$$

with a minimum of $0.5D_0$.

The same is applied to bolt B, so computing

$$F_{s,B} = \min\left\{F_{s,Btensile};\ F_{s,Bshear};\ F_{s,Bmixed}\right\}$$

Now if the part of the boundary of the convex hull considered as the failure path (ACB in Figure 8.24) is seen as the union of a number of straight lines $ns_{,ch}$, and if the angle of the generic straight line k with the vector $\mathbf{F_s}$ is $\alpha_{s,k}$, and the net length of the segment is $l_{s,k,net}$ then

$$F_{s,CH} = \sum_{k=1}^{n_{s,CH}}\left[\frac{\sigma_{lim}\cdot\tau_{lim}\cdot l_{s,k,net}}{\sqrt{\sigma_{lim}^2\cos^2\alpha_{s,k} + \tau_{lim}^2\sin^2\alpha_{s,k}}}\right]$$

It is now possible to evaluate a utilization factor for the subset s considered, as

$$U_{bs,s} = \frac{F_s}{F_{s,A} + F_{s,B} + F_{s,CH}}$$

The final utilization for block shear is computed as

$$U_{bs} = \max_{s=1,N}\left\{U_{bs,s}\right\}$$

That is, the maximum utilization is computed by considering all the possible bolt sets.

The method outlined is very general, but it has a drawback: if, to a force originally parallel to a row of holes, is added a slight perpendicular force, the failure path searched will not include the row of holes. Considering Figure 8.26, assume that applying a horizontal force, the block ABCLD is detached. Now if a slight vertical force is added, the resultant force will be inclined with a small angle to the horizontal. The two farthest bolts of the convex hull would be B and L, and the candidate block detached would be EBCLD. If the act of motion must be parallel to the direction of the resultant force, the block can no longer be ABCLD, as the "wall" AB would be found. However, with no correction, this would imply a sudden rise of the available strength, by adding a limited force.

However, this problem would disappear, and the solution would preserve its continuity, if, as suggested by Prof. Jönsson (Jönsson 2014 and Jönsson 2015) considering the experimental results, the gross area instead of the net one were to be used for the shear failure path. Using the following assumption:

$$A = \frac{A_g A_n}{\sqrt{A_g^2\sin^2\alpha_s + A_n^2\cos^2\alpha_s}}$$

that is, using an elliptic variation for the failure path area A, so that it would be equal to the gross one A_g, for $\alpha_s = 0$, that is, over the shear failure path, and equal to the net one, A_n, for $\alpha_s = 90°$, that is, over the tensile failure path.

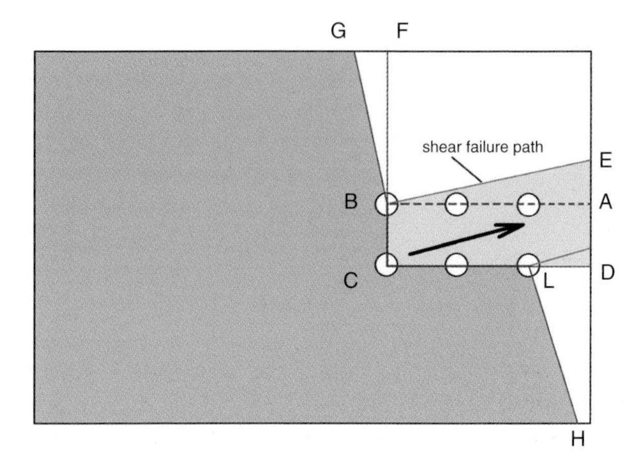

Figure 8.26 A slight change in the direction of the applied force implies much longer failure paths, if net area is to be used for shear failure path.

Inverting the reasoning, this can be a further argument for strengthening the idea that the gross area should be used for the shear part of the failure mode. At the time of writing (spring 2017), Eurocode 3 does not allow such an approach as it explicitly requires the use of the net area also for the shear failure path. However, referring to the block shear issue, the approach proposed by Eurocode 3 is simply unable to consider general situations and cannot be considered a usable tool when dealing with such problems.

Considering now the problem of the additional torque, it can be seen that the method outlined also includes failure paths for torsion. In order to explain why, the condition of pure torque will be considered, with reference to a simple 2 × 2 grid of bolts. According to Prof. Jönsson, a condition can exist of block shear under pure torsion, and using the static theorem, lower bounds have been proposed with reference to the C and L failure paths (Jönsson 2015).

Prof. Jönsson's approach, using static and kinematic theorem of limit analysis at given failure path (C or L shaped), gives us a very simple formula, that computing the block-shear limits related to shear only force V_R, axial force only N_R (two forces perpendicular one another presumed applied to the member by the bolt layout so N_R is the axial force in the member), and pure torque M_R sets the interaction limit condition as

$$\left(\frac{N}{N_R} + \frac{M}{M_R}\right)^2 + \left(\frac{V}{V_R}\right)^2 = 1$$

However, it is not clear why those specific C or L failure paths should be chosen, among all the possible ones. The failure path should be sensitive to the geometry, the distance from the boundary, and some kind of minimization such as the one outlined should probably considered. This is true for all the forces applied, considered separately, but it is particularly true for the torque-only loading condition, where there is no force resultant. It is also not clear to the author how a rotational act of motion implying a well defined movement against the assumed shear-only failure boundary, that is, with displacement components normal to it, could be matched with a shear-only stress state. The result is allowed by the static theorem but it seems not very compatible with the

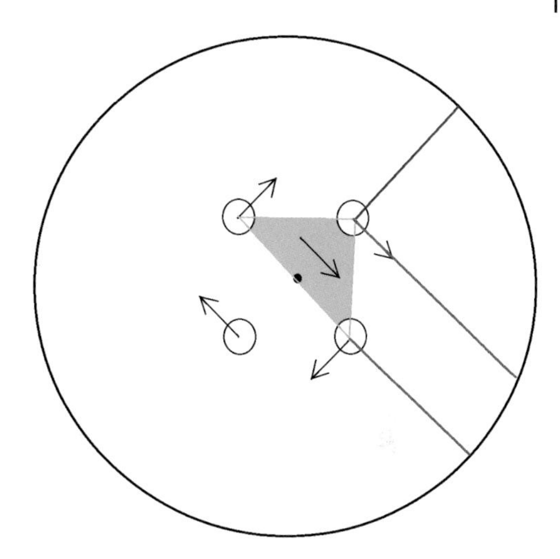

Figure 8.27 Block shear of a bolt layout under pure torsion. Considering the bolts' subsets, local failure modes are included.

real act of motion expected. This is an open problem. Also, it would be quite interesting to observe experimentally a pure rotational block shear failure mode, and to prove that it can be reached *before* bolt bearing failure. The issue is very interesting and more research is probably needed.

In the currently proposed approach, if a pure torque is applied to a bolt layout, the resultant obtained considering the set of all the bolts is null. However, when the subset of bolts is considered, for example for the 2 × 2 grid, the three subsets having three bolts, the resultant of the shear forces applied to the bolts of the subset is not null (Figure 8.27). Two of the three forces will cancel one another, but the third will be in place. This can be generalized to any situation, because the distribution of the shear forces in the bolts is considered not only as a whole but also when considering the local effects it provokes, thanks to the subsets. The shears in the bolts will directly also consider the effect of the applied torque.

Considering all the subsets, the one related to the maximum utilization will be stored and considered as reference. For the 2 × 2 grid under pure torsion, the three subsets will give different results considering the boundary of the bolted plate. If the plate is circular, as in this example, the three subsets will all give the same result.

8.7.3.2 Rupture in Single Cross-Sections

Sometimes the fracture line of a constituent is straight, so that a well defined plane of failure can be found. Traditionally the rupture of constituents weakened by cuts or by bolt holes has been investigated considering *net cross-sections*, that is, cross-sections obtained by ideally cutting the constituents with a plane. This is not to be confused with the Whitmore section, which is, for instance, a way to check for resistance gusset plates at a distance from the bolts.

For a given constituent, it is possible to automatically detect a huge number of such cross-sections, and to check them for resistance, including rupture. Each net section is related to a plane, whose equation in space is set considering the geometry of the faces of the constituent, and the geometry of the bolt layouts possibly applied to it.

For members, the planes are chosen perpendicular to the member axis (see Figure 8.28). The planes are chosen so as to pass through the sections reduced due to cuts or bevels, finding the weakened part (Figure 8.29). Net cross-sections defaulted to the gross-section of the member are discarded, as already covered by the member checks.

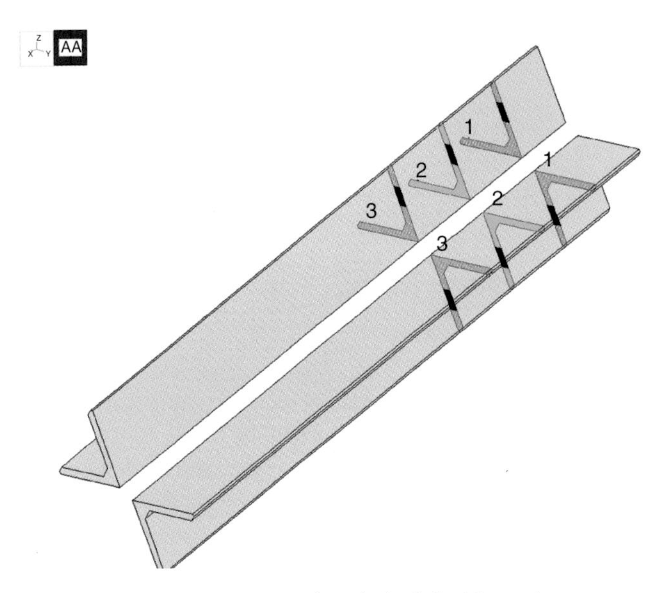

Figure 8.28 Net cross-sections for a bolted double angle.

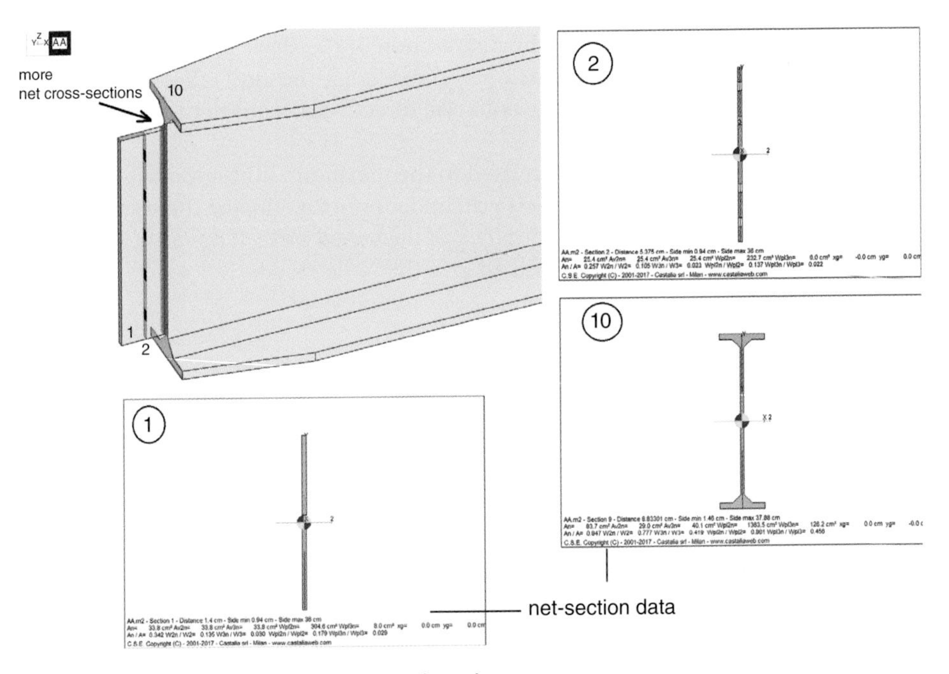

Figure 8.29 Cross-sections referring to weakened part.

Basically, once a cut-plane equation in 3D has been set, the collection of faces defining the BREP of the object is intersected to the plane, defining a vector of closed polygons laying over that plane. These polygons define the boundary of the needed cross-section and its possible holes.

The center, principal axes (2, 3), area, second area moment, elastic and plastic moduli of a net cross-section defined as a vector of closed polylines, and some possibly defining holes, can be obtained by integrating along the boundaries using Green's formula as has been described for instance in Rugarli 1998.

The failure is then checked by using a simple beam-like formula, and the resulting net cross-section area and moduli. Ideally six internal forces can be applied to such net cross-sections: an axial force, two shears, a torque and two bending moments.

The forces flowing into the generic net cross-section $(N, V_2, V_3, M_1, M_2, M_3)_{\text{cross}}$ are obtained by summing up simple terms, related to the forces applied to the constituent by single bolts and by single (parts of) welds. Only the bolts belonging to one of the two parts in which the constituent is divided by the plane are counted, and only the parts of single welds in the same part. A plane may well cut a weld in two parts: if this happens, only a subpart of the weld needs to be considered.

Some part of the internal forces can be discarded, if it is considered that they are taken by other resisting mechanisms, such as, typically, contact. So it should be possible to neglect weak axis bending (i.e. bending M_2 for the net cross-section number 7 in Figure 8.30), or torque, or in general any other component that is not considered meaningful. This decision should, however, avoid neglecting part of the forces if there is no good reason to do that, and is typically taken by the analyst.

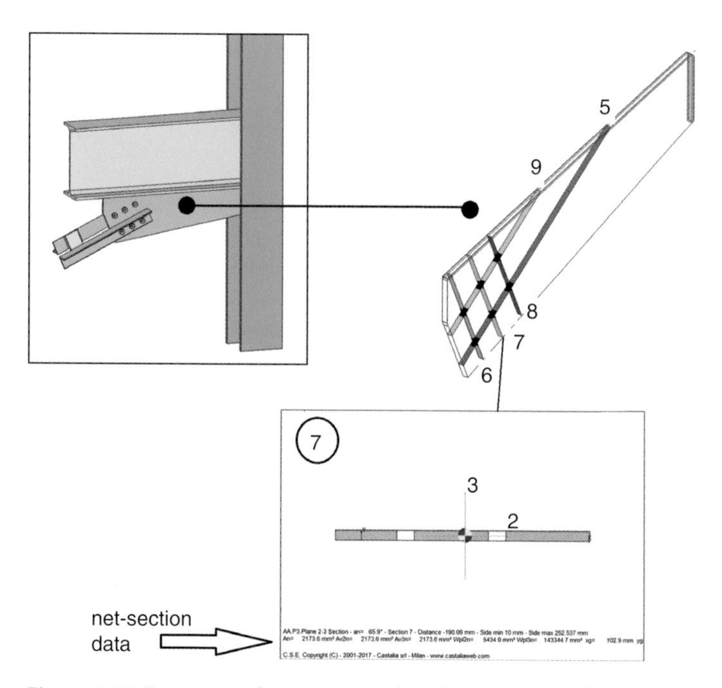

Figure 8.30 Extraction of net cross-sections from a gusset plate.

Using Eurocode 3, the resistance utilization U_{ncs} of the net cross-section can be found as follows.

If the axial force is compressive

$$U_{ncs} = \frac{\left|\frac{N\gamma_{M0}}{Af_y}\right| + \left|\frac{M_2\gamma_{M0}}{W_{2pl}f_y}\right| + \left|\frac{M_3\gamma_{M0}}{W_{3pl}f_y}\right|}{(1-\rho)}$$

where

$$\rho = (2U_V - 1)^2 \quad U_V > 0.5$$
$$\rho = 0 \qquad U_V \le 0.5$$

and

$$U_V = \max\{U_{V2}, U_{V3}\}$$

$$U_{V2} = \left|\frac{\sqrt{3}\cdot V_2}{A_{r2}f_y}\gamma_{M0}\right|$$

$$U_{V2} = \left|\frac{\sqrt{3}\cdot V_3}{A_{r3}f_y}\gamma_{M0}\right|$$

The terms A_{r2} and A_{r3} are suitable shear resistance moduli. These can be considered equal to the net area for rectangular net cross-sections, and must be evaluated by using more refined approaches for complex cross-sections. A possible way to compute these moduli is to use the Jourawskj formula, which is on the safe side.

If the term ρ is equal to one, or greater than one, then the shear applied is higher than the resistance available, and the utilization can be set equal to ρ.

If the axial force is tensile, then

$$U_{ncs} = \frac{\left|\frac{N\gamma_{M2}}{Af_u}\right| + \left|\frac{M_2\gamma_{M0}}{W_{2pl}f_y}\right| + \left|\frac{M_3\gamma_{M0}}{W_{3pl}f_y}\right|}{(1-\rho)}$$

That is, for the axial force term the ultimate factored stress is used.

If AISC is used, then according to section J4, the reference stress for tensile axial force is

- for tensile yielding 0.9 f_y
- for tensile rupture 0.75 f_u.

For compressive axial force, the reference stress is

- for compressive yielding 0.9 f_y.

For shear, the reference stress is

- 0.6 f_y for shear yielding
- $0.75 \cdot 0.60\, f_u = 0.45\, f_u$ for shear rupture.

The use of the rupture stresses should be restricted to net areas, while that of yield stresses is restricted to gross area. Given a "net" cross-section, if there are holes inside it then the rupture checks must be executed, but if there is no hole, then use the yield checks instead.

Net cross-section checks may help to execute resistance checks, but are rarely sufficient to prove the resistance of a constituent. In order to do that, constituent-global failure modes must be inspected.

8.7.4 Global Failure Modes

8.7.4.1 Yielding

Even if a constituent has not undergone any of the failure modes listed up to this point, it may still fail if a plastic mechanism is reached before the full application of the external loads.

A plastic mechanism implies the ability to increase the deformation at constant load. When a plastic mechanism is reached, the loads cannot be further increased. This condition is met sooner or later if a perfectly plastic constitutive law is used for the material, and the loads are increased. Using a hardening law, instead, the loads can be increased, but the ultimate stress will be reached in some part of the body of the constituent, signaling a failure. Moreover, although the loads can be increased, they will cause very high displacements, which can be compared to the perfect plastic flow of the perfect plasticity assumption. Usually if the aim is to get the limit loads, the perfectly plastic model is preferred.

The plastic mechanism will in general depend not only on the load level, but also on the load distribution over the constituent. So, for every distinct load combination, distinct plastic load multipliers and plastic mechanisms are expected. Considering isomorphically loaded constituents (see Chapter 5), the plastic mechanisms will be the same from load combination to load combination, and the load multipliers can be ordered, knowing the load level of each combination. For this very special case, running a unique analysis will be enough to know the plastic load multipliers of all the other combinations.

If the aim is merely to get a utilization index, that is, if the aim of the analysis is merely to assess that the applied loads are below the limit ones, then a static analysis with the applied loads as external loads may be sufficient. But if the aim is to determine the utilization factor, U_{pm}, then the procedure is much more complex, as the analysis must find the load multiplier α_{lim} that will take the constituent to its limit. Once this has been found, clearly

$$U_{pm} = \frac{1}{\alpha_{lim}}$$

It must be underlined that this in general should be done for every loading combination. The numerical procedures available nowadays, and the computing power available in modern personal computers, allow this task, especially if the computational effort is used as and when it is really needed.

In the frame of finite element analysis the goal can be reached by using a perfectly plastic constitutive law for the material, and increasing (or decreasing) the loads applied so as to find the limit loads, usually signaled by a lack of convergence. If displacement control or arc length methods are used, it is also possible to follow the post limit branch, and decrease the loads.

The analyst must have clearly in mind a tradeoff between the desired precision and the computational time needed to get it. So sometimes, very often indeed, simple elastic analyses related to the design loads applied can be used to get an idea of the stress state and to decide if this stress state is acceptable or not. These analyses run in few seconds, and must always be considered as the very first step to assess the safety of a constituent using finite elements.

If the aim is merely to get a utilization index, as the loads will probably be below the limit values, the convergence of a material-non-linear analysis is usually quick. All the combinations can be inspected, and reasonable information about the amount of utilization can be obtained by the Von Mises stress maps, and by the spreading of plasticity. A slow rate of convergence is normally related to the wide spreading of plasticity, and if this is the case, the designer has important information that can be used to improve the design.

If the aim is to get utilization *factors* for every loading combination, this may lead to a huge computational effort. However, the analyses done under the applied load levels may help the designer to choose the worst loading combinations, and, for those, they may wish to have a clear idea of the distance from the boundary. The limit load can be reached for instance by applying a factor 2 to the applied loads. If no limit is reached, then

$$U_{pm} < 0.5$$

If an elastic analysis has shown that nowhere in the constituent has the yield been reached, the non-linear analysis for that combination is clearly unnecessary, unless a limit load multiplier is needed.

So a possible ideal operative path, if only utilization indices are needed can be as follows:

1) Run elastic finite element analyses for all combinations, and see if and where the yield is crossed. By plotting the regions where the yield is crossed, some important information can be obtained. For each combination, the following reasoning should be followed.
2) If nowhere is the yield is crossed, an overestimate of the utilization factor can be obtained as $U_{pm} \leq (\sigma_{VMmax}/f_d)$, where σ_{VMmax} is the maximum Von Mises stress in the model and f_d is a design stress. This can be set equal to $f_d = f_y/\gamma_{M0}$ for Eurocode 3, and to $f_d = 0.9\, f_y$ for AISC.
3) If the yield is crossed but in very limited regions (Figure 8.31), in the vicinity of concentrated forces due to bolts or welds or in the vicinity of changes in geometry it can be reasonably assumed that these stress peaks will be redistributed in plastic analyses. Moreover, as the bolt bearing checks have already been done, or are done separately, near to the bolts the stresses obtained by finite element analysis are usually not very meaningful (also due to fem model scale issues). No measure of the utilization factor can be made, but it is assumed that the loads are below the limit.
4) If the yield stress is crossed in regions having some extent, say some square centimeters or square inches (Figure 8.32), it is reasonable to assume that in the plastic range these stresses will be redistributed, but it does not seem wise to assess whether the constituent has passed the check. In this situation, a non-linear plastic analysis is suggested for that combination, to be sure that the limit load has not been reached.

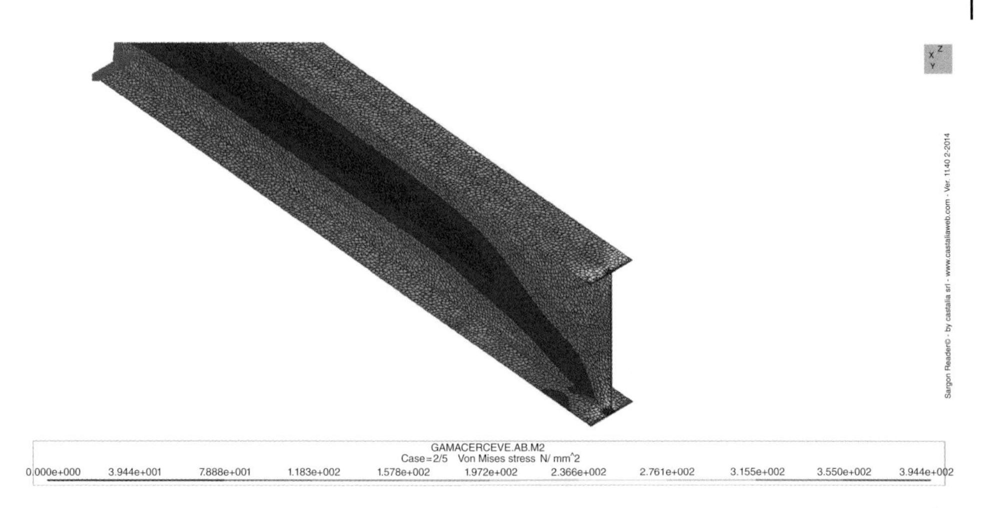

Figure 8.31 A very limited set of stress peaks due to welding.

Figure 8.32 An extended but still limited region where the yield has been crossed.

5) If the yield stress is crossed in wide regions (Figure 8.33), comparable to or not too far from the size of the constituent in at least one direction, then the probability that the limit load has been crossed is high, and once more a plastic analysis is needed. In this case, however, it may be wise, if the weight minimization is not an issue, to change the design accordingly.

If the utilization factor is needed, the plastic analyses should be run instead with proper load multiplier applied to the loading combination, so as to find a non-convergence condition (or a negligible current stiffness parameter – see Chapter 9).

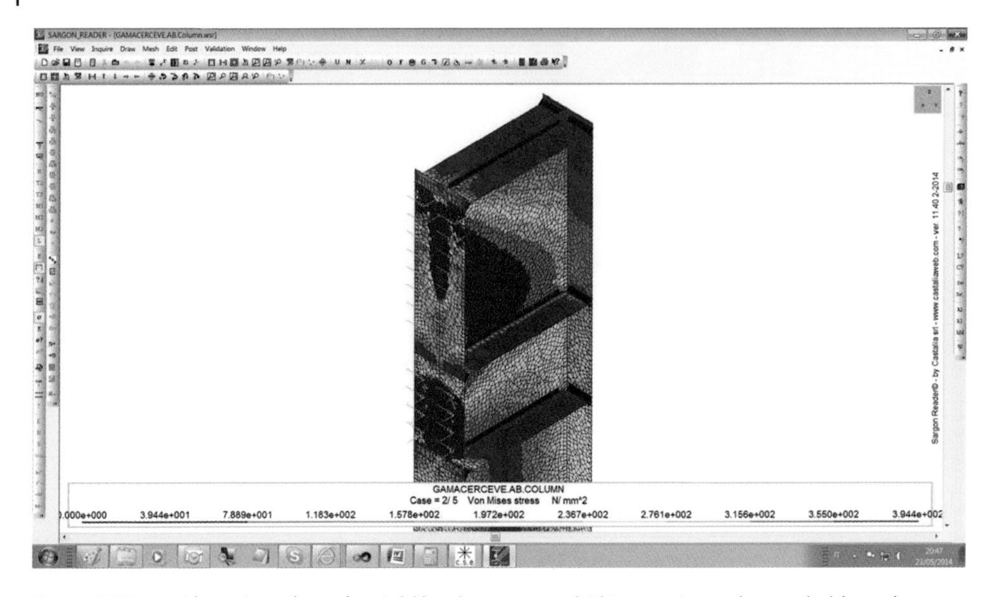

Figure 8.33 A wide region where the yield has been crossed. This constituent has probably undergone excessive loads.

If in the combination, the yield is not reached, and the maximum Von Mises stress is σ_{VMmax}, then a reasonable guess for the factor α to be applied to the loads could be

$$\alpha = 1.5 \cdot \frac{f_d}{\sigma_{VM\max}}$$

If the yield has been reached in the elastic analysis, the factor to be applied could be

$$\alpha = 1.5$$

If when applying α times the load the limit condition is not met, then the analyst can assume that

$$U_{pm} < \frac{1}{\alpha}$$

which very often is a sufficient information for design purposes.

If, instead, only a fraction κ $(0 < \kappa < 1)$ of the load α can be applied, and then the limit is reached, then

$$U_{pm} \approx \frac{1}{\kappa \cdot \alpha}$$

For instance if the design stress is $f_d = 355$ MPa, and the maximum Von Mises stress has been evaluated as $\sigma_{VMmax} = 280\,MPa$ by the elastic analysis, the guess-multiplier applied to the load would be $\alpha = 1.90$. If applying $\alpha = 1.90$ no mechanism is found then

$$U_{pm} < \frac{1}{1.9} = 0.52$$

If, instead, only a fraction $\kappa = 0.72$ can be applied, and then the limit is reached, then

$$U_{pm} \approx \frac{1}{1.9 \cdot 0.72} = 0.73$$

For very specific and well defined geometries, and for simple loading patterns, it is possible to get to reasonable estimates of the limit loads by means of simplified procedures. These may still be helpful, but in the general context of 3D analysis their use does not seem satisfactory, as they usually neglect some part of the loads, that is, some internal force component. Moreover, as already underlined in Chapter 1, they are not able to consider the superposition of more effects on the very same part. They should not, however, be dismissed, as they are useful both for pre-dimensioning and for cross-checks of results. As this argument is covered in depth by excellent publications (e.g. the *Green Books* by The Steel Construction Institute), the reader may wish to find there the needed simplified formulations.

8.7.4.2 Buckling

8.7.4.2.1 Generality

For the reasons explained in Chapter 5, the shelter offered by the static theorem of limit analysis does not fully cover against the buckling modes.

Buckling occurs under compressive stresses, and is strongly affected by the stiffness of the constituents, their possible degradation due to plasticity, the load distribution, the contact between parts with the complex patterns of pressure exchanged, and the level of the imperfections. As the load distribution that we use in our stress evaluations is almost always notional, and founded on the static theorem, there is an open problem to deal with. This is true for practically all the methods used in connection analysis, with the exception of pure fem models that model contact and friction properly, as will be described in Chapter 9. That is, this is also true for the simplified, non-fem approaches such as T-stub, Whitmore section, gusset plates uniform force method, and so on. It is, indeed, a common problem.

Finite element analyses are able to analyze the buckling problems with good accuracy, but if the loads applied to the constituent, or exchanged between them, are not “precise”, the buckling load multiplier will not be exact.

Despite this important consideration, the buckling analyses under balanced loading configurations may help to detect weaknesses of the design constituents, and to add material or tightening where it is needed. This is true especially if the analyses are repeated for all loading combinations, thus testing a huge set of load distributions. So, these buckling analyses should not be intended as a precise measure of the utilization with respect to the buckling of real-world designs, but as collateral information that may help to avoid unsafe designs.

As will be better explained in the next two chapters, there are several ways to evaluate the buckling phenomena using finite elements:

1) an eigenvalue analysis, directly measuring the buckling load multiplier α_{cr}
2) a step-by-step analysis, neglecting the effect of plasticity
3) a step-by-step analysis, including the effect of plasticity

The first and second approaches are similar, but the second one is more precise as it considers that the stresses are not linearly proportional to the applied loads, also in elastic range, if the geometrical effects are considered. Both approaches do not directly give the searched for *real* buckling multiplier as they do not take into account imperfections and residual stresses.

The third approach may well be unified with the limit-load search, getting to a utilization index or factor valid for both failure modes. Using the same techniques already outlined for the limit load search, the analyst can get useful estimates of the utilization index (greater than one or lower than one), and of the utilization factor against buckling, U_{bk}. If geometrical imperfections are added, modifying the initial geometry, these methods may also consider imperfections. Residual stresses need a specific approach in order to load the model with proper self stress fields. However, this approach is more complex.

A key problem is the choice of imperfections, and the number of different configurations for them. When combined with a high number of loading combinations, the requirements of this approach may be too high. Usually, a limited set of reputed meaningful combinations are analyzed.

Using the first approach, a critical multiplier α_{cr} for each loading combination can be found. This critical multiplier is not a measure of safety, even if it is higher than 1, approaching 3 or 4. As shown in Chapter 5, to get the "real" multiplier α_R, the effects of imperfections and residual stresses must be considered. A good way to do that is to use the general method of Eurocode 3, evaluating (also with a lower bound) a limit load multiplier α_L, and then a non-dimensional slenderness as

$$\bar{\lambda} = \sqrt{\frac{\alpha_L}{\alpha_{cr}}}$$

Using a proper buckling curve, a reduction factor χ can be found

$$\chi = \chi(\bar{\lambda})$$

and from that the needed multiplier α_R:

$$\alpha_R = \chi \cdot \alpha_L$$

that must be greater than one. The utilization *index* for buckling would then be

$$U_{i,bk} = \frac{1}{\alpha_R}$$

The buckling curve to be used depends on the problem at hand. No clear and general direction is given in Eurocode 3 Part 1.8, when dealing with connection constituents.

In the author's opinion rolled constituents welded in place can probably use buckling curve "b", while bolted in place rolled-constituents can use the buckling curve "c". Constituents created by welding, welded in place, may use the buckling curve "c", while if bolted, use the buckling curve "d". Safer buckling curves can be obtained by increasing the imperfection factor. There is no need and no real interest in sizing the constituents so as to avoid buckling for one single pound force, especially in seismic areas, where the sudden buckling of a stiffener may imply the collapse of the whole structure.

Eurocode 3 Part 1.8 explicitly mentions the stability of "compressed webs", and a reduction factor ρ, assuming the same role of χ, is evaluated as

$$\rho = \chi = 1 \qquad \bar{\lambda} \le 0.72$$

$$\rho = \chi = \frac{(\bar{\lambda} - 0.2)}{\bar{\lambda}^2} \qquad \bar{\lambda} > 0.72$$

(see §6.2.6.2(1) of Eurocode 3, Part 1.8), which means a value higher than that of the better possible curve, a_0. For the shear of web panels, an explicit limit to width-to-thickness ratio is set.

Referring to the stability of plates with or without stiffeners, Eurocode 3 Part 1.5 (§4.5.3) sets the following rules for "column-type" buckling behavior:

- Curve "a" must be used for unstiffened plates.
- The imperfection factor of curve "a" must be increased, depending on the geometry of the stiffeners, for stiffened plates.
- Curve "b" must be used for closed section stiffeners.
- Curve "c" must be used for open sections stiffeners.

AISC 360-10 explicitly mentions, in Section J4, the buckling of affected elements, setting no reduction if the "dimensional slenderness" is lower than 25. For higher values, the provisions of Chapter E must be applied. Here, considering that[5]

$$\bar{\lambda}^2 = \frac{f_y}{\sigma_{cr}}$$

The following reduction factor χ is prescribed:

$$\chi = 0.658^{\bar{\lambda}^2} \qquad \bar{\lambda} \le 1.5$$

$$\chi = 0.877 \cdot \frac{1}{\bar{\lambda}^2} \qquad \bar{\lambda} > 1.5$$

In AISC 360-10 explicit references are made (section J10) to web local crippling (J10.3), to web sideways buckling (J10.4), and to web compression buckling (J10.5). Specific formulae are provided which can be considered special cases and directly used in the checks. If the problem of stability is taken by an appropriate finite element model of the constituent (usually a member), these checks would be used as an additional tool.

It is important to consider a possible unified approach to all the problems of buckling, clearly identifying the buckling curves to be used for the different buckling checks. In light of this consideration, Eurocode 3 seems already able to give such an answer, provided that the pertinent buckling curve name is identified. In this way, approaching stability issues by using finite element techniques could be done well under the shelter of agreed provisions.

If the distribution of the forces exchanged between the constituents is evaluated with reasonable accuracy, the results obtained by buckling analysis should be considered meaningful for the analyst. However, the design of the constituents should be made considering that the load distributions effectively experienced by the constituent in the real

---

5 In AISC specification, the critical stress σ_{cr} is named F_e.

world might be different, and protecting the part against this possibility. This means sound proportions, and the need to test single constituents for a variety of different load distributions. This concept is indeed quite similar to that of cross-section classification, shared by both Eurocode and AISC provisions, and requires constituents that are able to resist to different loadings near yield, without significant reductions due to buckling.

The design rules used in the past for the sizing of force transferrers (e.g. the thickness of the stiffener must be the same as the stiffened part), were indeed a way of achieving this goal.

When dealing with members, the problem is caused by the fact that the loads locally applied to the members are not ordered and organized as the ones predicted by beam theory; the webs can be loaded out of plane, the flanges also, and in general the problems that will have to be faced are not covered by any simple theory.

With the aim of setting up reliable computational models, one useful tool would be to consider a constituent isolated from all the others, and apply to it a significant set of loadings, changing in distribution and direction, to test that the buckling is obtained for loads not too far from those leading to generalized yield. This procedure might be fully automated and could be used as an everyday tool by structural engineers.

8.7.4.2.2 Class 4 or Slender Cross-Section Connections

One important issue in steel connection analysis relates to cold formed members, where it is expected that local buckling may happen. Using gross section instead of the effective ones may lead to significant errors in the distribution of the stresses in the fasteners.

The problem would probably require inactivating the fasteners which are applied to the ineffective parts of the cross-section, so that only the fasteners in the effective region could be considered. The issue is not covered by Eurocode 3, where no specific provisions for cold formed member connections can be found in either Part 1.8, dealing with connection design, or in Part 1.3, with the problem of ineffective parts.

The effective section of a cold-formed member is usually evaluated (and this is allowed e.g. by Eurocode 3) by considering one internal force at a time. So, the effective area A_{eff} is related to compressive axial force, while the effective bending modulus W_{eff} is computed considering the cross-section under pure bending. When a combination of axial force plus bending is applied to the cross-section, the ineffective part strictly depends on the load level and on the mutual proportion of the two applied loads. If, also, the other bending is added (as in racks under seismic loadings, a condition which triggered important failures during the Italian Emilia earthquake of 2012), the effective section resulting can be quite complex to evaluate. And, it will change from load combination to load combination.

Considering these issues, the design of connections of class 4, slender members is a hard task. If the effective cross-section under pure compression is considered as a reference, the fasteners applied to the cross-section in order to connect it can be divided into two sets:

- the set of the fasteners applied to the effective part of the cross-section, which can be considered effective
- the set of the fasteners applied to the ineffective part of the cross-section that should be considered ineffective

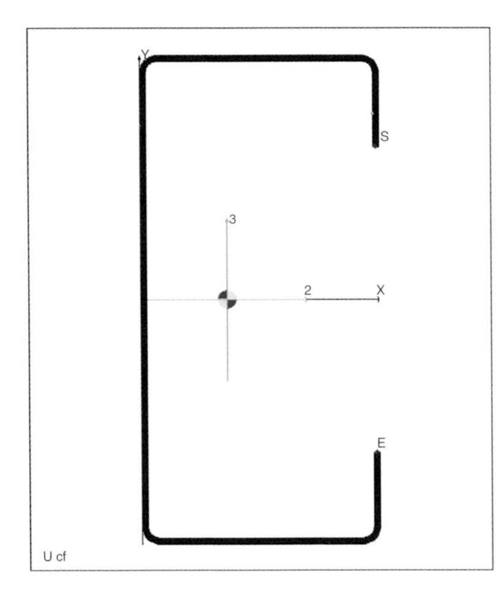

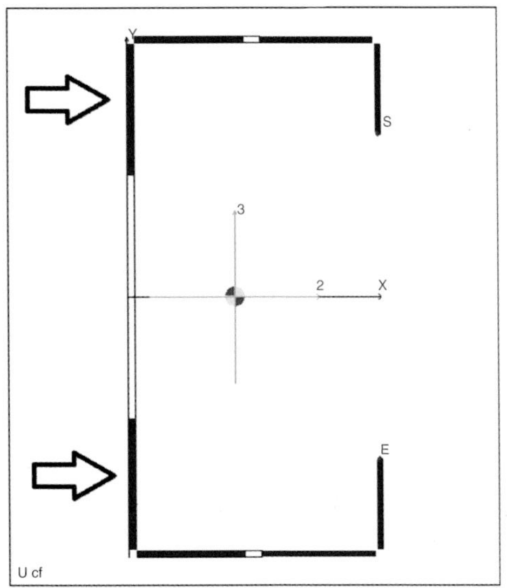

Figure 8.34 A clipped cold formed cross-section and its effective area under pure compression according to Eurocode 3. Note the shift in the centroid of the gross section. Ineffective parts are not filled with color.

However, the effective part of the cross-section (Figure 8.34) can be so reduced that little room is available for the "effective" fasteners.

It seems reasonable to state that the matter is still uncovered by the numerical procedures available, and remains an open question.

8.7.4.3 Excess of Displacement

In the context of the 3D analysis of a renode, the displacement check has two fundamental reasons.

The first one is that it is needed in order to ensure that the connectors have been properly put in place, that is, that not only are they in theory able to transfer the needed force packets, but also that the working mode choices that the analyst has made are fit for that particular connection, under the applied loads. If some connector working mode has been customized in such a way as to avoid some part of the structural stiffness, reducing it to negligible values, it is possible that the forces applied will find no other path to reach the destination, and this condition will be signaled by very high displacements. This may happen when declaring bolts or welds as shear only, or no-shear, or longitudinal shear only. It is a common error to misapply such flags, leading to connections that simply cannot take the loads (Figure 8.35).

Well before getting to the resistance failure checks, the analyst should be aware of the problem by looking at the displacements. It is not necessary, if this is the intent, to have accurate predictions, as the only thing that is relevant to the aim of controlling the good design of the system is the order of magnitude of the displacement.

The second reason why the analyst may wish to control displacements is that these must not reach values that would imply a loss of functionality or a change in the structural behavior. The best tool to obtain this kind of check is to consider the displacements

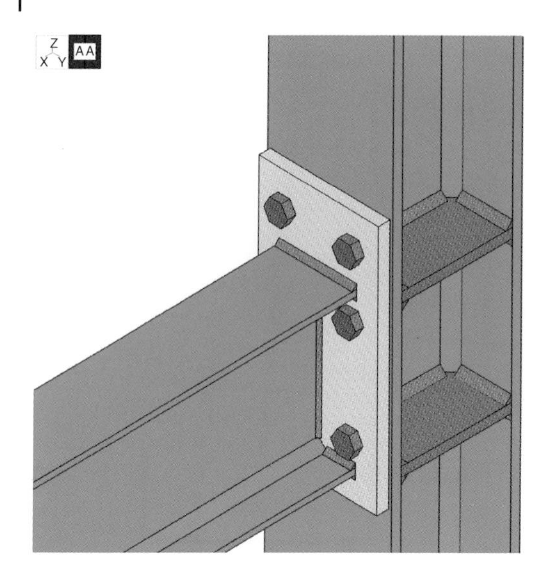

Figure 8.35 An example of a badly conceived renode: the bolts have all been declared shear only, but if the shear is delivered at the axis of the column, the bolts cannot be shear only. If the analysis is run, the beam will macroscopically rotate due to the lack of bending stiffness of the bolt layout.

evaluated by means of finite element models. Often high displacements are related to intrinsic limits of the models set up.

Displacement control is also needed to indirectly tackle the buckling problem. High displacements may be related to unwanted increases in the stresses, which by definition a first order analysis would not be able to take into account.

Finally, displacements are also useful to evaluate the rotational and translational stiffness, as will be seen in Chapter 10.

References

AISC 360-10, 2010, *Specifications for Structural Steel Buildings*, American Institute for Steel Construction.

Ballio, G. and Mazzolani, F.M. 1983, *Theory and Design of Steel Structures*, Chapman and Hall.

Bruneau, M., Uang, C.M. and Sabelli, S. 2011, *Ductile Design of Steel Structures*, McGraw Hill.

Driver, R.G., Grondin, G.Y. and Kulak, G.L. 2004, *A Unified Approach to Design for Block Shear*, Connections in Steel Structures V, Amsterdam, June 3–4 2004.

Eurocode 3 Part 1.5, 2006, *Design of Steel Structures – Plated Structural Elements*, EN 1993.1.8:2005, CEN.

Eurocode 3 Part 1.8, 2005, *Design of Steel Structures – Design of Joints*, EN 1993.1.8:2005, CEN.

Joints in Steel Construction – Simple Joints to Eurocode 3, P358/11, 2011, The Steel Construction Institute, ("Green Book").

Joints in Steel Construction – Moment Connections, P207/95, 1995, The Steel Construction Institute, ("Green Book").

Jönsson, J. 2014, *Block Failure in Connections Including Effects of Eccentric Loads*, in Proceedings of the 7th European Conference on Steel and Composite Structures, Eurosteel 2014, Naples.

Jönsson, J. 2015, *Generalized Block Failure*, in Proceedings of the Nordic Steel Construction Conference 2015, Tampere, Finland, 23–25 September 2015.

Kuhlmann, U., Wald, F. et al. 2014, *Design of Steel-to-Concrete Joints, Design Manual*, ECCS.

Kulak, G.L. and Grondin, G.Y. 2001, Block Shear Failure in Steel Members – A Review of Design Practice, Engineering Journal, **38**, 4, American Institute of Steel Construction.

Rugarli, P. 1998, *Proprietà Flessionali Elastiche e Plastiche. Calcolo Automatico di Sezioni Generiche*, Costruzioni Metalliche, **4**, 1998.

Schneider, P.J. and Eberly, D.H. 2003, *Geometric Tools for Computer Graphics*, Morgan Kaufmann Publishers.

Teh, L.H. and Clements, D.D.A. 2012, *Block Shear Capacity of Bolted Connections in Cold-Reduced Steel Sheets*, Journal of Structural Engineering, **138** (4), 459–467.

9

Analysis: Hybrid Approach

9.1 Introduction

Now that all the main single aspects of connection analysis have been properly discussed, it is possible to explain how the comprehensive goal of generic connections analysis can be accomplished by using general approaches.

Two main approaches are possible: the hybrid approach and the pure-fem approach (PFEM). This chapter deals with hybrid approach while the next chapter deals with PFEM approach.

The hybrid approach uses a simple initial renode finite element model (IRFEM) to compute the force packets flowing into the connectors, and then to check them. Then, it uses several possible approaches in order to check the other constituents, including the automatic FEM modeling of constituents. The term "hybrid" refers to the fact that there is no unique computational model but more models that are related one another by the third law and that all depend somehow on the IRFEM model, which does not use plate–shell elements. The hybrid approach is completely general and under the shelter of all the pillars of connection analysis (equilibrium, action and reaction principle, and static theorem of limit analysis) but for hyperconnected connections less precise than pure fem, albeit faster and much easier to implement.

Pure fem is used when *all* the connectors of the renode are explicitly modeled by suitable finite elements, so that both the external and internal indeterminacies are solved by the analysis itself. No assumption is made about the organization of displacements in any connector. If even one connector of the renode is replaced by the forces it exchanges with the connected constituents, as evaluated by IRFEM, the approach is named hybrid. So, the hybrid approach is also used when more constituents are modeled explicitly in the same finite element model and some connectors are modeled by specific elements, albeit these models have features in common with PFEM. In fact, at least a part of the applied forces will come from connectors not explicitly modeled inside the model.

Pure fem (Figure 9.1) is a "brute-force" approach and has pros and cons. Some failure modes are still very difficult to check by PFEM. Moreover, the computational effort in the non-linear range may sometimes be relevant. It may be considered the near future of connection analysis but it's already being used in practice.

Steel Connection Analysis, First Edition. Paolo Rugarli.
© 2018 John Wiley & Sons Ltd. Published 2018 by John Wiley & Sons Ltd.

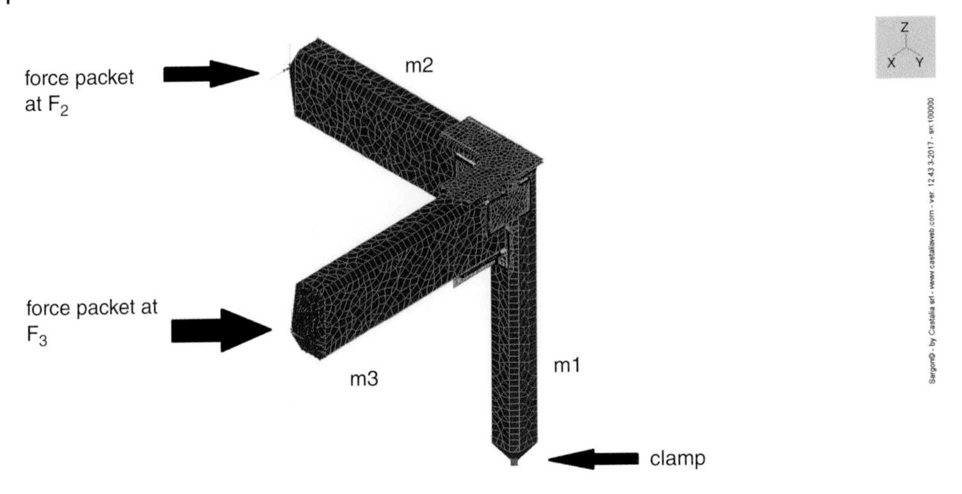

Figure 9.1 A pure fem model of a renode. Modification of an existing connection by adding reinforcements.(Courtesy Jason McCool, Robbins Engineering, Cabot, AR, USA.)

9.2 Some Basic Reminders About FEM Analysis of Plated-Structures

9.2.1 FEM Analysis as an Engineering Tool

Finite element analyses have for a long time been considered a specialist's tool, out of reach of normal practitioners.

Since the mid 1980s, however, the availability of software and hardware at low cost has made the use of finite element software widespread. Finite element models using beam and truss elements have become a regular tool in engineering companies. The use of the plate–shell, membrane and solid elements, however, is still not as diffused as it should, probably because the underlying theory is much more complex than that of beams and trusses.

The widespread use of computer software has also triggered a number of difficult problems that are for the most part still to be faced; often the software is used with no real understanding of the procedures involved, and incorrect models are considered good as long as attractive color maps deliver some meaning (Rugarli 2010, Rugarli 2014).

At the current stage of development, there is a very important issue almost completely uncovered – the use of computer software without the necessary knowledge of the theoretical basis behind it. However, this is not specific to connection analysis. It depends on the availability of cheap software, often advertised as a miracle-potion (Rugarli 2014), and on an unwillingness to introduce regulations. If the computer model is wrong it is very complex to find the errors. Some of them just cannot be found by the inexpert analysts because they need specific tools to be used (Rugarli 2014).

However, finite elements are nowadays a necessary tool to deal with the complexity of the analyses required by the standards and by the engineering practice. Much has to be done in order to teach at least the very basic concepts that are needed in order to use – with some hope of effectiveness – the available software. So, the good engineer or analyst dealing with structural engineering would have to take care of their knowledge, in order to work properly. This is particularly true in connection analysis, because very often the configurations needed in real engineering practice are far beyond the limits of the simple methods available, using computations by hand or by spreadsheet.

In the following sections some very brief reminders of basic concepts will be provided.

9.2.2 Linear Models

The most part of the steel constituents in a renode (not the connectors) can be successfully modeled by plate–shell elements.

These elements take into account both the membrane behavior (in plane tension, compression, and shear), and the plate behavior (out of plane bending, torque, and shears). By meshing constituents using plate–shell elements, very complex structural constituents, which have been cut, beveled, or that have holes, can be modeled in a reliable way, and so at least the geometrical description of the problem at hand can be done with a superior level of accuracy. Indeed, this is a very important result because simple formulations always rely on simple geometries, which often are not the ones really constructed.

There are both thin-plate and thick-plate elements, and the latter are also able to assess the out-of-plane shears. Simple elements having three or four nodes are most frequently used, and require meshes sufficiently refined. There are a number of finite element formulations, nowadays, which are the result of extensive studies carried on in the past few decades.

If the analysis is linear – that is, the material non-linearity, the geometrical non-linearity, and the contact non-linearity are not considered – the solution is fast, with no convergence issues for properly meshed models of properly connected renodes. Stress peaks well beyond the yield stress may appear, but they do not necessarily imply that the constituent is over stressed: in the plastic range those peaks would be redistributed, and in many cases the plastic region is very limited.

The Von Mises stress computed at the inner and outer surface and at mid-plane is the stress index usually considered, and compared to the (factored) yield stress. The plate–shell elements provide the membrane and plate generalized stresses, which, in the linear range, allow an evaluation of the stresses through the complete thickness. The stress state can be obtained by linear superposition of the membrane part and of the plate part – just as in a beam cross-section the normal stress is obtained by summing the effects of the axial force and of the bending moments.

Plotting the Von Mises stress map (Figure 9.2) a clear idea of the stress field can be obtained by – and this map must be judged by – a competent professional. Professional judgment must be used in order to decide if the stress peaks beyond the yield stress are so limited as to be neglected, or if they have an extension sufficient to consider the constituent over-stressed, or if a deeper (non-linear) analysis is suggested. In order to do that, color bands can be used where a specific color is assigned to Von Mises stresses higher than (possibly factored) yield. These maps will then allow an evaluation of the extent of the regions where the yield condition is crossed. The extent of the true plastic regions will obviously be greater, as the deformation energy will have to be redistributed to the surrounding still predicted elastic regions. However, there are cases where the peaks are so abrupt and narrow, and the predicted plastic regions so small, that plastic redistribution can be safely assumed able to absorb the peaks. However, this always has to be done with well designed details avoiding sharp corners and stress raisers, which might trigger fractures.

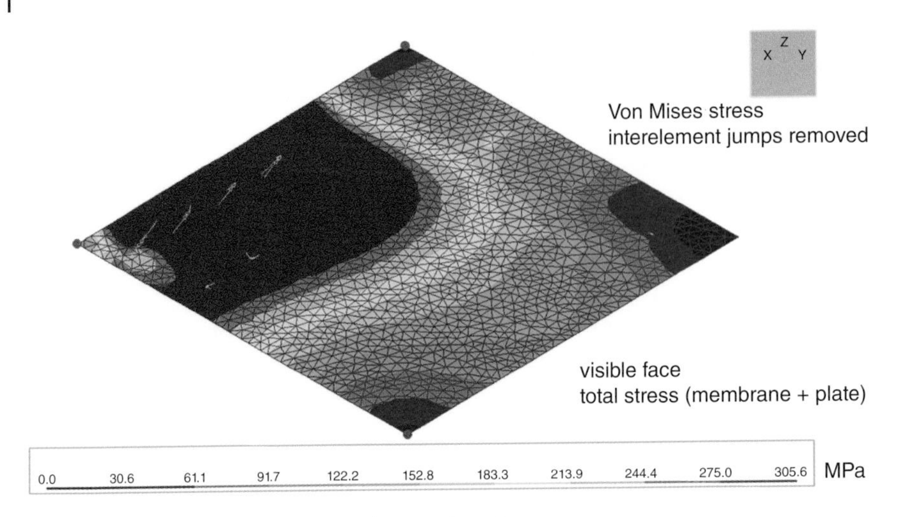

Figure 9.2 A Von Mises stress map of a constituent.

The stress state evaluation should be done for all the combinations to be checked. So it may be useful to consider the envelope of the Von Mises stress, obtained by considering the maximum value of Von Mises stress at every Gauss point, that is, the points where stresses are evaluated, considering all the combinations. This gives a quick snapshot of the health of the constituent.

Linear analysis should always be the first to be run. It quickly gives a huge amount of information, which can be very useful for the analysis in order to evaluate the health of the constituents, and the effectiveness of the model.

9.2.3 Linear Buckling Analysis

As mentioned in Chapter 5, the availability of the critical multiplier α_{cr} can be very useful to check a constituent, for every given combination.

There are two ways to compute such multipliers.

The first (Figure 9.3) uses the stress state evaluated with $\alpha = 1$, that is, considering the applied loads, and it assumes that the stresses are linearly variable with α. In the non-linear range, this is not generally true, but can be considered a useful approximation. If so, the critical multipliers for a given combination are found by solving the following eigenvalue problem:

$$(\mathbf{K_E} + \alpha \mathbf{K_G}) \cdot \mathbf{u} = 0$$

where:

- $\mathbf{K_E}$ is the elastic stiffness matrix of the constituent (or of the set of constituents) at hand
- $\mathbf{K_G}$ is the geometrical stiffness matrix (or stress matrix) computed with $\alpha = 1$
- α is an unknown multiplier
- $\mathbf{u}$ is the nodal displacement vector of the "structure" at hand; it may be a single constituent, a set of constituents, or the whole renode.

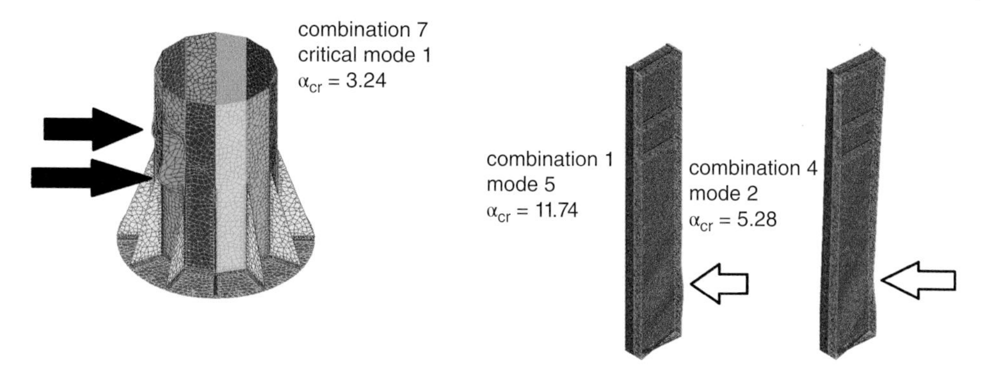

Figure 9.3 Buckling in a PFEM model (left) and in two single constituent plus stiffeners models (right).

The second method is to use a true non-linear step-by-step analysis, including the geometrical effects, trying to reach a no-convergence condition. This will be described in more detail in Section 9.2.5.

Considering the eigenvalue problem, what basically happens is that for compressed parts the geometrical stiffness $\mathbf{K_G}$ is negative, and so it diminishes the elastic matrix $\mathbf{K_E}$, up to a point where the determinant of the resulting matrix is null. If the order of the matrix is n, that is, there are n unknowns, n nodal displacements, there are n solutions to the polynomial equation imposing a null determinant to the square matrix

$$(\mathbf{K_E} + \alpha \mathbf{K_G})$$

Each of these solutions is a critical multiplier α_{cr} and it is an eigenvalue of the problem. The eigenvector is the critical shape, that is, the critical deformation of the structure arrived at buckling (Figure 9.3).

Usually the lowest positive eigenvector, that is, the lowest positive load multiplier $\alpha_{cr,min}$, is considered. This is the factor that, when applied to the loads, takes the structure to a buckling condition in the elastic range. However, for the reasons that will be better explained in the coming sections, sometimes some critical mode is not considered meaningful, and the next ones are examined and possibly considered.

All the first eigenvalues and eigenvectors are of great importance for the analysis of the constituent, and therefore should be carefully analyzed.

If the constituent, set of constituents, or renode is loaded by tens or hundreds of combinations, all the relevant critical multipliers must be considered. This means that it is important to consider all the applied combinations and not just some notional one, in general, because both the eigenvalue and the eigenvector depend on the *distribution* of the loads applied. A different load distribution may trigger some specific buckling condition that can remain unseen in other loading combinations.

As the number of combinations can be high, plotting the critical multiplier a_{cr} as a function of the combination number i, applying the general method of Eurocode 3 to all the available combinations (see Chapter 5), and plotting the final α_{Ri} obtained by for every combination i, can be very useful.

9.2.4 Material Non-Linearity

If the constitutive law used for the steel is non-linear, the superposition principle valid in the linear range is not valid anymore. This means that the solution is path-dependent, that is, it depends on how the loads are applied.

Usually the loads are grouped in load cases, considering that all the single loads embedded into a load case have a common physical origin. Then, in the linear range, linear combinations of the load cases are added. The superposition principle allows a run-time linear combination of the effects of single load cases.

In the non-linear range, several load paths should ideally be tested, adding the load cases one after another, and changing the sequence of the loads applied. However, this complex procedure is usually avoided by assuming that all the loads belonging to the different load cases are grouped together, and applied by means of a unique load multiplier α that increases from 0 to 1 (e.g. Eurocode 3 Part 1.5 clause C.7), and possibly getting into the plastic range (Figure 9.4).

The behavior of plates in the non-linear range is usually modeled by using a layer approach. The thickness of the plate is divided into a number of layers (Figure 9.5), where the problem solved is a plane stress one. The bending moments are computed by summing the effects of each layer and it is no longer true that the stresses in the thickness of the plate can be obtained by summing the effects of membrane and flexural behavior. The stresses at the inner and outer surfaces must then be computed and stored, as well as the stresses at mid thickness. To compute the needed integrals in finite element analyses, numerical integration techniques are used (for an introduction to these problems, see e.g. Rugarli 2010), transforming an integral sum into a simple sum of terms obtained by evaluating the integrand function in a number of *integration points*, and weighting the result by suitable weights. If material non-linearity is activated, the numerical integration needed to compute the stiffness matrix is no longer on the surface only (where the so-called *Gauss points* are used, see Figure 9.5), but also in the thickness. To this aim, a

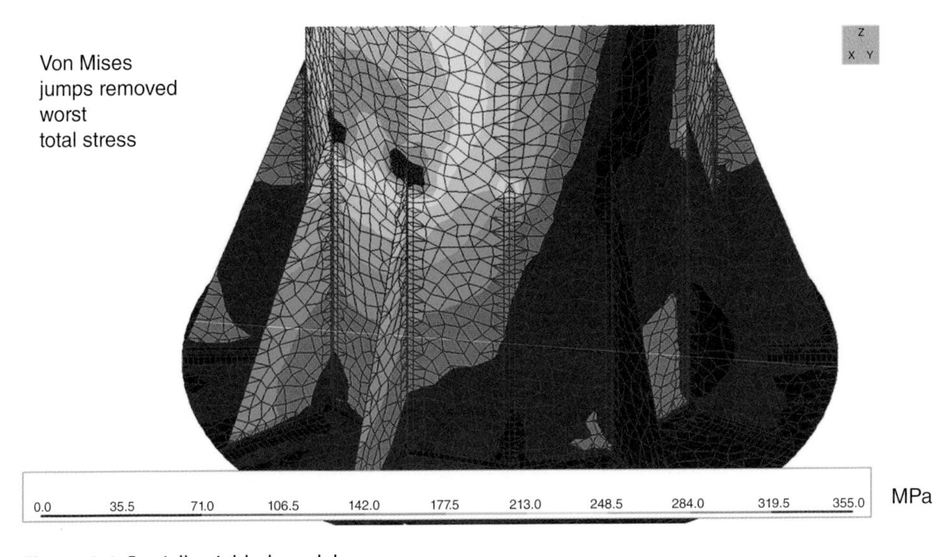

Figure 9.4 Partially yielded model.

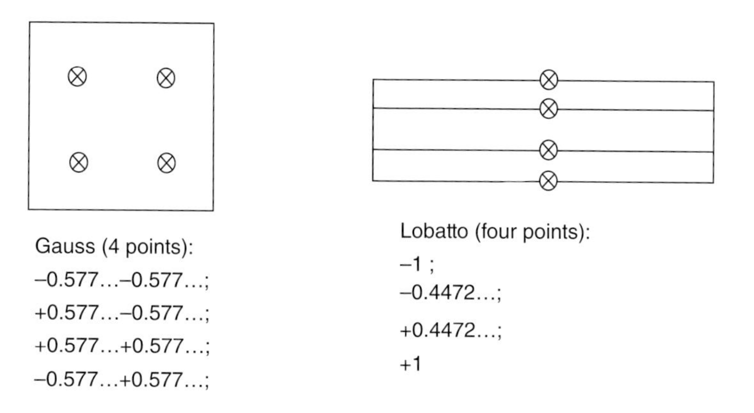

Figure 9.5 Integration points for surface and thickness.

Lobatto's integration rule is used, allowing the computation of the stresses at the extremity of the thickness domain, that is, at the inner and outer surface of the plate element (Figure 9.5).

There are many constitutive models applicable to steel, especially when the cyclic hysteretic behavior is considered. For monotonically increasing loadings, the stress–strain law can be (see Figure 9.6):

1) elastic perfectly plastic
2) hardening.

In elastic perfectly plastic model, once the yield stress is reached, the Young's modulus *of the particle* suddenly drops to zero. The stresses will then flow to the surrounding particles, that will be overloaded possibly propagating the plasticization and plastic flow effects. At the structure level, this implies that the loads cannot be increased indefinitely. There will be a load multiplier α_L that will imply the creation of a plastic mechanism, so that the structure is no longer able to resist the increasing loads. This is called the limit load multiplier, and the load level is the limit load. Limit load depends on the load

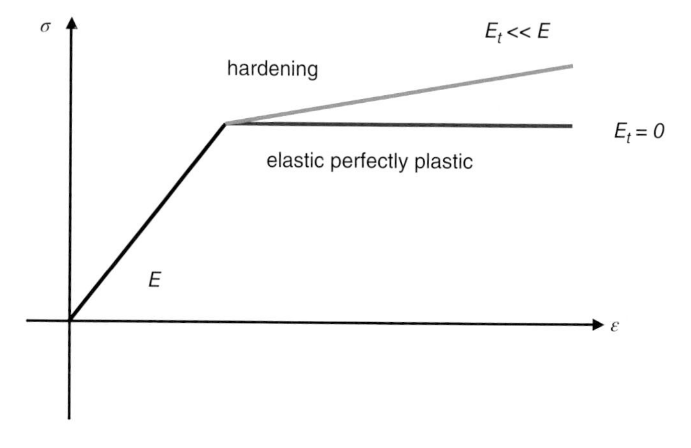

Figure 9.6 Uniaxial constitutive laws.

distribution, as well as the critical multiplier, so there will be a different limit load multiplier for every loading combination.

In non-linear analysis, if the limit load is reached during the analysis, usually this is signaled by a lack of convergence. Increased loads just cannot be applied, and if no special numerical formulation is used, the convergence – implying the satisfaction of equilibrium – cannot be reached. Special formulations like arc-length or displacement control (instead of load control), can be used in order to cross the so-called singular points. However, usually what is really needed is the limit load level, and this can be signaled by a simple lack of convergence. As lack of convergence may also be due to numerical issues, the analyst will carefully examine the displacements and stresses of the last converged step, in order to understand the physical meaning of the plastic mechanism, if any. Plastic mechanisms are announced by relevant displacements.

If the convergence is reached, despite the existence of plasticization, the load applied is below the limit load. Usually, in limit states design, ultimate limit states allow plasticization, provided that the loads can fully be applied: this means that it is not required to stay below yield, but it is required that the structure can still absorb more loads, albeit perhaps of very small amplitude.

It must be underlined that the elastic perfectly plastic model embeds a limitation on the strain. If strains are above a given limit, the particle undergoes rupture, a condition which can be simulated in numerical analysis by neglecting the contribution of a Gauss point to the stiffness matrix and to the computation of internal forces balancing the externally applied ones.

So also, if the load applied is below the limit one, a check on the strain must be accomplished, possibly embedding in the constitutive law of the single material grain a limit on the strain. Eurocode 3 Part 1.5 referring to plated structures, suggests a limit of 5% for the principal plastic strain (Eurocode 3 Part 1.5, Clause C.8).

If a hardening model is used, the stresses after yield may be increased up to a value which is related to the ultimate condition or rupture (e.g. for a S355 steel, the Von Mises stress reaches 510 MPa). So, using the hardening law approach, if no rupture condition is considered by the solver during the analysis, the loads can ideally be increased to any level, and the analyst will have to check that nowhere in the model has the ultimate Von Mises stress been reached. If instead the solver neglects the contributions of the particles that have reached rupture, then a lack of convergence will possibly be detected. Modeling fracture propagation, however, requires specialized finite element approaches and the introduction of additional nodes during the analysis. It is not presently an approach really needed, as fracture must itself be considered an unwanted condition for all the connections in real-world engineering.

Material non-linear analysis can be very useful to remove doubts about the ability of constituents analyzed in elastic range to carry the loads applied. Local plasticization does not usually imply convergence issues, if the loads applied are far from the limit ones. So it is the author's opinion that a material non-linear analysis using perfectly plastic constitutive law should become familiar to structural engineers, who should use it as an everyday tool.

9.2.5 Geometrical Non-Linearity

The so-called geometrical non-linearity arises when tensile and compressive forces cause additional stresses due to the displacement of the constituents: this is what is called P-Δ effect, and it may arise, in structures, at different scales.

To correctly take into account the problems related to geometrical non-linearity, the equilibrium should be stated with reference to the deformed configuration. As the deformation continuously changes during the loading, the problem is highly non-linear.

The additional stresses caused by the additional bending moments may trigger a failure in the constituents that is not otherwise predicted by other types of analyses, linear or maybe also non-linear, but only due to material constitutive law.

Several possible analytical approaches are available for considering the geometrical non-linearity. When the displacements are small, it is still possible to use the initial configuration as reference for the stiffness matrix of the system, provided that the so-called "second order" terms are added to the strain energy: this implies the addition of a new matrix, called the geometrical stiffness matrix. Other approaches use the so-called corotational approach in order to split the displacement of the bodies into a rigid rotation and a deformation, and also possibly large strains measures.

If displacements are considered relatively small, the incremental problem can be written in the following forms, elastic:

$$[\mathbf{K}_E + \mathbf{K}_G(\mathbf{u})]\cdot\boldsymbol{\delta}\mathbf{u} = \boldsymbol{\delta}\mathbf{p}$$

or elastic-plastic:

$$[\mathbf{K}_{EP}(\mathbf{u}) + \mathbf{K}_G(\mathbf{u})]\cdot\boldsymbol{\delta}\mathbf{u} = \boldsymbol{\delta}\mathbf{p}$$

where:

- $\mathbf{K}_E$ is the elastic stiffness matrix related to the original configuration
- $\mathbf{K}_G$ is the geometrical stiffness matrix, which is a function of displacements $\mathbf{u}$
- $\delta\mathbf{u}$ is the increment of nodal displacements
- $\delta\mathbf{p}$ is the increment of nodal forces, caused by the increment of the load parameter α
- $\mathbf{K}_{EP}$ is the elastic-plastic stiffness matrix, taking into account the spreading of plasticity in the system.

By increasing the loads, it may happen that the comprehensive stiffness matrix (in square brackets) becomes singular. This condition signals that a critical point has been reached. If the post critical behavior is not of interest, these critical points are related to the maximum load level that can be reached without collapse.

If the elastic problem is considered, and if a linear variation hypothesis for the geometrical stiffness is used, so that assuming a reference displacement vector $\mathbf{u}_0$, related to a load level $\alpha = 1$,

$$\mathbf{K}_G(\mathbf{u}) = \alpha\mathbf{K}_G(\mathbf{u}_0) = \alpha\mathbf{K}_{G0} \tag{9.1}$$

the critical point search can be done by means of an eigenvalue analysis, as already described:

$$(\mathbf{K}_E + \alpha\mathbf{K}_{G0})\cdot\boldsymbol{\delta}\mathbf{u} = \mathbf{0}$$

The critical multipliers α_{cr} are the values of α that make the determinant of the comprehensive stiffness matrix null.

If the full applied load can be reached, the system is able to carry the loads, but a clear distinction must be drawn between a problem where only geometrical non-linearity has been considered and a problem where both material and geometrical non-linearity have been taken into account.

If only geometrical non-linearity has been considered, the solution is correct if and only if nowhere has plasticity been reached. Otherwise, the interaction between the two phenomena must be taken into account, possibly by simplified tools such as the *general method* discussed in Chapter 5. This method evaluates a *real* maximum load multiplier α_R starting from the critical multiplier α_{cr}, that is, a load multiplier obtained considering only geometrical non-linearity, and from a load multiplier α_L, that is, a load multiplier obtained by considering only material non-linearity.

If both material and geometrical non-linearity have been considered, the load multiplier can be considered realistic as long as the initial conditions are meaningful, as the initial conditions affect the geometrically non-linear response of real-world systems. This means that a huge set of imperfect geometries and residual stress configurations should in theory be considered in order to assess the "real" load multiplier. This work is precisely the work done by the buckling curves adopted by all the standards.

Running a step-by-step geometrically non-linear analysis in the elastic range may be useful to evaluate the additional stresses caused by P-Δ effects and, indirectly, to evaluate the critical multipliers a_{cr} without introducing the simplifying hypothesis of Equation 9.1. In order to trigger the P-Δ effects, the initial geometrical configuration adopted for the system can be set considering the critical modes, properly tuned by a maximum reference displacement amount. This means that imperfect geometries are used at the beginning of the analysis. The initial deformed shape is that of the first critical multiplier. If nowhere is the yield crossed, then at the end of the analysis, the system is stable. If, instead, the yield has been crossed somewhere, then a fully non-linear analysis (material non-linearity and geometrical non-linearity) should be run in order to assess stability.

9.2.6 Contact Non-Linearity

Contact non-linearity is very important in connection analysis and usually considered by means of simplified tools. Probably, this is the most frequently neglected type of non-linearity in connection analysis, and with good reason. Using simplified approaches, only very simple and regular configurations can be analyzed, indirectly considering the displacement of plates such as that of beams (see the T-stub approach, whose merit is to be able to consider the contact non-linearity issue by adopting a simple geometry that assimilates the displacements of a half-flange of T-stub to that of a clamped beam).

Apparently, only finite elements using plate–shell elements, and directly modeling the contact non-linearity, are available to take into account with due generality the contact non-linearity problem.

Running a finite element analysis with contact non-linearity, however, is more difficult than to do it without; convergence might be difficult, and so more computational time is needed.

Contact non-linearity is due to the existence of one-sided constraints, that is, constraints that exist only when the displacements have a given sign. Considering two plates in contact, if the act of motion of one of the two is toward the other, some kind of constraint will be found. If instead the act of motion is in the opposite direction, then no constraint is found and the displacement is free. So, *the constraints depend on the displacements.* That is the source of contact non-linearity.

The displacement of the plates under the effect of the forces that load them is not easy to assess in general. Depending on the amount of displacement from point to point, it is possible that limited parts of the plates enter into contact with the adjacent plates, typically in bolted connections. This triggers additional contact forces that increase the loads in the bolts, and modify significantly the loading of the plates themselves, thus leading to different stress fields.

In this section it will be briefly outlined how the contact non-linearity issue can be tackled by a finite element model suitably prepared for it (see Yastrebov 2010 and Crisfield 1995).

Generally there are two mid surfaces modeling two plates in contact that may be subparts of more complex constituents (e.g. one "plate" could be the flange of the H cross-section member). The initial distance between the plates' mid surfaces is the sum of the thicknesses of the plates, divided by two (see Figure 9.7, right), and it is named the *initial gap*.

One of the two surfaces is named the contact surface, the other the target surface. At the beginning of the analysis, the plate–shell elements belonging to one of the contact surfaces are marked, as well as the twin plate–shell elements belonging to the related target surface: the two surfaces are those of the constituents decoded "connected". In this way, when checking the contact condition the search among the elements is only reduced to those potentially in contact.

Each node of the contact surface is considered, and the nearest nodes of the target surface are found (Figure 9.8). These non-aligned nodes lie over a plane and are those of a plate–shell element of the target surface, having three or four nodes. The projection of the contact node must lie inside the target element. The distance of the contact node under examination from the target nodes plane is the current gap, d. If the current gap is higher than the initial gap, the gap is "open" and there is no exchange of contact forces; if it is negative, the gap is "closed" and the contact is active. If this gap is higher than the initial gap, then no force is exchanged. If instead the distance is lower than the initial gap, contact forces are exchanged proportionally to the difference of the current gap to the initial gap. The amount of the forces is driven by a penalty constant K, which must be sufficiently high but not too high to trigger ill conditioning of the stiffness matrix.

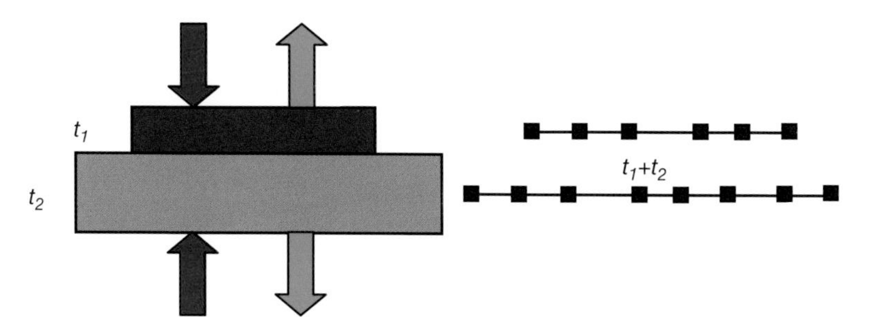

Figure 9.7 Two bolted plates and their plate–shell modeling. The limiting distance between the mid-surfaces is the sum of half the thicknesses of the two plates. It must be noted that the node positions of the two plates in general do not match.

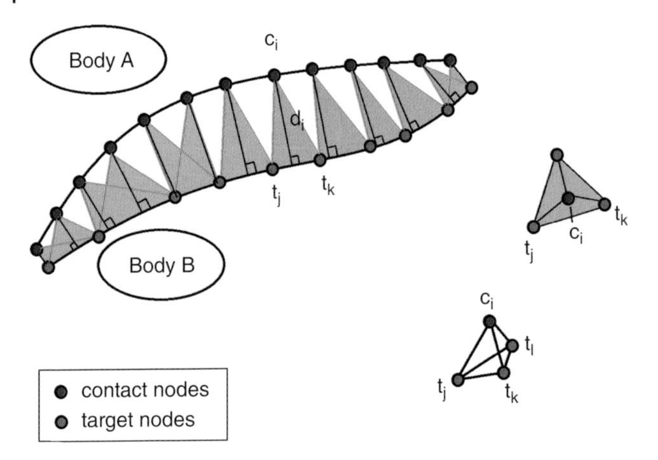

Figure 9.8 Target and contact surfaces.

Using a Boolean check is not efficient from the computational point of view, as it may trigger numerical instability and continuous switching between the two states "open" and "closed" gap. So, the contact elements are added and their "elastic modulus" is changed during the analysis as a function of the gap opening (this method is described in Crisfield 1997 and it is due to Zavarise).

$$t_N = \hat{t}_N + K\cdot(d-g) \quad \text{if } d<g$$

$$t_N = \hat{t}_N \cdot e^{\frac{K\cdot(d-g)}{\hat{t}_N}} \quad \text{if } d\geq g$$

where:

- d is the current gap
- g is the initial gap
- K is the penalty constant
- t_N is the normal force exchanged between the node of the contact surface and the three or four nodes of the target surface
- $\hat{t}_N$ is a suitable *negative* constant with the dimension of a force; this constant drives the speed of decrease of the exchanged force at gap open (see Figure 9.9).

It is also possible to define a friction coefficient, so that the tangential stiffness of the contact element can be dropped to zero if the tangential contact forces are higher than the normal forces times the friction coefficient.

By using contact non-linearity, complex interactions between the plates can be evaluated, thus leading to the analytical computation of the prying forces (e.g. see Figure 9.10). The main advantage of this analysis is that it has no geometrical limitations; it can be used in any structural configuration, leading to more realistic results.

9.2.7 Non-Linear Analysis Control

Non-linear analysis is more complex than linear analysis, and it is governed by a huge set of parameters and settings which can affect the results, so for this reason it is still an advanced tool that must be used with an especially competent professional judgment.

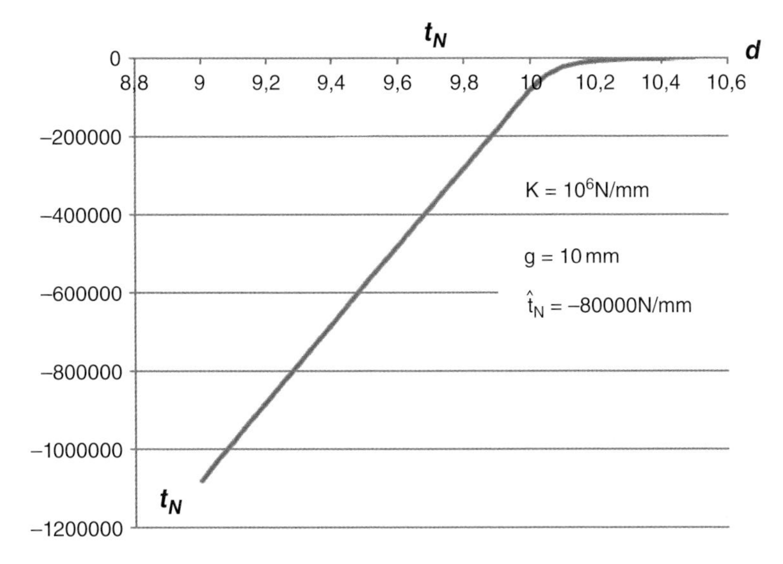

Figure 9.9 An example of the force contact force exchanged as a function of the displacement *d* (this choice is purely for display purposes).

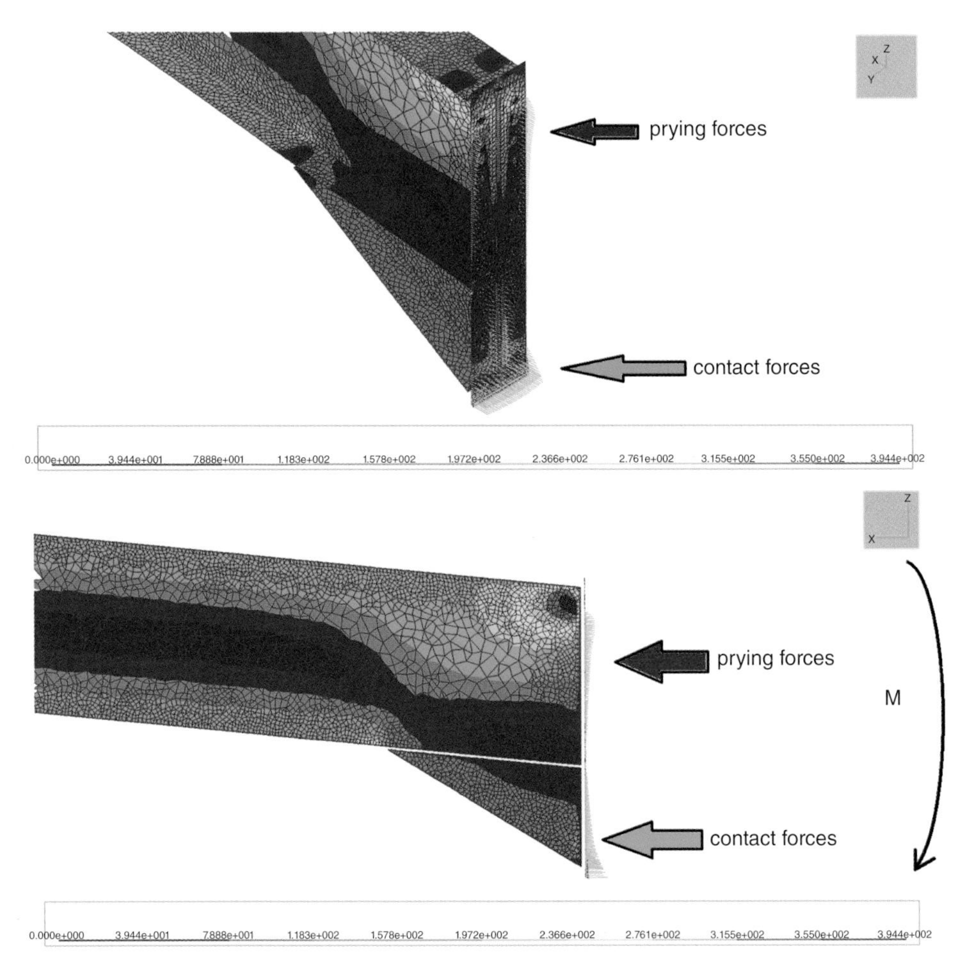

Figure 9.10 An example of PFEM solution considering also contact non-linearity.

However, it can no longer be said that non-linear analysis is only for researchers. Good structural engineers should be able to run non-linear analyses and should be aware of the meaning of the basic techniques, which would require specific studies.

At the moment, there are not many textbooks written with the aim of driving the basic ideas, hiding complex math. Non-linear analysis is explained in many excellent textbooks (Michael Anthony Crisfield's books are among the best), but usually they are complex and specialists' tools. In this section some very basic idea of the main settings used to drive a non-linear analysis are given, but the reader is invited to go deeper into the issues by consulting specific texts.

Usually the approach followed by non-linear solvers is said to be *incremental-iterative*. The load is applied by increments, and in order to apply each increment, an iterative procedure is followed (as done for the instantaneous center of rotation method in Chapter 6). At the end of the iterations, the load increment should have been fully applied, so that every node of the model is balanced under the effect of the internal and external forces. The internal forces are related to the strains, so the strain and possible plasticity status of each element will have to be evaluated in such a way that all the internal forces exerted by the elements at nodes are balanced with themselves and with the externally applied load.

Several numerical techniques are available to run non-linear analyses. The full Newton–Raphson method implies an upgrading of the stiffness matrix at each iteration. This is more precise and convergence is usually improved, but it has a computational cost related to the re-evaluation of the stiffness matrix. The modified Newton–Raphson method upgrades the stiffness matrix only at the beginning of each step. Initial stress methods use a constant stiffness matrix, created at the beginning of the analysis, which guarantees a well conditioned matrix but implies a greater number of steps and iterations.

The increments can be all equal, but this is not usually the best way to apply them; at the beginning of the analysis, far from critical points, usually the stiffness is higher, while in the vicinity of critical points the stiffness of the system is low. So the amount of the increments can be changed during the analysis, in such a way that roughly the same number of iterations is needed in order to get to convergence. This method has been described by Crisfield (1997), and it states that

$$\Delta\alpha_i = \Delta\alpha_{i-1} \cdot \left(\frac{I_d}{I_{i-1}}\right)^n$$

where:

- $\Delta\alpha_i$ is the increment of step i, and $\Delta\alpha_{i-1}$ is the increment at previous step
- I_d is the desired number of iterations
- I_{i-1} is the number of iterations needed to get to convergence at previous increment
- n is a suitable constant usually between 0.5 and 1.0.

In order to decide when the convergence has been reached, several possible error measures are available. These can be based on the displacement amount, the unbalanced forces (residuals) amount, or the work done by these unbalanced forces (residual multiplied by displacements).

What is usually compared is the norm of a vector related to the last iteration increment, and the norm of the total increment vector. The norm can be the square root of the sum of the squares, or the maximum absolute value of the vector. Usually the convergence is considered to have been reached when the ratio of the last iteration increment vector norm to the total step increment vector norm is lower than a suitable small value, say 10^{-4} or 10^{-3}. Smaller values lead to more precise results, but take more computational times. If the aforementioned ratio is higher than a preset value, such as 20 or 50, the numerical procedure can be declared divergent, and consequently the increment reduced.

In order to stop a numerical procedure that might not find an end, usually a maximum number of iteration per step can be set. So, if after say 20 or 50 iterations the convergence has not been reached, the increment is considered too high and can be reduced.

One every important computational parameter is the *current stiffness parameter*, which is used to evaluate the progressive loss of stiffness of the system. This numerical index (see once more Crisfield 1997), can be defined as

$$C_s = \frac{\dfrac{\Delta\alpha_s \mathbf{p}^{\mathrm{T}} \boldsymbol{\delta}\mathbf{u_s}}{\boldsymbol{\delta}\,\mathbf{u}_s^{\mathrm{T}} \boldsymbol{\delta}\mathbf{u_s}}}{\dfrac{\Delta\alpha_0 \mathbf{p}^{\mathrm{T}} \boldsymbol{\delta}\mathbf{u}_0}{\boldsymbol{\delta}\,\mathbf{u}_0^{\mathrm{T}} \boldsymbol{\delta}\mathbf{u}_0}}$$

where:

- $\mathbf{p}$ is a fixed load vector, equal to the total load to be applied
- $\delta\mathbf{u_s}$ is the displacement increment at step or iteration s
- $\Delta\alpha_s$ is the fraction of load $\mathbf{p}$ applied at step s
- $\delta\mathbf{u_0}$ is the displacement increment at the beginning of the analysis
- $\Delta\alpha_0$ is the fraction of load $\mathbf{p}$ applied at step 0.

At the beginning of the analysis, C is equal to 1 and decreases with the progressive weakening of the system. Special techniques such as *arc length* can be used to carry on the analysis when the current stiffness parameter is lower than a preset value. This technique is used up to cross-critical points where the stiffness matrix has null determinant, and can be used in several possible ways (e.g. spherical, cylindrical, Ramm and Fried). When arc length is switched on, what is relevant is the “length” of the load–displacement curve, not the increment of the applied loads. So, if the load–displacement curve we are interested in must also cross the critical points, arc length must be activated, possibly when the current stiffness parameter is lower than a suitable constant such as 0.2 or 0.3.

The evaluation of the stiffness matrix of the elements requires a numerical integration that is carried on using Gauss points or Lobatto points (for the integration along the thickness of plate–shell elements). A larger number of integration points usually means a more precise evaluation of the stiffness matrix, but it has a computational cost. For surface integrals and four-noded elements usually a 2×2 Gauss points integration scheme is used. For the integration in the thickness 4 to 10 Lobatto integration points are usually adopted. Setting more Lobatto points implies more precise results, but more computational time.

Within the incremental-iterative numerical procedure, some possible subproblems that may need an iterative solution can be found. One typical example is when the elastic perfectly plastic or hardening condition must be satisfied by a grain of material (an integration point) under the effect of a 3D stress state. The initially predicted stress increment usually violates the plastic condition, and so a specific numerical procedure, iterative, must be started in order to properly assess the increment of strains and stresses related to a given nodal displacement increment. Once more, several methods are available, such as the so-called radial return, that is, the return of the violating stress state over the plastic limit domain surface. Possible specific settings related to these procedures are the maximum number of iterations needed to return to the plastic locus, and the tolerance to be used to consider the plastic condition satisfied. When a part gets into the plastic range, the problem of returning to the plastic domain is important and is the main responsibility of the increment of the computational time related to material non-linearity.

9.3 IRFEM

9.3.1 Goal

Given a renode with all its members, force transferrers, stiffeners, possible constraint block, and connectors, it is assumed that the forces applied at member extremities are known by the solution of a related BFEM, as explained in Chapters 2 and 5. The force packets applied at each member k near-extremity $\mathbf{s_k}$ are by definition balanced in the BFEM node related to the jnode under examination.

The set of all the bodies which constitute the renode will have to be balanced under unknown force packets, and these force packets will have to satisfy equilibrium and the third law, i.e. the action reaction principle.

In order to analyze all the connections that the renode is made of, the force packets flowing into each connector i at extremity j, $\mathbf{c_{ij}}$, are needed (see Section 5.3.3). These force packets are vectors of three forces and three moments applied at each extremity of each connector, as defined in Chapters 6 and 7.

The goal of the initial renode finite element model (IRFEM) is to compute a set of force packets flowing into the connector's extremities which are balanced to the externally applied loads (i.e. the force packets at member extremities), and that make each constituent of the renode balanced.

IRFEM is a simplified finite element model, as it uses some important simplifying hypotheses, but it is able to allow the evaluation of the force packets applied to each constituent, regardless of the complexity of the scene of the renode, the number of members, constituents, connectors, or their layout, position and orientation in space. This means that IRFEM is able to tackle every renode configuration.

The results obtained by IRFEM are fully compliant with all the basic pillars of connection analysis:

1) equilibrium
2) action and reaction principle (third law)
3) static theorem of limit analysis

The main advantage of using IRFEM is that it is easy to be set up, it's general, and it allows the analysis of every possible renode, no matter what its geometrical complexity is.

IRFEM can be seen as the dual of BFEM. While in BFEM the members are explicitly modeled as deformable and the connectors often modeled assuming them rigid, in IRFEM the connectors are deformable and the members stiff. All the basic hypotheses needed to set up and use IRFEM are already normally used by engineers (see Chapters 6 and 7) and are deeply rooted in engineering practice. So IRFEM is just the rational extension and generalization of procedures that are already locally used.

The most important practical result of IRFEM is that it relieves the analyst of a sea of complex and cumbersome computations, which are often willingly avoided, very frequently leading to either the systematic neglect of important parts of the structural response, or its incorrect evaluation. IRFEM is a general tool, applicable to all the situations and that can immediately lead to a good part of the checks with no need for further finite element models. Its simplicity lets it also be developed by adapting software dealing only with beam and trusses.

Although it cannot be considered the most advanced method of analysis, because this role is played by PFEM, IRFEM's simplicity and speed of use and implementation makes it the natural candidate for everyday analyses of non-standard connections, as the results it delivers are totally rooted in computational mechanics, and no ad hoc trick is needed to use it in general.

9.3.2 Hypotheses

IRFEM can be constructed if some additional hypotheses are added to the pillars of connection analysis (which are laws and not hypotheses).

1) Members are considered rigid and are cut at their far extremities at a given distance from the node where the members in BFEM meet. This isolates the nodal region from the rest of the world (see Chapter 5).
2) Stiffeners and force transferrers are considered rigid, as well as the constraint block.
3) Weld layouts and bolt layouts are considered deformable and the displacements of subconnectors are considered organized, as discussed in Chapters 6 and 7.

Considering the constituents rigid, with the exception of the connectors, is a simplifying hypothesis which is not far from reality for many connections, where most of the deformability is actually confined to the bolts (especially due to bolt-bearing) or the welds (especially fillet welds).

Moreover, as explained in Chapter 5, and frequently recalled in Chapters 6 and 7, for the very much diffused and frequent isoconnected connections, *there is no error in force-packets computation when adopting this assumption*. The force packets flowing into constituents can be computed only on the basis of equilibrium. The IRFEM model is indeed a tool able to make such computation with all the needed generalizations.

For hyperconnected connections, where more than one connector is used to transfer the forces, or where more than one constituent in parallel shares some loads, each distribution of force packets is under the shelter of the static theorem of limit analysis, and better distribution of force packets can, if needed, be obtained by proper modifications of the flexibility indices of connectors. Here the force packet distribution is a function of the relative stiffness of connectors, and clearly with the flexibility of connectors must also be considered the flexibility of the surrounding parts. So, for instance, a connector may have a high flexibility index if it is attached to a flexible plate, loaded out of plane.

Also the working mode flags of single subconnectors can indirectly be used to finely tune the forces according to some preferred distribution.

Hypothesis 3, allows us to consider, as primary unknowns of the problem, the force packets: the three forces and moments transferred at each connector extremity. It is the existence of some organization of the displacements that allows the computation of the forces flowing into the subconnectors starting from the force packets applied to the layouts. This is, in principle, exactly like computing the stresses in "fibers" in a beam cross-section, once the bending moment and axial force are known.

IRFEM is the counterpart of BFEM and has deep analogies with it. It is commonly accepted practice to compute the member forces using elastic models, considering connections and connecting regions rigid, and then to check members using plastic method. In the same way it is possible to compute the force packets flowing into the connectors and check them using plastic and not elastic methods. Also, special elements considering the non-linear response of the connectors and their surrounding parts can be set up, thus leading to more precise but always simplified models.

The rational path to be followed once the force packets $\mathbf{c}_{ij}$ flowing in the connectors are known is the following:

1) Use the organization of displacements to evaluate the forces flowing into single subconnectors. For bolt layout using a bearing surface, evaluate the pressures exchanged at the interface. This was explained in Chapters 6 and 7.
2) Check the connectors using these locally computed forces or stresses (see Chapter 8).
3) Applying the third law directly, all the forces loading the constituents are immediately known. They are the forces stressing the subconnectors, with sign reversed. These forces are spread over the constituent, as they are applied at the subconnector level. *Force transferrers are self-balanced under these applied loads.* Members are self balanced under these applied loads *and* the force packets applied at the far extremities (Section 5.3.3 and Figure 5.13).
4) Using these forces, directly check the *local* failure modes, such as bolt bearing, punching shear, and block tear (Chapter 8).
5) Using these forces, check the constituents for the *global* failure modes, i.e. resistance and buckling (as explained in Chapter 8).
6) Using the displacements obtained by the IRFEM, evaluate the displacements of the constituents, checking for possible clear excesses that are usually the indicators of some improper design.

So, the availability of IRFEM and its solution is the key that opens the door of generic connection analysis in the simplest possible way.

9.3.3 Construction

Given a generic renode the following symbols are introduced:

- n_m, the number of members joined at the jnode; if a member is a compound member, all its subconstituents must be counted (e.g. for a double channel back-to-back member, two subconstituents)
- n_c, the number of connectors (weld layouts and bolt layouts)
- n_g the number of force transferrers

- n_s the number of stiffeners
- n_x the number of constraint blocks (1 or 0)
- m_i, the multiplicity of connector i
- m_{tot} the total multiplicity of connectors, i.e. the sum of all m_i for all connectors n_c
- m_m the total number of extremities of connectors connected to members in general; this may be higher than the number of members for two reasons: (i) a member can be connected by more than one connector to the other constituents, (ii) a member might have more than one constituent (compound members, see Section 5.3.2)
- $\mathbf{E_{ij}}$, a point in space related to the extremity j of connector i
- $\mathbf{N_k}$ a point in space related to the near extremity of member k. If no rigid offset is applied to member k, the point $\mathbf{N_k}$ coincides with the position of the node of the BFEM. If the finite element modeling in the BFEM of the extremity of member k (beam or truss) has rigid offsets, the point $\mathbf{N_k}$ does not coincide with the node of the BFEM. If a member is a compound member, more than one element is needed and more than one point $\mathbf{N_k}$ is defined.
- -$\mathbf{s_k}$ the force packets applied to the near extremity of member k, $\mathbf{N_k}$. These force packets, known by definition, and changing from load combination to load combination, are defined in global reference system of renode (X, Y, Z) and are applied to member (not to node), with the sign of the internal forces reversed, i.e. they have a minus sign – see the "dual" system of Figure 5.13. They can be obtained by considering the internal forces $-\boldsymbol{\sigma}_\mathbf{k} = \{-N, -V_2, -V_3, -M_1, -M_2, -M_3\}^T$ applied at member extremity, and transforming them from the local system of the member to the global reference system of the renode ($-\mathbf{s_k} = -\mathbf{T_k}^T\boldsymbol{\sigma}_\mathbf{k}$ – see Section 5.3.1). If a member is a compound member, the technique explained in Section 5.3.2 can be used to evaluate the force packets flowing into each submember.

The IRFEM is a finite element model obtained by assembling the stiffness matrices of connectors as explained in Chapters 6 (weld *layout*, i.e. not single weld seams) and 7 (bolt *layout*, i.e. not single bolts), and also using Timoshenko rigid beam elements. The rules of its construction depend on the jnode classification.

9.3.3.1 Hierarchical Jnodes (with No Constraint Block)

These are by far the most frequent jnodes. The IRFEM is a finite model has:

1) $(m_{tot} + n_m)$ nodes
2) $(m_{tot} - 1)$ flexible elements simulating the connectors
3) m_m rigid elements connecting the extremity $\mathbf{E_{ij}}$ of each connector i connected at extremity j to a member k, to the near extremity of the member k, $\mathbf{N_k}$
4) a number of rigid beam elements connecting all the points $\mathbf{E_{ij}}$, at equal i for all j. If a force transferrer i has m_i extremities of connectors connected, the number of rigid beam elements added for that force transferrer is $(m_i - 1)$. If the extremities of the connectors connected to i are A, B, C, D, the rigid elements connect: A to B, B to C, and C to D, that is, the elements AB, BC, CD are added.

9.3.3.1.1 Nodes and Nodal Constraints

The first m_{tot} nodes are the nodes related to the extremities of the connectors. For instance, a weld layout has two extremities, and will need two nodes. A bolt layout connecting three constituents, will have three extremities and will need three nodes.

These nodes are connected to the connectors, which are flexible, and to the constituents connected to the connectors, which are assumed rigid.

The next n_m nodes are the nodes related to member extremities. In the BFEM, assuming that no rigid offset is applied to members, all these nodes are merged into one single node (the node related to each instance of a jnode). Considering the IRFEM, each member "near" extremity has its node, *independent of the others*. The independence of the nodes is a necessary precondition to loading the connectors. It is like opening a flower whose petals are initially all closed together.

All the $\mathbf{N_k}$ nodes belonging to the master member, are clamped: all six degrees of freedom are fixed. They are just one node if the master member has no subconstituents. If the subconstituents are two, two nodes will be constrained, and so on. The application of a clamp to the nodes of the master near-extremities is fundamental as it ties the IRFEM down, allowing its solution. The other $\mathbf{N_k}$ nodes of the other members, will be free to move and will be directly loaded by the force packets – $\mathbf{s_k}$. Basically, these are the loads applied to the IRFEM. Also, the constrained nodes will possibly be loaded by force packets, but as the nodes are clamped, these force packets have no effect.

All the initial m_{tot} nodes are free to move and are attached to connectors layout finite elements providing the needed stiffnesses to avoid static indeterminacy.

9.3.3.1.2 Elements Simulating the Connectors

Each connector layout i having multiplicity m_i is modeled by $(m_i - 1)$ finite elements whose stiffness matrix was discussed in Chapters 6 and 7. In IRFEM, the secant stiffness matrix is usually adopted, as the IRFEM model is linear. We can also use the initial stiffness if the loads are low. Each finite element has a local reference system which is the principal reference system of the connector ($\mathbf{w_i}$, $\mathbf{u_i}$, $\mathbf{v_i}$). The stiffness matrix of the connector in the principal (local) reference system of the layout, $\mathbf{K_i}$, a $6m_i \times 6m_i$ square matrix discussed in Chapters 6 and 7, is transformed into the global reference system of the renode by the matrix operation

$$\mathbf{T_i^T K_i T_i}$$

where $\mathbf{T_i}$ is the orientation matrix. This is a $(6m_i \times 6m_i)$ square matrix, which has a block diagonal structure made by $2m_i$ identical (3×3) blocks, $\mathbf{T_i}^*$, placed in the diagonal. Each of these blocks is obtained by collecting by row the components of the unit vectors directed like the principal axes of the connector $\mathbf{w_i}$, $\mathbf{u_i}$, and $\mathbf{v_i}$, respectively:

$$\mathbf{T_i^*} = \begin{vmatrix} w_{ix} & w_{iy} & w_{iz} \\ u_{ix} & u_{iy} & u_{iz} \\ v_{ix} & v_{iy} & v_{iz} \end{vmatrix}$$

If the connector has just two extremities, such as weld layouts or bolt layouts drilling just two thicknesses, the matrix $\mathbf{T_i}$ has the form

$$\mathbf{T_i} = \begin{vmatrix} \mathbf{T_i^*} & \mathbf{0} & \mathbf{0} & \mathbf{0} \\ \mathbf{0} & \mathbf{T_i^*} & \mathbf{0} & \mathbf{0} \\ \mathbf{0} & \mathbf{0} & \mathbf{T_i^*} & \mathbf{0} \\ \mathbf{0} & \mathbf{0} & \mathbf{0} & \mathbf{T_i^*} \end{vmatrix} \tag{9.2}$$

The elements are then assembled in the usual way (see e.g. Rugarli 2010).

Embedded in the formulation of the stiffness matrix of the connector are both the working mode factors (efficiency factor for bolts, shear only, or no-shear flags for bolts and welds, longitudinal only flag for welds) and the flexibility index which, if needed, allows the tuning of the stiffness for hyperconnected renodes. For bolts in bearing the bearing stiffness is also included.

9.3.3.1.3 Rigid Elements Connecting Members Near Extremity to Connectors

These elements are rigid and mimic the members as rigid bodies. Moreover, connecting the point of application of forces applied to the member ends to the connectors, they transfer, with all the proper matrix and vector operations, the forces applied to the flexible connectors.

The use of near member extremity is allowed because of the fact that the members are modeled as rigid bodies (see Section 5.3.1 and the "dual" model).

These rigid elements can be thought of as rigid links. If the software used does not handle rigid links, rigid beam elements may be used instead. A reasonable model, which has been proved efficient, is to assume a Young's modulus E_{stiff} for steel 10^5 times higher than the typical value $E = 210$ GPa. The cross-section is dummy. A reasonable assumption is to use as area and second area moment for the rigid Timoshenko beam, the maximum value of the connectors A_i and J_i.

As the elements are rigid, their orientation is dummy.

9.3.3.1.4 Rigid Elements Internally Connecting Force Transferrers (and Stiffeners)

These elements are defined in a way very similar to that of the previous ones. They are rigid elements and their role is to mimic the non-member constituents as rigid bodies. All the extremities of the connectors related to that very constituent are connected by these elements, so the rigid body behavior of the constituent is guaranteed.

The rigid elements can be added following the same rules already seen for the previous elements.

9.3.3.1.5 Special Problems

When creating the IRFEM some special problems or conditions might be met.

Identical nodal position – It may happen that one of the extremities of the connectors $\mathbf{E_{ij}}$, is exactly in the same point where an extremity of a member, $\mathbf{N_k}$ is placed.This would imply a zero-length rigid element. As the element is rigid, however, there is no need to use its length, and its stiffness matrix can be defined with no reference to the element length.

Zero length flexible elements, i.e. zero length connector elements, cannot be found due to how the connector elements extremities have been defined (see Chapter 4).

Compound members – Compound members are members having more than one constituent, like double angle or double channel bracings. In the BFEM they are usually modeled by a unique truss or beam element, whose cross-section is composed by more than one profile.

When dealing with the analysis of a renode, the subconstituents of the members must be kept independent, and so there will be more than one $\mathbf{N_k}$ node for them (Figure 9.11). The operations needed for a single element member will have to be repeated as many times as the number of subelements. Connectors and force transferrers must be placed to simulate battened or castellated connections. Single subconstituents of members cannot be left without connections and must be balanced as the single members are.

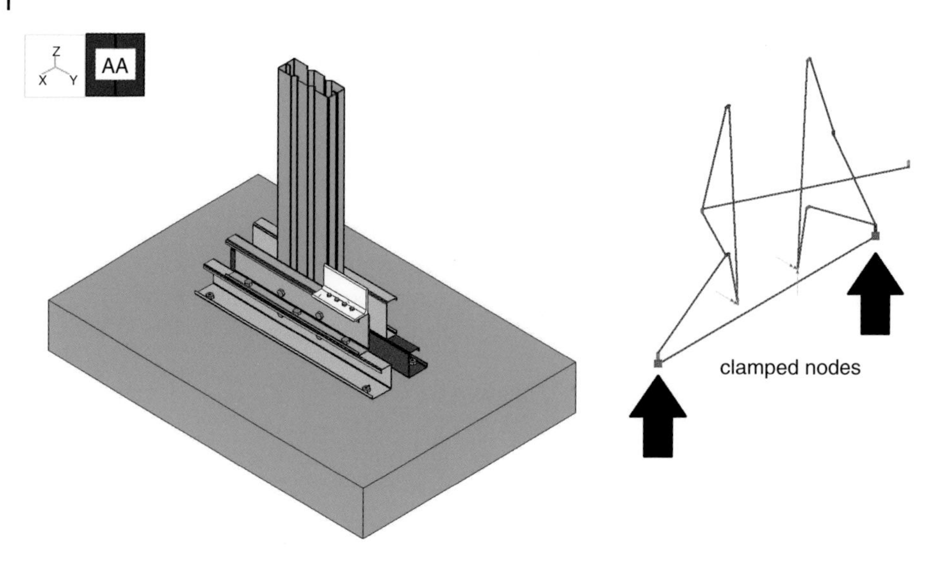

Figure 9.11 A renode with compound element and its loaded IRFEM model.

9.3.3.2 Central Jnodes (with No Constraint Block)

Central jnodes are very similar to hierarchical jnodes. The only difference is that a constituent (force transferrer) acts as master. So, the nodes to be constrained are the nodes of the connectors $\mathbf{E_{ij}}$ connected to that constituent, while no node $\mathbf{N_k}$ will have to be constrained.

9.3.3.3 Constraint Jnodes (with a Constraint Block)

If the jnode is a constraint, some modifications to the rules valid for hierarchical non-constraint jnodes must be applied.

The first modification is that the node to be constrained will not be the near extremity of the master member, $\mathbf{N_k}$, but all the nodes of the connectors connected to the constraint block, $\mathbf{E_{cj}}$ where c is the constraint block constituent number. The constraint block *must* be inserted in the scene if a constraint jnode is analyzed, so its existence must be guaranteed. Usually these connector extremities to be constrained are the second extremity of bolt layouts acting as anchors and embedded into concrete or anchored to proper devices embedded into concrete. However, ideally the constraint block might even be made of steel or other materials, and what is merely needed is that one or more connectors are connected to it.

The second modification is referred to the possible existence of shear keys (Figure 9.12). Shear keys are force transferrers applied to the bottom part of base plates by welding and that are embedded in concrete. Their role is to relieve the bolts from shears that are directly transferred to the concrete.

In the scene, the shear keys are modeled by cross-section trunks, usually fillet welded to the bottom face of base plates.

In order to be recognized as shear keys, these constituents must be properly flagged. The single finite element, which models the weld layout connecting the shear key to the plate, has two extremities, and one of these extremities, typically the second (i.e. not the

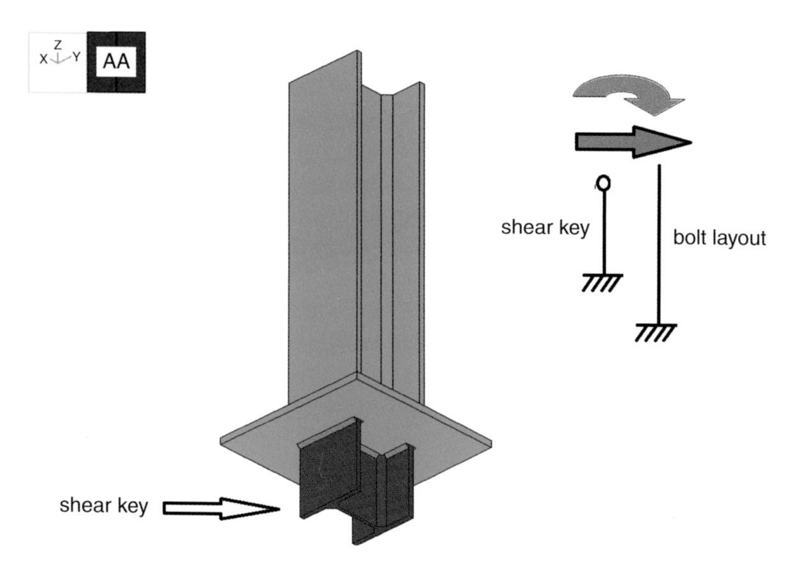

Figure 9.12 A shear key applied to a base plate bottom surface.

faying plane), is attached to the shear key. The related node must be clamped (as for the nodes of the IRFEM model attached to the constraint block), and the element extremity attached to that node, and simulating the weld layout must be released to

- translation in weld layout local **w** direction (axial for the weld layout, usually perpendicular to plate)
- rotation in weld layout local **u** and **v** direction, i.e. bending moments for the weld layout.

In this way, the beam element simulating the weld layout will only be loaded at the tip connected to the shear key by shear and torque, which is exactly the desired behavior: no bending moment but the one due to geometrical transfer will be carried.

Shear keys act in parallel to the anchoring bolts. The flow of (shear) forces flowing into the shear key is governed by the tuning of the flexibility indices of the bolt layout acting as anchor. Increasing this flexibility index, the shear will gradually flow in greater proportion to the weld layout connected to the shear key, while the bending moments will basically be carried by the anchor bolts and their bearing surface, as usually needed.

9.3.3.4 Cuspidal and Tangent Jnodes

These jnodes do not have a clear master. So there is no clear idea of which member or constituent must be constrained and considered as reference in the IRFEM. As explained in Chapter 2, it is always possible to avoid such jnodes by properly assigning the connection codes to the beam elements of the BFEM.

Even if it was possible to choose a master member by chance, it is preferred and cleaner from a logical point of view to keep these jnodes separated from the others and avoid their analysis.

9.3.4 Examples

IRFEMs have no familiar aspect (Figures 9.13 to 9.16). Their appearance is strange, because they are idealizations of the mechanical behavior of renodes, not a pictorial model of them. However, they are powerful tools for analyzing connections. At the time of writing, we have accumulated nine years of experience in using IRFEMs, and they have been proved with renodes to have very many different features, from very simple to very much complex: with many members, force transferrers, and connectors, in iso- and hyperconnected renodes.

IRFEM is a general tool, really applicable to most of the renodes and providing very important basic information, fully rooted in the basic principles of connection analysis, and of computational mechanics.

In order to better explain how the IRFEM model is constructed, the simple hierarchical renode of Figure 9.17 will be considered.

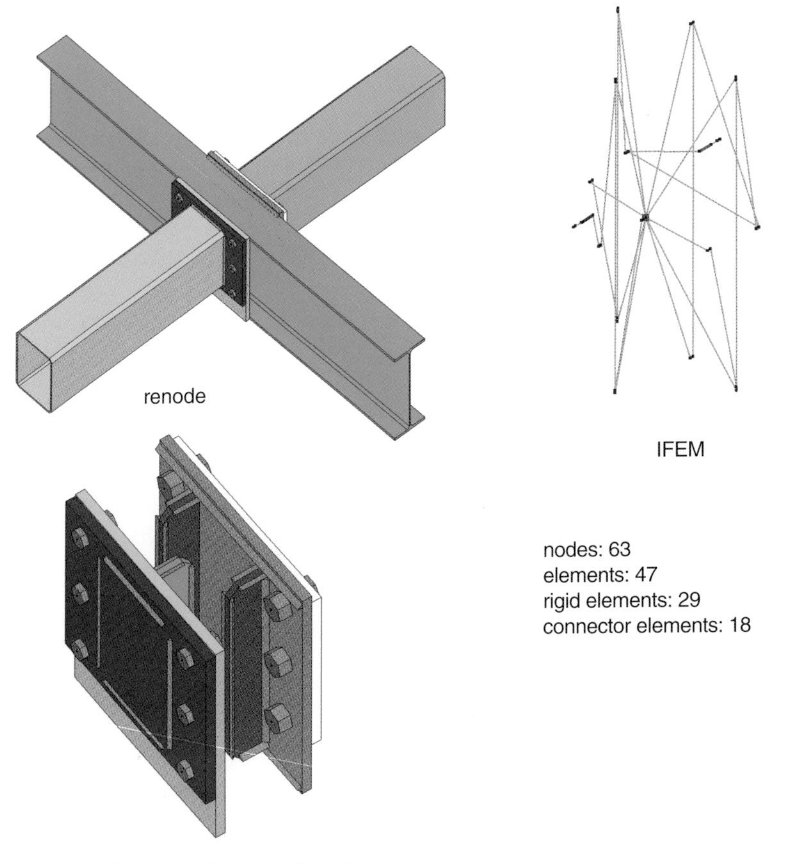

Figure 9.13 A renode and its IRFEM model (courtesy Jason McCool, Robbins Engineering, Little Rock, AR, USA).

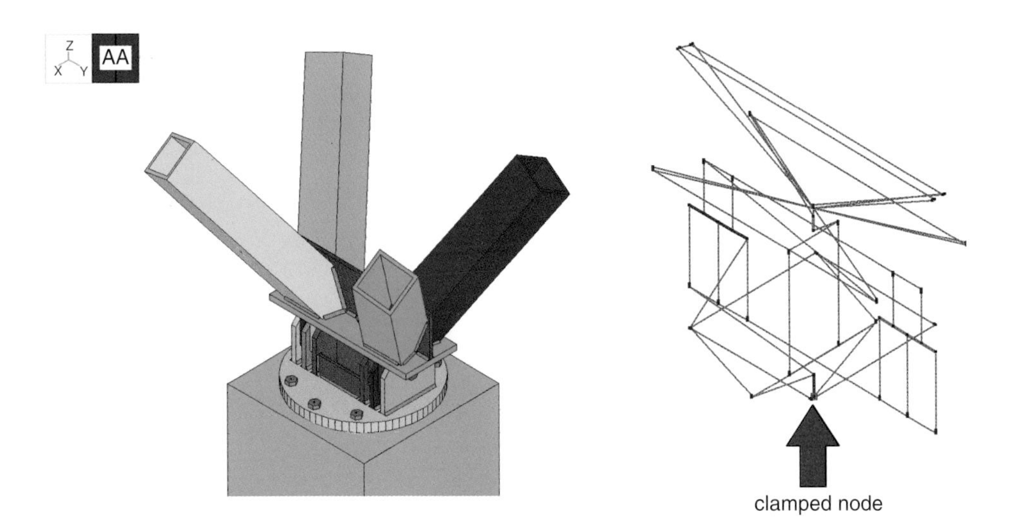

Figure 9.14 A complex, generic renode and its IRFEM (courtesy AMSIS srl, Rovato, Italy).

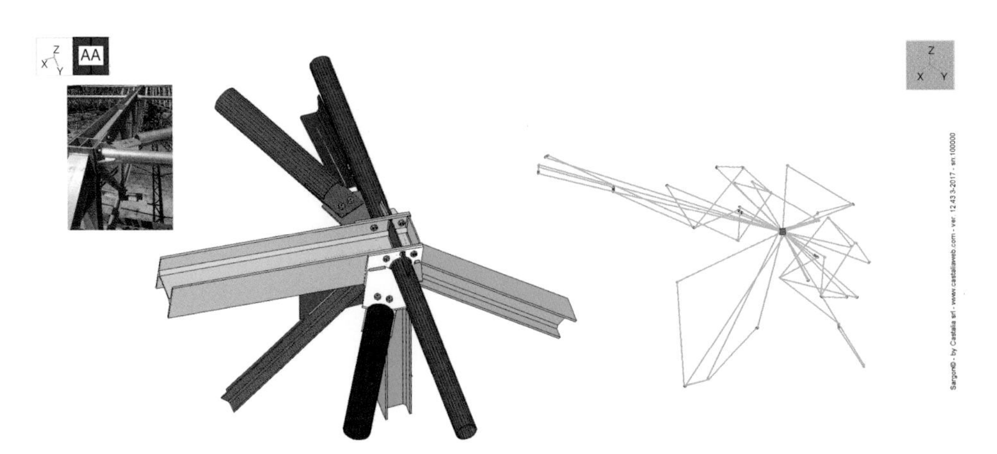

Figure 9.15 A complex generic rendode and its IRFEM model (courtesy CAVART srl, Villa di Serio, Italy). On the left the node built.

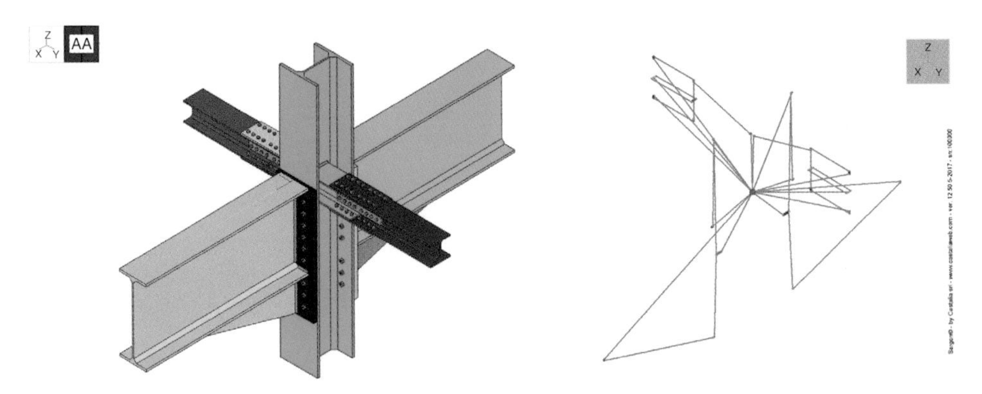

Figure 9.16 A generic renode and its IRFEM model (Courtesy L&P Technology Consulting, San Giorgio di Mantova, Italy).

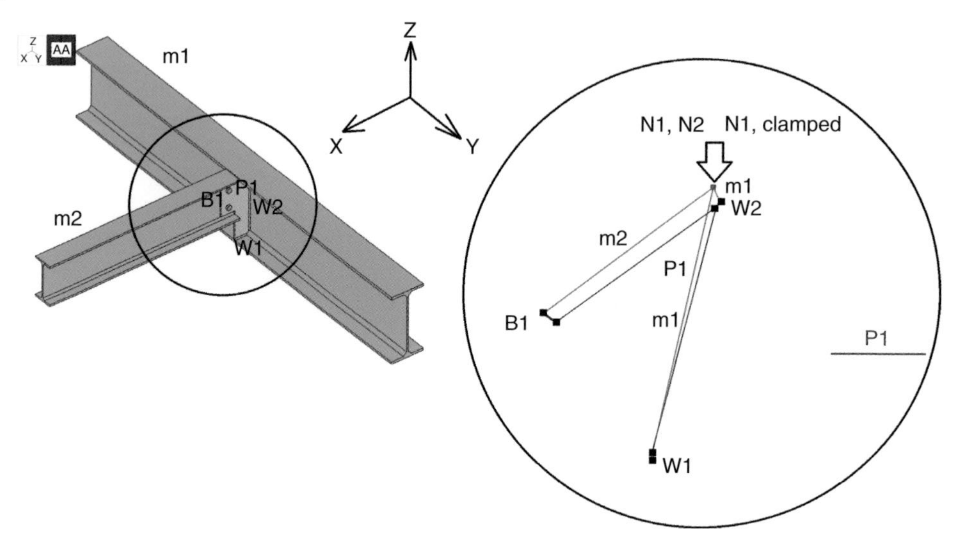

Figure 9.17 A simple beam-to-beam connection and its IRFEM model.

The renode has two members connected by a plate of thickness 7 mm. The plate is connected to the master member m1 (IPE300) by two weld layouts, each having two welds with $t = 4.9$ mm. The same plate is connected to the slave member m2 (IPE180, $t_w = 5.3$ mm) by a bolt layout. Global axis X is parallel to slave member, axis Y to the master member. The node of the BFEM is in the origin (0, 0, 0).

The IRFEM has eight nodes whose coordinates and constraints are in Table 9.1.

The elements and their meaning are listed in Table 9.2.

If the slave member is under a vertical load $V = 50$ kN, and delivers a downward force *toward* the master member, the node 8 will have to be loaded by a nodal force $\mathbf{F} = (0, 0, -50\text{ kN})$.

If the slave member is *pulled* by a force $N = 50$ kN, the node 8 will have to be loaded by a nodal force $\mathbf{F} = (50\text{ kN}, 0, 0)$.

Table 9.1 Nodes of the IRFEM model of Figure 9.17 (mm).

Node	X	Y	Z	Constraint
1	108.9271	6.15	60.0	Free
2	108.9271	0	60.0	Free
3	3.55	6.15	0	Free
4	6.024874	6.15	0	Free
5	46.775	6.15	−139.3	Free
6	46.775	6.15	−136.8251	Free
7	0	0	0	Clamped
8	0	0	0	Free

Table 9.2 The connectivity, type, and meaning of the elements of the IRFEM model in Figure 9.17. N1 is the first node, N2 is the second node. The meaning of the elements is the constituent of the renode the element is modeling.

Beam	N1	N2	Type	Meaning
1	1	2	Flexible	B1
2	3	4	Flexible	W2
3	5	6	Flexible	W1
4	2	8	Rigid	m2
5	3	7	Rigid	m1
6	5	7	Rigid	m1
7	1	4	Rigid	P1
8	4	6	Rigid	P1

These forces are opposite to the ones applied to the slave member, but are applied to its free node because the dual model is being used (see Section 5.3.1).

9.3.5 Results

The IRFEM is solved for every loading combination, very quickly, and it gives two very important results:

- nodal displacements
- force packets flowing into the connectors

The nodal displacements are very useful because they allow us to get an order of magnitude for the displacement of constituents. If some connector setting was not properly set (shear only or no-shear flags), or if the renode connections have not been all properly designed, the displacements will be high (Figure 9.18). Using a convenient displacement scale the constituents will appear grossly displaced from their original position. This is very important information, because if the scene and the applied member forces are such that a connector is requested to exert force packets, or components of the force packets, which it is not able to carry, the component of displacements related to the node of the connector will be high.

As in IRFEM the constituents are modeled as rigid bodies, it is enough for the displacement of one node of the IRFEM attached to a given constituent to rebuild the displacements of all the other points. If the vector of the three global displacements of a point **P** is $\mathbf{d_P}$, and the vector of three rotations in global reference is $\mathbf{r_P}$, then the displacements and rotations of another point **Q** of the same rigid body are

$$\mathbf{d_Q} = \mathbf{d_P} + \mathbf{r_P} \wedge (\mathbf{Q} - \mathbf{P})$$

$$\mathbf{r_Q} = \mathbf{r_P}$$

The BREP representation of the constituent is upgraded, adding the computed displacements $\mathbf{d_Q}$ to the original coordinates of the points **Q** defining each face. The object is

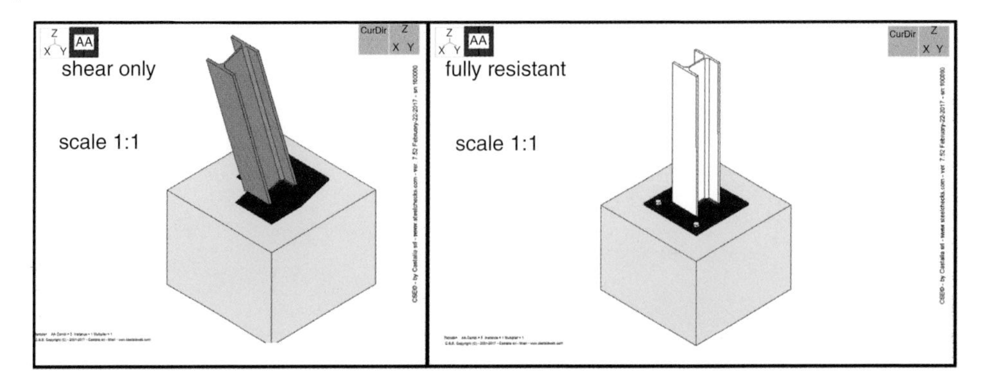

Figure 9.18 A base node where all the bolts have been flagged "shear only" (left), under a bending moment equal to 0.1 times the elastic limit for the cross-section (strong axis bending). Displacement scale 1:1. Clearly the connector is not able to take the loads applied. On the right, the same where all the bolts are "fully resistant".

then rendered in the usual way. The result is a map of displacement that may be very useful to detect unsound or contradictory designs, as when a shear-only bolt layout is requested to bear non-negligible bending moments.

The second important result provided by the IRFEM model is the evaluation of the force packets flowing into each connector. At each extremity of each connector, a force packet can be computed in the local reference system of the connector. This can be used to evaluate the forces and stresses flowing into each subconstituent of the connector (welds and bolts) and, for the case of contact and bolt layout using a bearing surface, the field of pressures exchanged at the contact part of the bearing surface. This contact surface is a subpart of the bearing surface and depends on the loading combinations.

Once the displacement vector $\mathbf{d_{gi}}$ of the nodes of a connector i is known, it can be transformed from the global reference system, that of the renode, to the local reference system, that of the connector, thus giving $\mathbf{d_i}$. This vector can be computed by multiplying the global vector by the orientation matrix $\mathbf{T_i}$ of the connector i, a block-diagonal matrix having order $6m_i$, where m_i is the multiplicity of the connector (see Equation 9.2, where the matrix valid for $m_i = 2$ can be found).

$$\mathbf{d_i} = \mathbf{T_i}\mathbf{d_{ig}}$$

Now, multiplying the stiffness matrix of the connector $\mathbf{K_i}$ by $\mathbf{d_i}$, the nodal forces applied to the extremities of the connector, in its local system are known (a vector of order $6m_i$):

$$\mathbf{c_i} = \mathbf{K_i}\mathbf{d_i}$$

Considering the weld layouts, the force packet applied to the second extremity, that is not that of the faying surface, may be considered equal to the internal forces applied to the weld layout, so

$$\{N, V_u, V_v, M_w, M_u, M_v\}_i^T = \boldsymbol{\sigma}_\mathbf{i} = \mathbf{c_{i2}}$$

These internal layout forces are then used to compute the stresses flowing into each weld.

For bolt layouts, the forces applied to extremities are different from the internal forces applied to the check sections used for the bolt checks (the planes where different plates meet), which are computed by Equation 7.2. The force packets $\mathbf{c_{ij}}$ at different extremities *j* are needed for bolt bearing and block shear checks, as well as for constituent checks.

Once the force packets applied to the connector extremities are known, all the needed steps to check connections can be made.

9.3.6 Remarks on the Use of IRFEM

9.3.6.1 Load Combinations

It must be pointed out that using the hybrid approach, as well as the pure fem approach of the next chapter, there is no need to squeeze the vector of the loading combination to a few units, because the checks can be done for all the applied combinations keeping into consideration *all the loads* applied at one time.

Each constituent, being it a force transferrer or a member, is loaded by a number of force packets coming from the connectors connected to it, and these are all balanced to the externally applied loads and take into consideration the effects of all the member forces at the same time.

Moreover, it must be underlined that if a jnode has more than one instance in the BFEM, the loading combinations to be checked are more than in the BFEM, since n_{combi}, but must be multiplied by the number of instances of the jnode of the renode considered, $n_{instance}$. As a generic instance of a jnode can be obtained by simply applying an orthogonal transformation to the first instance of the same jnode, the internal forces computed at the extremities of the members in each instance can be all considered into a single stack, which has as many final combinations as the $n_{combi} \cdot n_{instance}$.

The IRFEM can also be used if no internal forces have been computed for the BFEM, that is, if the member forces are unknown.

The member forces can be directly set by the analyst, typically considering single internal forces components for simple connections (a shear for a simply supported beam connection, an axial force for a tie, following, if wished, the simple approaches of the traditional methods) so as to test the effectiveness of the connections described in the scene. If this holds true, in general each combination will have to provide six internal forces at each member extremity (axial force, shears, torque, and bending moments), so for a n_m-member renode, $6n_m$ ordered values.

Clearly, if the internal forces are directly set by the analyst, all the instances will share the same values.

Another interesting method for checking connections without having internal forces coming from an analyzed BFEM, is to set limiting values, positive and negative, for each internal force component of each member, and generate a set of notional combinations using these values. Clearly, if wished, some values can be set equal to zero.

The limit forces can be taken as a fraction of the elastic limits of the cross-section, or as fractions of the plastic limits of the cross-section. This last approach is particularly useful for capacity design, where the connections must be designed for internal forces higher than the plastic ones, by considering proper over-strength factors.

An approach which has been proven useful through years of use is to generate 24 notional combinations for each member:

- The first six combinations use the positive limits specified by the analyst (i.e. the fraction of elastic or plastic limits of the member cross-section, $+N_{lim}$, $+V_{2lim}$, $+V_{3lim}$, $+M_{1lim}$, $+M_{2lim}$, $+M_{3lim}$).
- The second six combinations use the negative limits specified by the analyst.
- The other twelve combinations combine the limits for axial force and bending moments to the principal axes of the member cross-section according to Table 9.3.

These 24 notional combinations try to detach in all possible modes the member from the remaining part of the renode, and are, for this reason, an engineering tool to design connections for well determined member force limits.

The opposite problem to that of the absence of internal member forces is the excess of them. If a BFEM has a huge number of combinations, say some thousands, and the jnode of the renode under analysis has many instances, the number of loading combinations to be tested can be really too high.

It is then possible to find a reduced number of loading combinations, equal to $24n_m$ for each renode, finding for each of the n_m members connected in the renode the 24 worst stress indices related to the internal force combinations already listed.

That is:

1) Max N and concomitant internal forces
2) Max V_2 and concomitant internal forces
3) …
11) Min M_2 and concomitant internal forces
12) Min M_3 and concomitant internal forces.

Table 9.3 Possible notional combinations to test connections in a renode where no internal force is available. Combinations from 1 to 12 are obtained by adding the six limit values alone with plus (1–6) and minus sign (7–12). Twenty-four combinations for each member are so defined.

Combination	N_{lim}	$M_{2,lim}$	$M_{3,lim}$
13	0.5	0.5	0
14	0.5	−0.5	0
15	0.5	0	0.5
16	0.5	0	−0.5
17	−0.5	0.5	0
18	−0.5	−0.5	0
19	−0.5	0	0.5
20	−0.5	0	−0.5
21	0	0.5	0.5
22	0	0.5	−0.5
23	0	−0.5	0.5
24	0	−0.5	0.5

The other 12 combinations are found by searching the worst 12 values of the stress index

$$\sigma = \left| \frac{N}{A} + \frac{M_2}{W_2} + \frac{M_3}{W_3} \right|$$

where A is the cross-sectional area of the member, and W_2 and W_3 are the bending resistance moduli.

9.3.6.2 Renode Checks

IRFEM – and PFEM too – considers all the constituents of a renode in a unique model, and it is used to check all connectors, members, and force transferrers of the whole renode, for all the instances of the jnode to which it belongs.

So, even if the structural configuration is more complex than that of the typical simple nodes that are usually considered for the checks, perhaps considering subparts of a more complex renode, the gain obtained considering all in a single model is quite high, and may lead to dramatic time savings for the analyst.

9.3.6.3 Point of Application of Member Forces

If the analysis is to be coherent with the BFEM, the force packets applied to the member extremities $\mathbf{N_k}$, that is, the known loads in the IRFEM, must be applied exactly at the finite element extremities, otherwise some equilibrium condition is violated.

However, designers sometimes wish to neglect some eccentricities that would require a more refined BFEM (maybe using rigid offsets – see Figure 9.19) and perhaps some analysis recycling. Sometimes this is also needed because the offset amount depends on the connection design. As frequently discussed in this book, this shift of the point

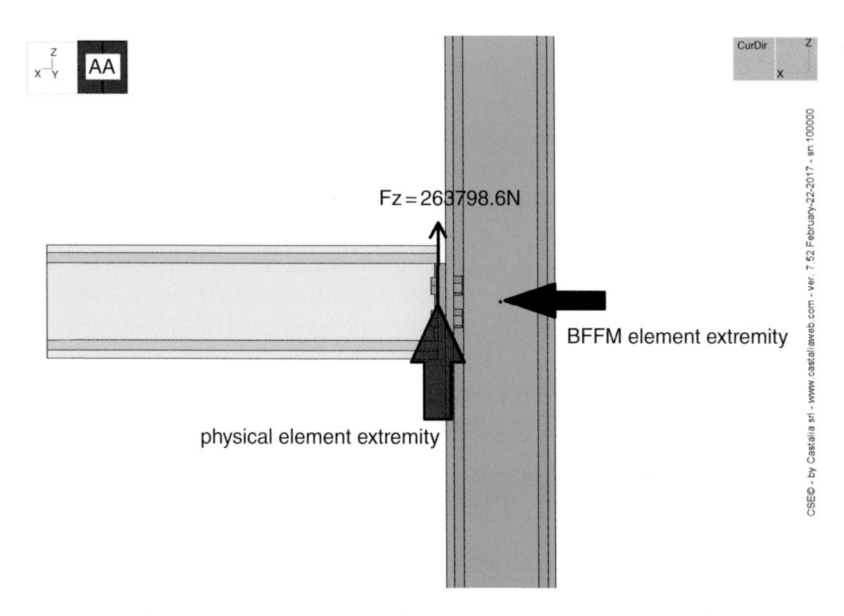

Figure 9.19 Example of the application of a shear force at the extremity of the physical member. The connection is unloaded, but the column should be considered in axial force plus bending, not under the effect of a simple axial force only.

of application of the forces introduces an incoherence in the analysis and might possibly determine severe errors in the checks.

Despite this general consideration, there are situations where the additional effects due to these eccentricities are really negligible, and so the possibility of applying the force packets at points different from the element extremities is attractive.

So it can become possible to allow some *connection-crime* by applying the force packets at points $\mathbf{N_k}$ that do not coincide with the extremities of the finite element in the BFEM, but that coincide with the extremities of the physical members as described in the scene, or with other points. This must be decided by the analyst.

9.3.6.4 Stiffeners

As the members and force transferrers are considered rigid by the IRFEM, the stiffeners, as defined in Chapter 4, are totally unloaded. In fact, as by definition they are only connected to a single constituent, and no deformation is experienced by it within the IRFEM model, the final result of the IRFEM is that their connectors are stress free.

This means that the connectors must be analyzed in a second analysis sweep, that referring to the constituents they belong to (see Section 9.6.2).

9.3.6.5 Pros and Cons of IRFEM

IRFEM is a simplified model that does not explicitly consider the deformability of the force transferrers and of the members. This deformability may lead to different force packets into the connectors, and to distribution of forces between the subconnectors of a connector which are not the simplified ones obtained by assuming a linear displacement field: this will be made clear in Chapter 10. Besides, IRFEM does not consider contact problems but indirectly, through bearing surface, so a direct evaluation of the possible existing prying forces is not possible, or it's hard (more research of simple approaches is needed to this end). A non-linear IRFEM would not be very meaningful, as the connectors are modeled notionally. A more refined IRFEM, considering the deformability of the connected parts explicitly, for instance by means of gridwork or trusswork modeling, would not be worthwhile; indeed, that path is better covered by a pure fem approach, using plate–shell elements.

These are the cons, but IRFEM has several important pros. The ratio of the benefit to cost for the IRFEM model is probably the highest of all the possible modeling approaches to generic renodes.

IRFEM is able to automatically compute the forces flowing into the connectors giving a balanced configuration, no matter what the complexity of the geometry is, so it relieves the analyst of a huge amount of work. If the currently widely adopted hypothesis of linear displacements of the subconnectors is to be taken coherently, then the force packets flowing into the connectors seen as *layouts* can be safely used to compute the forces flowing into the subconnectors, and so to check them. Moreover, some of the available methods of checking for connections, such as ICRM, implicitly rely on the rigid body displacement assumption, because the displacements of single weld seams can be obtained by the displacements of a single point.

As will be seen in the following sections, the IRFEM model results will open the path to a complete and coherent checking of all constituents. For isoconnected renodes or member connections, that are numerous, the force-packets obtained by the IRFEM are correct. These are very many design configurations. For hyperconnected joints, tuning the flexibility indices is easy, so as to get different force packets, balanced as well. For non-preloaded bolted connection in shear, the modification due to the bearing flexibility is already able to finely tune the forces, and usually no flexibility index is needed.

IRFEM, due to its simplicity and generality, and its coherence with all the most commonly used tools for checking bolts and welds, should in the author's opinion be considered the entry tool for dealing with renodes. Whatever the complexity of the renode and the number of loading combinations and jnode instances, IRFEM provides results fully coherent with the mechanical models currently adopted to check welds and bolts, and fully under the shelter of all pillars of connection analysis.

9.3.6.6 Contact Forces

IRFEM does not directly manage contact forces. The forces exchanged by the constituents all pass through the connectors, and the only way contact forces are taken into consideration is when bolt layouts using a bearing surface are used. For bolt layouts not using a bearing surface, the contact forces are modeled by the compressive axial forces in the bolt shafts, neglected for the checks of the bolts, but considered as such, concentrated compressive forces, in the checks of the constituents.

As mentioned in Chapter 6, it would also be possible to extend the bearing surface model to the welds, so that a cooperation of contact forces and tensile forces would carry the bending applied to the weld layout. If this is not done, the welds and not the contact forces carry the applied actions. So, for instance, if a fillet weld layout connects two plates to form a T (see also Figures 6.2 and 6.24), a possible vertical force in the web would first load the welds and then the flange, leading to a safe-side evaluation of the stresses that can sometimes be considered too severe. This problem can be removed, as mentioned, by extending the bearing surface concept also to welds. However, this is not yet allowed explicitly by the standards (see Section 6.7.6).

For the case that is frequently the most important – bolt layouts in (biaxial) bending and axial force – the contact forces are modeled by the linear strain model and are directly part of the balanced configuration of forces and stresses; the field of compressive stresses will be applied to the constituent exerting that pressure, and taken into account in the checks.

The contact forces due to the flexure of the plates in contact caused by the tensile forces in the bolts, the so-called *prying forces*, are not covered by the bearing surface plus linear strain field approach. They need specific considerations, as will be explained in Section 9.4.5.

9.3.6.7 Interdependent Jnodes

If a jnode interacts with another jnode, it must have one member in common with it. This member can be:

1) a master in both jnodes (this is the situation of seismic eccentric bracing frame, EBF)
2) a master in jnode A, and a slave in jnode B
3) a slave in both jnodes.

Configuration 1 can be dealt with by considering the single master receiving more slaves, coming from more jnodes. Only one node of the master will have to be constrained in the IRFEM.

Configuration 2 can be dealt with by considering as unique master the master of jnode A, and considering as slaves all the other members. The slaves of jnode B can be considered as slaves of jnode A not attaching directly to the master of jnode A, but attaching to it indirectly through other slave members.

Configuration number 3 can be solved by assuming that the unique master is the master of one of the two jnodes, and treating as slave all the other members. Clearly, only this member extremity-node would have to be constrained.

9.4 Connector Checks

9.4.1 Weld Checks

The availability of the internal forces flowing into each weld layout, at the faying plane, immediately allows the computation of the stresses flowing into each weld by means of the methods described in Chapter 6. If the instantaneous center of rotation method is used, then the incremental-iterative procedure for each weld layout can be started, leading to a utilization factor valid for the combination checked.

If the instantaneous center of rotation method is not used, the application of the methods of Chapter 6 is straightforward. The standards usually provide closed check formulae for the welds using n, t_{par}, and t_{per}, and as these can be immediately computed for each loading combination, and for each weld, and so the checks are immediate.

Using false colors, the utilization index or factor can be mapped to color ranges, and the checks can be displayed for an immediate evaluation (Figure 9.20).

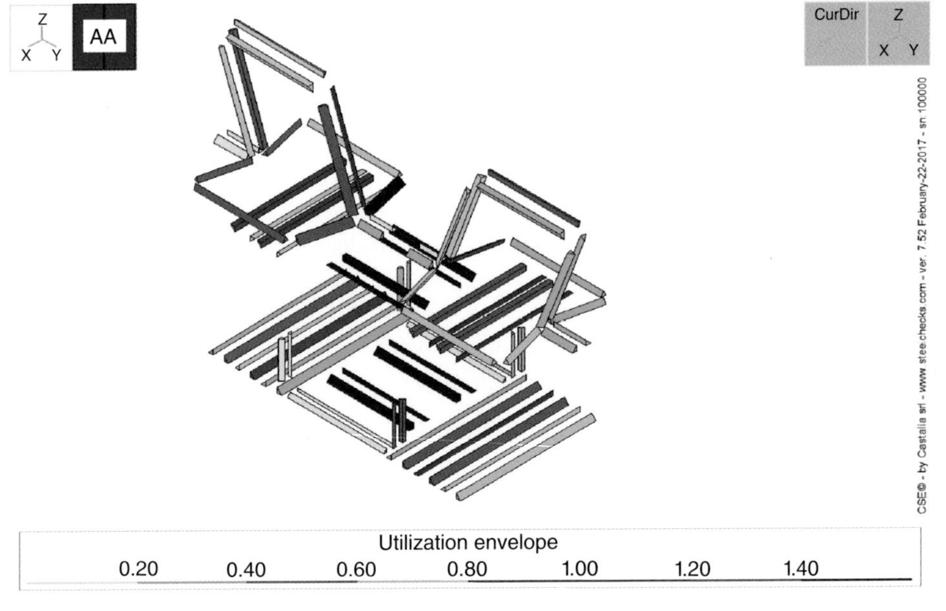

Figure 9.20 False colors representing the utilization factor for single welds of several weld layouts (see also Figure 9.14).

9.4.2 Bolt Resistance Checks

The forces flowing into each check section of each bolt can be computed from the layout i check-section s internal forces $\boldsymbol{\sigma}_{\mathbf{is}}$ using the methods described in Chapter 7.

Once a shear V_{isb} and an axial force N_{isb} are computed for each bolt b of bolt layout i at check section s, the rules provided by the standards can be directly applied (see Chapter 8) and the bolt checked against them, giving the bolt resistance checks, bolt by bolt.

Following this path, as well as for the single welds, the bolts can be colored according to their utilization, and a quick look can be taken at the results, both for each combination and for each bolt.

9.4.3 Pull-Out Checks

Once the tensile force N of each single anchor bolt has been computed, the pull-out checks according to the standard used can be immediately applied, with no need to change the reference system, as they are quite local checks.

9.4.4 Slip Checks

Once the axial force N and shear V of each single bolt have been computed, the slip checks according to the standard used can be immediately applied.

9.4.5 Prying Forces

Prying forces are secondary forces arising when the plates loaded by the concentrated tensile forces of bolts, bend in such a way that, at some distance from the loaded bolts, edges, parts of edges or parts of plate surfaces of the bent plate find a bearing in the connected part.

Prying forces are secondary forces because they are not needed for equilibrium, and so they are statically indeterminate; given a load level there is no unique solution for prying forces (Thornton 1985). Their evaluation can be only be made by introducing some stiffness assumption for the bearing, or, at limit, when the plate forms yield lines, and/or the bolts reach their limit. The amount of the prying forces depends on a number of factors, such as the length and area of the bolts (whose elongation increases the distance of the bent plate from the potential support), the geometry of the connection (i.e. the relative position of the bolts, of the plate edges, and of the applied tensile loads), and the thickness of the plate.

Prying forces increase the tensile force of the bolts, and unload the bent plate that finds a support. So it can be generally stated that it is unsafe to neglect them in the check of the bolts, while it is generally on the safe side to neglect them when checking the plate.

Prying forces may typically develop when the bolts are robust and the plate is relatively thin. This is unfortunately the condition also needed to ensure ductile behavior of sheared bolts. When the bolts are weak and the plate is thick, the flexure of the plate is modest, and the first failure to happen is that of the pulled bolts. This mode of failure, however, is not the most frequently designed, due to its lack of ductility.

When considering the problem of prying forces in general, it is clear that the problem is very complex, and that no simple accurate solution exists. Depending on the geometry and flexibility of the plate, complex configurations of displacements can be found. To face such a complex problem for what it is, only finite element models using contact non-linearity can cope.

If the geometry is made simple by several additional hypotheses, a closed form solution is available, namely that of T-stub. This is the most useful reason why the T-stub model is so diffused.

A pulled T-stub connected to something else by bolts placed over the flange, can be modeled like a T-frame attached to two elements simulating the bolts (Figure 9.21). Bolts can be modeled hinged or clamped to the flange (they are usually assumed hinged). Depending on the distance of the bolts from the web and from the free edge of the T-stub, the thickness of the T-stub flange, its width, and the stiffness of the bolts, the flexure of the flange (Figure 9.21, first row) may be so relevant that the connected body might be hit. If this happens (Figure 9.21, second row) prying forces Q develop. These forces imply a relevant change in the distribution of bending in the flange (Figure 9.21, third and last rows), and an increase in the axial force N carried by the bolts (see circular detail of Figure 9.21).

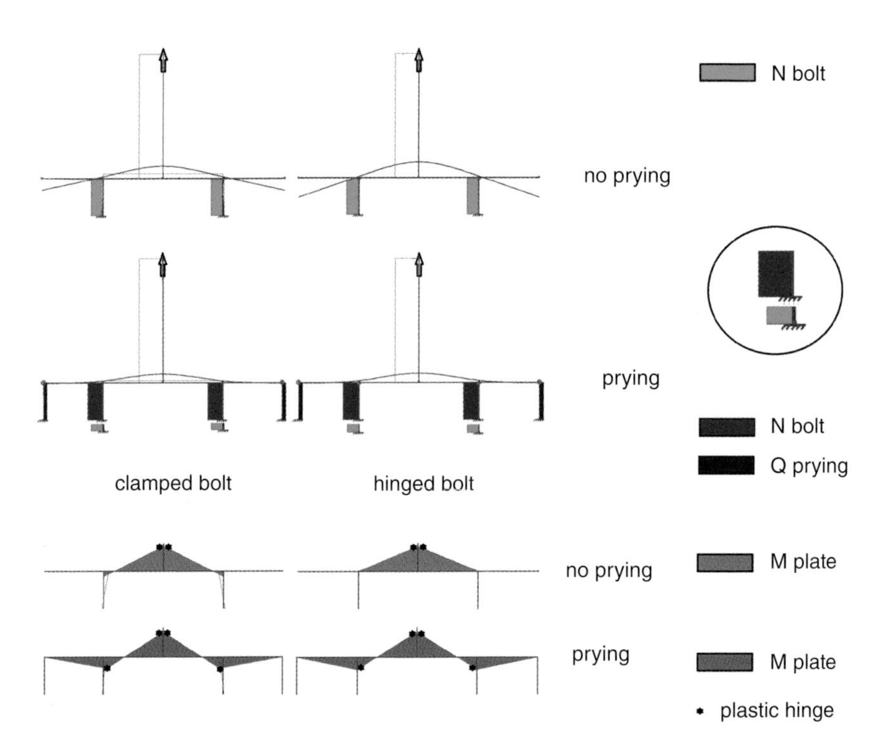

Figure 9.21 A pulled T-stub modeled beam-like in elastic range. Bolts are hinged to the plate (on the right column of images) or clamped (on the left). Axial forces in bolts are higher if prying is active (see round detail). In plastic range, assuming the plate thin, plastic hinges will form leading to a mechanism (see black stars).

Considering the simple, regular, simply pulled T-stub connected by two sets of bolts, at equal distance from the free edge, three types of failures are usually described:

1) Mode 1, implying the formation of four plastic hinges and the developing of prying forces. Two plastic hinges are near to the bolts, two at the connection of the flange to the web (Figures 9.22 and 9.23).
2) Mode 2, implying the failure of the bolts by excessive tension, *and* the formation of only two plastic hinges at the connection of the flange to the web.
3) Mode 3, implying only the failure of the bolts, and no plastic hinge in the flange.

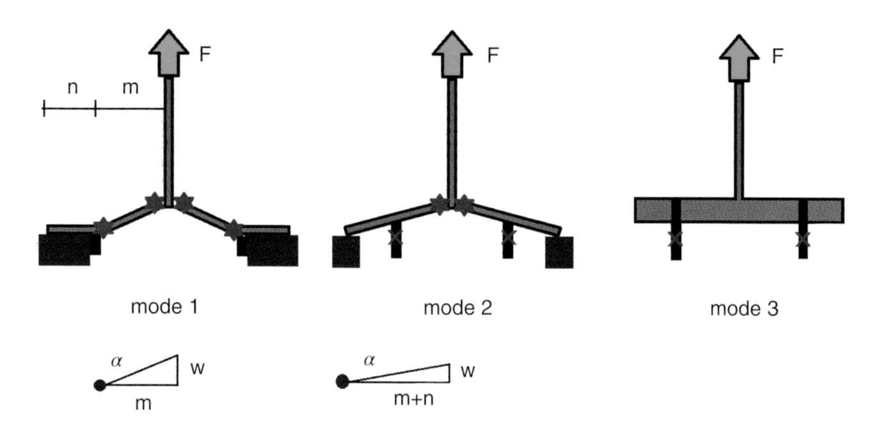

Figure 9.22 The three typical failure modes of a pulled T-stub according to Eurocode 3. Plastic hinges as marked by a red star. Red crosses assigned to bolts mean failure.

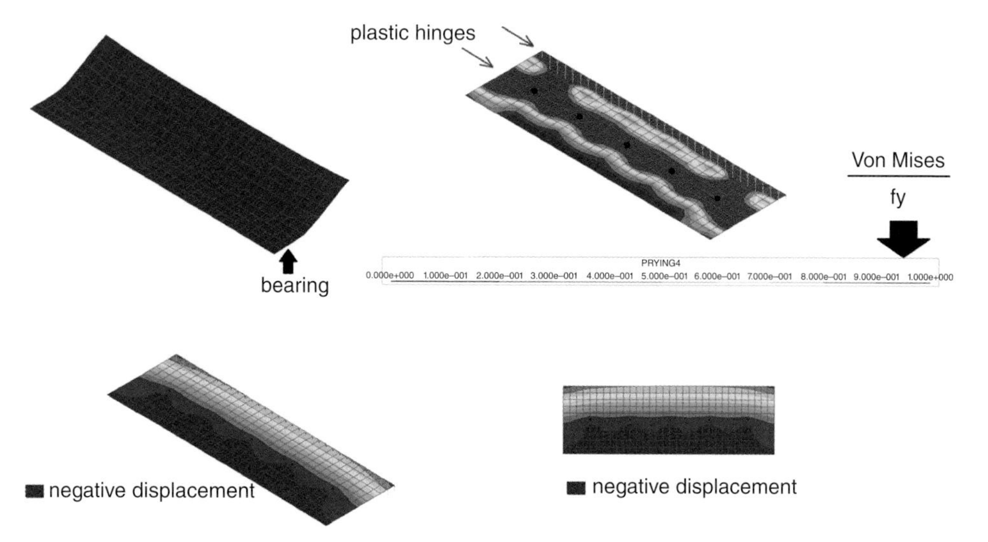

Figure 9.23 Plasticity distribution at limit for half T-stub pulled by increasing forces (right long side) and modeled by plate–shell elements. Black dots are in the bolt positions. Bolts modeled by truss elements. The half T-stub rests over a bed of no-tension springs. The red parts are plastic.

The model assumes implicitly that the bolts are equally loaded, disposed in such a way that what is actually the bending of a plate can be predicted by the formulations valid for beams.

At limit, if the plastic moment referred to the yield lines at failure is $4M_{pl}$ for mode 1 failure (four plastic hinges, two per side) and $2M_{pl}$ for mode 2 failure (two plastic hinges, one per side), it can be written equating the external to the internal work:

Mode 1 (see Figure 9.22):

$$F\delta w = 4M_{pl}\delta\alpha$$

$$F = \frac{4M_{pl}}{m}$$

Mode 2:

$$F\delta w = 2M_{pl}\delta\alpha + \left(\sum_{i=1}^{nrows} 2F_{\max,\, bolt}\right)\cdot n\delta\alpha$$

$$F = \frac{2M_{pl} + \left(\sum_{i=1}^{nrows} 2F_{\max,\, bolt}\right)\cdot n}{m+n}$$

Mode 3:

$$F = \sum_{i=1}^{nrows} 2F_{\max,\, bolt}$$

If the bolts were continuous over the whole length L of the T-stub, then

$$M_{pl} = \frac{1}{4}Lt^2 f_y$$

where t is the thickness of the flange, and f_y the yield stress. Since the bolts are applied at given distances one to another, and do not cover the whole length of the T-stub, complex yield lines patterns will appear (in Figure 9.23 they are evaluated numerically by a plate–shell model), so that the plastic moment could be expressed as

$$M_{pl} = \frac{1}{4}L_{eff}t^2 f_y$$

where L_{eff} is a suitable “effective length”.

The problem is more complex if the bolts are not equally loaded. Assuming that each “row” consisting of two bolts is such that the two bolts are equally loaded, single T-stub “strips” can still be analyzed in order to find the maximum F of a single bolt row, or of a group of rows, exploring yield line patterns matching one another (this is the method explained by *Green Book, Moment Connections,* SCI 1995, and derived from Eurocode 3). Using this method the maximum N and the maximum M of bolted T-stubs can be found and, *drawing a linear limit domain,* checks can be made.

The equal load assumption of the bolts of a row (i.e. the need for an axis of symmetry, because otherwise no equilibrium with externally applied load would be easy to ensure), forbids using biaxial bending, and limits the possible loading conditions to the 1D ones (axial force N, *or* strong axis bending moment M_u, *or* weak axis bending moment M_v). Linear domains must be used. If there are more than two bolts over the flange, the distances are uneven, the pattern of bolts irregular, or no T-stub is actually present, then the method simply cannot be applied. However, as there is no other available method, T-stub is often extended to situations where it should not be used.

As has been seen, there is no unique solution for a given load level, referring to the simple T-stub. Solutions are all found *at plastic limit,* that is, assuming that some plastic mechanism is triggered. Moreover, these solutions refer to a subset of the possible loads acting over the T-stub, considered one by one. So T-stub is used in answering the following questions:

1) What are the maximum simple loads, one by one, *N or M*, that the T-stub can carry? This requires the evaluation of simple loading configurations but assuming complex yield line patterns, applicable only for those very simple loading conditions (see *Green Book, Moment Connection*, SCI, 1995).
2) Is the stress point inside or outside the *linear* limit domain drawn using the simple actions limits?

Designers are implicitly forced to evaluate the simple plastic limits of the T-stub, also if they are not interested in evaluating a utilization *factor,* that is, also if they are only interested in a utilization *index.* In turn, this safety measure – if buckling, slip, fatigue, and fracture do not come into the discussion – is considered safe due to the convexity of the limit domains ensured by plasticity theory. Moreover:

- as this method is used at plastic limit, when serviceability loads are to be checked, and it must be ensured that nowhere plasticity has been reached, they are not very helpful
- as has already been explained in Chapter 1, this safety measure obtained by linear domain is safe under well delimited assumptions, but it is far from *exact,* as it forgets the coupling of the internal forces and the true shape of the limit domain (the true yield line patterns under $\alpha_L \cdot (N, M_u, M_v)$).

In the real world, designers may be interested in a possibly much simpler question: is the loading condition (N, M_u, M_v) (or better, $N, V_u, V_v, M_w, M_u, M_v$) *inside or outside the limit domain*? And the answer to this question comes from examination of the displacements and stresses *related to the given load condition* $(N, V_u, V_v, M_w, M_u, M_v)$, that is, the one related to α, not to α_L. Not all designers want to design with nominal loads taking the constituent at its limit, perhaps to save 2 mm of steel. Other considerations are also important: the time spent analyzing, the availability of constituents, the ease of fabrication and control, and so on. Of course this does not negate the validity of the issue "find the limit loads"; it merely states that there is another way to look at the problem. Moreover, very often computing a utilization factor is wishful thinking, because there are loads that are only known notionally, such as the seismic ones.[1]

---

1 After much study, and many discussions with advanced seismologists (see Rugarli 2014, Chapter 9), namely Prof. G.F. Panza and his colleagues, and the latest earthquakes that hit Japan and Italy, the author is convinced that the approach currently used by the standards to get the seismic input, probabilistic seismic hazard assessment (PSHA), is totally flawed. Much more convincing is the less presumptuous neo deterministic seismic hazard assessment, and scenario-earthquake approach based on the numerical simulation of the seismic wave propagation, as done by Prof. Panza's group. The seismic input that structural engineers normally use for their computations, if based on PSHA is unfit to protect constructions, as it is not on the safe side. Since seismic loads are among the main ones that a structure must withstand in a good part of the world, including USA, South America, Japan, Asia, and Southern Europe, it seems that the request to evaluate a utilization *factor* is wishful thinking, if internal coherence has to be strictly guaranteed.

The question referring to the safety of the loads applied (inside or outside the limit domain?) can be better answered by finite element models, that will show the stresses and displacement obtained by considering all the internal forces applied at a time, related to load level α.

However, the problem of prying forces must be addressed. This can be done in the following three ways:

1) assuming a PFEM approach and contact non-linearity
2) increasing the tensile forces of the bolts by a safe percentage, for the check of the bolts, and neglecting the prying forces when considering the stress analysis of the constituents
3) evaluating by means of simplified approaches the amount of the prying forces, and adding them to the computational fem model in proper positions.

Approach number 1 is the most correct one and it's available in PFEM models. In the long run it will be used routinely.

Approach number 2 is safe as long as the increase of the tensile forces of bolts considered for the checks is high enough. If it is true that the fem models of the constituents will not embed the prying forces, so leading to a more severe stress state, it is also true that the models will take into account all the forces applied at the same time, and will not use any linearization of limit domain. So if from one side something is lost, the description of the prying forces, from another much is gained, as all the loads are considered at the same time with no need to draw a straight line. The stress maps will have to be considered in elastic range, or if convergence in the plastic range is reached, then

$$\alpha < \alpha_{L,\,noprying} < \alpha_{L,\,prying}$$

That is, if convergence in the plastic range is reached under load level α, which is the one of the applied loads, the applied loads are below the limit loads not considering the prying forces, $\alpha_{L,no\ prying}$ which in turn are lower than the limit loads of the constituent, considering the prying forces, $\alpha_{L,prying}$. This can be proved by considering that prying forces can be seen as the forces exerted by additional constraints. So, the fact that

$$\alpha_{L,\,noprying} < \alpha_{L,\,prying}$$

is equivalent to assuming that, at equal externally applied loads, the limit load multiplier related to a structural configuration A, is lower than that related to a structural configuration B, obtained by adding more constraints to the system. This can be proved by the kinematic theorem of limit analysis.

All the possible plastic mechanisms of configuration B, including the true plastic mechanism for this configuration, are also kinematically admissible for configuration A. So the true plastic mechanism of configuration B is a kinematically admissible one for configuration A. As by kinematic theorem of limit analysis all the kinematically admissible mechanisms of A are related to a load multiplier greater than or equal to the true limit load multiplier for configuration A, then

$$\alpha_{L,\,noprying} < \alpha_{L,\,prying}$$

In approach 2, then, the problem is to get a realistic increase of the tensile forces for the bolt checks. Usually the bolts must be stronger to avoid their brittle fracture.

Their tensile resistance is grossly proportional to the square of the diameter. An increase of Δ(%) of the tensile force due to prying implies an increase of 0.5Δ in the diameter. However, rarely are the bolts sized to take exactly the tensile forces predicted by the analysis, because their failure is not ductile and it's usually unwanted. This means that, realistically, more than Δ is often already available for other reasons. If the safety margin against bolt failure is to be kept constant also after prying force evaluation, then the 0.5Δ increase will have to be applied.

In order to predict a reasonable amount for the fraction Δ, simple models can be used, including the T-stub "strip" model currently used by the T-stub based methodologies. Here the "strip" is a constant width of plate, and the two distances m and n can be evaluated from the geometry of the problem (Figure 9.24).

According to the work of Thornton (1985), the increasing factor K_{pr} to be applied to the axial forces in the bolts can be evaluated as follows

$$K_{pr} = 1 + \frac{\delta\alpha}{(1+\delta\alpha)} \cdot \rho$$

where

$$\delta = 1 + \frac{D_h}{p}$$

$$\rho = \frac{m'}{n'}$$

$$m' = m - 0.5D$$

$$n' = n + 0.5D$$

$$\alpha = \left(\frac{8N_{\max,b}m'}{pt^2 f_y} - 1\right) \cdot \frac{1}{\delta \cdot (1+\rho)}$$

and

- p is the pitch, i.e. the distance between bolts
- t is the thickness of the flange
- f_y is the yield stress

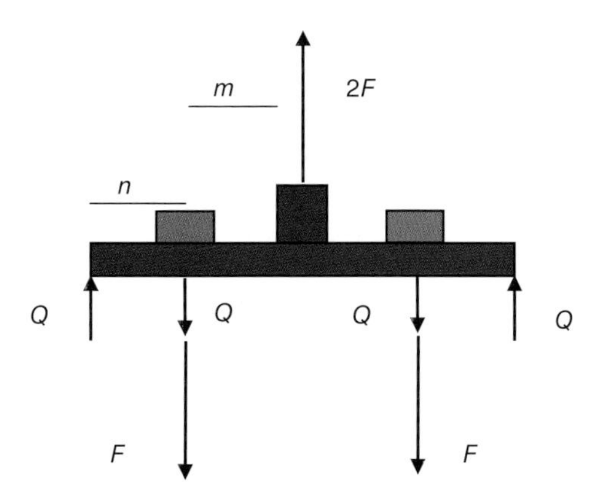

Figure 9.24 Simple Thornton's model for prying forces.

Table 9.4 Prying forces factor in a T-stub according to Thornton (1985) for several design situations. Bolts are M16 class 8.8.

n (mm)	*m* (mm)	*p* (mm)	*t* (mm)	K_{pr}
40	30	48	20	1.26
40	30	64	20	1.24
40	30	48	10	1.26
40	30	64	10	1.25
40	30	48	30	1.13
40	30	64	30	1.05
60	30	48	10	1.18
60	30	64	10	1.18
60	30	48	30	1.09
60	30	64	30	1.03

- D is the bolt diameter
- D_h is the bolt hole diameter
- $N_{max,b}$ is the maximum force that can be taken by the bolt.

If the flange is relatively thin, and the distance of bolts from the flange-clamp high, the values of K_{pr} may rise, usually an increase up to 20–30%, or lower can be expected in the axial forces in the bolts for normal design situations. In Table 9.4, some sampling results are presented, using class 8.8 bolts, with $D = 16$ mm, and $f_y = 235$ MPa.

More refined T-stub models are available, also taking into account the stiffness of the bolts and their elongation. Long bolts tend to increase their lengths more than short ones, and so the detachment of the T-stub from the contact surface is made easier. However, all these models consider a simple or simplified geometry, and so they are not applicable to generic contexts.

Approach number 3 is an open front of research.

9.5 Cleats and Members Non-FEM Checks

9.5.1 Action Reaction Principle

By using the third law, action and reaction principle, the forces loading the subconnectors with sign reversed are immediately the forces loading the constituents.

The force transferrers are self balanced (see Figure 9.25) under:

- the axial force, shears and possibly bending moments delivered by each single bolt connected. These forces are diffused over the body of the constituent, no longer concentrated at the connector extremity as when dealing with layouts (Figure 9.26).

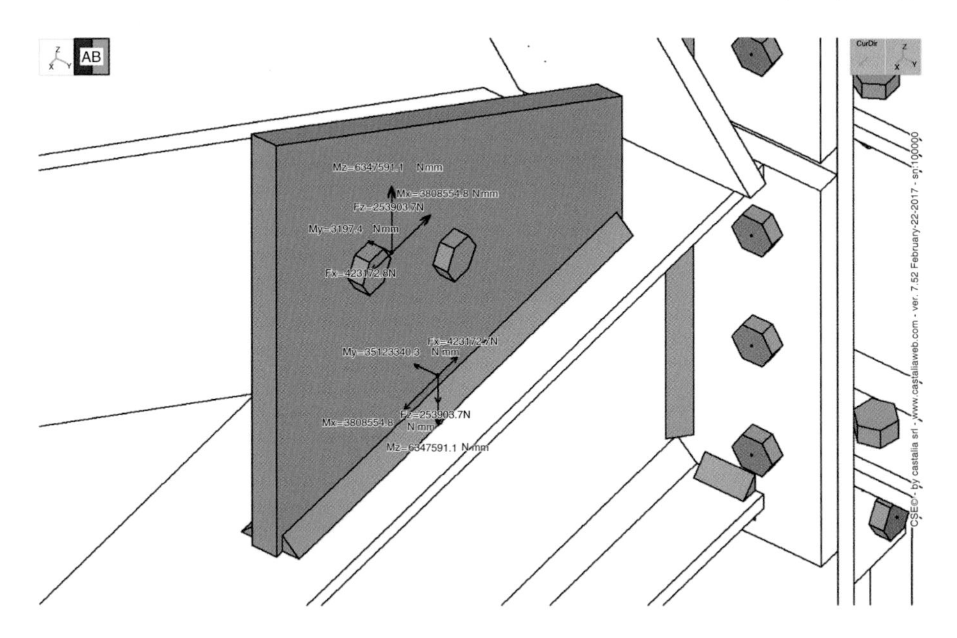

Figure 9.25 The force packets loading a force transferrer at connector extremities are self balanced.

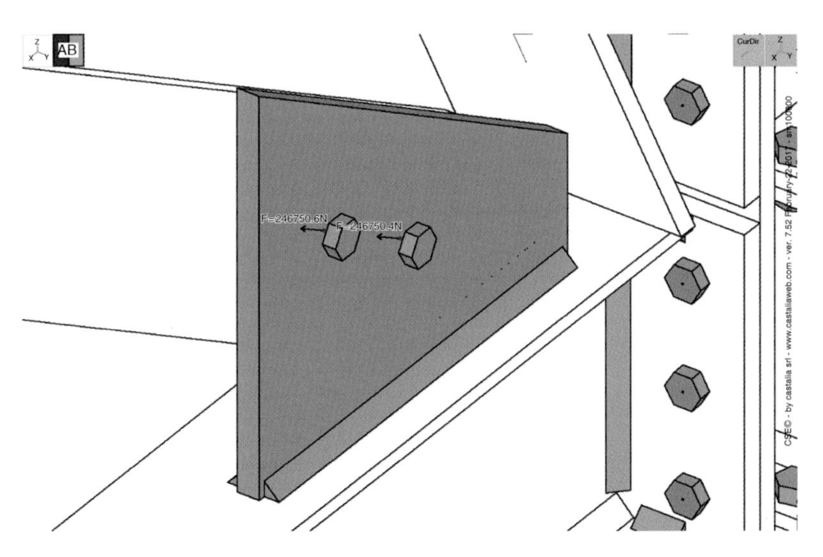

Figure 9.26 The forces exerted by the bolts over a constituent.

- the pressure fields exchanged at the subset of the bearing surface reacting, in the specific loading combinations adopted, for every bolt layout using a bearing surface connected to the constituent. These pressures are also diffused.
- the stresses (collected into nodal forces), exerted by the single welds of fillet and penetration weld layouts connected to the constituent. These nodal forces are diffused over the body of the constituent.

So, the fem analysis of each force transferrer can be done by applying dummy constraints to it (six elementary constraints to block all the six rigid free body motions).

The members are balanced under:

- the axial force, shears and possibly bending moments delivered by each single bolt connected
- the pressure fields exchanged at the bearing surface subset reacting, in the specific loading combinations adopted, for every bolt layout using a bearing surface connected to the member
- the stresses (collected into nodal forces), exerted by the single welds of fillet and penetration weld layouts, connected to the member
- the force packet applied at the far extremity **F** of the member stump, a force packet that is already known by the BFEM, or that can be recomputed assuming a clamp at the far extremity itself (for non-passing members).

And so also the fem model of members can be used, applying a clamp at the far extremity **F** of the member.

9.5.2 Bolt Bearing

In order to check bolt bearing, the forces taken by each bolt at each extremity $\mathbf{c_{ij}}$ (not to be confused with the internal forces at the check sections of the bolt, which are slightly different), are applied with sign reversed to the bolted thickness, $-\mathbf{c_{ij}}$ (Figure 9.26). As mentioned in the previous section, this means applying the third law, the action reaction principle.

Then, the methods outlined in Chapter 8 can be used to check each constituent bolted against this local failure mode. If the multiplicity of a bolt layout is m_i, m_i constituents will be checked, for each bolt b, leading to *nbolts* different utilization factors for each constituent and each bolt layout connected. At each extremity j, of the bolt layout i, a different value of the applied bolt bearing force V_{ijb} is found, in general.

9.5.3 Punching Shear

The axial force taken by each bolt can be directly used to check punching shear. So, also this failure mode can be immediately checked with no particular extra effort. In this case, in order to execute the check, it is enough to consider the forces in the principal system of the bolt layout, with no change of reference system.

9.5.4 Block Tearing

Block tearing can be checked following the methods outlined in Chapter 8, once the forces exerted by each bolt to the thicknesses bolted are known. In general, a constituent will receive forces from more than one bolt of more than one bolt layout. These forces are known in direction and modulus for each loading combination considered.

9.5.5 Simplified Resistance Checks

The resistance of a constituent to the loads applied must be ensured, not only for the already enumerated local failure modes, but also as a whole.

In principle this means considering the whole constituent, usually having a complex shape that does not allow the use of simple theories, and assessing the limit load multiplier for it. As the task can be extremely complex in by hand computation, traditionally a number of simplified approaches have been used.

Also, if the preferred way to check a constituent is nowadays the plate–shell finite element analysis of it, traditional approaches still have an importance, for these reasons:

1) They can often (but not always) be evaluated with limited computational effort, and with no need for advanced software.
2) They can be easier to understand for an average analyst.
3) They can be used as a means of getting to a preliminary sizing or as a means of cross-checking fem results, provided that the problems studied are the same (same loads considered). If this is not true, comparisons are difficult.

The simplified methods, however, are rarely sufficient to test the resistance of a constituent, especially when realistic loading conditions are considered. So, they should be used with extreme care, being well aware of their field of applicability.

Among the most diffused simplified methods to check resistance, the following can be enumerated:

- Strut and tie. These are ideally cut inside complex bodies, and simple forces are assigned to them in order to ensure the force transfer. Balanced configurations are searched, and accepted under the shelter of the static theorem of limit analysis.
- Whitmore section. It is assumed that the stress flow follows paths open at 30° from the point of application of a concentrated force. At a suitable distance, a cross-section is considered and checked beam-like.
- Net cross-sections (Figure 9.27). Cutting the constituent by suitable planes, "net" cross-sections of the constituent can be obtained. These also take into account the width or depth reductions due to cuts and bevels, and the existence of the holes

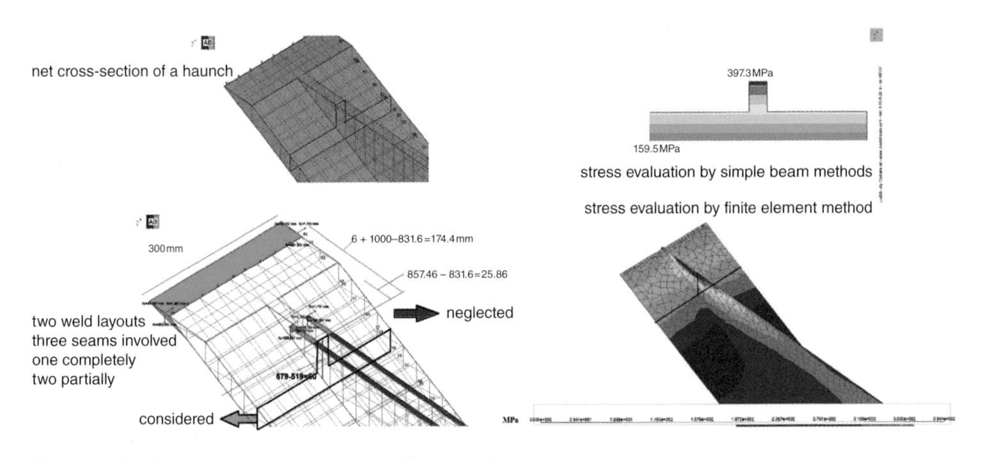

Figure 9.27 A net cross-section check of a haunch.

due to bolting. The net cross-sections of a BREP-described constituent can be obtained automatically, including holes, if any. Considering the solid divided into two parts by the cutting plane, and summing the effects due to elementary sets of forces related to bolts and welds of one part (e.g. the green part of Figure 9.27), it is possible to compute the forces and moments resultant applied to each net cross-section, and check this cross-section by beam-like methods (top right, Figure 9.27). This is a very useful approach, and sometimes it leads to results equal to those of a true plastic analysis (e.g. when the failure is due to the plasticization of a plate bent out of plane along a well delimited and unique yield line).

- Ad hoc formulae. These are simple formulae to be used to check very specific failure modes, related to very specific subconstituents or subparts of constituents, in the context of typical connections. The Steel Construction Institute's so-called *Green Books* are the champion references for these approaches.

It is not within the scope of this book to discuss in detail these methods, which are already explained by many textbooks. It is assumed that the reader is already familiar with them, and the focus is placed instead on the finite element techniques.

9.6 Single Constituent Finite Element Models

9.6.1 Remarks on the Finite Element Models of Single Constituents (SCOFEM)

Once the IRFEM has been solved, the force packets exchanged at connector extremities are known. Using the methods explained in Chapters 6 and 7, the forces exchanged by the subconnectors, and the pressure field related to the bolt layouts (and possibly also weld layouts) using a bearing surface are known.

The third law is applied at this level, that is, at each single subconnector and at each ideal point of the bearing surfaces where a contact pressure is exchanged (Figure 9.28).

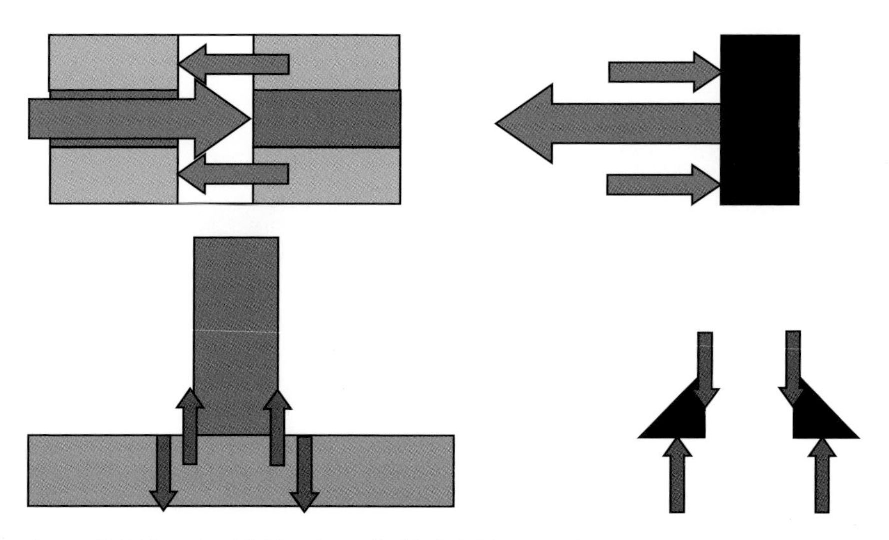

Figure 9.28 How the third law is applied in hybrid approach.

In this way, the forces loading the constituents are immediately known. So, the path is open for their stress analysis using fem methods. The finite element model of a single constituent (a force transferrer or a member) will be here named SCOFEM (single constituent finite element model).

If a force transferrer is considered, the set of all the elementary forces and pressures loading it are certainly balanced. This means that there is no need of constraint reactions to guarantee equilibrium: six dummy iso-static simple constraints can be added, so as to eliminate the six rigid body motions of the constituent.

If a member is considered, the forces applied by the subconnectors are not balanced; to get a balanced condition, the force packet acting at the far end of the member **F** must also be added (see also Chapter 5).

The finite element models of single constituents are made of plate–shell finite elements, summing the membrane and the bending stiffness and stresses (Figure 9.29). The finite elements to be used must take into account, in the highest possible detail, the correct shape of the constituent, as resulting after the application of several work-processes such as cuts and bevels (as explained in Chapter 4, about *dual geometry*). This requires general meshing tools with the capability of respecting complex geometries. Moreover, the points of application of the forces exerted by the subconnectors must be respected, in order to avoid violations of the third law. In turn, this means the capability of respecting the existence of well defined positions in the mesh: the so-called *hard-points*, points where a node of the model must necessarily exist.

Usually three or four noded finite elements are used, and possibly referring to thick plate theory, which also enables the formulation for the out-of-plane shears.

The finite element models of single constituents of a renode are not compatible with one another (Figure 9.29). This means that the displacement of a constituent A obtained by analyzing a load combination, is not the same as that of a constituent B when the boundary between A and B is considered in the two models. This is to be expected. Under the shelter of the static theorem of limit analysis, a balanced configuration is also not compatible.

The reason why the finite element model of a single constituent is prepared and run is to prove that a balanced configuration can be carried without violating the plasticity

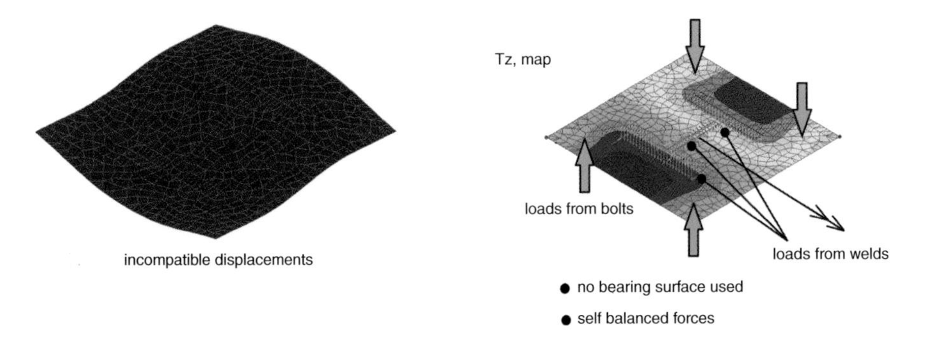

Figure 9.29 Displacement of a base plate SCOFEM under pure strong axis bending of the column (not shown). No bearing surface is used. The deformation is clearly incompatible with that of the base column cross-section. A compatible displacement would require near-zero displacement in the whole footprint of the column end.

limit, that is, in order to ensure that the applied load multiplier is lower than the limit load multiplier (see also Chapter 5). If ideally all the single constituents of a renode have been modeled by finite elements, and if all of them can carry a load configuration coherent with that of their neighbors – i.e. not violating the third law – without reaching the plastic limit, then the externally applied loads are below the limit load for all constituents. However, as discussed in Chapter 5, this does not finally prove that all the constituents do not fail for other reasons. Hybrid analysis of connection involved the following steps:

1) Prepare and run an IRFEM.
2) Check connectors using elementary forces obtained from layout force packets, as computed by the IRFEM.
3) Check constituents for the "local" failure modes, using the third law.
4) Automatically prepare and run the (plate–shell) finite element models of ideally all the constituents of the renode (SCOFEMs), one by one, under balanced and third-law compliant load patterns. This is done in order to prove that nowhere the plastic limit load has been crossed (generic resistance checks).

Note that the automatic preparation and running of the fem model of a constituent is a matter of a few seconds in the linear range, and very frequently a matter of some tens of seconds or maybe three or four minutes for non-linear analysis, when the loads applied are far from the plastic limit (which is the most frequent situation).

If some constituent does not comply with generic resistance checks, there are several possible alternative paths:

1) Improve the constituent by increasing thicknesses or changing widths.
2) For hyperconnected constituents, change the internal force packet distribution by modifying the flexibility indices of the connectors.
3) Change the connectors internal stress distribution (e.g. modifying the bearing surface amount or the constitutive law of the bearing surface, or the working flags of subconnectors).
4) Use models of sets of constituents, i.e. more refined finite element models.

The last choice is useful for reducing the compatibility problems at the interface between single constituents, as the interface connectors are modeled explicitly (see Section 9.7). However, it requires some specific action to preserve coherence.

9.6.2 Stiffeners

If the model of a single constituent is asked for, and this single constituent embeds one or more stiffeners, then the SCOFEM must include the stiffeners of the constituent and all the pertinent connectors. Otherwise the stress state of the stiffener would be wrong. This step, in the hybrid approach, is also useful for assessing the forces and moments loading the stiffeners, especially in view of their buckling analysis.

This problem was faced by the author in 2008, keeping the meshes of the constituent and of the stiffeners independent, and adding special finite elements to model each subconnector of the connectors that connect the stiffeners to the constituent under examination (usually: weld layouts). For each bolt or weld segment, two nodes (hard points in the respective meshes) were found and related to the two extremities (Figure 9.30). The finite element simulating the bolt or weld segment was then the "bridge" between the two

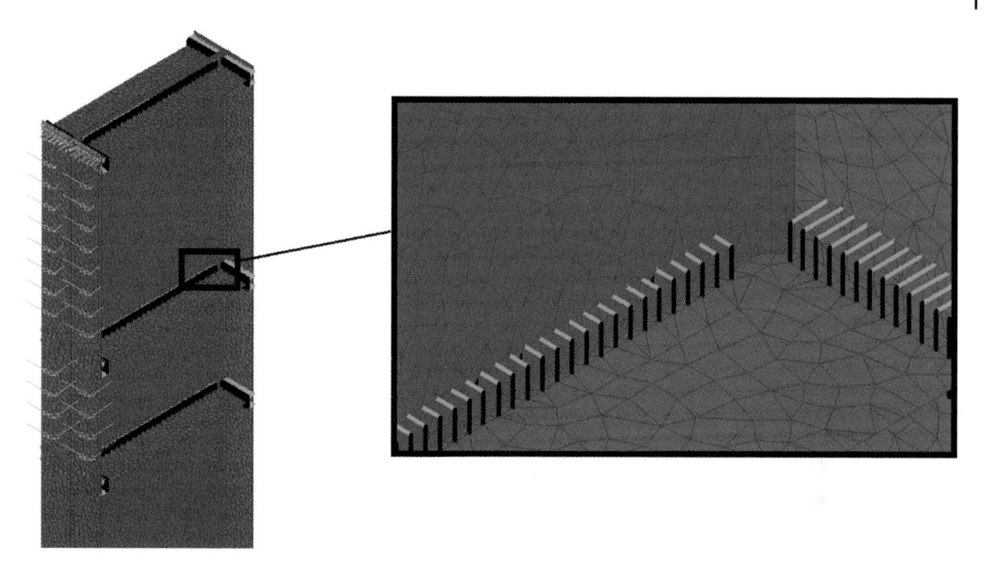

Figure 9.30 A column with its stiffeners as a unique SCOFEM. The "bridge" elements simulating the welds are visible. Plate–shell elements are at middle plane of the plates.

meshes, responsible for the structural continuity between the stiffened constituent and its stiffener.

Although referring to a very special problem, limited to a detail of single constituents, this approach later proved itself to be extremely useful for automatically building the finite element models of sets of constituents (2009), and of the whole renode (2012).

The issue of how to model these "bridges" will be treated in detail in Chapter 10.

9.6.3 Meshing

9.6.3.1 Meshing Phases

Meshing operations make use of two important concepts: mesh *hard points* and mesh *segments*.

A hard point is a point that the final mesh must respect, meaning that a node must be in the 3D position of the hard point. The mesh must then "pass" through the hard point.

A segment is a straight line joining two hard points, and no element side of any plate–shell element must intersect the segment. Segments are boundaries between two regions A and B that the mesh must respect, so that no element has part of it inside A and part inside B (see Figure 9.31).

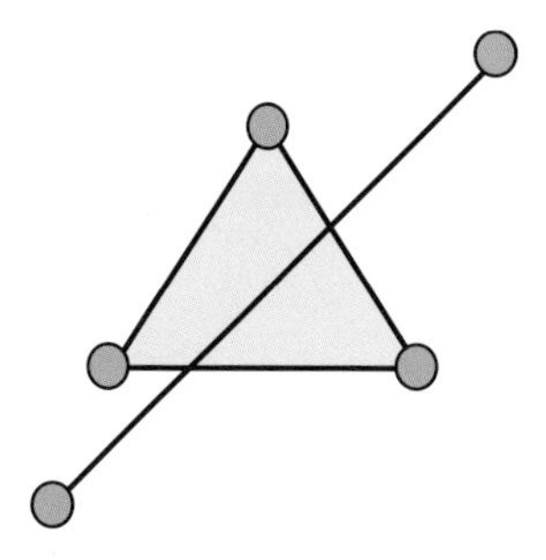

Figure 9.31 Segment violation in a mesh is not allowed.

The meshing of a constituent is carried on in three different steps.

1) Draw basic construction segments and add proper hard points.
2) Fill the closed regions with plate–shell elements by means of suitable meshing tools.
3) Remove the plate–shell elements filling the holes.

The first two steps allow the creation of a *sketch model* (Figure 9.32). This is not yet the mesh of the constituent, but lists all its main features. Sometimes it might be useful to have a look at the segments and hard nodes, with no element filling, in order to detect possible difficult regions where, due to the feature of the problem, meshing can be complex or problematic. Sometimes, shifting a weld a little implies no substantial modification to the mechanical resistance of the weld layouts, but allows the creation of the mesh in a much easier way.

The first sweep of the first step is adding a number of closed polylines, each referring to a subplate of the constituent (see Chapter 4, dual model), as resulting after the application of the needed work-processes (see Figure 9.32). Each side of each polyline is divided into a number of equal segments, each having approximately the size requested for the mesh. Among these closed polylines there may also be closed polylines referring to holes.

The second sweep of the first step is to add the closed polylines needed to define the bearing surfaces possibly needed (see Figure 9.33). These closed polylines are needed in order to have better geometrical compliance with the part of the model where pressures

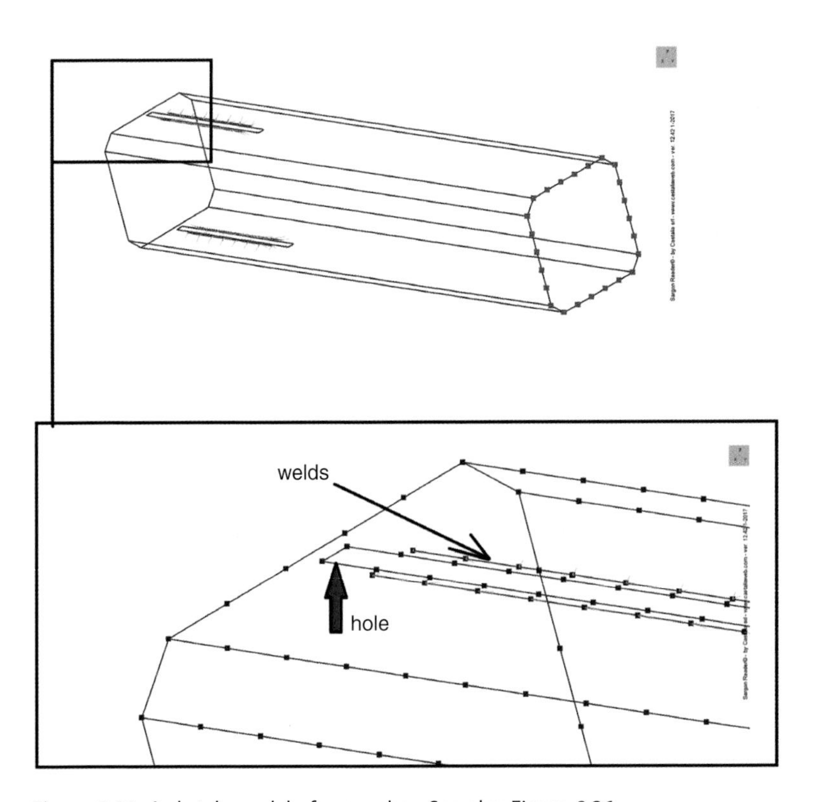

Figure 9.32 A sketch model of a member. See also Figure 9.36.

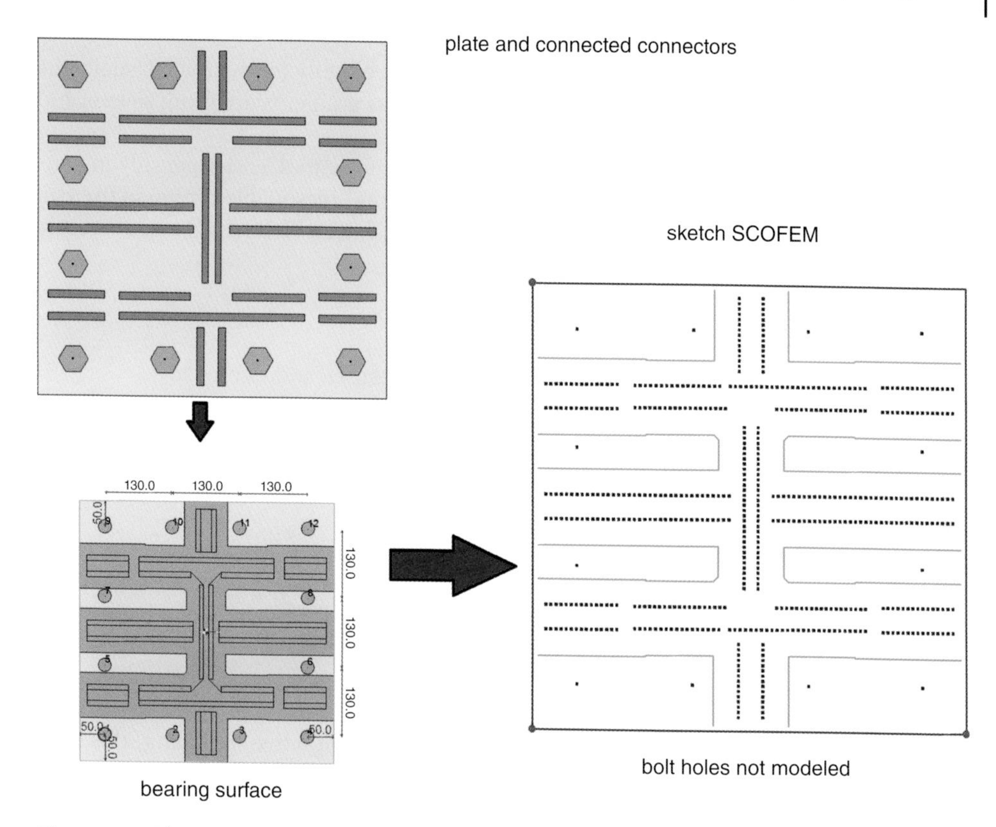

Figure 9.33 The polyline referring to a complex bearing surface in the sketch model of a plate. To avoid segment intersection the bearing surface polyline is not closed. The nodes of the bolts and of the welds (hard points) are also displayed. No bolt hole modeling is used in this example.

are exchanged. Lacking these segments, the application of the nodal forces referring to the pressures would have to be less precise. As these are boundaries independent of the loading combinations, they should be respected by the mesh.

The third sweep of the first step is to add segments referring to weld seams. These are segments, divided in such a way as to get the typical required size for the mesh near to the welds (Figure 9.32).

The second step, typically carried on by suitable meshing tools, is to fill the closed regions with plate–shell elements respecting hard point positions. Several algorithms and methods are available, and the description of these methods is beyond the scope of this book. Usually a Delaunay triangulation is the first sweep, and then refining algorithms improve the initial triangulation. Other approaches use paving, for example the progressive fill of the region by layers of (usually quad) elements.

The author has implemented a mesher using several algorithms and usually adopts Ruppert's algorithm after a Delaunay triangulation (Ruppert 1995). However, the literature on the issue is endless.

The third step is to remove the elements filling the holes, so as to respect the regions where no material is available.

9.6.3.2 Mesh Size and Regularity

The mesh size of plate–shell finite element models must be sufficiently small. Usually the mesh size must be smaller in the regions where the stress gradients are more severe, and therefore typically near to welds and bolts.

If considering the finite element method as an engineering tool, and especially in connection analysis, there is no need to push precision beyond reasonable limits, so the size of the mesh must be good but not excessive. Default values which have been averagely proved good are:

- 10 mm or 0.5″ for the segments of the closed polylines
- 10 mm or 0.5″ for the segments of the welds
- 30 mm or 1″ for the elements unconnected to boundaries and connectors.

Also the minimum angle of triangular elements is an indirect measure of the mesh quality. Using Ruppert's algorithm, a minimum angle of 19.8° guarantees convergence. Better angles, approaching 30° require more nodes and lead to heavier-to-compute meshes.

9.6.3.3 Near Hard Points and Tolerance

Sometimes due to the feature of the renode under examination, hard points are positioned very near to one another. If this happens there may be more problems in meshing due to the need for decreasing abruptly the size of the elements to respect minimum angle requirements. This may, for example, happen when a weld seam does not exactly end at the extremity of a side, or when a corner of a bearing-surface polyline is very near to a bolt center, or to a weld extremity. Sometimes, in interactive modeling, this may happen. The availability of the sketch model helps us to understand where and why these problems may appear.

Usually a tolerance value of 0.1 mm (0.004″) is enough for most problems, but in special cases this might be reduced to 0.05 mm or even 0.01 mm (0.002 or 0.004″, respectively).

9.6.3.4 Bolt Holes

Modeling bolt holes in SCOFEM poses a problem.

On the one hand the resistance of the constituent is weakened by the existence of the bolt-holes. This is especially important when considering failure modes involving a cross-section where holes are present, or when considering block shear.

On the other hand, bolt hole modeling requires more nodes and finer meshes, and for compressive membrane stresses it is a too severe model, as it neglects the stiffening effect of the bolt shaft, unless special formulations are used.

If the bolt holes are not modeled, the model will appear stiffer and stronger, and single nodal forces will be applied to the (ideal) center of the hole, where a node will be placed.

Stress peaks will possibly appear near to the concentrated nodal forces (Figure 9.34), so as to coarsely model the effect of bolt bearing. Since bolt bearing failure mode is already taken into account as a local failure mode, these stress peaks, when localized near to a bolt center, can be discarded. Also, block tearing is considered explicitly and checked by other means. So, only if the plastic mechanism is affected by holes, might its evaluation be unsafe.

If the bolt holes are modeled explicitly, to take into account the stiffening of the bolt shaft, stiff elements can be added connecting the node of the hole center to the nodes at

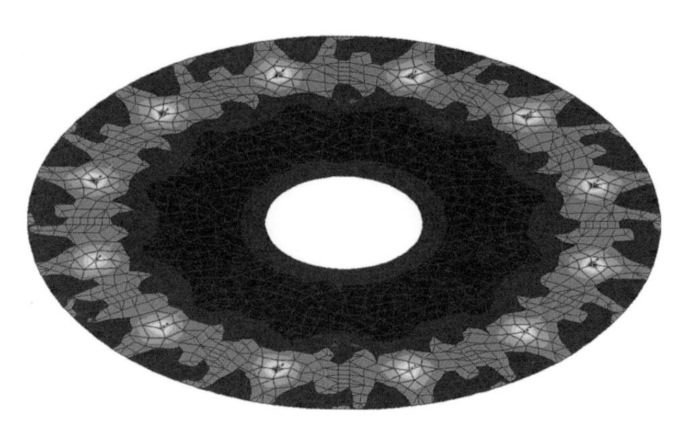

Figure 9.34 Example of stress peaks near to nodes simulating the bolts.

the hole boundary (and using say a 45° angle to model the circle). If their application is correct for compressive membrane forces, it is questionable for membrane tensile forces, due to the contact non-linearity between bolt shaft and hole edge. In the non-linear range the issue can be tackled by using no-tension springs (see Chapter 10).

9.6.4 Constraints

9.6.4.1 Force Transferrers

In order to solve the SCOFEM it must be constrained to avoid the rigid body motions. Generally speaking, a structure loaded is under the effect of stresses which depend on the constraints applied. However, there is an important exception, which is directly applicable to force transferrers: if and only if the structure is under self balanced loads, then every statically determinate set of constraints implies null constraint reactions.

Indeed, the force transferrers are only loaded:

- by the forces equivalent to the force packet exchanged at the pertinent connected extremity *j*, when bolt layouts *not* using a bearing surface, connected to the constituent under examination are considered
- by the forces *and pressures* equivalent to the force packet exchanged at the pertinent extremity *j*, when bolt layouts using a bearing surface, connected to the constituent under examination are considered
- by the stresses equivalent to the force packet exchanged at the pertinent connected extremity *j*, when weld layouts connected to the constituent under examination are considered.

So all the elementary forces and stresses loading the constituent are statically equivalent to the force packets computed by the IRFEM, and that ensures the equilibrium. In turn, this means that the SCOFEM of a force transferrer is self-balanced, and that six elementary constraints can be applied to it, exerting null constraint reactions.

From a numerical point of view, the nullity of constraint reactions is reached when the (not in general numerically null) constraint reactions exerted by the constraints can be considered negligible. Sometimes this means some tens or even hundreds of newtons, and is thus negligible when tonnes or tens of tonnes are being exchanged. It must be considered, in fact, that the nodal forces applied to the SCOFEM (see next section) are balanced to the external pressures and stresses with some possible minor violation due to the nodal lumping (neutral axis cuts the elements, as it changes from load combination to load combination). Moreover, if bolt holes have been modeled explicitly, it might happen that part of the bearing surface pressures gets lost, unless the bearing surface is properly decreased by the surface amount of the holes.

In the absence of bearing surfaces and neglected bolt holes, the constraints reactions are indeed very small: 10^{-8} to 10^{-6} times the forces exchanged is the typical order of magnitude, in those cases (Figure 9.35).

If an excess of reactions is found, it can be a good choice to decrease the size of the mesh. Decreasing mesh size, the elements crossed by the plastic neutral axis have on average a lower surface, and this ensures a decrease of the unbalanced nodal forces.

The choice of constraints is free. The six constraints should be such as to avoid rigid body motions. A good choice might be to find three non-aligned nodes sufficiently far from one another and set constraints depending on the plane of the nodes. For a plane of the three nodes having normal parallel to the Y axis, for instance (Figure 9.35):

- global translations X, Y and Z fixed for node 1
- global translation Y fixed for node 2
- global translations X, Y fixed for node 3.

The maximization of the distance would ensure the minimization of the constraint reactions, and therefore the minimum possible perturbation of the stresses.

The deformation of a self-balanced body respecting a set of *minimum* constraints A, can be obtained by the deformation of the same body under the same system of self-balanced loads, but respecting a set of *minimum* constraints B, by adding a rigid body motion. In fact, the difference of displacements must imply null strain energy, as the constraint reactions are null, and this can only be achieved by adding a rigid body motion, by definition.

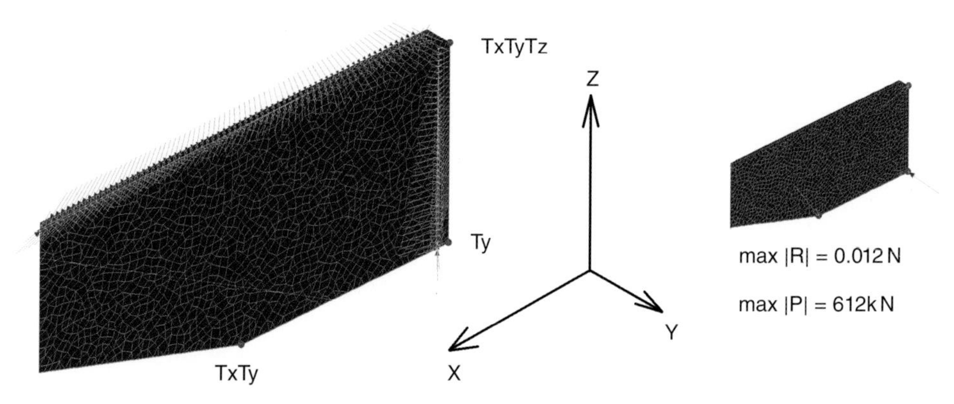

Figure 9.35 Dummy constraints application for a plate. Loads due to welds are visible. The maximum reaction modulus is about 0.012 N, each weld layout of the two connected exchange a force **P** = 612 kN.

9.6.4.2 Members

Members are *not* self balanced under the effect of the force packets delivered *by the connectors*. So, the force packet of the far extremity **F** must be added, that is, at the extremity of the member stump.

If constraints are to be applied, it can be accepted to clamp a dummy node rigidly connected to the nodes of the far extremity of the member considered **F**, by means of suitable rigid elements (see Figure 9.1). This clamped node, placed on the member axis, will exert the forces and moments needed to ensure equilibrium, and basically identical (a change of axes is needed) to the internal forces and moments (N, V_2, V_3, M_1, M_2, M_3) of the beam or truss element used in BFEM to model that very member, at the same distance from the near extremity **N** of the element, that is, from the near extremity of the member.

As an alternative, all the nodes of the member belonging to the far extremity **F** can be clamped (Figure 9.36).

These nodal forces will imply some perturbation only near to the far extremity **F** of the member, a region where the analyst is not interested in getting detailed information of the stress field. In fact, there, the stress field is already "covered" by the checks done in the BFEM. Using the rigid elements and a unique clamped node, a lower perturbation is usually detected. It must be underlined that possible stress concentrations in the zone of the member very near these clamped nodes (at far extremity **F**) must *not* be considered meaningful by the analyst, unless they comply with Euler–Bernoulli distribution of stresses. If so, they may mean that the checks of the member need to be reviewed.

9.6.5 Loading

9.6.5.1 Load Cases vs Combinations

The number of load cases to be studied for each SCOFEM depends on the strategy used to define the member forces.

If the member forces are those of the BFEM, then the number of load cases referred to each SCOFEM is the number of load *combinations* of the BFEM, times the number of

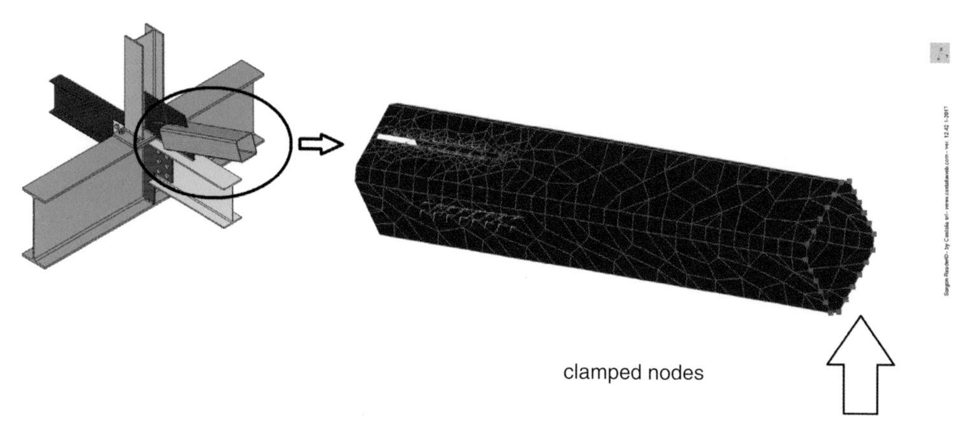

Figure 9.36 A member constrained at the far extremity **F**, under the effect of the elementary forces delivered by the (weld) connectors. A double cut has been applied to insert the gusset plate. The elementary nodal forces transferred by the connectors are visible. The mesh is midsize. Finer elements are near to the welds.

different instances of the jnode of the renode under examination. It is important to stress that the load cases in the SCOFEM are the load combinations in the BFEM. This is for two reasons:

1) It is not of interest to check for BFEM load cases, as taken alone they in general do not physically model possible loading situations (e.g. only wind).
2) If the BFEM has been run in the non-linear range, only combinations (usually dummy, i.e. related to unity diagonal matrices) are of interest, and not load cases. These "load cases" are equivalent to the combinations of BFEM in the linear range.

Moreover, since each jnode can occur more than once, in different positions in the structure, the number of load cases in the SCOFEM must be increased accordingly. As by definition the physical construction of the renode is identical for all jnode instances, the member forces of different jnode instances can be taken as such, with no modifications. So if the load combination of the BFEM is i, if the total number of combinations in the BFEM is $ncombi$, and if the number of instances of the jnode is nin, then the jth instance of the jnode becomes the load case number lc, where

$$lc = ncombi \cdot (j - 1) + i$$

and the total number of load cases in the SCOFEM, $nlcase$, is

$$nlcase = nin \cdot ncombi$$

If on the other hand the member forces are notionally related to 24 simplified loading conditions for each member (as explained in Section 9.3.6.1), and use the factored elastic or plastic limits of the member cross-sections, then it is of no interest to duplicate the same loading conditions for the instances of the jnode higher than the first one. In this case, the total number of load cases in the SCOFEM is 24 times the number of members of the renode, and usually as the master is taken as the reference, the first 24 imply null stress state (a master cannot be detached by itself). Load case 25 will be the tensile axial force detachment of the member 2 (the first slave), load case 31 will be the compressive axial force detachment of the first slave, and so on.

Finally, if the member forces have been directly set by the analyst, then the number of load cases of the SCOFEM is exactly the same as specified by the analyst, as once more there is no need to duplicate the same information for jnode different instances.

The IRFEM has solved exactly the same number of load cases of each SCOFEM, so each SCOFEM load case inherits the (diffused) connectors force packets of the pertinent load case of the IRFEM.

9.6.5.2 Bolts

Each single bolt center, at the extremity of the bolt layout corresponding to the constituent under examination, implies one hard point. That hard point, after meshing, will become a node of the SCOFEM, and that node will receive the forces and moments exchanged by the bolt with the constituent due to the third law. It must be stressed that these are the forces exchanged at the extremity of the connector and not the internal forces used to check bolts at bolts check-sections.

The compressive forces in the bolt shafts, for bolt layouts not using a bearing surface, will be added to the SCOFEM, as they are needed to ensure equilibrium and to model as lumped the contact forces.

Bending moments of the bolt shaft are applied with sign reversed to the constituent, and are sometimes responsible of stress concentrations.

Possible stress concentrations nearby the bolt nodes are usually not to be considered, as they are related to bolt-bearing checks, already covered by the closed formulae of the standards (see Chapter 8).

9.6.5.3 Bearing Surface

Bolt layouts (or also possibly weld layouts) using a bearing surface exchange not only concentrated forces but also a field of contact pressures. In the local reference system of the bolt layout, for each load case, the strain is assumed linear (Section 7.5.2.2.3) so that

$$\varepsilon = au + bv + c$$

The related contact pressure is computed by the strain using the constitutive law of the bearing. The parameters *a*, *b*, and *c*, are computed by solving the non-linear problem described in Section 7.9.2, for each bolt layout using a bearing surface, for each extremity, and for each load case of the IRFEM. The step is necessary in order to apply the checks of the bolts as described in Sections 8.5.1–3 and 9.4.2, so when a SCOFEM is prepared, the results are already available.

Each point inside the bearing surface of the bolt layout will exert a pressure or – for no-tension constitutive laws – a null stress. Only the bearing surfaces of preloaded bolt layouts, if used, can exert "tensile" stresses (Section 7.6.3). The coordinates of a point in the local reference system of the bolt layout (**w**, **u**, **v**), can be transformed to the global reference system, and vice versa. So, for each node of the SCOFEM, defined in the global reference system, the coordinates of this node in the local system of the bolt layout can be evaluated. If the node is outside the bearing surface (this problem only uses u and v), the contact stress is null. If the node is instead inside the bearing surface, then the strain and contact stress can be evaluated.

Considering the plate–shell finite elements of the SCOFEM, the evaluation of the stress can be repeated for each node. Quad elements can be seen as two triangular elements, and so the problem is reduced to the evaluation of a number of triangles. The following situations are possible (see Figures 9.37 and 9.38):

1) All the three nodes of the triangle are unloaded. This may happen because all three nodes are outside the bearing surface or because all the three nodes have null contact stress if the element is inside the bearing surface. It must be stressed that by the definition of the initial segments, an element cannot cross the boundary of a bearing surface, and so it cannot be part inside and part outside the bearing surface itself (Figure 9.31).
2) All three nodes of the triangle are loaded. Then the three nodal forces balanced and equivalent to the field of pressures can be assigned to nodes. This is like nodal force vector evaluation of distributed pressures, in the finite element modeling literature.
3) One node is loaded, the other two are unloaded. Assuming a linear variation for the stresses, the equivalent nodal forces applied to the three nodes can be evaluated.

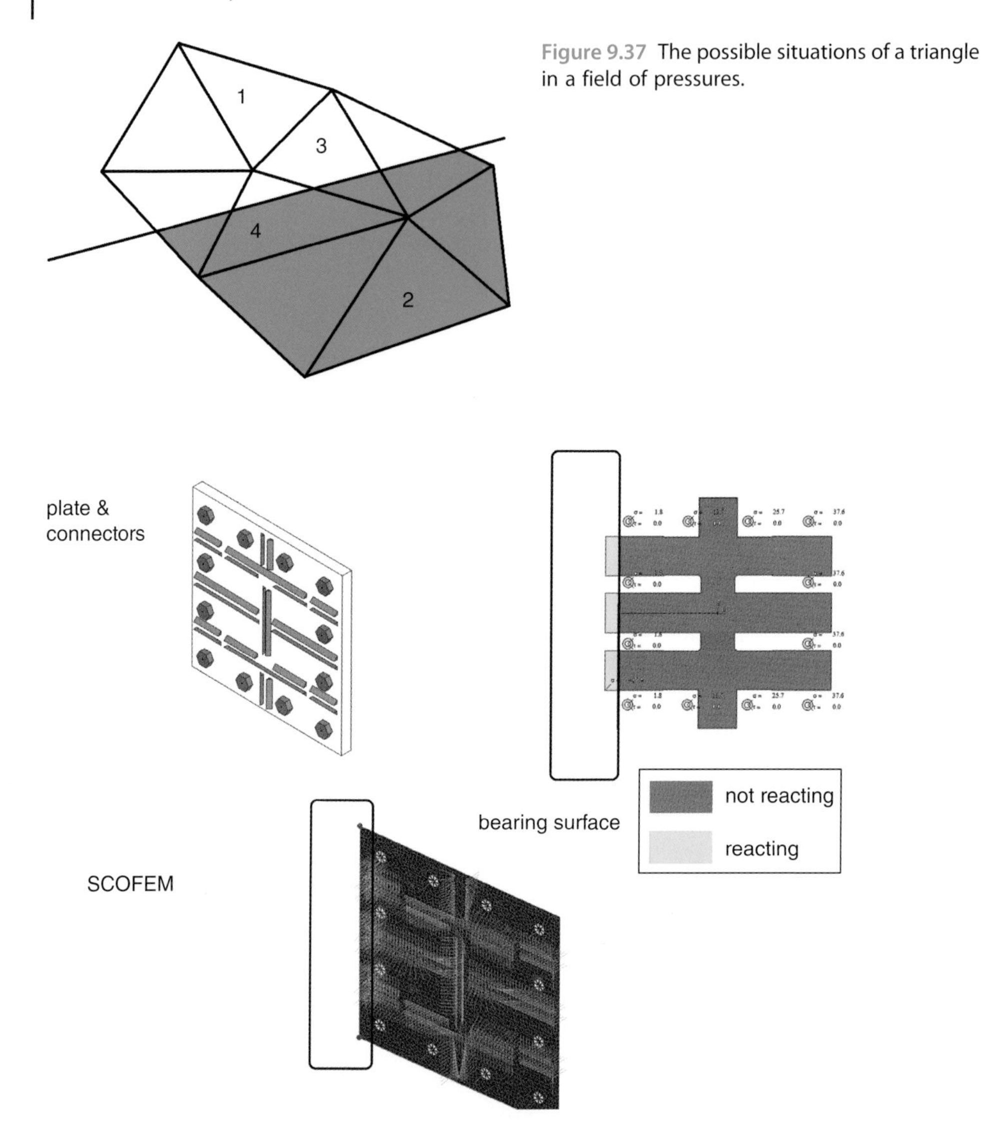

Figure 9.37 The possible situations of a triangle in a field of pressures.

Figure 9.38 Finite element model of a single constituent loaded by the bolt forces, weld forces, and nodal forces related to the pressures exchanged at the bearing surface.

The bolt layout neutral axis found by the non-linear procedure described in Section 7.9.2 can change from load combination to load combination. So the same element can be cut in one combination and uncut in another.

4) Two nodes are loaded, and the last one is unloaded. As for the previous case, it is possible to find the nodal forces equivalent to the applied pressure, and assign them to the pertinent nodes.

The set of all the nodal forces applied to the nodes loaded by pressure, are statically equivalent to the field of pressure evaluated by means of the non-linear method described in Section 7.9.2. The locally adopted linear *stress* hypothesis is not always correct, because if a non-linear constitutive law has been adopted in the compression branch, the linearity of the strains does not imply linearity of stresses. Also for this reason, if the size of the elements is not small, some equilibrium violation may be inserted in the SCOFEM, if the linear stress assumption is adopted. However, if mesh size is reduced, these violations are usually negligible.

9.6.5.4 Welds

Using the IRFEM model, the stresses n, t_{par}, and t_{per} flowing into each weld seam are known, in every load case. If each weld seam is divided into a proper number of equal segments, the forces resulting by the integral summation of these stresses in the area of each segment can be computed and assigned with sign reversed to the nodes of the connected constituent. The equally spaced points defining the equal segments of each weld seam are all hard points of meshing and therefore a node is available in the SCOFEM for each of them.

This node is then loaded by the equivalent force computed considering the stresses n, t_{par}, and t_{per} acting over the weld seam. The sum of all the nodal forces applied to the weld seams nodes are equivalent to the forces exerted by the weld layouts over the constituent under examination.

9.6.5.5 Far End Member Force Packets

These forces are not necessary if the SCOFEM of a member is to be analyzed, because the far end of the member is constrained. However, when considering finite element models of sets of components it may be necessary to consider members with no constraints, loaded at their far extremity with the aim of loading connections. These loads are equivalent to the internal beam-like forces computed at some distance from the near extremity **N**, and applied to the center of the member cross-section.

Usually the force packet can be computed considering the near end member forces, transferred to the far extremity (this implies no other load applied to the member stump, or that if this local load is present its effects are negligible. If the local load is not negligible it can be added to the SCOFEM, and the far extremity internal forces must take into account its presence). In order to apply the forces to the member, a single notional node can be loaded by the needed force packet (three forces and three moments), and this node can then be connected to the far end member cross-section nodes by rigid links, or rigid beam elements (see Figures 9.1 and 10.2).

9.6.6 Members: Deciding Member-Stump-Length

The length of the member stumps directly affects the size of the finite element models of the members, that is, the number of nodes and so the number of degrees of freedom. Therefore, the analyst has a specific interest in reducing the size of the member stumps to a minimum.

On the other hand, the size of the member stumps must not be reduced too much (Figure 9.39), or possible stress concentrations or peaks can be missed. The rule to follow is that the distribution of the Von Mises stresses in the member stump, must reach that of the De Saint Venant's prism. In turn, this means that the stress distribution computable with beam finite element models must be reached, far from the nodal region.

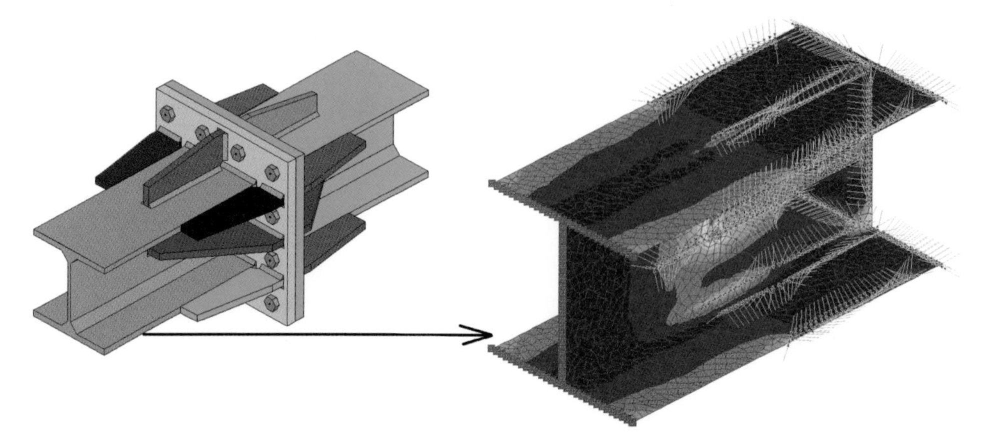

Figure 9.39 An example of a perhaps too short member stump.

This distribution is that computed and checked in the BFEM. If it is true that the finite element model of a member will also provide useful information for the member checks, this is not the tool commonly used to check members. However, sometimes it happens that also in the regions far from the node of the BFEM, and governed by De Saint Venant's principle, the member plate–shell finite element model provides information useful for the member checks. This happens when the placing of the members in the renode is not coherent with the position of the members in the BFEM, or when the checks done in the BFEM have been too coarse or are even missing.

9.6.7 Compatibility Issues

The deformation of an SCOFEM is not compatible with that of the neighbor component. This is expected, as the static theorem ensures balanced configurations generally not compatible. However, in some cases, when high compressive contact forces are exchanged, the compatibility can be improved by some simple modification to the SCOFEM.

A particularly clear example is that of base plates (Figure 9.29). These constituents are balanced under the loads applied, but their deformation is clearly incompatible with that of the column end face, as the plate bends under the footprint of the column, losing contact with it. If a particularly simple loading condition is considered, that of a purely compressed column, several issues can be cleared. In the analytical model of the weld layouts, contact non-linearity is usually neglected, so the compression will flow into the welds.

A pure compression load will imply constant n stresses on the faying surface (let us imagine neglecting the effect related to weld thickness differences). These stresses are balanced by the pressure field exerted by the foundation, arriving from below. No bending stresses are locally applied at the column plates, as they would imply a change of stress in the weld thickness, while the computational model for the welds usually neglects these stresses.

However, if a high permanent compression is applied, and if the column is initially in contact with the plate, the force is transferred directly by contact, and the welds will not be loaded.

Moreover, this initial assumed very high compressive stresses in the thicknesses of the end column cross-section will be able to locally absorb and limit the base plate bending under the cross-section flanges and webs (generally speaking the plates comprising the column cross-section) simply reducing and increasing their values, *but always remaining compressive*. If a high pre-compression is applied, and it never vanishes, there is no need of other connectors.

So, for base plates under permanent high compressive forces, these effects can be modeled:

- assuming that the weld nodes (loaded) are fully fixed in the SCOFEM of the base plate; this means considering the footprint of the welds as clamps and no extra constraints will then be applied to the SCOFEM
- neglecting the compressive n stresses while keeping only t_{par} and t_{per} when checking for the welds; if the shear forces exchanged at the interface are below the sliding ones, considering a friction-resisting mechanism, the checks of the welds connecting the base plate to the column can be totally discarded
- verifying that the reactions at the clamped nodes do not imply tensile stresses normal to the faying plane of the weld layout, as these stresses would necessarily flow in the welds, and not be transferred by contact.

So, weld layouts can be marked as "using contact" when the *permanent* compressive stresses on the faying surface are expected high, so that in no combination is it expected to reach tensile n stresses in the welds. If this is done the checks of these welds can be modified, discarding the compressive n, and checking the resultant tangential stresses t, ($t^2 = t_{par}{}^2 + t_{per}{}^2$) against a friction limit (typically $\mu|n|$ where μ is the friction coefficient and n the compressive stress).

If there is not sufficient confidence in the permanent existence of these high compressive stresses, this is an unsafe approach.

9.7 Multiple Constituents Finite Element Models (MCOFEM)

9.7.1 Goal and Use

In this section the typical problems arising when using a partial modeling of the renode – that is, a kind of modeling between SCOFEM and PFEM – will be described. This means that using at least in part the force packets computed by the IRFEM, a multiple constituent finite element model, MCOFEM, is set up in order to get results nearer to those of PFEM, at a lower computational cost. This is needed when the distribution of stresses computed by IRFEM is not considered realistic enough by the analyst, and then, still avoiding the PFEM approach, more refined results in a well delimited area of the renode are searched. As more than one constituent is inside the MCOFEM, some connector internal to the MCOFEM will be *saturated*, that is, it will have to be partially or totally modeled by specific subconnector finite elements (therefore, no longer *layout* finite elements as in IRFEM).

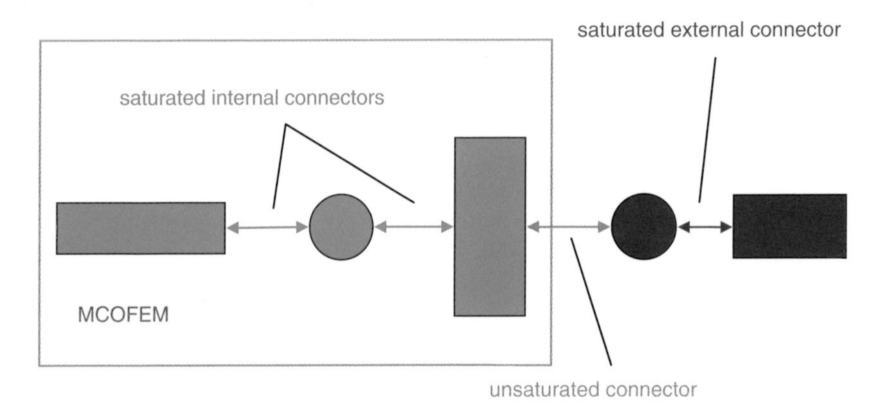

Figure 9.40 Saturated and unsaturated connectors.

A connector (Figure 9.40) is said to be *saturated* if all its extremities are connected to constituents which are *all* part of the MCOFEM (*internal saturated connectors*), or to constituents which are *all* outside of the MCOFEM (*external saturated connectors*).

A connector is said to be *unsaturated*, if it is connected at least at one of its extremities to one constituent that *is not* included in the set of constituents modeled by the MCOFEM, and if it is also connected at least at one extremity to a constituent that *is* included in the set. The need of cutting at least two parts the renode to extract the MCOFEM implies the existence of at least one unsaturated connector.

An internal saturated connector must be explicitly modeled by specific finite elements, related to its subconnectors (and this allows us to relax the hypothesis of linear displacement field).

An external saturated connector is not part of the MCOFEM.

An unsaturated connector is modeled by the forces it exerts on the MCOFEM. The elementary forces loading the MCOFEM are obtained by the IRFEM, and for this reason a MCOFEM is partly related to the results of the IRFEM; it is not yet pure fem, but it is still a hybrid approach.

The "bridge" elements simulating the single bolts and weld seam segments, i.e. the saturated internal connectors, have in great part already been described in Chapter 6 and 7. The discussion of their use in plate–shell finite element models is in Chapter 10.

9.7.2 Mesh Compatibility Between Constituents and Connector Elements

Constituents are meshed independently of one another. This greatly improves the speed and ease of meshing. As already mentioned for stiffeners, the author has found that the modeling of connected components does not require the merging of nodes, but the addition of "bridge" elements simulating single subconnectors. This idea, implemented in 2008, was of great help and opened the path to the automatic modeling of the whole renode with a unique, standardized approach.

9.7.3 Saturated Internal Bolt Layouts and Contact Non-Linearity

If the analysis of MCOFEM is run in the non-linear range, using contact non-linearity, then the bolt layouts using a bearing surface can be modeled explicitly by single-bolt finite elements. The bearing surface will then be one of the unknowns of the problem, and the field of pressures exchanged will be found by analysis.

However, if contact non-linearity is not included in the analysis, then these spreading pressures cannot be found by the analysis, modeling the single bolts with single finite elements. If the bolts are considered both in tension and compression, this would be equivalent to adopting the AISC-allowed approach of using the bolt shafts to absorb the needed contact pressures (Section 7.5.2.1).

If an intermediate approach is chosen, so as to avoid the computational cost of a contact non-linear analysis, but modeling the pressures exchanged more realistically, then the bolt layouts using a bearing surface can be replaced by the forces *and pressures* computed by the linear strain field hypothesis (Section 7.9.2). So, two sets of forces and pressures will be applied, one loading part A, and one, with sign reversed, loading part B (Figure 9.41). As no finite element connecting the two parts is added, *one more body will have to be constrained*. The constraints, typically applied to members, will always be

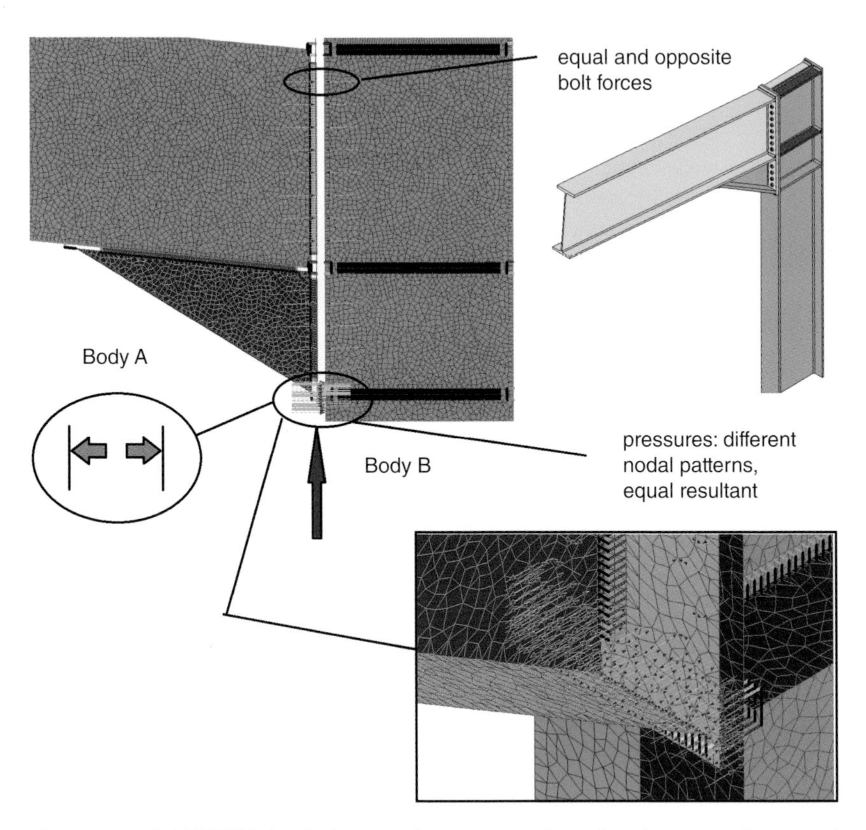

Figure 9.41 A MCOFEM detail, showing the two sets of equal and opposite forces exchanged at an internal saturated bolt layout using a bearing surface.

applied to their far extremity, and so no perturbation of the stress state *in the nodal region* is to be expected, provided the member stump is sufficiently long. If a saturated bolt layout using a bearing surface is in the MCOFEM, and if contact non-linearity is *not* used, the MCOFEM is split into two bodies, because a break of connection continuity will be related to that particular bolt layout.

9.7.4 Constraints

Depending on how the MCOFEM is prepared, the needed constraints should be applied to one or more than one constituent.

If the MCOFEM includes only one member, usually that member is to be constrained, at its far extremity.

If more than one member is included in the MCOFEM, and all the slave members are connected to the master by connectors explicitly modeled by single subconnector finite elements, then only one equivalent body is in the MCOFEM, and the master member is typically the constituent to be constrained.

If more than one member is modeled, and n slave members are connected to the master by bolt layouts not explicitly modeled with finite elements for the reasons explained in the previous section, then all these n slave member have to be constrained as well as the master member. The other members normally connected by saturated connectors, must not be constrained. They will be loaded at their far extremity, and their loads will flow into the connectors, and from there to the other force transferrers and members connected.

If no member is modeled in the MCOFEM, and at least one force transferrer, then at least one of the constituents included in the MCOFEM must be constrained. The model will be self-balanced, and the constraints applied notionally. The same considerations already applied to the members are also valid for force transferrers, if one or more bolt layouts using a bearing surface are included in the MCOFEM, and the contact non-linearity is not tackled by the analysis.

Single stiffeners cannot be a SCOFEM, and sets of stiffeners cannot be, as such, a MCOFEM.

9.7.5 Stabilizing Springs and Buckling of Members

If more than one member is included in the MCOFEM, and this member is a truss in BFEM, in order to avoid the spurious cantilever-like buckling of this possibly compressed member in buckling analyses, two stiff stabilizing springs can be applied at the far extremity **F** node of the member in the MCOFEM. These stabilizing springs are directed normally to the member axis and are themselves perpendicular. They should be stiff enough to prevent the cantilever buckling of the member.

Due to the existence of bending and shear, this trick cannot be applied to a member simulated by beam elements in BFEM, as the stress state would be perturbed: part of the loads applied at far extremity would flow in the springs.

This means that if a buckling analysis is done by means of an eigenvalue analysis, some critical multipliers will have to be discarded. It may happen for instance that the first critical multiplier is related to this spurious buckling mode. If this happens the analyst has two choices:

1) Consider the second, third, and successive critical multipliers until a critical shape related to a realistic buckling mode is found. The analyst should discard those critical shapes related to the existence of member stumps. From this side, shorter member stumps are preferred to long ones.
2) Adopt the first critical multiplier even if it is not realistic, because it is certainly lower than the critical multipliers related to meaningful critical modes. This is a heuristic valid when it is not needed to get higher critical multipliers in order to prove that the constituent is to be considered safe (according to the general method, see Chapter 5).

If a true non-linear step-by-step analysis including geometrical non-linearity is run, there is no way to prevent the model diverging following spurious buckling modes. Or the load multiplier is sufficiently high to be considered safe (as once more the meaningful mode are related to higher load levels), or the analysis must be discarded and re-run with shorter member stumps. Of course, if the member is only axially loaded, the stabilizing springs method can be used.

9.7.6 Need for Rechecks

In general, the saturated connectors of the MCOFEM will be loaded by forces different from those estimated by the IRFEM. In order to keep coherence and avoid violations of the static theorem, the checks of the connectors should be rerun using the elementary internal forces computed by the MCOFEM. Also the local checks of the components connected must be rerun (bolt bearing, punching shear, block shear). The MCOFEM can be seen as an additional model to better evaluate the internal distribution of the sub-connector forces in the saturated connectors or, for hyperconnected connections, as a system to evaluate differently the force packets flowing into the saturated connectors.

Both the IRFEM force packets and internal saturated connectors force distribution, and the MCOFEM ones, are under the shelter of the static theorem of limit analysis, as both are balanced to the externally applied loads. However, if the MCOFEM has been used, it means that the analyst prefers the MCOFEM as considered more realistic and nearer to the probable force distribution of the real system.

9.8 A Path for Hybrid Approach

The following is an outline of the path to check a renode by hybrid approach:

1) Set the working mode of each connector (fully resistant shear-only, no-shear, or longitudinal-only flags, bearing surface extent and constitutive law, flexibility indices, prying forces factors). This is a part where the analyst's judgment is fundamental.
2) Prepare and run the IRFEM for all loading combinations and all loading instances (automatic).
3) Check the connectors using the forces and stresses flowing into the subconstituents, computed from the force packets evaluated at step 2 by the IRFEM. Modify the axial forces in the bolts to take into account prying forces (automatic).

4) Reverse signs and check for bolt bearing, punching shear, block tear (automatic).
5) Check the constituents for resistance and stability using the following tools:
   a) Traditional approaches
      i) simplified resistance checks (net cross-sections, Whitmore sections; can be done automatically, possibly "teaching" the sections to use.
      ii) closed formulae checks; can be done automatically
   b) Finite element models of single constituents using plate–shell elements (the analyst may decide which constituents, of course also all).
      i) linear static analysis – automatic, but needs expert's judgment
      ii) non-linear static analysis – automatic, but needs expert's judgment
      iii) linear buckling analysis – automatic, but needs expert's judgment
   c) Finite element models of sets of constituents using plate–shell elements.
      i) linear static analysis – automatic, but needs expert's judgment
      ii) non-linear static analysis, possibly also using contact non-linearity – automatic, but needs expert's judgment
      iii) linear buckling – automatic, but needs expert's judgment
      iv) non-linear buckling (geometrically non-linear analysis, possibly also using contact non-linearity) – automatic, but needs expert's judgment
      v) recheck the saturated connectors of the model of the sets, repeating the steps 3 and 4. (automatic); this is to allow checks fully coherent with stress evaluation.
6) Possibly change the settings set at step 1, and re-execute.

The hybrid approach should be used first, and very often does not need further improvements.

If all the constituents are checked, this will be sufficient, because the solution is balanced, fully compliant to the third law, and fully compliant with plastic resistance of constituents. If the constituents are not checked, or if the sizing should be improved, it may be wished to use the PFEM approach, that is, to directly use a more complex modeling of the renode. This will be outlined in the next chapter.

References

Crisfield, M.A. 1997, *Non-linear Finite Element Analysis of Solids and Structures Volume 2 Advanced Topics*, John Wiley & Sons.

Eurocode 3 Part 1.5, *Design of Steel Structures – Plated Structural Elements*, EN 1993-1-5:2006, CEN.

Moment Connections, 1995, SCI Steel Construction Institute, 207/95 ("Green Book").

Rugarli, P. 2010, *Structural Analysis with Finite Elements*, Thomas Telford, London.

Rugarli, P. 2014, *Validazione Strutturale Volume 1: Aspetti Generali*, EPC Libri, Rome.

Ruppert, J. 1995, *A Delaunay Refinement Algorithm for Quality Two-Dimensional Mesh Generation*, Journal of Algorithms. **18** (3): 548–585.

Thornton, W.A. 1985, *Prying Action – a General Treatment*, Engineering Journal, American Institute of Steel Construction, Second Quarter 1985.

Yastrebov, V.A. 2010, *Introduction to Computational Contact Mechanics – Part 1 Basis*, WEMESURF short course on contact mechanics and tribology, Paris, France, 21–24 June 2010, Centre des Matériaux, MINES ParisTech.

10

Analysis: Pure FEM Approach

10.1 Losing the Subconnector Organization

By using the name pure fem (PFEM), in this book we are referring to a finite element model dealing with the problem of steel connection analysis that does not rely on any basic assumptions regarding the behavior of the connector subcomponents.

In the previous chapters, connectors have been seen as *layouts*, i.e. sets of organized subcomponents, obeying some (usually kinematic) assumption.

If these hypotheses are relaxed, and each single subconnector is modeled independently of the others, and if all the connectors existing in the renode are modeled in this way, then the resulting plate–shell (or also solid) finite element model is called pure fem.

Relaxing the basic assumptions of the organized kinematic or static behavior of subconnectors, immediately implies that most of the simple rules usually adopted for computing the stresses in the welds or the forces in the bolts lose their effectiveness. More precisely, as the traditional computational rules used to compute connectors are indirectly based on a rigid body movement assumption (that is the linear strain or linear displacement hypotheses), these rules will be violated more thoroughly when the rigid-body movement hypothesis is further violated, that is, with thin components in the absence of stiffeners.

So, one common issue when the flexibility of the components is correctly taken into account, is that the forces and stresses flowing into the connectors, and the stress states that the constituents undergo, are no longer the simple ones we are accustomed to.

In order to show how this is true even for the simplest joints, an FR beam-to-column joint will be considered, with no stiffeners (Figure 10.1). Both the beam and the column have a HEB 200 cross-section. The beam is fillet welded to the column. Connections of this type (but using full penetration welds) have been extensively used in seismic areas of the world, and were discussed in depth after the Northridge earthquake.

In their original version, they used bolted web and welded flanges (full penetration), and were designed to transfer the plastic moment of the beam in steel structures using MRF frames. These structures (see Bruneau et al. 2011) had been progressively weakened by reducing redundancy, that is, the number of MRF, and by avoiding the application of stiffeners to the column in continuity with the beam flanges, in the years from before the earthquake, so that a number of existing steel structures using a very limited number of such MRF frames, two for each direction, with no column stiffeners, were in place at the

Steel Connection Analysis, First Edition. Paolo Rugarli.
© 2018 John Wiley & Sons Ltd. Published 2018 by John Wiley & Sons Ltd.

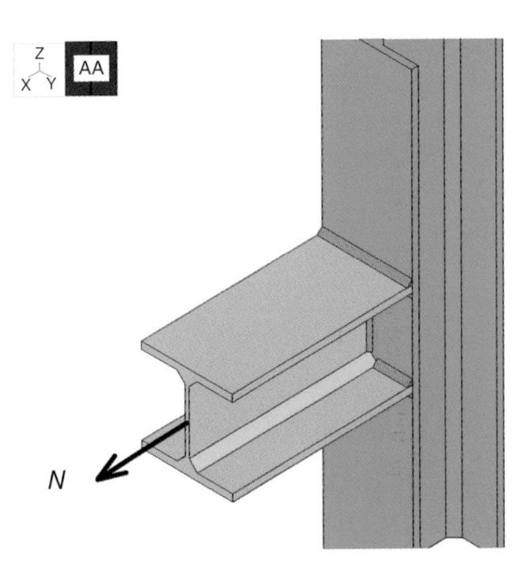

Figure 10.1 Beam fillet welded to a column under a tensile force. Both cross-sections are HEB 200.

moment of the earthquake. After the Northridge earthquake, it became clear that these connections had severe failures and an extended examination of them was carried out in the Northridge area. Many of them were found to have severe cracks, also spreading to the column flange and to the column web. Among the several reasons that were proposed in order to explain the cracks, the uneven stress distribution in the beam flange welds was discussed. It is the author's opinion that this must be considered as one possible key issue, because the evaluation of the weld stress made by means of the traditional approaches is completely inapplicable if the column flange is flexible enough.

In order to explain the issue, and to see its central relevance to the discussion that is being carried on, a simple tensile force applied to the beam is considered. Equivalently, a strong-axis bending moment applied to the beam can also be considered. Both imply a tensile force applied to the beam flange, and this tensile force must be transferred to the column flange by the fillet or penetration weld.

Now if the traditional approaches are used, and a tensile force N is applied to the beam, the stress flowing in the beam flange weld *is constant from point to point of the weld* and equal to N/A_w, where A_w is the area of the weld (a similar reasoning would also apply for the strong axis bending). This implicitly assumes that the weld experiences a rigid body displacement, so that all its points displace in the very same manner. This is a special case of the linear displacement hypothesis. If a force N equal to 0.3 times the elastic cross-section limit is applied as tensile force, the utilization factor computed in the weld layout of Figure 10.1 by the traditional approaches using for example Eurocode 3 is 0.671. The normal stress is 156.8 MPa, constant from fillet to fillet.

The linear displacement hypothesis can be verified in a numerical experiment, where the whole renode is modeled with a PFEM (Figure 10.2). In this example, the weld has been considered as a fillet weld, but the same results would be obtained by considering it a penetration weld. In order to ensure the rigid body movement of the weld, the plate–shell elements of the beam and the column are considered rigid, by assigning a very stiff material.

All the weld segments (Figure 10.2) displace in the very same manner and an even stress distribution in welds is found. The normal stress found in each weld segment is exactly the same as would have been predicted using traditional approaches: a constant stress in the welds. Also the utilization ratios computed by the forces flowing in the weld

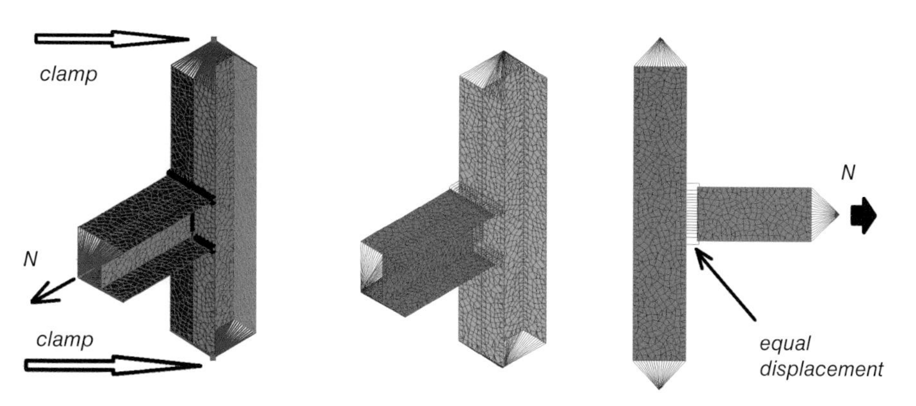

Figure 10.2 Pulled beam. The beam and the column are rigid. The displacements of the weld segments are all equal. This is a simulation of what traditional approaches do.

segments are exactly the same as those obtained by the simplified methods, using a "weld layout".

However, if a PFEM model is set up with the right flexibility for the beam and column plate–shell elements, in the linear range, the distribution is totally different (Figure 10.3).

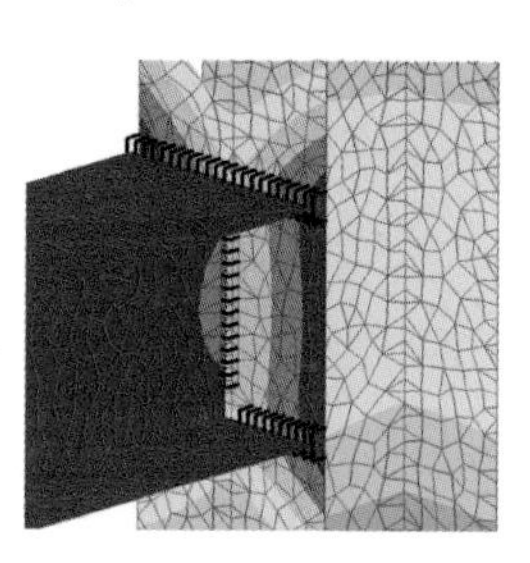

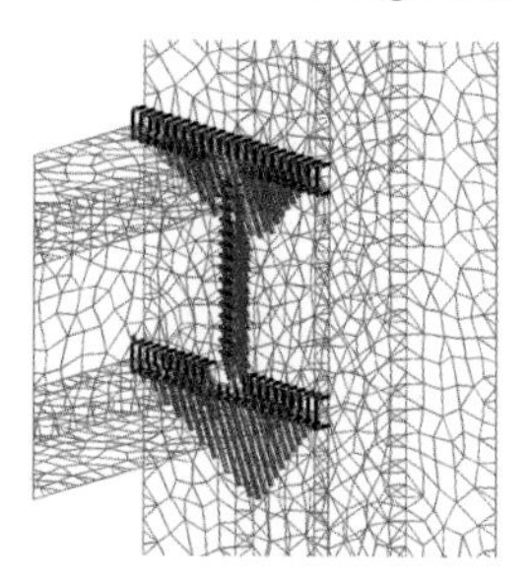

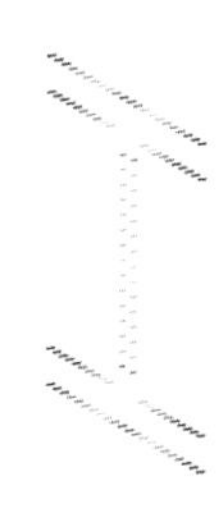

Figure 10.3 Pulled beam. The beam and the column are modeled with their flexibility resulting from the thickness and the material. The displacements of the weld segments are totally different. A stress concentration at mid beam flange is observed. Top left: deformed view. Top right: false color map of the displacement normal to the column flange. Bottom left: distribution of axial forces in the weld segments. Bottom right: false color map of the normal stresses in weld segments.

In particular, note the effect discussed by several authors (Bruneau et al. 2011), that is, the stress concentration in the mid part of the beam flange weld. The maximum predicted normal stress in the weld would be (locally, at mid flange, in some weld segments) 329.5 MPa, instead of 156.8 MPa: more than double its value. So, evaluating the stresses by using the traditional approaches would not be safe.

If 2 + 2 stiffeners are added to the column (of the same thickness as the column flange, 15 mm, encased in the column and in line with beam flanges, see Figure 10.4), the out-of-

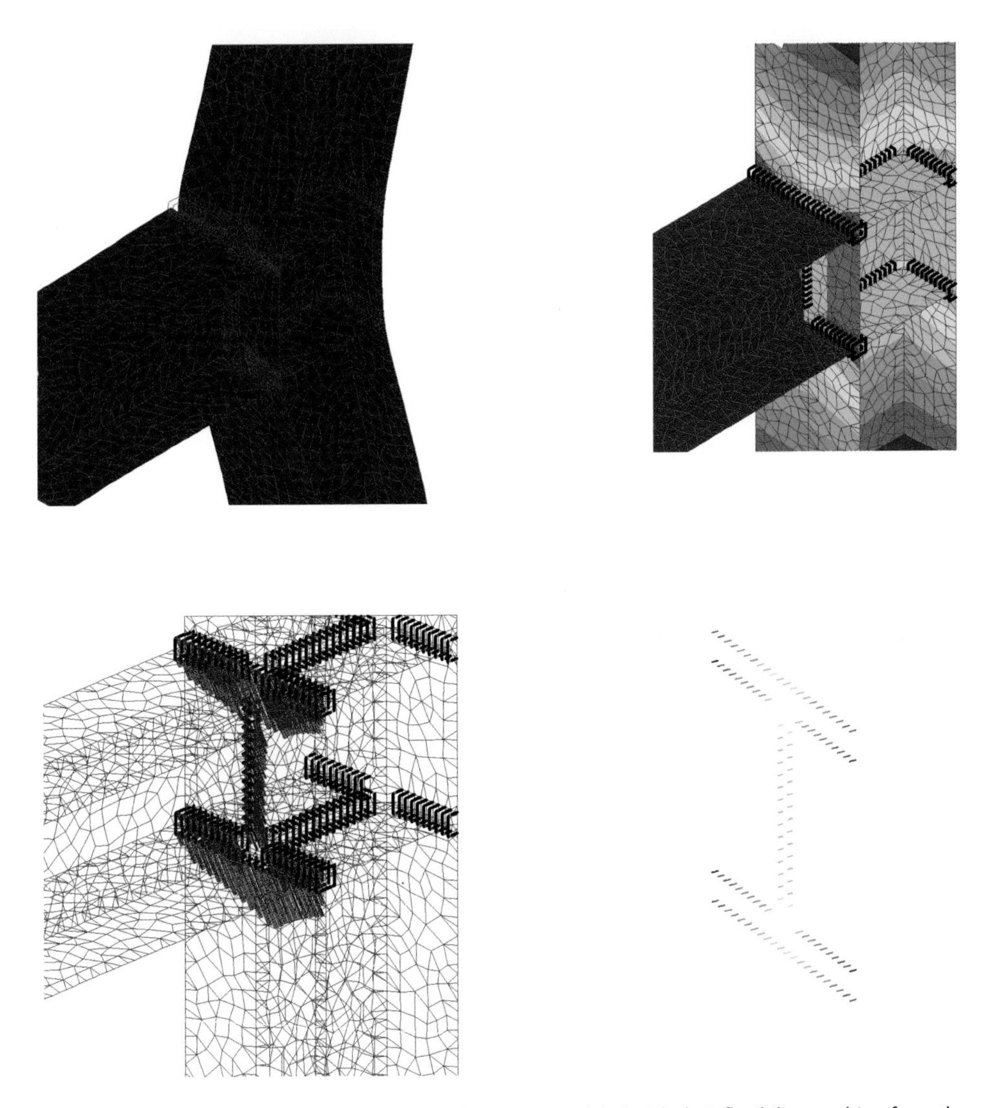

Figure 10.4 Pulled beam. The beam and the column are modeled with their flexibility resulting from the thickness and the material. Stiffeners added to the column. The displacements of the weld segments are different. A stress concentration at mid beam flange is observed. Top left: deformed view. Top right: false color map of the displacement normal to the column flange. Bottom left: distribution of axial forces in the weld segments. Bottom right: false color map of the normal stresses in weld segments.

plane deformability of the flange is greatly reduced but it is not made zero. The deformation is qualitatively similar to that of the no-stiffeners case. The peak of normal stress is still at the center of the flange weld, but now it is 248.8 MPa, still much higher than the value predicted by the simplified approaches.

It is the author's opinion that these results require a major rethinking of the methods normally used to evaluate the stresses in the connectors and the safety margin of steel structures. Analyzing the connection with more realistic models, a whole new world, almost completely unexplored and unknown, is opened in front of the analyst. A small indication of the importance of this kind of issue has already been provided by the described structural failures, so care needs to be taken to properly consider all this in the next generation of standards.

10.2 Finite Elements for Welds

10.2.1 Introduction

In the context of PFEM, or when MCOFEMs are prepared, the saturated weld layouts are modeled dividing each weld seam into a number of weld segments, and modeling each of these segments by a suitable finite element.

The nodes of the weld segments are placed over the mid-surface of the plates connected, but the extremities of the elements are on the surfaces welded. This implies that rigid offsets are applied at the two element nodes (Figure 10.5). The weld segment element axis is always orthogonal to the faying surface. The length of the weld segment is t for penetration welds, and $t/2$ for fillet welds.

In order to have an appropriate modeling, the number of weld segments should be high enough. This number is usually related to the mesh size, by dividing each weld seam into a number of equal elements having a size comparable to that required by the analyst.

The finite elements modeling the weld segments do not use bending moments or torque, that is, they use only the translational degrees of freedom. This does not imply free rotations, because the plate–shell elements to which the weld segments are attached have the necessary stiffness to avoid such a problem. The bending moment M_{vi} and the torque

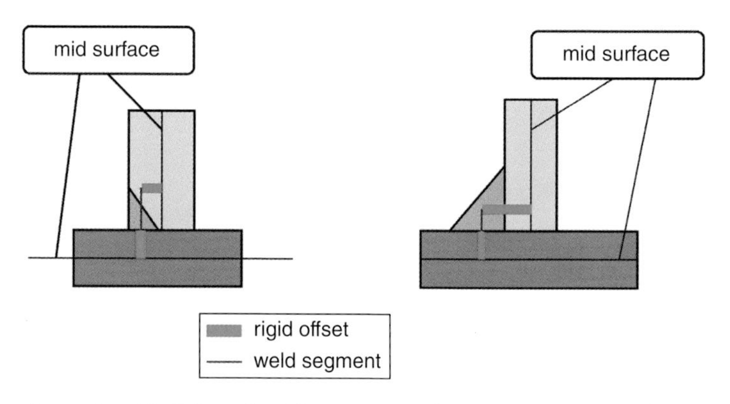

Figure 10.5 Modeling of weld segments. Left: penetration. Right: fillet.

M_{wi} taken by each weld seam i will be related to the normal forces F_{wij} and to the shears F_{vij} taken by the weld segments j. No bending M_{ui} is expected.

The cross-section used for each weld segment is a rectangular section and the area of each segment is related to the tributary area of each node. The weld segments at the extremities of the weld seam have half the area of the other weld segments, all having the same area. In this way, the nodes correctly respect the extension of the weld seam (Figure 10.6).

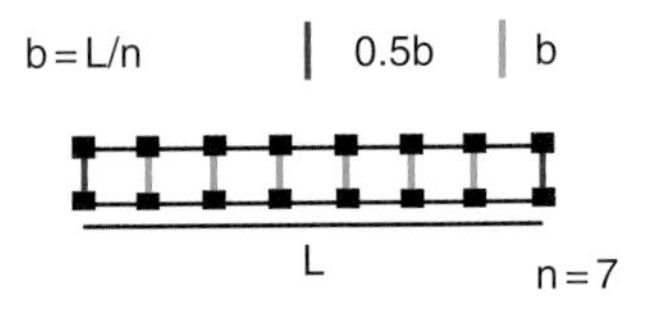

Figure 10.6 Tributary areas for weld segments.

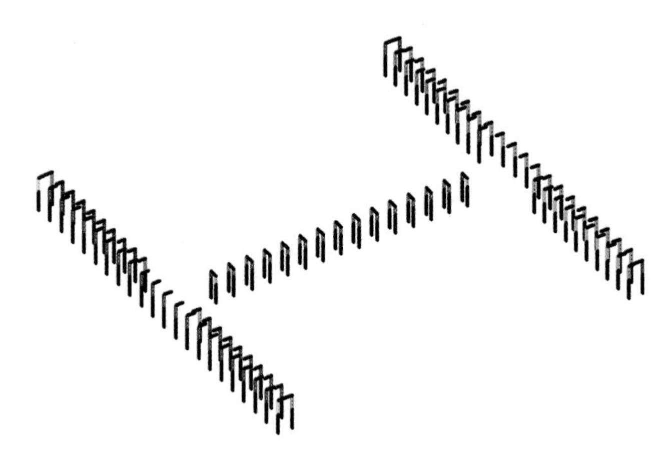

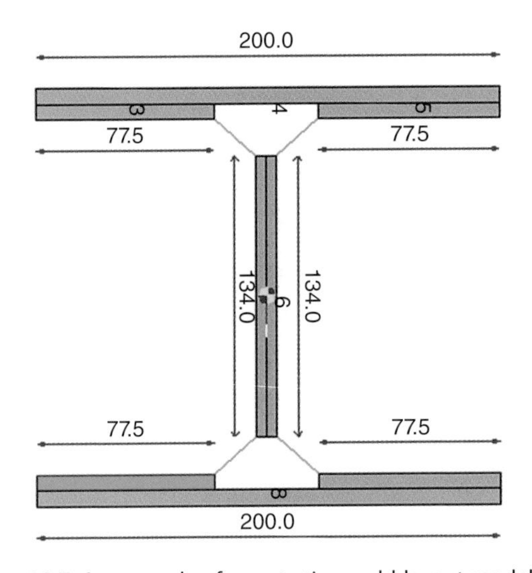

Figure 10.7 An example of penetration weld layout modeled in PFEM.

The penetration weld segments are not connected longitudinally (see Figure 10.7). They correctly react to all the elementary translational acts of motion. Namely axial (direction **wi**), longitudinal (direction **ui**), and transverse (direction **vi**).

For fillet welds, the nodes of weld seams placed on the two surfaces of the same thickness may have different distance one another, and different positions, but all lie over the mid-surface, therefore it is wise to locally protect the plate elements (especially in the non-linear range) by a line of fictitious beam elements to be kept linear, and having a square cross-section of size a, that is, the throat size of the fillet. This protects the plate by the delivery of uneven forces normal to its mid-surface, and by the related spurious bending.

It must be stressed that the evaluation of the stress state of the connected plate just at the welds should be considered as out of scale: to get such a stress, solid elements at a greater detail scale would be necessary.

The lines follow the **ui** direction of each weld seam. These beam elements will also partly stiffen the plate for out-of-plane bending, so as to partially reduce the apparent flexibility.

10.2.2 Penetration Welds

10.2.2.1 Linear Element

Each weld segment can be modeled by an element having the stiffnesses k_N, k_{Vu}, and k_{Vv} placed in the following positions in the 6 × 6 stiffness matrix using the displacement vector $\mathbf{d} = \{d_{w1}, d_{u1}, d_{v1}, d_{w2}, d_{u2}, d_{v2}\}^T$:

$$\mathbf{K} = \begin{vmatrix} k_N & 0 & 0 & -k_N & 0 & 0 \\ 0 & k_{Vu} & 0 & 0 & -k_{Vu} & 0 \\ 0 & 0 & k_{Vv} & 0 & 0 & -k_{Vv} \\ -k_N & 0 & 0 & k_N & 0 & 0 \\ 0 & -k_{Vu} & 0 & 0 & k_{Vu} & 0 \\ 0 & 0 & -k_{Vv} & 0 & 0 & k_{Vv} \end{vmatrix} \tag{10.1}$$

If the stiffness of the elements is considered proportional to the area only, and not to the thickness of the single weld seam, the following choices apply:

$$k_N = \frac{E\mu_{ni}A}{t_{ave}}$$

$$k_{Vu} = \frac{G\mu_{ti}A}{t_{ave}}$$

$$k_{Vv} = \frac{G\mu_{li}\mu_{ti}A}{t_{ave}}$$

where E is Young's modulus, G is the shear modulus, A is the area of the weld segment, and t_{ave} is the average thickness of the weld seams of the weld layout. The working mode modifiers μ_{ni}, μ_{li}, μ_{ti} of the weld seam i to which the weld segment belongs are also used.

10.2.2.2 Non-Linear Element

In the non-linear range the element is kept equal to the linear one, but a resistance check is made using appropriate limiting conditions. The normal stress σ can be evaluated as

$$\sigma = \frac{F_w}{A}$$

while the tangential stress τ is

$$\tau = \sqrt{\frac{F_u^2 + F_v^2}{A}}$$

10.2.2.3 Example

The same test examined in Chapter 6 (Section 6.6.6) has been studied using an MCO-FEM. A column is welded to a plate and it is then bolted to a constraint block. The plate–shell fem models the HEB 200 column, the 300 mm × 300 mm plate having a thickness of 20 mm (rather flexible), and all the penetration welds of the weld layout used to connect the column to the plate. The bolts are not saturated, and so only the forces they exert are applied.

The model is an MCOFEM because the loads coming from below due to the bolts are evaluated using the results of the IRFEM.

If the analysis is done by considering the flexibility of the plates, the distribution of the stresses in the welds is uneven, and cannot be compared to the distribution obtained by traditional approaches (see Figure 10.8, no matter the thinness of the plate and the severity of the loading condition).

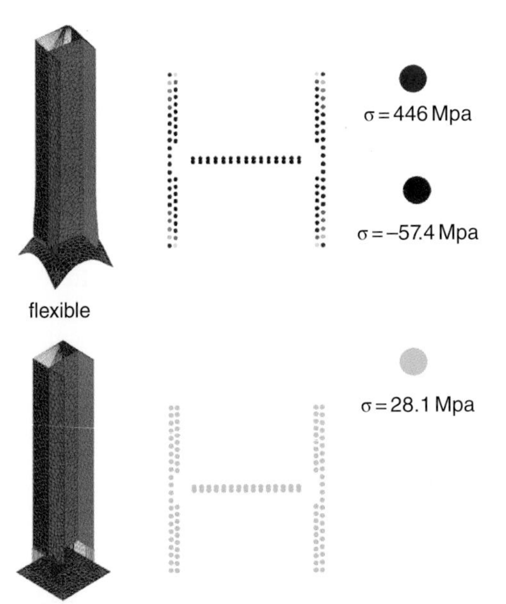

Figure 10.8 Flexible and rigid MCOFEMs using penetration welds in pure tension (different scale used).

In order to establish a direct comparison, the plate–shell elements are made rigid, so as to simulate the IRFEM and the results that would be obtained by traditional approaches.

If the results from the pure tensile force condition with the flexible and the stiff models are compared, very stark differences are found. Using the rigid model the stress distribution is even, as predicted by the simple traditional methods. If the plate is flexible, the distribution of the stresses in the welds is totally different. Prying forces are not taken into account in this example.

All four tests of Tables 6.1–6.4 have been reproduced using rigid MCOFEMs, giving results identical to those obtained in Chapter 6 using the layout approach and closed

Table 10.1 Comparison of the main results obtained using IRFEM and the closed formulae discussed in Chapter 6, and the results obtained by rigid MCOFEMs, simulating the IRFEM and the range of applicability of the traditional methods.

Meaning	Reference table for IRFEM	Load case	IRFEM value (MPa)	Rigid MCOFEM (MPa)
n_{per}	6.1	1	28.1	28.1
t	6.1	2	11.1	11.1
t	6.1	3	5.2	5.2
t	6.1	4	0.8	0.8
n_{per}	6.1	5	26.9	26.9
n_{per}	6.1	6	23.7	23.6
n_{per}	6.2	1	34.5	34.5
t	6.2	2	11.1	11.1
t	6.2	3	5.2	5.2
t	6.2	4	0.8	0.8
n_{per}	6.2	5	27.9	28.0
n_{per}	6.2	6	23.7	23.6
n_{per}	6.3	1	34.5	34.5
t	6.3	2	13.6	13.6
t	6.3	3	27.8	27.8
t	6.3	4	1.1	1.1
n_{per}	6.3	5	27.9	28.0
n_{per}	6.3	6	23.7	23.6
n_{per}	6.4	1	74.0	74.0
t	6.4	2	23.8	23.8
t	6.4	3	27.9	27.9
t	6.4	4	1.9	1.9
n_{per}	6.4	5	56.2	56.2
n_{per}	6.4	6	46.2	46.0

formulae (see Table 10.1 where, for each analysis, the maximum stress is compared). This means that the finite elements used for the weld segments are equivalent to the closed-formulae approach, as long as the plate elements are rigid. It also means, once more, that the evaluation of the state of stress of connectors of flexible parts cannot rely safely on the simple and traditional approaches, an issue that deserves a deeper consideration by the standards.

10.2.3 Fillet Welds

10.2.3.1 Linear Elements

For fillet welds in the linear range the same stiffness matrix as Equation 10.1 is used, with the replacements of Tables 10.2–10.5.

The secant stiffness for the orthotropic approach is the modified one (Section 6.7.1).

Table 10.2 Terms of the stiffness matrix for isotropic fillet weld segment and initial stiffness.

k_N	k_{Vu}	k_{Vv}
$546\cdot\frac{A\cdot f_{uw}}{t_{ave}}\cdot\mu_{ni}$	$546\cdot\frac{A\cdot f_{uw}}{t_{ave}}\cdot\mu_{ti}$	$546\cdot\frac{A\cdot f_{uw}}{t_{ave}}\cdot\mu_{ti}\cdot\mu_{li}$

Table 10.3 Terms of the stiffness matrix for isotropic fillet weld segment and secant stiffness.

k_N	k_{Vu}	k_{Vv}
$3.58\cdot\frac{A\cdot f_{uw}}{t_{ave}}\cdot\mu_{ni}$	$3.58\cdot\frac{A\cdot f_{uw}}{t_{ave}}\cdot\mu_{ti}$	$3.58\cdot\frac{A\cdot f_{uw}}{t_{ave}}\cdot\mu_{ti}\cdot\mu_{li}$

Table 10.4 Terms of the stiffness matrix for orthotropic fillet weld segment and initial stiffness.

k_N	k_{Vu}	k_{Vv}
$2793\cdot\frac{A\cdot f_{uw}}{t_i}\cdot\mu_{ni}$	$546\cdot\frac{A\cdot f_{uw}}{t_i}\cdot\mu_{ti}$	$2793\cdot\frac{A\cdot f_{uw}}{t_i}\cdot\mu_{ti}\cdot\mu_{li}$

Table 10.5 Terms of the stiffness matrix for orthotropic fillet weld segment and modified secant stiffness.

k_N	k_{Vu}	k_{Vv}
$18.3\cdot\frac{A\cdot f_{uw}}{t_i}\cdot\mu_{ni}$	$10.14\cdot\frac{A\cdot f_{uw}}{t_i}\cdot\mu_{ti}$	$18.3\cdot\frac{A\cdot f_{uw}}{t_i}\cdot\mu_{ti}\cdot\mu_{li}$

10.2.3.2 Non-Linear Elements

The fillet weld segment in the non-linear range should be computed using the instantaneous center of rotation method, and in particular using the tangent stiffness matrix of Appendix 2, section A2.1. This tangent stiffness matrix does not have the form of Equation 10.1 because it is not diagonal. In Appendix 2 there is the formulation of the tangent stiffness matrix in closed form, for both the orthotropic and the isotropic approach.

In a non-linear analysis, if a weld segment has reached the maximum strain $p = 1$, it is declared to be broken and its stiffness matrix made null.

Non-linear analyses in which the welds also have a non-linear behavior require more iterations to converge. So, the maximum number of iterations before declaring the lack of convergence should be increased from 20 to 40 or 60. If automatic stepping is used, it is wise to declare a "wanted" higher number of iterations, so as to avoid steps that are too short. This is because each single weld segment acts as a non-linear subsystem. Moreover, the tangent stiffness matrix is not symmetric if the orthotropic model is used. So, the convergence of isotropic fillet welds is easier and the non-linearity is weaker. It is also possible to run non-linear analyses where the fillet welds are computed by a linear model (isotropic or orthotropic and using the secant stiffness), while the plate–shell elements are computed using a plastic material model.

10.2.3.3 Example 1

In order to compare the results obtained by the non-linear solver using the tangent stiffness of Appendix 2, and the results from the layout approach of Chapter 6, using the method to build the tangent stiffness matrix described in Section 6.7.4.1, the second test described in Section 6.7.4.3 has been reapplied using an MCOFEM similar to that of the example of Section 10.2.2.3.

The weld layout is computed using the IRCM and the orthotropic approach. The loading conditions are those described in Table 6.7 and the results obtained by the non-fem method are listed in Table 6.8.

The same three loading conditions have been run using an MCOFEM where the plate elements have been made rigid, so as to have a comparable situation.

The MCOFEM analysis does not apply the 0.75 safety factor of AISC, and so the utilization ratios should be 0.75 times lower than those obtained by layout approaches. This means that dividing by 0.75 the utilization from MCOFEM, should give the utilization obtained by IRFEM. The non-linear analyses have been run applying a load case 10 times the nominal one, so as to generate a no-convergence condition. The true load multiplier α_{LIM} is then evaluated by multiplying by 10 the limit multiplier reached by the analysis using loads increased 10 times. So, a utilization can be evaluated, that can later be converted by applying the 0.75 factor (see Table 10.6).

The comparison shows that the agreement is very good, with differences less than 2%.

10.2.3.4 Example 2

A single fillet weld segment with $f_{uw} = 500$ MPa, having throat thickness $a = 10$ mm, length $l = 10$ mm, and thickness $t = 10$ mm, has been loaded with several loading conditions as explained in Table 10.7. No safety factor has been applied. The approach used is that of orthotropic fillet welds.

The results obtained by using the tangent stiffness matrix of Appendix 2 are shown in Table 10.8.

Table 10.6 Comparison between (a) the utilizations ratios for several generic loading conditions computed using IRFEM (see also Table 6.8) and the methods of Section 6.7.4.1, and (b) the utilizations computed using a non-linear rigid MCOFEM.

Load	U_{IRFEM}	$\alpha_{LIM,MCOFEM}$	U_{MCOFEM}	Δ%
1	0.306	0.443	$\frac{1}{10 \cdot 0.443 \cdot 0.75} = 0.301$	1.6
2	0.536	0.253	$\frac{1}{10 \cdot 0.253 \cdot 0.75} = 0.527$	1.7
3	0.588	0.230	$\frac{1}{10 \cdot 0.230 \cdot 0.75} = 0.580$	1.4

Table 10.7 Single fillet weld segment loading conditions.

Load	F_w (N)	F_u (N)	F_v (N)
1	−2000	−14000	−15000
2	0	0	−60000
3	0	−60000	0
4	−600000	0	0
5	−12000	−12000	−12000

Table 10.8 Results obtained for the loads applied to a single fillet weld segment (the loads are those of Table 10.7). Displacements d_w, d_u, d_v; angle θ; functions h(p) and q(θ). Forces resulting, to be compared with those of Table 10.7.

Load	d_w (mm)	d_u (mm)	d_v (mm)	θ (deg)	h(p)	q(θ)	Fw (N)	Fu (N)	Fv (N)
1	−3.64e − 3	−2.55e − 2	−2.73 e − 2	47.22°	0.523	1.31	−2,000	−14,002	−14,997
2(*)	0.	0	−0.4738	90°	0.999	1.5	0	0	−44,955
3(*)	0.	−1.669	0.	0°	0.999	1.0	0	−29,997	0
4(*)	−0.4738	0	0	90°	0.999	1.5	−44,955	0	0
5	−1.851e − 2	−1.851e − 2	−1.851e − 2	54.735°	0.506	1.369	−11,998	−11,998	−11,998

*) At the last converged step. The full load was not applied.

10.2.3.5 Elements Using Contact

The use of finite elements for a fillet weld segment simulating the contact non-linearity has been outlined in Chapter 6 and Appendix 2.

The addition of the contact non-linearity related to fillet welds stiffens the connections, because the compressed weld segments do not react as transversely loaded fillets, but as stiff bearings. This may help to make stiffeners more effective and to get nearer to the

Table 10.9 Results obtained for the loads applied to a single fillet weld segment (the loads are those of Table 10.7). Displacements d_w, d_u, d_v; angle θ; functions h(p) and q(θ). Forces resulting, to be compared with those of Table 10.7.

Load	d_w (mm)	d_u (mm)	d_v (mm)	θ (deg)	h(p)	q(θ)	F_w (N)	F_u (N)	F_v (N)	Status
1	−4.762e − 4	−2.533e − 2	−2.714e − 2	46.97°	0.521	1.312	−2000[†]	−13,993	−14,993	Locked sliding
2[*]	0	0	−0.4678	90°	0.998	1.5	0	0	−44,910	Unlocked
3[*]	0	−1.673	0	0°	0.999	1.0	0	−29,999	0	Unlocked
4	−1.428e − 2	0	0	—	—	—	−59,976[†]	0	0	Locked Not sliding
5	−2.857e − 3	−1.449e − 2	−1.449e − 2	45°	0.436	1.297	−11,999[†]	−11,995	−11,995	Locked sliding

*) At the last converged step.
†) Using the contact stiffness $EA/(0.5\,t) = 4.2 \times 10^6$ N/mm.

rigid body assumption that is behind the traditional methods of computing the connections.

If the same fillet weld segment of the example of Section 10.2.3.4 is loaded in the same way, but using the "contact" flag, the results of Table 10.9 are obtained.

10.3 Finite Elements for Bolts

10.3.1 Introduction

When considered alone, in the context of a 3D finite element analysis where the constituents have been modeled by plate–shell elements, bolts can be modeled in a number of different ways, which depend on the following:

- if the bolts act in bearing or if they are preloaded
- if the bolt hole is modeled explicitly or not
- if the analysis is linear or not linear, and, if it is linear, if secant stiffness or initial stiffness has to be used
- if the bending moments in the bolt shaft are to be computed or not
- if the bolt is shear only or no shear
- if contact non-linearity is activated or not.

In the following sections, an overview of the numerical formulations proposed and tested is presented. These formulations follow strictly what has been explained and described in detail in Chapter 7.

Once the organization between the bolts ensured by the kinematic hypothesis of rigid body movement is lost, the distribution of the forces in the bolts strictly depends on the coupling of the stiffness of the single bolts and of the surrounding regions. Contact forces play a very important role in the mechanical behavior of bolts, and if the contact non-linearity is not considered, this issue should be tackled by proper assumptions.

10.3.2 Bolts in Bearing: No Explicit Bolt-Hole Modeling

If the bolt hole is not explicitly modeled, a single finite element must model both the stiffness of the bolt and the stiffness of the bearing.

As was shown in Chapter 7, the coupling of the three stiffnesses k_s, k_{be1}, and k_{be2} can be considered by static condensation, getting to an equivalent shear stiffness k_V which basically takes into account three springs in series, for a bolt drilling two thicknesses.

The stiffness k_s is

$$k_s = \frac{8D^2 f_{ub}}{D_{M16}}$$

In the linear range the stiffnesses k_{be1} and k_{be2} can be the secant ones using 0.149 K_{ini} (for medium to high loads) or the initial ones using K_{ini}, if the loads are expected to be low, where (D is in mm)

$$K_{ini} = 120 \cdot t \cdot f_y \cdot \left(\frac{D}{25.4}\right)^{0.8}$$

The axial stiffness is k_N is $k_N = EA_{res}/L$, the torsional stiffness is null.

The bending stiffness and the coupling terms between translation and rotation are managed by the terms R_{mod}, K_{mod}, and $k_V L/2$. If it is not wanted to consider the bending moments in the bolt shaft, the 12 × 12 stiffness matrix can be reduced to a 6 × 6 stiffness matrix that uses only the translational degrees of freedom, as explained in Chapter 7.

If the modifiers related to the shear-only or no-shear flags are used, as explained in Chapter 7, the factors μ_n and μ_t are applied respectively:

- μ_n to the stiffness k_N and k_M (which is part of R_{mod} and K_{mod})
- μ_t to the stiffness k_V.

As the effect of the bearing of the bolt is already considered in the stiffness of the element, the bolt-bearing checks can be done using the computed shear V. This means that it would not be meaningful to consider the Von Mises stress of the plate–shell element immediately near the bolt. These elements exist because the bolt hole has not been modeled, but their stress state would not be very meaningful, as the checks are done at the bolt-element level.

So, for this reason, a number of eight fictitious stiff beam elements can be added (Figure 10.9), connecting the node of the bolt-plus-bearing element to a number of nodes placed over a circle having radius equal to the radius of the hole, at angles of 45°. These radii act axially and bending out-of-plane and protect the plate region surrounding the bolt by membrane forces, to an extent equal to the diameter of the bolt hole, in all directions. The fictitious elements are released at both extremities to the rotation about axis 3, so in plane they act as trusses, and out of plane as rigid elements.

These elements are to be kept elastic also in non-linear analyses so as to help avoid the local plasticization of the nearby plate–shell elements simulating the plates bolted, due to the shear transmitted by the bolt-plus-bearing element. At a greater distance, no

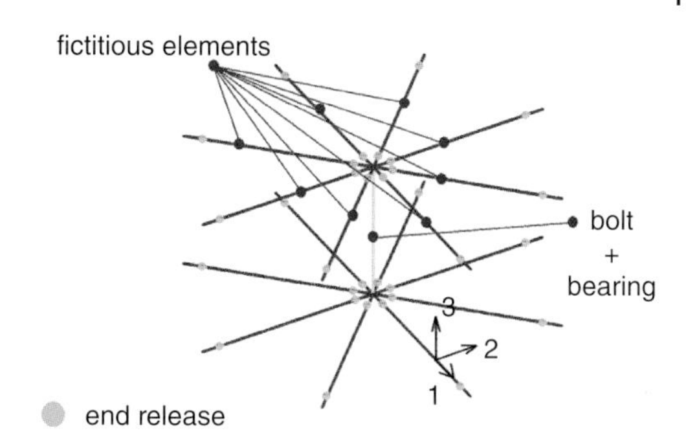

Figure 10.9 The bolt-plus-bearing element and the 8 + 8 fictitious beam elements, with end releases applied.

protection is exerted by the elements and the plates can coherently go into the plastic range. If the plasticization occurs very near to the bolt shaft it must still be considered related to the bolt bearing and not considered dangerous, provided that the bolt bearing checks are passed. The plasticization avoided is already taken into account by the bolt plus bearing element.

In the non-linear range, this element can keep the linear stiffness k_N, k_M, and k_s, up to a point where the bolt shaft is plastic (and the plastic condition to be checked using N, V, and M is that of Equation 7.31).

In the non-linear range the stiffnesses k_{be1} and k_{be2} must be upgraded during the analysis so as to follow the Rex and Easterling curve. Once the total shear displacement of the element is known (sum of the displacements of the three springs obtained by subtracting the shear displacement of node 1 from the shear displacement of node 2), the shear is evaluated by applying the computed total displacements in small increments, and adding the related small shear increments.

Each increment is obtained by multiplying the tangent *total* shear stiffness k_V times the displacement increment δd, up to the point where all the computed displacements d have been considered. After the end of each increment, the tangent stiffness of each element is upgraded. It must be recalled that once a shear $V = R$ is known, the r factor can be evaluated ($r = R/R_{be}$), and thanks to Equations 7.5 the tangent stiffness k_{be}, at first and second extremity, can be computed and then used at the next iteration or step.

10.3.3 Bolts in Bearing: Explicit Bolt-Hole Modeling

If the hole is to be modeled explicitly, a ring of nodes placed at a distance $D_0/2$ from the bolt center can be added, being D_0 the bolt-hole diameter. The angular span can be 45°, which is a reasonable compromise between precision and speed. No plate–shell element is added inside the ring, so there is a hole in the mesh (Figure 10.10). This hole can be obtained by adding during meshing fictitious truss elements along the circle, to define a closed loop. The fictitious trusses are segments (see Chapter 9) which cannot be crossed by the plate–shell elements. All the shell elements inside closed loops are then discarded, at the end of the meshing phase.

Two sets of elements are added in order to simulate the bolt and the bearing.

The first set is made up of one single element that models the bolt. This element has two nodes, at the two extremities of the bolt, placed in the bolt-shaft center (the element is normal to the page in Figure 10.10).

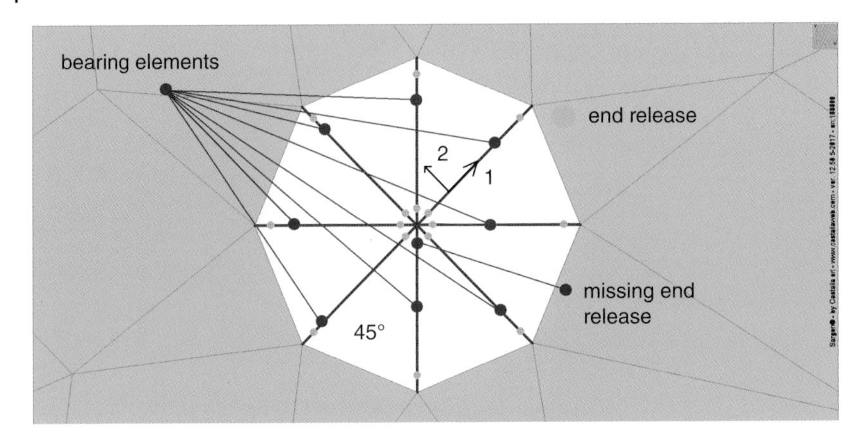

Figure 10.10 The bolt hole modeled. Bolt element is normal to the page. The eight bearing elements with their releases are visible, and the local reference system for one of them (axis 3 is normal to page, pointing toward the viewer).

The shear stiffness of this element is $\mu_t{\cdot}k_s$ Remember that μ_t may be equal to 1 or 0.0001, depending on the flag state (inactive, or active) and considers only the shear stiffness of the bolt. The axial stiffness is $\mu_n{\cdot}k_N$ (if the shear-only working flag is used). The bending stiffness is related to $\mu_n{\cdot}k_M$, and uses the terms R and K as defined in Chapter 7.

The second set is made up of eight elements at each extremity of the bolt (so a total of sixteen elements for one bolt with multiplicity 2), connecting the bolt center to the nodes of the ring. These eight elements simulate the bearing, and their rotation R_2 is released at both extremities, with the exception of one element of the group of eight, which is released only at the first extremity. The reason why one element is not released is to avoid torsional lack of restraint in the bolt-element. The end releases of the radial elements:

- allow the transfer of axial forces
- allow the transfer of forces in the direction of the bolt axis, i.e. normal to the plane of the bolted plate
- allow the transfer of bending moments with vector in the plane of the plate
- do not allow the transfer of forces normal to each radius
- do not allow the transfer of bending moments with vector normal to the plane of the plate.

Thanks to these settings, the in-plane stiffness globally exerted by the eight elements is four times the axial stiffness of each single radius. This means that whatever the direction of application of a force to the central node common to all eight elements, the apparent stiffness will always be four times the axial stiffness of one radial element. So, if the axial stiffness of each radius is set such that

$$k_N = 0.25{\cdot}k_{be}$$

the global effect of all radii will be that of a bearing of stiffness k_{be}.

The out-of-plane shear stiffness of each radius must be high, in order to transfer the axial forces of the bolt to the nodes of the ring. The bending stiffness of each radius k_M can be set null, so that $R = k_V L^2/3$, and $K = k_V L^2/6$.

In the non-linear range, there are two possible approaches. The first is to consider all radii acting both in tension and in compression. The second approach considers the radii reacting only in compression. This would help to simulate more realistically the forces applied.

If the first approach is adopted, then once more

$$k_N = 0.25 \cdot k_{be}$$

while if the second one is used then

$$k_N = 0.5 \cdot k_{be}$$

A problem that arises in the non-linear range is how to relate the axial force N in a generic radii, during the analysis, to the force R globally applied to the bearing, and partially embedded into other elements. This is necessary in order to properly evaluate the tangent stiffness of the bearing. If the angle between the axis line of the generic radius and the in-plane displacement vector is α (Figure 10.11), it can be seen that if the elements act both in tension and compression then

$$R = 4\frac{N}{\cos\alpha}$$

while if the elements act only in compression then

$$R = 2\frac{N}{\cos\alpha}$$

where N is the axial force of the generic radius.

Once the total R is known, the value of $r = R/R_{be}$ is computed, and from this the new tangent axial stiffness of each radius

$$k_{N,tan} = 0.25 k_{be,tan}$$

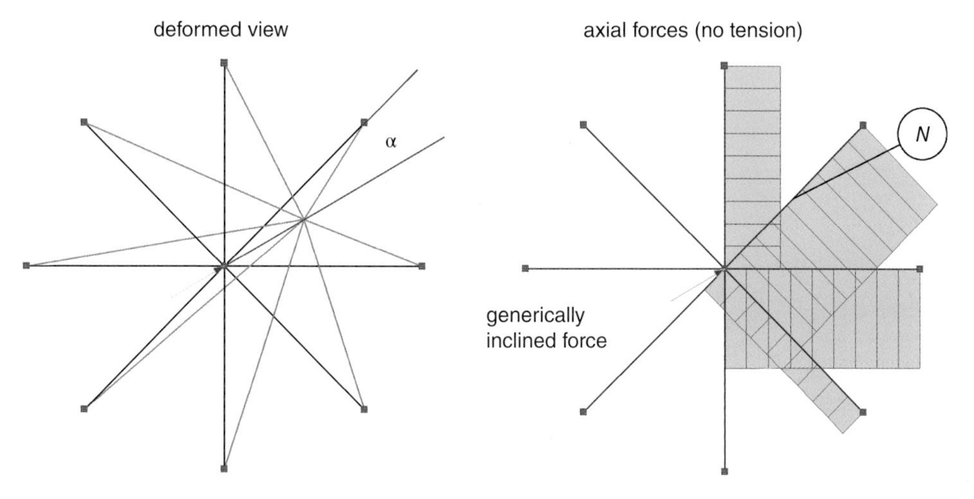

Figure 10.11 How to rebuild the overall force R applied to the bearing from the axial force of one radii under generic loading conditions.

or if the no-tension model is assumed

$$k_{N,tan} = 0.5k_{be,tan}$$

During the non-linear analysis, if one of the bearings, or the bolt, reaches its limit, the related stiffness (k_N for the bearing elements, k_s k_N, and k_M for the bolt element) is set to null, and the bolt is no longer active, implying a redistribution of the forces.

10.3.4 Preloaded Bolts: No Explicit Bolt-Hole Modeling

If a bolt is preloaded, its apparent axial stiffness is greater, and so is its bending stiffness. Using the "full cylinder" option described in Chapter 7, the bending and axial stiffness of the preloaded bolt can be obtained by applying properly the factor Q^*. Precisely, it has been seen that

$$k_N = \left(\frac{EA_{res}}{L} + \frac{Q^{*2}EA}{L}\right)\cdot\mu_n \approx \frac{EA}{L}\cdot\left(1+Q^{*2}\right)\cdot\mu_n$$

$$k_M = \frac{\pi ED^4}{64L}\left(1+Q^{*4}\right)\cdot\mu_n$$

$$k_s = \frac{10GA}{L}\cdot\mu_t$$

where E is Young's modulus and G is the tangential modulus of steel, A_{res} and A are the threaded and total area of the bolt, D is its diameter, and L is the length of the bolt element.

The protective radii already seen for the bolts in bearing are to be applied also for preloaded bolts, as there is no bearing effect.

In the non-linear range, two different conditions must be controlled.

The first is slip. If slip occurs, there are two possible strategies: declare the shear stiffness null or replace it with the bearing stiffness plus the bolt stiffness as for bolts in bearing.

The second is the bolt rupture. True bolt forces N_{true} and M_{true} can be computed by the forces stressing the preloaded bolt by the following rules:

$$N_{true} = F_{P,C} + \frac{N}{1+Q^{*2}}$$

$$M_{true} = \frac{M}{1+Q^{*4}}$$

where $F_{P,C}$ is the preload applied. These true forces can then be used to check for bolt rupture using the limit condition of Equation 7.31. If the bolt is broken, all its stiffnesses are set to zero.

10.3.5 Preloaded Bolts: Explicit Bolt-Hole Modeling

If the bolt-hole is explicitly considered, the bolt can be modeled as if the hole were missing, and the radii can receive a high stiffness, to transfer the shear to the ring. All the radii will have to be reacting both in tension and in compression.

10.3.6 Effect of the Bending Moments in Bolt Shafts

In Chapter 7, the formulation of the stiffness matrix of the single bolt and of the bolt layout has been carried on taking into account the bending stiffness of the bolt shaft. It has also been explained that in order to totally neglect the bending moments in the bolt shafts, instead of a 12×12 stiffness matrix, using three translations and three rotations at each extremity of each bolt segment, a 6×6 matrix using only translations can be used.

The addition of the bending moments in the bolt shaft or their removal is related to a number of quite complex issues.

It was explained in Section 5.1.1, that a shear-only bolt layout cannot exist in principle because the transport moment due to the translation of the shear force from the sliding surface to plate mid-thickness implies a bending that must be absorbed somehow. If two shears are applied to the extremities of a bolt, a bending moment *must* arise to ensure equilibrium. In the real world, the bending of the plates caused by this extra moment is counteracted by contact pressures if the bending implies a movement against the sibling bolted plate. So, the extra-bending loads the bolts which bend the plate which (usually but not always) finds a constraint to its bending due to contact. The contact, in turn, implies the exchange of normal pressures that load the bolted parts. As these pressures are usually considered low, they are not taken into account.

For this reason the bending in the bolt shaft is often totally neglected in the analysis and checking of the bolts. The standards do not even mention it.

However, if the bending of the plate implies its going off from the sibling, no constraint is found and no contact pressure is exchanged. In bolted splice joints this might happen to the flange plates when compressive forces are applied to the member. Usually, however, the members are assumed to be in contact, and the compression does not imply loading the bolts. So, once more, for all this chain of reasons, the bending of the plate related to the bending moment in the bolt shaft is neglected. However, if the members are not in contact (non-bearing type), the compression in the members implies a bending in the plate.

In general, it cannot be assumed that sheared bolts are not loaded by additional bending, and it cannot always be assumed that the extra bending of the plates is absent. For instance, if *two* plates that are connected by bolts are pulled (as happens to a channel bracing and its gusset plate), there must also be a bending moment in the plates, and not just membrane forces. In compression, this helps the buckling of the gusset plate. Usually this bending is neglected.

In the following section, an interesting case is discussed in depth, showing the importance of all these issues.

10.3.7 Example: A Bolted Splice Joint Using PFEM

In this section an interesting simple case will be discussed using PFEM as the analysis tool. A HEB 200 cross-section member is connected by a splice joint to an identical member. It is assumed that there is an unfilled gap between the two members which are not in contact (Figure 10.12).

The members are connected by four plates: two for the flanges, upper and lower, having thickness 15 mm, and two for the web, having thickness 6 mm. The material grade is S275.

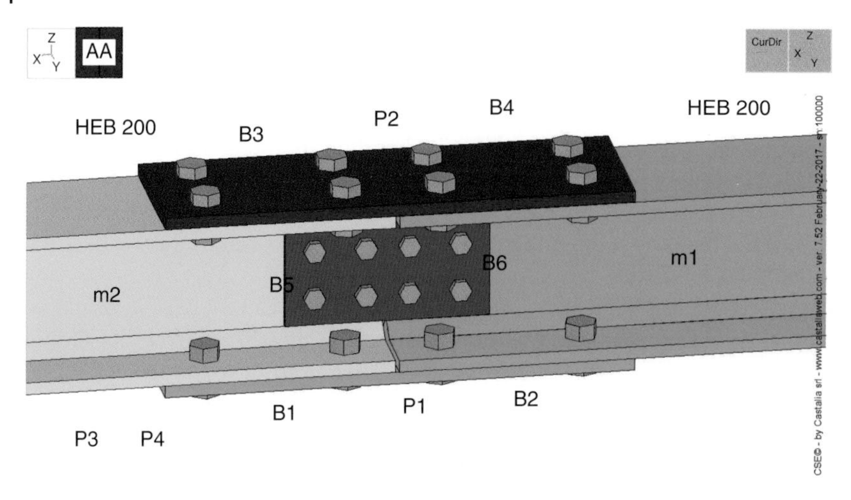

Figure 10.12 A bolted splice joint.

The flange plates are connected by means of a four-bolt layout, each using 4 M24-8.8. The webs are connected by means of two bolt layouts having multiplicity 3, and using 4 M16-8.8 each. The ultimate stress f_{ub} for class 8.8 bolts is 800 MPa.The dimensions and sizing are dummy – it is just a numerical experiment.

The properties of an HEB 200 cross-section are in Table 10.10. The renode is hyper-connected, because there are more connectors than those strictly needed to ensure the transfer of the forces. In particular, the proportion of the shear flowing in the web bolt versus that flowing in the flange bolt depends on the relative stiffness of the connectors. This, in turn, depends on several parameters, and should take into account the bearing stiffness at the bolt–plate interface.

The loading condition analyzed is that of a tensile axial force equal to 0.5 times the yield limit of the cross-section. That is $N = 1{,}073{,}600$ N. Before presenting the results obtained by finite elements, some by hand computation will be prepared, for comparison purposes.

There are basically two bearing interfaces: the interface between the M24 bolt and two 15 mm thick plates (the flange and the top or bottom plate) and the interface between the M16 bolt with a thickness of 6 mm (one of the two web plates) and with a thickness of 4.5 mm (half the web thickness, $h = 0.5$).

The limit bearing force for the first interface is

$$R_{be1} = R_{be2} = 2.5{\cdot}15{\cdot}24{\cdot}430 = 387{,}000\,\text{N}$$

Table 10.10 HEB 200 cross-section properties.

A (mm^2)	H (mm)	B (mm)	t_w (mm)	t_f (mm)
7808	200	200	9	15

The limit bearing forces for the second interface are

$$R_{be1} = 2.5 \cdot 6 \cdot 16 \cdot 430 = 103,200\,\text{N}$$

$$R_{be2} = 2.5 \cdot 9 \cdot 16 \cdot 430 = 154,800\,\text{N}$$

These interfaces lead to the following stiffnesses:

Flange stiffnesses

Initial, bearing: $k_{be1,ini} = k_{be2,ini} = 120 \times 15 \times 275 \times (24/25.4)^{0.8} = 473{,}050$ N/mm.

Secant, bearing: $k_{be1,sec} = k_{be2,sec} = 0.149 \times 473{,}050 = 70{,}484$ N/mm

Bolt $k_s = 8 \times 24^2 \times 800/16 = 230{,}400$ N/mm

The total resulting stiffness for each bolt is

$$k_{V,ini} = \frac{473,050 \cdot 473050 \cdot 230,400}{2 \cdot 473050 \cdot 230,400 + 473,050 \cdot 473050} = 116,711\,\text{N/mm}$$

$$k_{V,\sec} = \frac{70,484 \cdot 70,484 \cdot 230,400}{2 \cdot 70484 \cdot 230,400 + 70,484 \cdot 70484} = 30,566\,\text{N/mm}$$

Web stiffness

Initial, bearing,plate: $k_{be1,ini} = 120 \times 6 \times 275 \times (16/25.4)^{0.8} = 136{,}802$ N/mm.

Initial, bearing, web: $k_{be2,ini} = 120 \times 4.5 \times 275 \times (16/25.4)^{0.8} = 102{,}602$ N/mm.

Secant, bearing,plate: $k_{be1,sec} = 0.149 \times 136{,}802 = 20{,}383$ N/mm

Secant, bearing, web: $k_{be2,sec} = 0.149 \times 102{,}602 = 15{,}288$ N/mm

Bolt $k_s = 8 \times 16^2 \times 800/16 = 102{,}400$ N/mm

The total resulting stiffness for each bolt *segment* is

$$k_{V,ini} = \frac{136,802 \cdot 102,602 \cdot 102,400}{136,802 \cdot 102,602 + 136,802 \cdot 102,400 + 102,602 \cdot 102,400} = 37,283\,\text{N/mm}$$

$$k_{V,\sec} = \frac{20,383 \cdot 15,288 \cdot 102,400}{20,383 \cdot 15,288 + 20383 \cdot 102,400 + 15,288 \cdot 102,400} = 8049\,\text{N/mm}$$

Now, assuming the members to be rigid, the total stiffness of the system can be evaluated as

$$K_{ini} = \frac{116,711 \cdot 8 + 37,283 \cdot 8}{2} = 615,976\,\text{N/mm}$$

$$K_{\sec} = \frac{30,566 \cdot 8 + 8049 \cdot 8}{2} = 154,460\,\text{N/mm}$$

This is because on each side there are eight springs in the flanges, and eight springs in the web. The two sides of the splice are connected in series, and are identical, so their final stiffness is half.

It is now possible to predict the elongation assuming the rigid body hypothesis:

$$d_{ini,rigid} = \frac{1,073,600}{615,976} = 1.743\,\text{mm}$$

$$d_{\sec,rigid} = \frac{1,073,600}{154,460} = 6.951\,\text{mm}$$

The elongation due to the elasticity of the members (having each a length of 1000 mm), is readily computed:

$$d_{members} = \frac{1,073,600 \cdot 2,000}{210,000 \cdot 7808} = 1.31\,\text{mm}$$

So, the final displacement would be

$$d_{ini} = 1.74 + 1.31 = 3.05\,\text{mm}$$

$$d_{ini} = 6.95 + 1.31 = 8.26\,\text{mm}$$

The shears in the bolt layout are distributed proportionally to the stiffness. So, in each bolt:

Flange

$$V_{M24,\,ini} = 1,073,600 \cdot \frac{4 \cdot 116,711}{2 \cdot 4 \cdot 116,711 + 2 \cdot 4 \cdot 37,283} \cdot \frac{1}{4} = 101,709\,\text{N}$$

$$V_{M24,\,\sec} = 1,073,600 \cdot \frac{4 \cdot 30,566}{2 \cdot 4 \cdot 30,566 + 2 \cdot 4 \cdot 8049} \cdot \frac{1}{4} = 106,227\,\text{N}$$

Web

$$V_{M16,\,ini} = 1,073,600 \cdot \frac{2 \cdot 4 \cdot 37,283}{2 \cdot 4 \cdot 116,711 + 2 \cdot 4 \cdot 37,283} \cdot \frac{1}{8} = 32,490\,\text{N}$$

$$V_{M24,\,\sec} = 1,073,600 \cdot \frac{4 \cdot 2 \cdot 8049}{2 \cdot 4 \cdot 30,566 + 2 \cdot 4 \cdot 8049} \cdot \frac{1}{8} = 27,973\,\text{N}$$

Several PFEMs have been prepared automatically by the software, CSE, and the results compared with that of by hand computation, in order to test the correct behavior of the finite elements proposed. Table 10.11 explains the features of each model. The finite elements described in Chapter 7 and in this chapter have been used to model the bolt, the bearing, and the hole. Only the shear and axial force of one bolt of each layout, always the same, is listed. Slight differences in the bolt shears are due to local effects.

The non-linear models do not use initial or secant stiffness, but follow the Rex and Easterling curve. So, it is expected that the displacement should be between the displacement obtained by with initial stiffness and that obtained by with the secant stiffness. In modeling the bolt hole, the bearing elements have been assumed to react both in tension and in compression. The constitutive law for the members is elastic perfectly plastic, or, in the simulations considering the members as rigid bodies, elastic and rigid constitutive law ($E = 1.0 \times 10^{11}$ MPa, instead of 2.1×10^{5} MPa).

A first set of analyses has been done considering all the bolts shear only, and considering the deformability of the members. Results are in Table 10.12.

The models have about 5600 nodes, 33,000 degrees of freedom, 6500 plate–shell elements, and 600 bolt and member-extremity elements (there are slight differences in these numbers between the models). The typical mesh size of 10 mm has been adopted. Each model is created automatically in about 7 seconds, and runs 48 loading configurations in the elastic range in about 18 seconds using a workstation with processor Intel™ Xeon™ 3.2 GHz, in the Win 32 subsystem of Windows 7 Professional™. So, in about 25 seconds all results are available, for 48 different loading conditions. Linearly comparable times are obtained by with more complex models.

Table 10.11 Features of the PFEM models. "Y" stands for "Yes", "N" stands for "No".

Model	**Hole Modeling**	**Initial Secant Non-linear**	**Bending in shafts**
A	N	Initial	Y
B	Y	Initial	Y
C	N	Secant	Y
D	Y	Secant	Y
E	N	Non-linear	Y
F	Y	Non-linear	Y
G	N	Initial	N
H	Y	Initial	N
I	N	Secant	N
L	Y	Secant	N
M	N	Non-linear	N
N	Y	Non-linear	N

Table 10.12 Main results obtained by analyzing the different models whose features are explained in Table 10.11. All models use shear-only bolts. All models consider the members deformable. The shear V and the axial force N of a single bolt are listed, as well as the total axial displacement d.

Model	d **(mm)**	V_{M24} **(kN)**	N_{M24} **(kN)**	V_{M16} **(kN)**	N_{M16} ***(kN)***	**Flange cover plate stress status**
A1	3.27	95.57	0.80	32.62	0.37	KO (pulled & bent)
B1	3.31	95.68	0.81	31.58	0.37	KO (pulled & bent)
C1	8.49	104.1	0.85	28.1	0.32	KO (pulled & bent)
D1	8.52	104.0	0.86	28.1	0.32	KO (pulled & bent)
E1	—	—	—	—	—	No convergence-KO
F1	—	—	—	—	—	No convergence-KO
G1	2.99	103.0	0.003	31.51	≈ 0	OK (pulled)
H1	3.03	103.1	0.003	31.44	≈ 0	OK (pulled)
I1	8.20	106.4	0.003	27.86	≈ 0	OK (pulled)
L1	8.23	106.4	0.003	27.85	≈ 0	OK (pulled)
M1	3.65	106.1	0.003	28.21	≈ 0	OK (pulled)
N1	3.64	106.0	0.003	28.30	≈ 0	OK (pulled)

The models with and without holes give very similar results, meaning that the two modeling systems are equivalent for this specific type of loading.

The analyses G–N do not consider the bending moments in the bolt shafts. This implies null axial forces, and null bending moments. We observe some redistribution due to the deformability of the web vs that of the flange for the stiffer bolt-bearing interfaces, those with the initial stiffness (G1, H1). Using the secant stiffness (I1, L1) the results obtained are identical to those predicted using the stiff-member assumption.

Analyses A–F do consider also the bending in the bolt shafts. Results for the bolts are similar, but the axial force and bending moments in the bolts bend the cover plates, and this implies an additional stress field, unacceptable for the plate. Considering the four shears applied to the flange plate equal to 106 kN, the additional bending moment in the plate would be

$$M = 4{\cdot}106{,}000{\cdot}7.5 = 3{,}180{,}000\ \text{Nmm}$$

and considering the elastic modulus of the plate

$$\sigma = \frac{M}{W} = \frac{6{\cdot}3{,}180{,}000}{200{\cdot}15^2} = 424\ \text{MPa} \gg 275\ \text{MPa}$$

So the plate is overstressed. Usually this additional bending moment is neglected considering that the bending of the plate is avoided by the contact pressures that it receives at the interface with the member flange (Figure 10.14). If the analysis is run in the linear range, or in the non-linear range neglecting the contact non-linearity, this contact effect is not taken into consideration.

If the bolt moments are not considered, the deformation is totally different (Figure 10.13), and there is no bending. This is wrong, in principle, but may become correct provided that: there are contact forces, these are low, and they do not imply overstressing of the flange.

However, if the members are *compressed*, and there is a gap, then the bending moves the plate far from the contact, and the additional bending would have to be considered (Figure 10.15). If the gap is low, it can be assumed that the contact of the two end

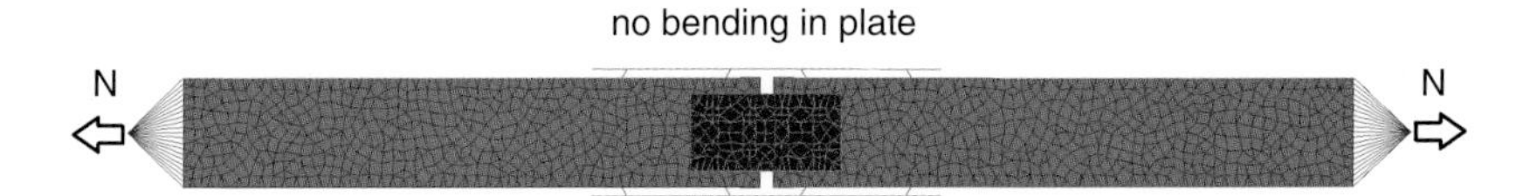

Figure 10.13 Deformation of the models G–N with no bending moments in the bolt shafts. The end-cones of rigid elements used to apply the loads are visible.

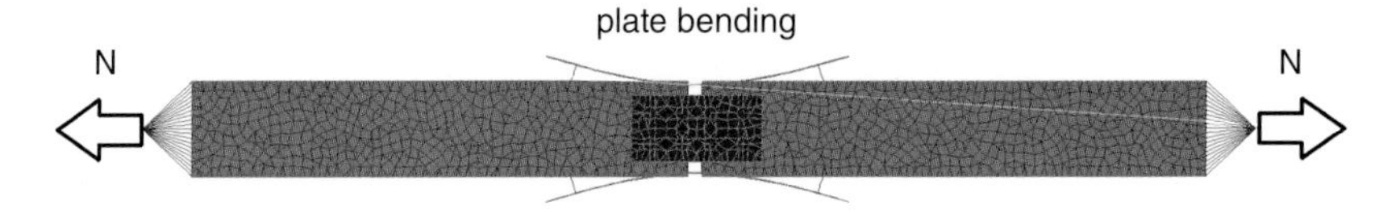

Figure 10.14 The bending of the cover plate for tensile forces when the bolt moments are considered (models A–F). No contact pressure assumed.

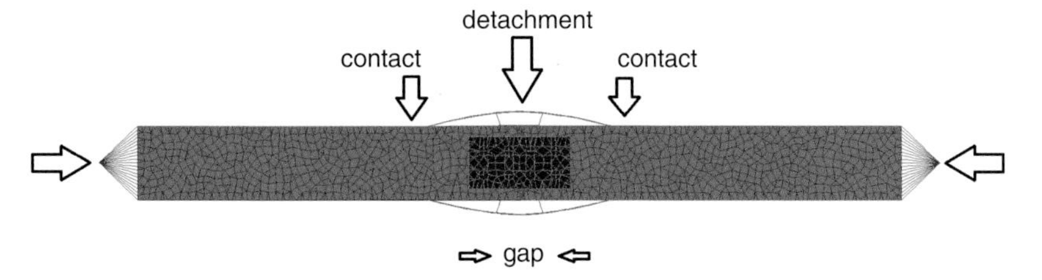

Figure 10.15 If the elements are compressed, and there is an initial gap, bending moment in the cover plates may arise. The contact at the ends acts as an end constraint.

cross-sections will be activated, and so the moment is reduced. But if the gap is not low, the bending moment may indeed load the plate. In the present example, the gap should be lower than about 3 mm to protect the plates.

The non-linear analyses run with the bolt bending moments do not converge. This means that the limit load has been reached. In turn, this is related to the bending of the cover plate for the reasons explained.

If the bending moments in the bolts are not computed, then the non-linear analyses converge to results almost identical for the model with and without the hole. The displacement is lower than that obtained using the secant stiffness, and higher than that predicted using the initial stiffness. Indeed, the non-linear curve proposed by Rex and Easterling has been followed, instead of using the secant or initial linear stiffnesses. In order to prove that the result is exact, the first step is to subtract the displacement due to the deformability of the members. So the net displacement due to the local deformability of the bolts and at the bolt-bearing interface can be evaluated as

$$d_{net} = 3.63 - 1.31 = 2.32\,\text{mm}$$

The force at the M24 bolt interface is (see M1 in Table 10.12)

$$R = 106{,}100\,\text{N}$$

so

$$r = \frac{R}{R_{be}} = \frac{106{,}100}{387{,}000} = 0.274$$

The displacement related to such r in the Rex and Easterling curve is

$$d_{be} = \frac{R_{be}}{k_{be1,ini}} 0.444 = \frac{387{,}000}{473{,}050} \cdot 0.444 = 0.363\,\text{mm}$$

The displacement due to the sheared bolt is

$$d_{bolt} = \frac{R}{k_s} = \frac{106{,}100}{230{,}400} = 0.460\,\text{mm}$$

So, considering that there are two systems in series, the final expected displacement is

$$d_{total} = 2 \cdot (0.363 + 0.460 + 0.363) = 2.37\,\text{mm}$$

which is very near that obtained (2.32 versus 2.37, a difference of 2.1% due to the local deformability of the plate–shell elements and to the approximation of the non-linear analysis).

The same displacement should be found also at the other interface, the M16 one. There it results, for each bolt segment

$$R = 28{,}220\,\mathrm{N}$$

so

$$r_1 = \frac{R}{R_{be,1}} = \frac{28{,}220}{103{,}200} = 0.273$$

$$r_2 = \frac{R}{R_{be,2}} = \frac{28{,}220 \cdot 2}{154{,}800} = 0.365$$

The displacements related to the r in the Rex and Easterling curve are

$$d_{be,1} = \frac{R_{be,1}}{k_{be1,ini}} 0.441 = \frac{103{,}200}{136{,}802} \cdot 0.441 = 0.333\,\mathrm{mm}$$

$$d_{be,2} = \frac{R_{be,2}}{k_{be2,ini}} 0.739 = \frac{154{,}800}{2 \cdot 102{,}602} \cdot 0.739 = 0.557\,\mathrm{mm}$$

The displacement due to the sheared bolt segment is

$$d_{bolt} = \frac{R}{k_s} = \frac{28{,}220}{102{,}400} = 0.276$$

so the final expected displacement is

$$d_{total} = 2 \cdot (0.333 + 0.557 + 0.276) = 2.33\,\mathrm{mm}$$

which can be considered equal to the previous one and to the value expected, considering the rounding errors, and the possible local deformability of the members. The resulting Von Mises stress map is plotted in Figure 10.16.

If the analysis is repeated considering all the bolts fully resistant, i.e. by removing the shear-only flag, different results are obtained if the bolt bending moments are kept. The results are given in Table 10.13.

The main difference is that now the bolt *layouts* have a bending stiffness that clamps the cover plate. Axial forces arise in the bolt shafts and the stress-state of the plate changes significantly. Now, the plate bending is limited by the stiffness of the bolts, which act in tension and compression as a contact interface. The shears in the bolts do not change much; there is a slight scatter between the shears of the bolts of the same layout.

The results obtained considering the no-bending working mode with fully resistant bolts (G2–N2) are identical to those obtained by with the shear-only flag and the no-bending working mode (G1–N1).

If the analysis E is repeated using contact non-linearity, the results are listed in Table 10.14. The axial force in the bolts increases due to the prying forces.

Finally, if the bolts are all marked as preloaded, the stiffness is much higher, as can be seen in Table 10.15.

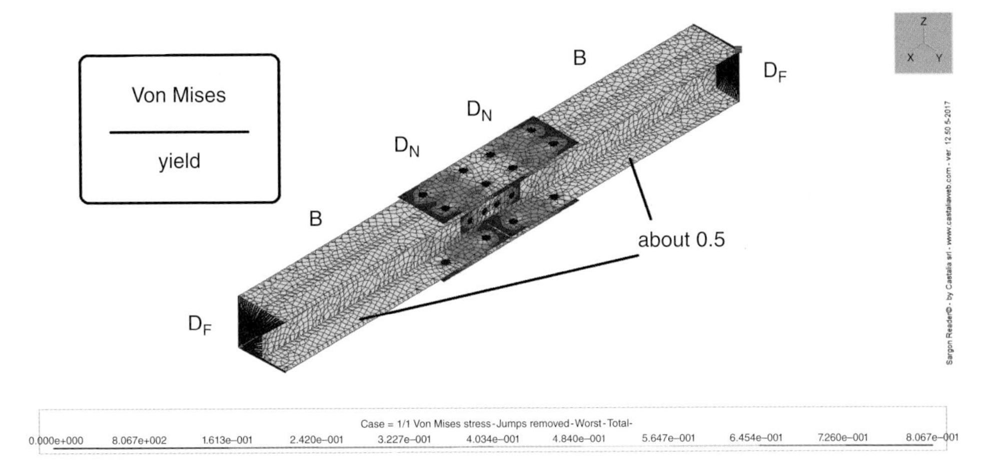

Figure 10.16 Non-dimensional Von Mises stress map for model N of Table 10.13. The disturbed (near and far) and Bernoulli regions of the two members are also marked.

Table 10.13 Main results obtained by analyzing the different models whose features are explained in Table 10.11. The shear V and the axial force M of a single bolt are listed, as well as the total axial displacement d. All models use fully resistant bolts. All models consider the members deformable.

Model	d (mm)	V_{M24} (kN)	N_{M24} (kN)	V_{M16} (kN)	N_{M16} (kN)	Flange cover plate stress status
A2	3.04	101.8	5.76	32.3	1.05	OK (pulled & bent)
B2	3.08	101.9	5.76	32.3	1.04	OK (pulled & bent)
C2	8.25	106.1	6.00	28.08	0.91	OK (pulled & bent)
D2	8.29	106.1	6.00	28.09	0.90	OK (pulled & bent)
E2	3.71	105.4	6.03	28.62	0.93	OK (pulled & bent)
F2	3.70	105.4	5.97	28.74	0.92	OK (pulled & bent)
G2	3.00	103.0	0.003	31.51	≈ 0	OK (pulled)
H2	3.03	103.1	0.003	31.44	≈ 0	OK (pulled)
I2	8.20	106.4	0.002	27.86	≈ 0	OK (pulled)
L2	8.23	106.4	0.002	27.85	≈ 0	OK (pulled)
M2	3.65	106.1	0.002	28.21	≈ 0	OK (pulled)
N2	3.64	106.0	0.002	28.30	0.002	OK (pulled)

Table 10.14 Results obtained for model E, with bearing and contact non-linearities activated. Bolts fully reacting, no explicit bolt hole modeling.

Model	d (mm)	V_{M24} (kN)	N_{M24} (kN)	V_{M16} (kN)	N_{M16} (kN)	Flange cover plate stress status
E3	3.70	105.5	14.00	28.77	7.09	OK (pulled & bent)

Table 10.15 Results obtained by with preloaded bolts: model O without holes; model P with holes, linear range; model Q without hole, model R with hole, non-linear range. Bolts fully resistant. All models consider the members deformable.

Model	d (mm)	V_{M24} (kn)	N_{M24} (kN)	V_{M16} (kN)	N_{M16} (kN)	Flange cover plate stress status
O	1.30	103.0	5.12	32.18	0.019	OK (pulled & bent)
P	1.31	103.0	5.14	32.31	0.19	OK (pulled & bent)
Q	1.22*	96.17*	4.85*	30.3*	0.18*	$\alpha_L = 0.937$
R	1.23*	96.12*	4.88*	30.4*	0.18*	$\alpha_L = 0.937$

*) at the α_L reached.

The tests use the following data:

- $F_{P,C,M24} = 0.7 \times 353 \times 800 = 197{,}680$ N; $V_{lim,M24} = 0.5 \times 1 \times 197{,}680 = 98{,}840$ N
- $F_{P,C,M16} = 0.7 \times 157 \times 800 = 87{,}920$ N; $V_{lim,M24} = 0.5 \times 1 \times 87{,}920 = 43{,}960$ N

The displacement is practically only due to member extension, so the preloaded connection is quite stiff. It is observed that the distribution of the shears is now different, as it does not depend anymore on the bearing stiffness, but only on the area of the bolts, on the thickness bolted, and on the local deformability of the web and of the flange, which, being lower now than the shear stiffness of the bolt, plays an important role.

Neither of the non-linear analyses (row Q and R in Table 10.12) reaches the applied load. The shear in the M24 bolts has reached its sliding limit and this is signaled by the analysis (which does not use factored resistance). The values at which the analyses were arrested are very near the slip limit of M24 bolts, one fraction of step before the limit:

$$\frac{96{,}170}{98{,}840} = 0.973$$

$$\frac{96{,}120}{98{,}840} = 0.972$$

10.4 Loads

10.4.1 PFEM

If PFEM is used, all the members are connected by elements simulating the connectors and so the loads are applied at the far extremity of the members. There are several ways to do that: one way which has been proved effective, is to add a cone of fictitious stiff elements connecting the nodes of the far extremity of each member to a single node (Figure 10.2), placed over the ideal member axis. To this connected node are then applied the forces and moments read at the pertinent distance in the BFEM elements simulating the member.

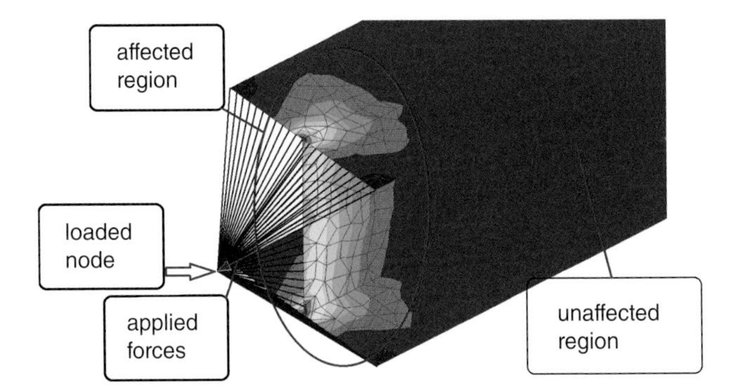

Figure 10.17 Stress concentrations in the loaded region at the far extremities of members.

If the effect of the loads applied locally to the member is negligible (usually distributed loads), then the forces applied at the loaded node can be computed by using equilibrium, starting from the internal member forces at the near extremity.

Although the fictitious elements distribute the applied forces to several nodes, sometimes it is observed that a stress concentration near the loaded region (Figures 10.16 and 10.17). In this disturbed region, stresses are not meaningful because they depend on how the loads have been applied. Given a generic member, starting from its near extremity and going to the far extremity, three regions are in general met:

1) a disturbed region at the near extremity D_N – this is where the stresses are interesting, as they depend on the application of the forces due to connectors
2) a Bernoulli-beam undisturbed central region, B, where the stresses follow the rules of simple beam theory – here the checks have usually already been done in the BFEM software package, or when the members were designed
3) a disturbed region at the far extremity D_F. – not meaningful as the stress concentrations depend on how the loads have been applied

Sometimes the region D_F merges into B.

10.4.2 MCOFEM

The loading of MCOFEMs differs from that of PFEM because some interface connector-constituent has been broken. At that interface, the forces exchanged between the constituents must be applied explicitly.

If the connector is a weld layout, or a bolt layout not using a bearing surface, the forces exchanged are nodal forces, simulating the internal forces in the weld segments, and the internal forces in the bolts.

If the connector is a bolt layout using a bearing surface, also the pressures exerted by the reacting part of the bearing surface must be added. This is done in the same way already seen for SCOFEMs.

10.5 Constraints

10.5.1 PFEM

In a PFEM model of a hierarchic renode that is not a constraint, the unique constituent to be constrained is the master member. If this is interrupted, the constraints will be applied as a clamp to the node at the far extremity of the member. If the master member is passing, there are two options.

If proper loads are applied to the far extremities of the member, balanced to the externally applied loads, then the system will be self-balanced and a single clamp can be applied to one of the two far extremities. This clamp should exert null or negligible reactions.

If these loads are not applied properly, both the nodes can be clamped. This implies a perturbation of the stress state of the master, because the system is redundant.

In the PFEM of a constrained renode, the only part to be constrained is the constraint block. This can be modeled by a block of stiff brick elements clamped at the far surface.

In the PFEM of a central renode, one of the slaves can be clamped at the far extremity.

10.5.2 MCOFEM

The only difference between the constraint of a MCOFEM and those of the related PFEM is that since one or more connectors have been removed, there will possibly be more than one equivalent body to be constrained.

Applying the rules already seen for PFEM, one or more bodies will be left free to move in space: these are usually the members whose connectors have been removed. So, a clamp will have to be applied at the far extremities of these free members. This will not imply any perturbation of the stresses in the D_N region of the member, which is the one interesting.

If some connector is removed but the constituent preserves its link to the remaining part, because it is hyperconnected then there is no need to apply more constraints.

10.6 Checking of Welds and Bolts

In the context of MCOFEM or PFEM, the checks of the internal saturated welds and of bolts can be done by directly reading the internal forces flowing into the finite elements used to model the weld segments and the single bolts (N, V, and M for bolts, F_w, F_u, F_v for weld segments). The no-slip condition can be checked directly. If using preloaded bolts, the checks of the bolt shaft should be done by evaluating the true forces flowing in the bolt, as explained in Chapter 8. For weld segments using contact, we need to check the existence of a compression and the no-slip condition.

During a non-linear analysis, the checks can be embedded into the analysis so as to make the unchecked simple connector ineffective.

10.7 Checking of Components

The execution of a global analysis, embedding many different parts, has a cost (very low if executed in the linear range or if searching for buckling modes), but it has huge revenue: all the constituents can be checked at one go for a good part of the failure modes.

It must be stressed once more that the creation of automatic fem models usually takes just a few seconds, and that the analysis in the linear range for many loading conditions can take the same order of magnitude of time. Using buckling analysis for all the design combinations is a quick tool. So, in a few minutes at worst – usually less – the analyst has a huge amount of interesting information, at almost no cost. The time of the analyst should be spent in understanding the physical problem and the working mode of the connections.

Finite element models can be precious for checking the constituents for generic resistance, buckling, and fatigue, and to better assess the forces flowing into the connectors, if and when this is needed. These forces in turn allow us to check the subconnectors (single bolts and weld segments), and the local failure modes (punching shear and bolt-bearing – see Chapter 8).

As has been seen, the checks of the connectors and the checks of the local failure modes can be embedded into a non-linear analysis, flagging as “broken” the connector involved in these checks.

The fracture failure modes – namely the block shear check, but also partly the net cross-section checks – are more difficult to tackle directly by fem analyses and so they should be tackled using the results of the fem analyses and applying other methods.

The availability of comprehensive stress maps like that of Figure 10.18 greatly helps the analyst, because it allows the delivery of a huge piece of information at one go. Moreover, fem analyses in the linear range (and also buckling checks) can help to find the most dangerous loading conditions, so as to reduce the number of loading conditions that need to be analyzed in the non-linear range. Even if tens or hundreds loading

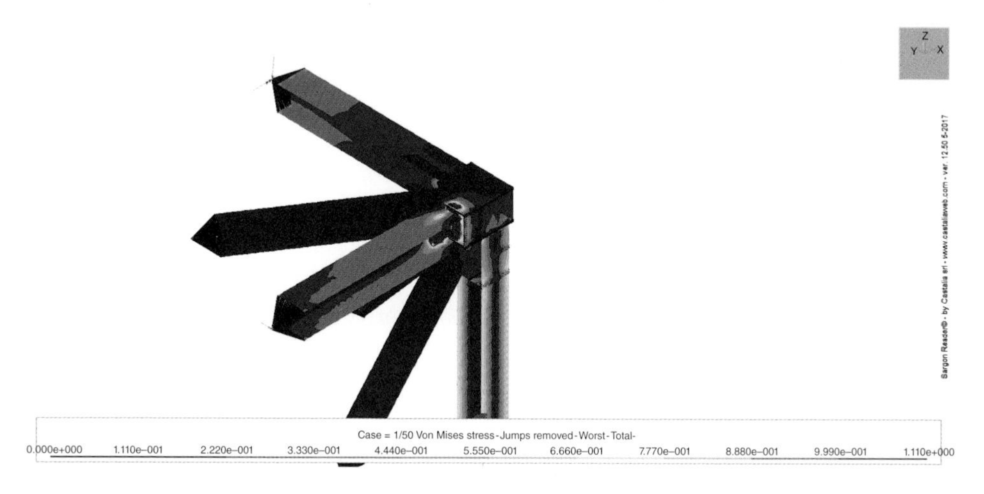

Figure 10.18 A Von Mises stress map of a complex PFEM.

combinations have to be checked, Von Mises envelope maps can help to quickly look at the most stressed components.

If the general method is automated (computing for each combination the critical and limit multiplier, or a safe lower bound for it), it may be used to quickly focus the most dangerous loading combinations. Then proper design actions can be carried on, improving the size or improving the analysis.

When the configuration of the nodal region is complex, this way to proceed is very much better than dividing the problem into a number of simplified ones, needing special tools to be solved. Not only is the single problem split into tens, but the simplifications run the risk to forgetting the total frame.

So, even if there are still limitations, problems to be solved, and methods to be improved, the author has no doubt that the fem approach is the simplest for the skilled analyst (albeit the much complex for the designers of the software).

10.8 Stiffness Evaluation

Using global PFEMs or MCOFEMs, it becomes natural to assess the stiffness of the connections. With this aim, if the far-extremity loaded-node of the members is flagged, so as to store its displacements during the non-linear analysis, then a load–displacement curve will be available, and this curve can be used to get the required stiffness. Clearly, if the analysis is linear the final displacement can be directly read.

Usually, due to the deformability of the members, a part of the displacement is related to member flexibility and this part should be neglected.

For instance, if the splice joint of Section 10.4.7 is considered, it has been seen that

1) the displacement related to the extension of the members was 1.31 mm (= NL/EA)
2) the total displacement using the initial stiffness was 3.05 mm
3) the total displacement using the secant stiffness was 8.26 mm
4) the total displacement using non-linear analysis was 3.63 mm.

So, the following stiffness of the connection can be predicted by the three analyses:

$$S_{j,ini} = \frac{1{,}073{,}600}{(3.05-1.31)} = 617\,\text{kN/mm}$$

$$S_{j,\text{sec}} = \frac{1{,}073{,}600}{(8.26-1.31)} = 154\,\text{kN/mm}$$

$$S_{j,nl} = \frac{1{,}073{,}600}{(3.63-1.31)} = 463\,\text{kN/mm}$$

For bending or shear loads, the displacement to be subtracted can be evaluated by means of simple beam theory. For instance, the displacement d_{beam} of a cantilever of span L under a constant bending moment M (the typical situation considered meaningful in order to evaluate the rotational stiffness of a joint), can be evaluated as

$$d_{beam} = \frac{ML^2}{2EJ}$$

So if the total displacement read at the extremity of the member is d_{tot}, then the rotational stiffness of the joint is

$$S_j = \frac{ML}{(d_{tot} - d_{beam})}$$

If the analysis is not linear, special considerations can be made in order to choose a suitable representative value of stiffness. Non-linear analyses can be illuminating. For instance, if the non-linear curve of the displacement of the load tip is plotted, considering the splice joint of the previous example, it can be seen that the non-linearity is weak (Figure 10.19, left). If the analysis is repeated increasing the load by 50%, and so getting to 0.75 N_{pl}, the non-linearity remains low, and a linear approximation may still be used (Figure 10.19, right).

The same considerations also apply to complex connections.

Sometimes, the stiffness of the connection depends on the loads applied, typically when contact non-linearities are activated. So, it might be necessary to execute a preliminary step loading the parts, and then moving with the pertinent loads applied. This can be done if the analysis, instead of considering each loading condition separately, considers the loading condition $(i + 1)$ applied at the end of the application of the loading condition i. This means activating a *load path*, which is a normal operating mode for non-linear solvers.

So, it is clear the PFEM and MCOFEM allow a prediction of the stiffness of the connections potentially that is much more accurate than that obtained by simple methods, especially when the geometry is complex and does not fall within the range of applicability of the simple methods. Moreover, sometimes there is no simple method available.

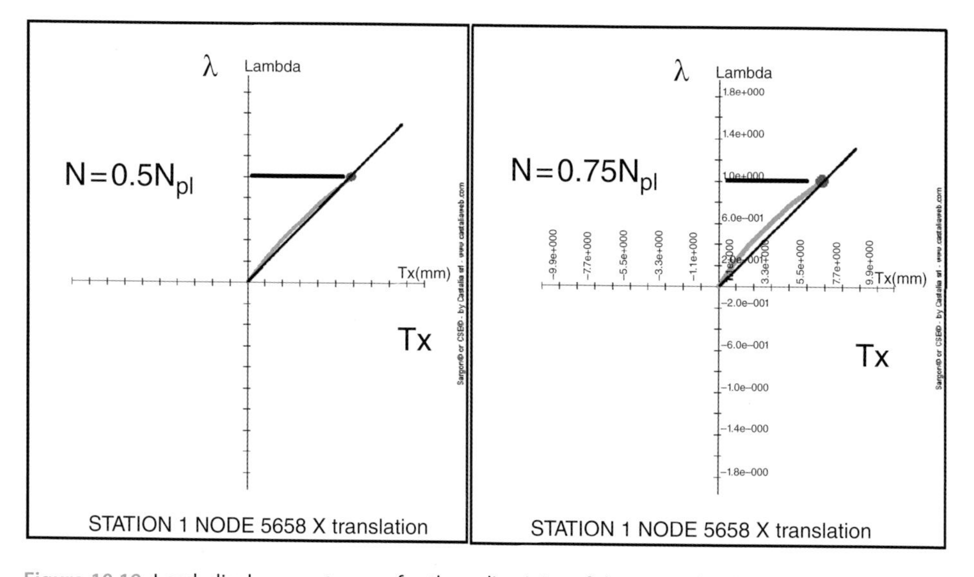

Figure 10.19 Load–displacement curve for the splice joint of the example. The load is applied using a load parameter λ, ranging from 0 to 1. On the left, is applied a total load equal to 0.5 times the yield axial force N_{pl} of the member. On the right the load is increased to 0.75 N_{pl}, and the analysis re-executed. The non-linearity increases, but it remains low.

10.9 Analysis Strategies

There are broadly speaking different levels of refinement and different levels of complexity related to PFEMs and MCOFEMs. The analyst should choose a good tradeoff between the need of refined analyses and the need to save time and money.

The analysis tools available are those listed below, with a judgment about difficulty and cost, according to the author's experience. It must be stressed that the cost of an analysis depends on how close the loads are to the limit, because the number of iterations needed will be different.

1) static linear analyses – easy and quick, very low cost
2) buckling analyses using eigenvalues – easy and quick, very low cost
3) general method of Eurocode 3 using 1 and 2 – easy and quick, very low cost
4) non-linear static analyses using material non-linearity only – easy, and quick if sufficiently far from limit loads, but slower if the applied loads are near the limit ones; it is suggested that when the elastic analysis has signaled that it is questionable, to use Von Mises stress maps
5) non-linear static analyses using geometrical non-linearity only – useful to estimate the differences with the critical multipliers obtained using the eigenvalue analysis approach
6) non-linear static analyses using contact non-linearity only – suggested when it is expected that contact forces may significantly change the stress state of constituents – usually runs quickly; these analyses have medium cost but may have a huge return
7) non-linear static analyses using material and geometrical non-linearity – can be used as an alternative to the general method, if an important dependence on imperfections is not expected; usually it has a meaning on well defined loading combinations that other methods have signaled interesting or dominant
8) non-linear static analyses using material and contact non-linearity only – it is one step forward from (6), and can be useful if (6) has signaled questionable Von Mises stress maps
9) non-linear analysis using all three non-linearities – may be used for specific loading conditions after all the features of the problem are well understood by the analyst

Reference

Bruneau, M., Uang, C.M., and Sabelli, R. 2011, *Ductile Design of Steel Structures*, 2nd ed., McGraw Hill.

11

Conclusions and Future Developments

There are more things in Heaven and Earth, Horatio,
Than are dreamt of in our philosophy.
Hamlet, I,v

11.1 Conclusions

The work presented in this book has been carried on – not continuously – over a period of 17 years. Some of the methods assembled in this work had been proposed by other authors in other contexts, and have been used here in the frame of the work presented: for instance, the extended experimental studies of fillet welds or bolts, or preloaded bolts, have been a great help to improve the modeling of connectors.

Some of the terms, approaches, and methods presented in this book, as far as the author knows, have been discovered by the author; they are here proposed, for the first time, for the attention of colleagues and scholars in the field. These in particular are:

- some new and much-needed terms: *jnode, renode, prenode, chain*
- the jnode concept, jnode analytics (hierarchical, central, simple, tangent, and cuspidal jnodes) and jnodes automatic search in BFEM, plus indices related to connection complexity and to the amount of work expected
- methods to assess by purely geometrical considerations if the bodies are connected or not, considering their geometry only
- methods to automatically extract the *chains* from the scene and analyze them in order to check that the renode is coherent and well-posed
- some logical rules assumed for the connectors and connected components and their classification into *members, force-transferrers, stiffeners, connectors,* and *constraint blocks*
- classification of renodes into isoconnected, hypoconnected, and hyperconnected, in a parallel way to framed structures, using the concept of the force-packet
- unified approach to the modeling and checking of connectors (bolts and welds) using special Timoshenko-like finite elements for single subconnectors. This approach can be used in PFEM, and, adding hypotheses to the kinematics of the connectors,

Steel Connection Analysis, First Edition. Paolo Rugarli.
© 2018 John Wiley & Sons Ltd. Published 2018 by John Wiley & Sons Ltd.

coherent to the rigid-body assumption implied by the traditional rules adopted to check single bolts and single welds of a group; this approach has also been used in order to model the mechanical behavior of *groups* of simple connectors – here called layouts. The approach has been described in linear and non-linear ranges, with several possible options.

- application of the instantaneous center of rotation method to the checks of weld groups loaded by 3D, *not 2D*, loading conditions (three forces and three moments), explained and tested, and a non-linear numerical procedure
- 3D tangent stiffness matrix of orthotropic fillet weld segments
- discussion of the difference between utilization index and factor, and of the danger related to the confusion between the two concepts
- new general method to tackle the problem of block shear
- *hybrid approach* to the analysis of steel connection, introducing the IRFEM as fundamental tool. IRFEM is proposed as the workhorse of steel connection analysis, in a way comparable to that of linear beam and truss models for structural analysis. Also, the use of automatically generated SCOFEMs related to the IRFEM is new
- automation of the creation of detailed plate–shell finite element models of single constituents (SCOFEMs) and of assembly of constituents (MCOFEMs or PFEM) in a way fully coherent with all the steps of the analysis
- fully automatic creation of PFEMs referring to generic renodes, developed by the author for the first time in 2008–2012, although a pure fem approach as a tool to analyze whole generic renodes is not new and has been used many times in the past decades. However, the *Connection Study Environment* (CSE) is the first software in the world to automate the creation of PFEMs, SCOFEMs, and MCOFEMs fully coherent with the pillars of connection analysis from 2008 to 2012 (MSC 2012, Rugarli 2013). In this book, PFEM inherits the finite element formulations of single connectors and it is seen as a generalization or IRFEM due to the relaxation of the hypotheses of rigidity.

Practically all the methods described in this book are already part of CSE, which has been used in real-world engineering since 2008.

11.2 Final Acknowledgments

11.2.1 Reasons of This Project

This research was begun at the end of the 1990s, because the author's customers and colleagues needed software able to check *generic* connections. At the time, the author tried to apply the component method of Eurocode 3 to a general context, finding it inapplicable to *generic* connections. The author is still convinced that the component method is intrinsically inapplicable to generic connections under generic loading conditions.

The component method specific innovation is not related to considering the bolts and welds as springs in series or in parallel. This had already been done, particularly with bolts, decades before the component method was proposed and named as such (e.g. Tate and Rosenfeld did it in 1946). Instead, the brilliant idea of the component method is to extend the use of springs to all the parts of a renode, more specifically dividing into "components" the bodies that create the scene. This fruitful concept can be very useful when applied to specific, well defined connections, using specific cross-sections, and specific loading conditions. However, this method cannot be applied to a general context.

The approaches presented in this book differ from the component method precisely in the way the bodies are modeled (which is the specific aim of the component method) for the following reasons:

- When using the IRFEM, a simple very useful method, it is assumed that the bodies are rigid, and later they are analyzed by using flexible plate–shell finite element models created automatically, and loaded by balanced loading configurations fully bearing on all the three pillars of connection analysis (equilibrium, the action and reaction principle, and the static theorem of limit analysis).
- When using PFEM and MCOFEM, the bodies are directly fine-modeled by plate–shell finite elements, and the forces flowing into the connectors are evaluated, partially (MCOFEM) or totally (PFEM), by finite element models.

So, the approach presented here is the complement of the *component method*. The component method transforms the (parts of the) bodies into uniaxial springs. Here this has only been partially done for connectors (that are modeled by finite elements, not by uniaxial springs), and this had already been done before the component method was developed. The bodies are initially rigid and later stress-analyzed one by one by fem, as flexible, in the simple *hybrid* approach, or are accurately modeled as flexible from the start and linked by connectors, in the more refined approaches (MCOFEM and PFEM).

The author first arrived at the hybrid approach, which was implemented and has been used since 2008. The path was then open for the automatic modeling of the whole renode. This was reached at the end of 2012 and announced in February 2013: the automated PFEM, was the natural extension of the models created to check constituents using the hybrid approach (Rugarli 2009a, Rugarli 2009b, MSC 2012, Rugarli 2013).

When this work was started it would have been a bet to believe that the *creation* and running of pertinent finite plate–shell finite element models would be quick *and totally automated*. Years of practice (since 2008) have shown that this bet has been won. It is the author's opinion that these approaches should be considered as the future normal mode to check connections, exactly as has already happened for framed structures.

11.3 Future Developments

At the time of writing, the main open fronts for research consistent with the approach here proposed are, in the author's view (not in order of importance):

1) improving the block shear checks to better taking into account load eccentricity (torque)
2) improving and testing the mechanical model of welds also using contact
3) finding a simple, more general model than T-stub for evaluating in a simplified way prying force locations and amounts
4) improving the speed and robustness of contact non-linearity algorithms for PFEM
5) extending the instantaneous center of rotation method for bolts, also for loading conditions including axial force and bending

6) exploring the issues related to uneven stress distribution in welds, when the implicitly assumed rigid body assumption is removed
7) improving MCOFEM and PFEM to avoid or reduce spurious buckling modes
8) extending the methods described here to Class 4 (slender) member cross-sections
9) extending this approach to wood connections.

If it can be frustrating for the scholar to consider that this kind of research is never-ending, some of the famous last words of *Candide*, are indeed of help:

> *"You must certainly have a vast estate," said Candide to the Turk.*
> *"I have no more than twenty acres of ground," he replied, "the whole of which I cultivate myself with the help of my children; and our labor keeps off from us three great evils – idleness, vice, and want."*
>
> *Candide, Voltaire*

References

MSC 2012, *Cool Software – Choose your own Connection*, What's Cool in Steel, Modern Steel Construction, August 2012: https://www.aisc.org/globalassets/modern-steel/archives/2012/08/2012v08_cool.pdf

Rugarli, P. 2009a, *CSE. A Brief History of the Project*, Castalia srl: http://www.steelchecks.com/CONNECTIONS/history.asp

Rugarli, P. 2009b, *Calcolo Automatico di Collegamenti con Posizionamento Libero (Software Computation of Steel Joints Built Up Freely Placing Components)*, Acta XXII Congress of Collegio Tecnici Acciaio, *L'Acciaio per un Futuro Sostenibile*, Padova, September 28–30 2009, ACS ACAI Servizi srl, Milano.

Rugarli, P. 2013, *CSE Press Release – Automatic FEM Modeling of Steel-Structures Connections*, February 2013: http://www.steelchecks.com/connections/press1.asp

Appendix 1

Conventions and Recalls

A1.1 Recalls of Matrix Algebra, Notation

A *matrix* **T** is a table of numbers organized in rows and columns. In this book, bold capital letters will be used for matrices. The size of a matrix having n rows and m columns, can be written (n, m).

A *vector* **v**, is a list of numbers. In this book lowercase letters in bold are used for vectors. Vectors can be row $(1, n)$ or columns $(n, 1)$. Usually they will be columns in this text, unless otherwise stated.

A matrix having the same number of rows and columns is said to be *square*. The matrix obtained by interchanging rows and columns is the *transpose matrix* of the original one. The transpose of a row vector, is a column vector, and vice versa. Transpose matrices and vectors are identified by the superscript "T", for example, $\mathbf{T}^{\mathrm{T}}$, $\mathbf{v}^{\mathrm{T}}$.

The elements of a matrix **T** can be identified by two indices, i for the row and j for the column: t_{ij}. The order of a square matrix or of a vector is the number of rows or columns.

A matrix is said to be diagonal if it is square and if for each $i \neq j$ $t_{ij} = 0$. Diagonal matrices have diagonal terms not null, and extra-diagonal terms null.

A diagonal matrix having all the diagonal terms equal to 1 is the unity matrix and its symbol in this text is **I**.

The dot product between two vectors of the same order will be identified using the raised dot: $\mathbf{v} \cdot \mathbf{t}$. Its result is a number. The square root of the dot product of a row vector with its transpose is the norm of the vector and is a number:

$$\sqrt{\mathbf{v}^{\mathrm{T}} \cdot \mathbf{v}} = \|\mathbf{v}\| = norm$$

The vector product between two vectors of order 3 is still a vector of order 3. The vector product will be identified in this text using "$\wedge$", for example, $\mathbf{s} = \mathbf{v} \wedge \mathbf{t}$.

A matrix **A** can be multiplied by a matrix **B** (possibly a vector) if the number of columns of **A** is equal to the number of rows of **B**. The result will be a matrix having the number of rows of **A** and the number of columns of **B**. The product **C** is written as $\mathbf{C} = \mathbf{AB}$. The transpose of a matrix **C** which is the result of the multiplication of two matrices **AB**, can be written $\mathbf{C}^{\mathrm{T}} = \mathbf{B}^{\mathrm{T}}\mathbf{A}^{\mathrm{T}}$. The dot product can be seen as the product between a row vector $(1, n)$ and a column vector of the same order $(n, 1)$. The result is a $(1, 1)$ matrix, i.e. a number.

Steel Connection Analysis, First Edition. Paolo Rugarli.
© 2018 John Wiley & Sons Ltd. Published 2018 by John Wiley & Sons Ltd.

If a matrix or a vector is multiplied by the unity matrix, it is left unchanged, that is **Iv** = **v**, **vI** = **v**, **IA** = **A**, **AI** = **A**.

A linear system of equations can be written in short as **Ax** = **b**, where **A** is a square matrix, **x** is an unknown vector, and **b** is a vector of known terms.

The *inverse* of a square matrix **A** is written $\mathbf{A}^{-1}$ and is such that $\mathbf{A}\mathbf{A}^{-1} = \mathbf{I}$.

A square matrix **T** is said to be *orthogonal* if its inverse is equal to its transpose, so that $\mathbf{T}\mathbf{T}^{\mathrm{T}} = \mathbf{T}^{\mathrm{T}}\mathbf{T} = \mathbf{I}$. It can be easily shown that multiplying an orthogonal matrix times a vector, the norm of the vector is left unchanged:

$$(\mathbf{Tv})^{\mathrm{T}} \cdot (\mathbf{Tv}) = \mathbf{v}^{\mathrm{T}}\mathbf{T}^{\mathrm{T}}\mathbf{Tv} = \mathbf{v}^{\mathrm{T}}\mathbf{v}$$

This means that the vector has rotated at equal norm. In this book, orthogonal matrices will be used frequently.

Points are also vectors of order 3, in space. For this reason, they will sometimes be identified by capital letters in bold. This is an exception to the rule that lowercase letters are used for vectors, and capital for matrices. The vector connecting point **P** to point **Q**, is written (**Q-P**).

A1.2 Cross-Sections

In this book, the term *cross-section* is mainly used to describe the *member cross-section,* which, in a prismatic member, does not change from point to point along the axis line. Member cross-sections are obtained by considering the underlying truss or beam element cross-sections.

A cross-section is usually referred to its geometrical (gravity) center, and two reference orthogonal axes can always be drawn from that center: the cross-section *principal axes.* These two orthogonal axes always imply a null mixed second area moment for the cross-section. Beam and truss element local reference systems can be obtained by considering the axis 1, which is the element axis from first extremity to second extremity, and the axes 2 and 3, which are the cross-section principal axes.

So, here, the cross-section principal axes will be named axis 2 and axis 3, and the variables x_2 and x_3 will be used to identify points over a cross-section. The variable x_1 will be used to identify a point over the member or finite element axis. In Eurocode 3 and in AISC 360-10 a different convention is used. Table A1.1 summarizes the relationship between the different axis names.

Table A1.1 Different conventions for cross-section principal axes.

This book	Eurocode 3	AISC
1	x	Z
2	y	X
3	z	Y

It is common usage, for ordinary C, T, I, and H shaped cross-sections, to name as "strong axis" the axis normal to the web. For those cross-sections, this axis is usually here named *axis 2*. The weak axis is therefore *axis 3*.

However, this is not a general rule. There are cross-sections whose principal axes are not parallel to any plate, and these principal axes will also be named axis 2 and axis 3. Given a cross-section, its principal axes will always be found with a unique rule, which will always be used for that cross-section type.

If the cross-section has a symmetry axis, that axis is always one of its principal axes.

A cross-section made up by constant-thickness plates, which are all parallel or perpendicular to the principal axes 2 and 3 will be said to be *aligned*. Ordinary IPE, HEA, HEB, M, W, UB, and UC cross-sections are all aligned.

However, an angle L, is not aligned. If all the plates of a cross-section are parallel or perpendicular to each other, the two axes passing through the center of gravity and parallel or perpendicular to these plates are the *axes of construction*, and will sometimes be identified here by the letters (y, z). If the cross-section is aligned, the axes of construction are also principal axes.

It is assumed that the axes (1, 2, 3) are a right-hand regular set.

Cross-sections have standard geometrical data, which will be here referred to by using the symbols listed in Table A1.2.

A cross-section is in general subjected to six internal forces: an axial force, two shears, one torque, and two bending moments. These internal forces are usually referred to the principal axes of the cross-section. Table A1.3 summarizes the symbols that will be used here to refer to these internal forces.

The internal forces applied to a cross-section can be organized in a vector, $\boldsymbol{\sigma}$. This column vector has the following form:

$$\boldsymbol{\sigma}^{\mathrm{T}} = [N \;\; V_2 \;\; V_3 \;\mid\; M_1 \;\; M_2 \;\; M_3]$$

Table A1.2 Main cross-section symbols and their meaning.

Symbol	Meaning
A	Area
J	Second area moment
W_{el}	Elastic modulus
W_{pl}	Plastic modulus
G	Center of gravity
t	Thickness
t_w	Web thickness (if any)
t_f	Flange thicknesses (if any)
h	Depth
b	Width
r	Fillet radius

Table A1.3 Symbols used for cross-section internal forces in this book.

Symbol	Meaning
N	Axial force (positive if tensile)
V_2	Shear parallel to axis 2
V_3	Shear parallel to axis 3
M_1	Torque
M_2	Bending moment: axis 2
M_3	Bending moment: axis 3

A1.3 Orientation Matrix

A set of three mutually orthogonal vectors having unit norm define a reference system. If the three local axes $\mathbf{v}_1$, $\mathbf{v}_2$, $\mathbf{v}_3$ are defined in such a way that

$$\mathbf{v}_1 = \mathbf{v}_2 \wedge \mathbf{v}_3$$

$$\mathbf{v}_2 = \mathbf{v}_3 \wedge \mathbf{v}_1$$

$$\mathbf{v}_3 = \mathbf{v}_1 \wedge \mathbf{v}_2$$

then the set follows the "right hand rule". In this book all the reference systems used follow the right hand rule.

The three unit vectors can be arranged into a square 3 × 3 matrix $\mathbf{T}$. Here, this is done according to this scheme:

$$\mathbf{T} = \begin{bmatrix} \mathbf{v}_1^{\mathrm{T}} \\ \mathbf{v}_2^{\mathrm{T}} \\ \mathbf{v}_3^{\mathrm{T}} \end{bmatrix} = \begin{bmatrix} v_{1x} & v_{1y} & v_{1z} \\ v_{2x} & v_{2y} & v_{2z} \\ v_{3x} & v_{3y} & v_{3z} \end{bmatrix}$$

This matrix is also called the *orientation matrix*, as it can be used to specify how an object in space is positioned. As the three vectors have unit norm and are orthogonal, it can be seen easily that

$$\mathbf{T}\mathbf{T}^{\mathrm{T}} = \mathbf{I}$$

i.e. the orientation matrix is always orthogonal.

The proof follows:

$$\mathbf{T}\mathbf{T}^{\mathrm{T}} = \begin{bmatrix} \mathbf{v}_1^{\mathrm{T}} \\ \mathbf{v}_2^{\mathrm{T}} \\ \mathbf{v}_3^{\mathrm{T}} \end{bmatrix} \begin{bmatrix} \mathbf{v}_1 & \mathbf{v}_2 & \mathbf{v}_3 \end{bmatrix} = \begin{bmatrix} \mathbf{v}_1^{\mathrm{T}}\cdot\mathbf{v}_1 & \mathbf{v}_1^{\mathrm{T}}\cdot\mathbf{v}_2 & \mathbf{v}_1^{\mathrm{T}}\cdot\mathbf{v}_3 \\ \mathbf{v}_2^{\mathrm{T}}\cdot\mathbf{v}_1 & \mathbf{v}_2^{\mathrm{T}}\cdot\mathbf{v}_2 & \mathbf{v}_2^{\mathrm{T}}\cdot\mathbf{v}_3 \\ \mathbf{v}_3^{\mathrm{T}}\cdot\mathbf{v}_1 & \mathbf{v}_3^{\mathrm{T}}\cdot\mathbf{v}_2 & \mathbf{v}_3^{\mathrm{T}}\cdot\mathbf{v}_3 \end{bmatrix} = \begin{bmatrix} 1 & 0 & 0 \\ 0 & 1 & 0 \\ 0 & 0 & 1 \end{bmatrix} = \mathbf{I}$$

A1.4 Change of Reference System

A vector **v** expressed with reference to the global reference system can be expressed with reference to a new system, provided that the orientation matrix **T** of the new system, relative to the global system, is known. If the vector expressed with reference to the global system is $\mathbf{v_g}$ and the vector expressed with reference to the local system is $\mathbf{v_l}$, the following relationships hold true:

$$\mathbf{v_l} = \mathbf{T}\mathbf{v_g}$$

$$\mathbf{v_g} = \mathbf{T}^{\mathrm{T}}\mathbf{v_l}$$

A1.5 Pseudocode Symbol Meaning

The following special symbols – coherent with C/C + + programming language, will be used in pseudocode:

+ +	increment operator
||	or
&&	and
! =	is different?
= =	is equal?
array[0]	the first element of an array
array[*n*-1]	the *n*-th element of an array

Appendix 2

Tangent Stiffness Matrix of Fillet-Welds

A2.1 Tangent Stiffness Matrix of a Weld Segment

Given a weld segment *j* of a weld seam *i* of a weld layout, it is useful to be able to write the tangent stiffness matrix of this weld segment.

The following symbols and definitions will be used:

- f_{uw} is the ultimate stress of weld. If needed, it can be set applying a reduction factor like $\phi = 0.75$ in AISC or by dividing by the product of β_w times γ_{M2}, as in Eurocode 3 (see Chapter 8, section 8.4.3).
- t_i is the thickness of the weld *i*.
- a_i is the throat thickness of the weld *i*.
- l_i/n_{si} is the weld segment length, obtained by dividing the weld seam length l_i by the number of weld segments of that weld seam.
- d_{wij}, d_{uij}, d_{wij} are the displacements of the weld segment centroid in the direction normal to the faying surface, d_{wij}, parallel to weld seam *i* (d_{uij} longitudinal), and perpendicular (d_{vij} transverse). For brevity, in this section the indices *i* and *j* will be omitted.
- *d* is the total displacement: $d = \sqrt{d_w^2 + d_u^2 + d_v^2}$.
- $\theta = \arccos(d_u/d)$ is the angle (in radians) of the displacement vector to the longitudinal axis of weld, **ui**. If $\theta > \pi/2$, θ is set equal to $\pi - \theta$.
- *p* is the strain measure defined as $p = d/d_{max}$, and d_{max} is the maximum displacement function of θ (see below).
- $q(\theta) = [1 + 0.5\ (\sin\theta)^{1.5}]$
- $h(p) = [p(1.9 - 0.9p)]^{0.3}$. The correction described in Section 6.7.4.1 is used, that is

$$
\begin{aligned}
h(p) &= 152.6p & p &\le 0.001 \\
h(p) &= [p\cdot(1.9-0.9p)]^{0.3} & 0.001 < p &\le 0.95 \\
h(p) &= 1.0503357p & p &> 0.95
\end{aligned}
$$

The maximum displacement d_{max} is defined as

$$
d_{max} = 0.209t\cdot\left(\frac{180}{\pi}\cdot\theta + 2\right)^{-0.32}
$$

where θ is in radians.

Steel Connection Analysis, First Edition. Paolo Rugarli.
© 2018 John Wiley & Sons Ltd. Published 2018 by John Wiley & Sons Ltd.

The internal force F carried by the weld segment j is a function of its displacements, according to the following rule:

$$F = 0.6\frac{a_i l_i}{n_{si}} f_{uw} \cdot h(p) \cdot q(\theta)$$

and it is directed as $\{d_w, d_u, d_v\}^{\mathrm{T}}$. Its components are F_w, F_u, F_v:

$$F_w = F\frac{d_w}{d}$$

$$F_u = F\frac{d_u}{d}$$

$$F_v = F\frac{d_v}{d}$$

The tangent stiffness matrix of the weld segment is a complex function of displacements, and its main property is that given an increment of displacement the increment of forces can be predicted, or vice versa:

$$\begin{vmatrix} \delta F_w \\ \delta F_u \\ \delta F_v \end{vmatrix} = \mathbf{K_t} \cdot \begin{vmatrix} \delta d_w \\ \delta d_u \\ \delta d_v \end{vmatrix} = \begin{vmatrix} \frac{\partial F_w}{\partial d_w} & \frac{\partial F_w}{\partial d_u} & \frac{\partial F_w}{\partial d_v} \\ \frac{\partial F_u}{\partial d_w} & \frac{\partial F_u}{\partial d_u} & \frac{\partial F_u}{\partial d_v} \\ \frac{\partial F_v}{\partial d_w} & \frac{\partial F_v}{\partial d_u} & \frac{\partial F_v}{\partial d_v} \end{vmatrix} \cdot \begin{vmatrix} \delta d_w \\ \delta d_u \\ \delta d_v \end{vmatrix}$$

that is,

$$\mathbf{K_t} = \begin{vmatrix} \frac{\partial F_w}{\partial d_w} & \frac{\partial F_w}{\partial d_u} & \frac{\partial F_w}{\partial d_v} \\ \frac{\partial F_u}{\partial d_w} & \frac{\partial F_u}{\partial d_u} & \frac{\partial F_u}{\partial d_v} \\ \frac{\partial F_v}{\partial d_w} & \frac{\partial F_v}{\partial d_u} & \frac{\partial F_v}{\partial d_v} \end{vmatrix} \tag{A2.1}$$

Taking the partial derivatives of F_w, F_u, and F_v, the following relationships are obtained.

First row of tangent stiffness matrix:

$$\frac{\partial F_w}{\partial d_w} = \frac{\partial F}{\partial d_w} \cdot \frac{d_w}{d} + \frac{F}{d} - \frac{F d_w d_w}{d^3}$$

$$\frac{\partial F_w}{\partial d_u} = \frac{\partial F}{\partial d_u} \cdot \frac{d_w}{d} - \frac{F d_w d_u}{d^3}$$

$$\frac{\partial F_w}{\partial d_v} = \frac{\partial F}{\partial d_v} \cdot \frac{d_w}{d} - \frac{F d_w d_v}{d^3}$$

Second row of stiffness matrix:

$$\frac{\partial F_u}{\partial d_w} = \frac{\partial F}{\partial d_w} \cdot \frac{d_u}{d} - \frac{F d_w d_u}{d^3}$$

$$\frac{\partial F_u}{\partial d_u} = \frac{\partial F}{\partial d_u} \cdot \frac{d_u}{d} + \frac{F}{d} - \frac{F d_u d_u}{d^3}$$

$$\frac{\partial F_u}{\partial d_v} = \frac{\partial F}{\partial d_v} \cdot \frac{d_u}{d} - \frac{F d_u d_v}{d^3}$$

Third row of stiffness matrix

$$\frac{\partial F_v}{\partial d_w} = \frac{\partial F}{\partial d_w} \cdot \frac{d_v}{d} - \frac{F d_w d_v}{d^3}$$

$$\frac{\partial F_v}{\partial d_u} = \frac{\partial F}{\partial d_u} \cdot \frac{d_v}{d} - \frac{F d_u d_v}{d^3}$$

$$\frac{\partial F_v}{\partial d_v} = \frac{\partial F}{\partial d_v} \cdot \frac{d_v}{d} + \frac{F}{d} - \frac{F d_v d_v}{d^3}$$

Noted that

$$\frac{\partial d}{\partial d_w} = \left(\frac{1}{2\sqrt{d_w^2 + d_u^2 + d_v^2}} \right) \cdot 2 d_w = \frac{d_w}{d}$$

$$\frac{\partial d}{\partial d_u} = \frac{d_u}{d}$$

$$\frac{\partial d}{\partial d_v} = \frac{d_v}{d}$$

The partial derivative of F to displacements are

$$\frac{\partial F}{\partial d_w} = \frac{\partial F}{\partial p} \cdot \frac{\partial p}{\partial d_w} + \frac{\partial F}{\partial \theta} \cdot \frac{\partial \theta}{\partial d_w}$$

$$\frac{\partial F}{\partial d_u} = \frac{\partial F}{\partial p} \cdot \frac{\partial p}{\partial d_u} + \frac{\partial F}{\partial \theta} \cdot \frac{\partial \theta}{\partial d_u}$$

$$\frac{\partial F}{\partial d_v} = \frac{\partial F}{\partial p} \cdot \frac{\partial p}{\partial d_v} + \frac{\partial F}{\partial \theta} \cdot \frac{\partial \theta}{\partial d_v}$$

and

$$p \le 0.001 \Rightarrow \quad \frac{\partial F}{\partial p} = \frac{0.6 \cdot f_{uw} a l}{n_s} \cdot q(\theta) \cdot 152.6$$

$$0.001 < p \le 0.95 \Rightarrow \quad \frac{\partial F}{\partial p} = \frac{0.6 \cdot f_{uw} a l}{n_s} \cdot q(\theta) \cdot 0.3 \cdot [p \cdot (1.9 - 0.9p)]^{-0.7} \cdot (1.9 - 1.8p)$$

$$p > 0.95 \quad \frac{\partial F}{\partial p} = \frac{0.6 \cdot f_{uw} a l}{n_s} \cdot q(\theta) \cdot 1.0503357$$

The partial derivative of F to θ is

$$\frac{\partial F}{\partial \theta} = \frac{0.6 \cdot f_{uw} \cdot a l}{n_s} \cdot h(p) \cdot 0.75 \cdot \cos\theta \cdot \sqrt{\sin\theta}$$

Now recalling that

$$\theta = \arccos\left(\frac{d_u}{d}\right)$$

and that if $\theta > 90°$ then it must be replaced by $180 - \theta$, the following function can be defined:

$$\text{tsign}(\theta) = +1 \quad \text{if}(\theta \le 90°)$$
$$\text{tsign}(\theta) = -1 \quad \text{if}(\theta > 90°)$$

So, the derivatives of θ to the displacements are

$$\frac{\partial\theta}{\partial d_w} = \text{tsign}(\theta)\cdot\frac{d_w d_u}{d^2\cdot\sqrt{d^2 - d_u^2}}$$

$$\frac{\partial\theta}{\partial d_u} = -\text{tsign}(\theta)\cdot\sqrt{\frac{d^2 - d_u^2}{d^2}}$$

$$\frac{\partial\theta}{\partial d_v} = \text{tsign}(\theta)\cdot\frac{d_u d_v}{d^2\cdot\sqrt{d^2 - d_u^2}}$$

The partial derivatives of the strain p to the displacements are

$$\frac{\partial p}{\partial d_w} = \frac{d_w}{d\cdot d_{max}} - \frac{d}{d_{max}^2}\cdot\frac{\partial d_{max}}{\partial d_w}$$

$$\frac{\partial p}{\partial d_u} = \frac{d_u}{d\cdot d_{max}} - \frac{d}{d_{max}^2}\cdot\frac{\partial d_{max}}{\partial d_u}$$

$$\frac{\partial p}{\partial d_v} = \frac{d_v}{d\cdot d_{max}} - \frac{d}{d_{max}^2}\cdot\frac{\partial d_{max}}{\partial d_v}$$

Finally, the partial derivatives of d_{max} to the displacements are

$$\frac{\partial d_{max}}{\partial d_w} = -0.32\cdot 0.209\cdot t\cdot\left[\frac{180}{\pi}\cdot\theta + 2\right]^{-1.32}\cdot\frac{180}{\pi}\cdot\frac{\partial\theta}{\partial d_w}$$

$$\frac{\partial d_{max}}{\partial d_u} = -0.32\cdot 0.209\cdot t\cdot\left[\frac{180}{\pi}\cdot\theta + 2\right]^{-1.32}\cdot\frac{180}{\pi}\cdot\frac{\partial\theta}{\partial d_u}$$

$$\frac{\partial d_{max}}{\partial d_v} = -0.32\cdot 0.209\cdot t\cdot\left[\frac{180}{\pi}\cdot\theta + 2\right]^{-1.32}\cdot\frac{180}{\pi}\cdot\frac{\partial\theta}{\partial d_v}$$

and the partial derivatives of θ to the displacements have already been found.

Substituting the expressions found back into the proper positions, the tangent stiffness matrix of a weld segment, as a function of the displacements d_w, d_u, and d_v, can be found.

If the weld segment j is modeled as a single finite element, with two nodes, considering the translational degrees of freedom the tangent stiffness matrix can be expressed as

$$\begin{vmatrix} \delta F_{w1} \\ \delta F_{u1} \\ \delta F_{v1} \\ \hline \delta F_{w2} \\ \delta F_{u2} \\ \delta F_{v2} \end{vmatrix} = \begin{vmatrix} \mathbf{K_t} & -\mathbf{K_t} \\ -\mathbf{K_t} & \mathbf{K_t} \end{vmatrix} \cdot \begin{vmatrix} \delta d_{w1} \\ \delta d_{u1} \\ \delta d_{v1} \\ \hline \delta d_{w2} \\ \delta d_{u2} \\ \delta d_{v2} \end{vmatrix}$$

Most of the bending and torsional stiffness of a weld seam is obtained by assembling the translational stiffnesses of the weld segments, and is due to the lever of segments relative to the weld seam center. At the segment level, the rotational stiffnesses are low, and would depend on a variation of the stresses along the length or throat-thickness, so they can be neglected.

If it is wished to consider an isotropic behavior for the weld segment (i.e. neglecting the effect related to the direction of loading, and considering always as if $\theta = 0$), then the following replacements must be applied to the tangent matrix:

$$q(\theta) = 1$$

$$d_{max} = 0.1674t$$

$$\frac{\partial F}{\partial \theta} = \frac{\partial \theta}{\partial d_w} = \frac{\partial \theta}{\partial d_u} = \frac{\partial \theta}{\partial d_v} = \frac{\partial d_{max}}{\partial d_w} = \frac{\partial d_{max}}{\partial d_u} = \frac{\partial d_{max}}{\partial d_v} = 0$$

A2.2 Modifications for Weld Segments Using Contact

If a weld segment uses contact, three states can be detected examining the displacements and the forces (see Chapter 6).

1) unlocked – this is when $(d_{w2} - d_{w1}) \geq 0$
2) locked and not sliding – this is when $(d_{w2} - d_{w1}) < 0$ and $\sqrt{F_u^2 + F_v^2} < \mu \cdot |F_w|$, where μ is a friction coefficient
3) locked and sliding – this is when $(d_{w2} - d_{w1}) < 0$ and $\sqrt{F_u^2 + F_v^2} \geq \mu \cdot |F_w|$, where μ is a friction coefficient

The tangent stiffness matrix for the unlocked state was described in the previous section.

For the locked no-sliding state, the elementary block of the tangent stiffness matrix can be written as

$$\mathbf{K_t} = \begin{vmatrix} \frac{EA}{t} & 0 & 0 \\ 0 & \frac{GA}{t} & 0 \\ 0 & 0 & \frac{GA}{t} \end{vmatrix}$$

For the locked-sliding state, the matrix can be written as

$$\mathbf{K_t} = \begin{vmatrix} \frac{EA}{t} & 0 & 0 \\ 0 & \frac{\partial F_u}{\partial d_u} & \frac{\partial F_u}{\partial d_v} \\ 0 & \frac{\partial F_v}{\partial d_u} & \frac{\partial F_v}{\partial d_v} \end{vmatrix}$$

The partial derivatives should be computed using the expressions found in the previous section, and applying the following replacements:

$$d = \sqrt{d_u^2 + d_v^2}$$

$$\frac{\partial(\cdot)}{\partial d_w} = 0$$

A2.3 Tangent Stiffness Matrix of a Weld Layout for the Instantaneous Center of Rotation Method

The displacements of a weld segment j are related to the displacement of a weld seam i center by the following matrix relation

$$\begin{vmatrix} d_{wij} \\ d_{uij} \\ d_{vij} \end{vmatrix} = \begin{vmatrix} \mathbf{I} & \mathbf{S_{ij}} \end{vmatrix} \cdot \begin{vmatrix} d_{wi} \\ d_{ui} \\ d_{vi} \\ \hline r_{wi} \\ r_{ui} \\ r_{vi} \end{vmatrix}$$

where $\mathbf{I}$ is the unity matrix, and

$$\mathbf{S_{ij}} = \begin{vmatrix} 0 & 0 & -u_{ij} \\ 0 & 0 & 0 \\ u_{ij} & 0 & 0 \end{vmatrix}$$

u_{ij} is the distance with sign, from the centroid of the weld segment j of weld seam i, to the weld seam i center (Figure 6.30).

The contribution of the weld segment j to the forces and moments acting at the weld seam i level is

$$\boldsymbol{\delta\sigma}_{\mathbf{ij,loc}} = \begin{vmatrix} \delta N_i \\ \delta V_{ui} \\ \delta V_{vi} \\ \delta M_{wi} \\ \delta M_{ui} \\ \delta M_{vi} \end{vmatrix} = \begin{vmatrix} \mathbf{I} \\ \mathbf{S}_{\mathbf{ij}}^{\mathbf{T}} \end{vmatrix} \cdot \begin{vmatrix} F_{wij} \\ F_{uj} \\ F_{vj} \end{vmatrix} = \begin{vmatrix} \mathbf{I} \\ \mathbf{S}_{\mathbf{ij}}^{\mathbf{T}} \end{vmatrix} \cdot \mathbf{K}_{\mathbf{tij}} \cdot \begin{vmatrix} \mathbf{I} & \mathbf{S}_{\mathbf{ij}} \end{vmatrix} \cdot \begin{vmatrix} \delta d_{wi} \\ \delta d_{ui} \\ \delta d_{vi} \\ \delta r_{wi} \\ \delta r_{ui} \\ \delta r_{vi} \end{vmatrix} = \begin{vmatrix} \mathbf{K}_{\mathbf{tij}} & \mathbf{K}_{\mathbf{tij}}\mathbf{S}_{\mathbf{ij}} \\ \mathbf{S}_{\mathbf{ij}}^{\mathbf{T}}\mathbf{K}_{\mathbf{tij}} & \mathbf{S}_{\mathbf{ij}}^{\mathbf{T}}\mathbf{K}_{\mathbf{t}ij}\mathbf{S}_{\mathbf{ij}} \end{vmatrix} \cdot \delta\mathbf{d}_{\mathbf{i}}$$

where $\mathbf{K}_{\mathbf{tij}}$, a 3 × 3 square matrix, is the tangent stiffness matrix of the segment j of weld seam i (see Equation A2.1).

The sum of the contributions of all weld segments gives the increment of forces carried by weld seam i, in its reference system (loc):

$$\boldsymbol{\delta\sigma}_{\mathbf{i,loc}} = \left(\sum_{j=1}^{nsi} \begin{vmatrix} \mathbf{K}_{\mathbf{tij}} & \mathbf{K}_{\mathbf{tij}}\mathbf{S}_{\mathbf{ij}} \\ \mathbf{S}_{\mathbf{ij}}^{\mathbf{T}}\mathbf{K}_{\mathbf{tij}} & \mathbf{S}_{\mathbf{ij}}^{\mathbf{T}}\mathbf{K}_{\mathbf{t}ij}\mathbf{S}_{\mathbf{ij}} \end{vmatrix} \right) \cdot \boldsymbol{\delta}\mathbf{d}_{\mathbf{i}}$$

As already seen in Equations (6.27–6.29) the displacements of the centroid of the weld seam i can be expressed as a function of the displacements of the weld layout centroid G as follows.

$$\mathbf{d}_{\mathbf{i}} = \begin{vmatrix} d_{wi} \\ d_{ui} \\ d_{vi} \\ r_{wi} \\ r_{ui} \\ r_{vi} \end{vmatrix} = \begin{vmatrix} \mathbf{T}_{\mathbf{i}} & \mathbf{S}_{\mathbf{i}} \\ \mathbf{0} & \mathbf{T}_{\mathbf{i}} \end{vmatrix} \cdot \begin{vmatrix} d_w \\ d_u \\ d_v \\ r_w \\ r_u \\ r_v \end{vmatrix} = \mathbf{Q}_{\mathbf{i}} \cdot \begin{vmatrix} d_w \\ d_u \\ d_v \\ r_w \\ r_u \\ r_v \end{vmatrix} = \mathbf{Q}_{\mathbf{i}} \cdot \mathbf{d}$$

where (see Equation 6.29)

$$\mathbf{T}_{\mathbf{i}} = \begin{vmatrix} 1 & 0 & 0 \\ 0 & c_i & s_i \\ 0 & -s_i & c_i \end{vmatrix}$$

$$\mathbf{S}_{\mathbf{i}} = \begin{vmatrix} 0 & v_{Gi} & -u_{Gi} \\ Y_i & 0 & 0 \\ X_i & 0 & 0 \end{vmatrix}$$

and

$$X_i = v_{Gi}s_i + u_{Gi}c_i$$
$$Y_i = -v_{Gi}c_i + u_{Gi}s_i$$

$$c_i = \cos(\alpha_i - \beta)$$
$$s_i = \sin(\alpha_i - \beta)$$

The tangent stiffness matrix of the whole layout can then be found as

$$\mathbf{K_t}(\mathbf{d}) = \sum_{i=1}^{nw} \left[\mathbf{Q_i^T} \cdot \left(\sum_{j=1}^{nsi} \begin{vmatrix} \mathbf{K_{tij}} & \mathbf{K_{tij}S_{ij}} \\ \mathbf{S_{ij}^T K_{tij}} & \mathbf{S_{ij}^T K_{tij} S_{ij}} \end{vmatrix} \right) \cdot \mathbf{Q_i} \right] \tag{A2.2}$$

This matrix can be used to evaluate the stresses flowing into welds, by the instantaneous center of rotation method (see Section 6.7.4).

Appendix 3

Tangent Stiffness Matrix of Bolts in Shear

A3.1 Tangent Stiffness Matrix of a Bolt

Given a bolt i of a bolt layout, it is useful to write the shear-related tangent stiffness matrix of this bolt, according to the rules of AISC 360-10, in view of a general application of the instantaneous center of rotation method.

The following symbols and definitions will be used:

- $R_{ult,i}$ the ultimate shear force of a bolt i. – in general it is different from bolt to bolt
- d_{ui}, d_{vi} the displacements of the bolt centroid in the direction parallel to principal axis **u** (d_{ui}), and parallel to principal axis **v** (d_{vi}) – for brevity in this section the index i will be omitted
- d the total displacement: $d = \sqrt{d_u^2 + d_v^2}$

The internal shear force V carried by the bolt i is a function of its displacements, according to the following rule (d is in millimeters)

$$V = R_{ult} \cdot \left(1 - e^{-0.3937d}\right)^{0.55}$$

and it is directed as $\{d_u, d_v\}^T$. Its components are V_u, V_v, and

$$V_u = V\frac{d_u}{d}$$

$$V_v = V\frac{d_v}{d}$$

The tangent stiffness matrix of the bolt is a function of displacements, and its main property is that given an increment of displacement the increment of forces can be predicted, or vice versa:

$$\begin{vmatrix} \delta V_u \\ \delta V_v \end{vmatrix} = \mathbf{K_t} \cdot \begin{vmatrix} \delta d_u \\ \delta d_v \end{vmatrix} = \begin{vmatrix} \dfrac{\partial V_u}{\partial d_u} & \dfrac{\partial V_u}{\partial d_v} \\ \dfrac{\partial V_v}{\partial d_u} & \dfrac{\partial V_v}{\partial d_v} \end{vmatrix} \cdot \begin{vmatrix} \delta d_u \\ \delta d_v \end{vmatrix}$$

Steel Connection Analysis, First Edition. Paolo Rugarli.
© 2018 John Wiley & Sons Ltd. Published 2018 by John Wiley & Sons Ltd.

that is,

$$\mathbf{K_t} = \begin{vmatrix} \dfrac{\partial V_u}{\partial d_u} & \dfrac{\partial V_u}{\partial d_v} \\ \dfrac{\partial V_v}{\partial d_u} & \dfrac{\partial V_v}{\partial d_v} \end{vmatrix} \tag{A3.1}$$

Taking the partial derivatives of V_u and V_v, the following relationships are obtained.
First row of tangent stiffness matrix:

$$\frac{\partial V_u}{\partial d_u} = \frac{\partial V}{\partial d_u} \cdot \frac{d_u}{d} + \frac{V}{d} - \frac{V d_u d_u}{d^3}$$

$$\frac{\partial V_u}{\partial d_v} = \frac{\partial V}{\partial d_v} \cdot \frac{d_u}{d} - \frac{V d_u d_v}{d^3}$$

Second row of tangent stiffness matrix:

$$\frac{\partial V_v}{\partial d_u} = \frac{\partial V}{\partial d_u} \cdot \frac{d_v}{d} - \frac{V d_u d_v}{d^3}$$

$$\frac{\partial V_v}{\partial d_v} = \frac{\partial V}{\partial d_v} \cdot \frac{d_v}{d} + \frac{V}{d} - \frac{V d_v d_v}{d^3}$$

Noted that

$$\frac{\partial d}{\partial d_u} = \left(\frac{1}{2\sqrt{d_u^2 + d_v^2}} \right) \cdot 2 d_u = \frac{d_u}{d}$$

$$\frac{\partial d}{\partial d_v} = \frac{d_v}{d}$$

The partial derivative of V to displacements are

$$\frac{\partial V}{\partial d_u} = \frac{\partial V}{\partial d} \cdot \frac{\partial d}{\partial d_u}$$

$$\frac{\partial V}{\partial d_v} = \frac{\partial V}{\partial d} \cdot \frac{\partial d}{\partial d_v}$$

and

$$\frac{\partial V}{\partial d} = 0.55 \cdot R_{ult} \cdot \left(1 - e^{-0.3937d}\right)^{-0.45} \cdot 0.3937 \cdot e^{-0.3937d} \equiv A$$

Substituting back the expressions found in the proper positions, the tangent stiffness matrix of a bolt, as a function of the displacements d_u and d_v, can be found:

$$\mathbf{K_t} = \frac{1}{d^3} \begin{vmatrix} A d d_u^2 + V d^2 - V d_u^2 & A d d_u d_v - V d_u d_v \\ A d d_u d_v - V d_u d_v & A d d_v^2 + V d^2 - V d_v^2 \end{vmatrix} \tag{A3.2}$$

If the bolt i is modeled as a single finite element, with two nodes, considering only the translational shear degrees of freedom, the tangent stiffness matrix can be expressed as

$$\begin{vmatrix} \delta V_{u1} \\ \delta V_{v1} \\ \hline \delta V_{u2} \\ \delta V_{v2} \end{vmatrix} = \begin{vmatrix} \mathbf{K_t} & -\mathbf{K_t} \\ -\mathbf{K_t} & \mathbf{K_t} \end{vmatrix} \cdot \begin{vmatrix} \delta d_{u1} \\ \delta d_{v1} \\ \hline \delta d_{u2} \\ \delta d_{v2} \end{vmatrix}$$

A3.2 Tangent Stiffness Matrix of a Bolt Layout for the Instantaneous Center of Rotation Method

The displacements of a bolt i are related to the displacement of the bolt layout center by the following matrix relation:

$$\mathbf{d_i} = \begin{vmatrix} d_{ui} \\ d_{vi} \end{vmatrix} = \begin{vmatrix} 1 & 0 & -v_i \\ 0 & 1 & u_i \end{vmatrix} \cdot \begin{vmatrix} d_u \\ d_v \\ r_w \end{vmatrix} = \mathbf{S_i d}$$

The contribution of the bolt i to the increment of forces and torsional moment acting at the bolt layout level is

$$\boldsymbol{\delta\sigma}_\mathbf{i} = \begin{vmatrix} \delta V_{ui} \\ \delta V_{vi} \\ \delta M_{wi} \end{vmatrix} = \begin{vmatrix} 1 & 0 \\ 0 & 1 \\ -v_i & u_i \end{vmatrix} \cdot \begin{vmatrix} \delta V_{ui} \\ \delta V_{vi} \end{vmatrix} = \mathbf{S_i^T K_t \delta d_i} = \mathbf{S_i^T K_t S_i \delta d}$$

where $\mathbf{K_{ti}}$, a 2×2 square matrix, is the tangent stiffness matrix of the bolt i (Equation A3.2).

The sum of the contributions of all bolts gives the increment of forces carried the bolt layout

$$\boldsymbol{\delta\sigma} = \left(\sum_{i=1}^{nb} \mathbf{S_i^T K_{ti} S_i} \right) \cdot \boldsymbol{\delta}\mathbf{d} = \mathbf{K_t} \cdot \boldsymbol{\delta}\mathbf{d}$$

and the shear-torsional part of the tangent stiffness matrix of the bolt layout, a 3×3 square matrix, is defined as

$$\mathbf{K_t} = \sum_{i=1}^{nb} \mathbf{S_i^T K_{ti} S_i} \qquad \text{(A3.3)}$$

This matrix can be used to evaluate the forces flowing into bolts, by the instantaneous center of rotation method (see Section 7.9.3).

Symbols and Abbreviations

Symbols

a member identifier
a throat size
a parameter
A point
A set identifier
A area
$\mathbf{A}$ stiffness matrix subblock
AS all shear
b width
b member identifier
b parameter
B point
B set identifier
B bolt layout identifier
$\mathbf{B}$ stiffness matrix subblock
c parameter
c cosinus
$\mathbf{c}$ force packet at connector extremity
C point
C set identifier
C constituent identifier
$\mathbf{C}$ modifying matrix
C capacity
C couple or moment
d translational displacement
d amplification factor
d total translational displacement
D point
$\mathbf{D}$ distribution matrix
D distribution of internal forces
D diameter
D demand

Steel Connection Analysis, First Edition. Paolo Rugarli.
© 2018 John Wiley & Sons Ltd. Published 2018 by John Wiley & Sons Ltd.

D	bolt diameter
$\mathbf{D}$	stiffness matrix subblock
e	edge distance
E	point
E	Young's modulus
E	extremity point
$\mathbf{E}$	stiffness matrix subblock
f	stress
$\mathbf{f}$	force vector
f	flexibility index
F	force
F	far extremity of member
FR	fully resistant
G	center
G	shear modulus
h	height
h	intermediate weld resistance factor
$\mathbf{I}$	unity matrix
J	second area moment
k	stiffness
k	imperfection factor
k	scaling factor
k	hole factor
k	bolt bearing factor
$\mathbf{K}$	stiffness matrix
l	length
L	length
LO	longitudinal only
m	member identifier
$\mathbf{m}$	moment vector
m	non-dimensional bending
m	ratio of Young's moduli
M	bending moment, applied moment
n	number of items
n	normal stress
$\mathbf{n}$	unity vector
n	non-dimensional axial force
N	axial force
N	node identifier
N	near extremity of member
NS	no shear
$\mathbf{o}$	offset vector
O	origin or point
p	weld strain
p	load per unit length
p	pitch
P	applied load

P	point
$\mathbf{P}$	point vector
P	plate identifier
q	intermediate weld resistance factor
Q	point
Q	prying force
$\mathbf{Q}$	point vector
$\mathbf{Q}$	generic matrix
Q	factor for preload bolt compressed area
r	rotation
r	radius
r	polar distance
r	ratio of bolt bearing applied force R to maximum R_{be}
r	ratio
R	resistance
$\mathbf{s}$	force packet
s	sinus
S	closed polyline
$\mathbf{S}$	transformation matrix
t	thickness
t	tangential stress
T	applied force
$\mathbf{T}$	transformation matrix
u	coordinate
$\mathbf{u}$	nodal displacement vector
U	utilization
$\mathbf{v}$	unitary vector
v	coordinate
v	non-dimensional shear
V	shear
w	displacement of axis line in beams
w	coordinate
w	size of fillet weld
W	weld layout identifier
W	bending modulus
x	coordinate
y	coordinate
z	internal lever arm
z	coordinate
Z	check section for bolt

Greek Letters

α	load parameter
α	angle

α bolt bearing factor
α angle of weld seam to reference axis 1 or x
β angle between principal axis u and reference axis 1 or x
χ reduction factor
Δ displacement
ε strain
φ intermediate factor
φ safety factor
γ angle
γ angle between fusion faces
γ safety factor
λ slenderness
λ load parameter
$\bar{\lambda}$ non-dimensional slenderness
λ non-dimensional abscissa
μ friction coefficient
μ suitable small value
ν Poisson's ratio
Π failure path
θ angle
ρ shear reduction factor
σ normal stress
$\boldsymbol{\sigma}$ internal force vector
Σ subset
Σ bearing surface
τ tangential stress
Ω safety factor

Subscripts

0 permanent
0 hole
1 axis 1
1 variable
2 principal axis 2
3 principal axis 3
A point A
b bolt
b bearing
B point B
be bearing
bm base metal
C compression
CH convex hull
com compression

cr critical
cr crushing
d design
e bolt head diameter
e extremity
eff effective
el elastic
f factor
fr fracture
g gross
G geometrical
i index
IC instantaneous center
ini initial
j index
k bolt head height
k step
k index
l longitudinal
L limit
lim limit
m mean
m bolt nut height
m maximum
M material
max maximum
mod modified
n net
n normal
n normalized
nom nominal
p pin
p preload
P point P
P preload
par parallel
per perpendicular
pl plastic
pm plastic mechanism
Q point Q
R real
res threaded part
s weld seam
s hole factor
s subset
s bolt-head distance from faces
s shear

s	secant
sec	secant
t	tensile
t	transverse
u	ultimate
u	principal axis u
v	shear
v	principal axis v
VM	Von Mises
w	principal axis w
w	weld
y	yield

Index

a

action reaction principle 111, 124, 400, 402, 430, 432, 440
  and checks 426
  applied to connectors 319
  at interfaces 113
  enunciation 111
  violation 112
AISC 16, 28, 142, 197
alignment
  by strong 39
  by weak 39
  of axes 38
  splice 38
alignment code 64
analytic geometry 91
analytical model 29
  check 30
  need of the 30
anchor
  failure modes 345
  friction resistance 278
  permanent stiffness 277
  preloaded 278
  stiffness 277
arc-length method 392, 399
assembly matrix 144
assumptions, appropriate 19
automatic increment evaluation 398
automatic stepping 461
automation of finite element model creation 29
average surface 84
axes, principal *see* principal axes
axial force 3, 71

b

base metal, utilization factor, AISC 335
base plates 444
battened connection 405
beam element 27
beam theory 116
beam, principal 42
beam, secondary 42
bearing bolt layout
  action reaction principle 250
  axial stiffness of bolt 289
  bearing stiffness 244
  Crawford and Kulak curve 247
  elastic distribution 249
  force-displa cement relation 246
  initial stiffness 246
  non linear analysis 248
  plastic distribution 249
  pressure exchanged 250
  Rex and Easterling curve 246
  secant stiffness 248
  secant vs initial stiffness 245
  shear stiffness 244
  shear stiffness of bolt 289
  shear stiffness, secant and initial 291
  shear stiffness, tangential 291
  two springs in series 245
  using only axial forces in bolts 249
bearing forces, constitutive law 441
bearing stiffness 470

Steel Connection Analysis, First Edition. Paolo Rugarli.
© 2018 John Wiley & Sons Ltd. Published 2018 by John Wiley & Sons Ltd.

bearing surface 402, 412, 416
  model, description 252
  and mesh 434
  as result of Boolean operations between polygons 256
  boundary integrals 303
  candidate 252, 260
  constitutive law 252, 262, 263, 302
  crushing stress 255
  definition 251
  elementary operations 259
  example 264
  extent 253, 302
  for slip resistant bolt layouts 276
  intersection of polygons 258
  iterative procedure 305
  linear strain model 261
  nodal forces exchanged 441
  plastic neutral axis 260
  pressure as loads 255
  pressure exchanged 254
  pressures, true distribution 264
  reaction region 254
  stiffness 263
  union of polygons 257
  used for welds 417
  using crushing stress 358
bending moment, in bolt shaft 469
bending moment, in plate-shell 390
bending stiffness, of bolt shaft 469
Bernoulli beam 28, 173
bevel *see* work processes
bevels 84, 329
BFEM (Bar Finite Element Model) 28, 40, 41, 45, 51, 55, 56, 61, 73, 77, 78, 113, 115–117, 400, 415, 439, 444
BFEM, combinations 439
block shear *see* block tear
block tear, definition 21
block tear 148, 436
  and convex hull 364
  bolt layout subset 360, 363
  eccentric loading 360
  equivalent stress 362
  Eurocode 3 method, limits 368
  general model 365
  general problem description 363
  limit condition 361
  need of a general model 361
  net lengths 367
  pure rotational failure 369
  standard model 360
  tests 360
  utilization, AISC 360
  utilization, Eurocode 3 359
bolt
  nut, contact 79
  point, geometrical conditions 87
  resistance, utilization formulae 325
  row 12, 16, 159, 422
  row, inner 352
  segment 471
  shear stiffness, of no-slip bolt layout 292
  efficiency factor 262, 282
  efficiency factor, example 314
  European, grades, table 239
  European, table 239
  flexibility index 289
  fully resistant 476
  limit domain (M, N) 294
  limit domain (N, M, V) 295
  limit domain (N, V) 294
  limit domain (N, V) 324
  maximum distance, AISC 244
  maximum distance, Eurocode 3 244
  maximum edge distance, AISC 244
  maximum edge distance, Eurocode 3 244
  minimum distance, AISC 244
  minimum distance, Eurocode 3 244
  minimum edge distance, AISC 244
  minimum edge distance, Eurocode 3 244
  non linear analysis 293
  preload limit, Eurocode 3 342
  preloaded, model for non linear analysis 296
  shear behavior 287
  translational stiffness 285
  USA, grades, table 240
  USA, table 240
bolt axial stiffness, of no-slip bolt layout 292

bolt bearing
failure mode 436
hole modeling 465
modeling with finite elements 288
non linear modeling 465
stiffness evaluation, example 471
bolt efficiency factor 289
bolt group *see* bolt layout
bolt head 79
contact 79
bolt hole
modeling 465
modeling, for preloaded bolts 468
modeling in PFEM 463
radial elements 466
bolt layout
center 87
angle of principal axes 241
as a single object 88
bearing type and slip resistant 237
bearing type *see* bearing bolt layout
center 240
center 282
check planes 242
differences with weld layout 237
displacements and rotations 282
drilling direction 242
entry face 241
equivalent properties 299
extremities 88
extremity 240
finding connected constituents 89
force packets 241
generalized resistance moduli 302
in weak bending 105
insertion 79
insertion point 88
internal force vector 242
internal forces at the check sections 297
minimum geometrical information 80
multiplicity 88
no rows or columns 351
no slip (*see* slip resistant bolt layout)
overall internal forces 243
preloaded *see* slip resistant bolt layouts
principal axes 88, 283
principal system 241
reference system 240
regular grid 241
representation 79
shear only, cannot in principle exist 469
typical problems 238
using a bearing surface, computation of forces in bolts 302
boundary surface 74
BREP (Boundary Representation) 73, 84, 411
brick elements 84
bridge elements 432, 446
British Standard 28
brittle fracture 6
buckling 6, 13, 127, 129, 131, 136, 152, 348, 382, 389, 481
analysis 388, 432, 448, 481, 484
curve 141, 378
curve choice, for plates, Eurocode 3 379
curve, choice 134
depends on load distribution 389
and imperfections 147
length 134
load, factors affecting 131
local 380
modes 14
multiplier, real 378
premature 7
rules of good practice 132

c

capacity 320
capacity design 71, 129, 413
castellated connections 405
CBF (Concentric Bracing Frames) 61
chain 94, 126
automatic search 96
broken 95
closed loop 95
data structure 97
invalid 95
maximum number of objects 96
strings 100
check
ad hoc 348
ad hoc formulae 430
anchors 345

check (*cont'd*)
  and PFEM 481
  base metal, AISC 334
  block tear 114, 349, 428
  block tear 358–362
  block tear, utilization factor 367
  bolt bearing 350–352, 428, 441
  bolt bearing, AISC 355
  bolt bearing, bolt layout with no rows or columns 353
  bolt bearing, Eurocode 3 353
  bolt bearing, needed or not for preloaded bolts 344
  bolt bearing, using PFEM 464
  bolt resistance 337, 419
  bolt resistance, AISC 341
  bolt resistance, Eurocode 3 338
  bolt resistance, Eurocode 3 340
  bolt resistance, utilization factor, general formula 339
  bolt rupture, using PFEM 468
  buckling 132, 142, 147, 349, 377, 378
  buckling, utilization index 378
  cross section rupture 369
  crushing 357
  displacement control 349
  excess of displacement 381
  fillet weld, AISC, stress-strain curve 336
  fillet welds, AISC 334
  fillet welds, Eurocode 3 332
  fillet welds, Eurocode 3, directional method 333
  fillet welds, utilization factor, AISC 335
  fracture 349
  generic resistance 12, 15, 432, 481
  generic resistance, need of finite element method 348
  global 349
  local 349
  need of redoing 449
  net cross section 429
  net cross section, AISC 372
  net cross section, Eurocode 3 372
  no slip 468
  penetration weld 328
  penetration weld, rupture, AISC 331
  penetration weld, shear yielding, AISC 331
  penetration weld, tensile rupture, AISC 331
  penetration weld, utilization factor, AISC 331
  penetration welds, yield, Eurocode 3 330
  penetration welds, rupture, Eurocode 3 330
  pin bearing 356
  pin, AISC 347
  pin, Eurocode 3 346
  preloaded bolt resistance 345
  pull out 345, 419
  punching shear 357, 428
  simplified approaches 348
  simplified resistance 429
  slip 9, 342, 419
  slip, AISC 344
  slip, Eurocode 3 342
  threaded shaft 338
  welds and bolts using PFEM 480
  weld 418
  welds, effective area 330
  yielding 349, 373
  yielding, strategy 374
  yielding, utilization factor 373
  yielding, utilization factor evaluation 375
chosen face 162
clashes 89
class 4, 380
cleats 65
CNC (Computer Numerical Control) 3
codes 14
coherence 156
cold formed members 380
collapse 147
color bands 387
color maps 386
combination, number of 389
  notional in IRFEM 413
  of Eurocode 9
  realistic 10
combined loadings 6
comparisons, between IRFEM and MCOFEM 459

comparisons, PFEM and rigid MCOFEM, for fillet welds 462
complex math 156
component method 17, 18, 65, 158
  apparent ease 19
  infinite stiffness 18
  stiffnesses 158
  typical components 158
component *see* constituent
components, elementary 17
compound member 83, 403, 405
concrete block 254, 357
concrete slab 71, 151
connected main elements 65
connection
  analysis, features 1
  analysis, general path 153
  analysis, scope 29
  automatic detection 86
  check, basic questions 30
  classification 70
  classifications, limits of 72
  code 33, 37, 49
  code, color 34, 42, 60
  code, setting 34
  codes, using groups 33
  complexity 8
  considered one by one 7
  and cusps 37
  design, main problem 20
  direct 75
  failures 1
  FR (fully restrained) 71
  FS (full strength) 71
  how many different in a structure? 22
  hyperconnected, no-slip 151
  identical 40
  logic 66, 95
  meaning 69
  no-slip and shear 151
  no-slip and torque 151
  PR (partially restrained) 71
  preliminary conditions 85
  PS (partial strength) 71
  semi-rigid, design 71
connectivity list 101
connector 66, 76
  external 66
  extremities 66
  internal 66
  multiple 66
  multiplicity 66, 70
  placement 85
  rules 69
  saturated 445
  simple 66
CONNECTORDATA 94, 96
constituent
  common 8
  vs. component 65
  definition 65
  identical in parallel 144
  identification 77
  isomorphically loaded 143, 144
constitutive law 390
  perfectly plastic 373
constitutive models 391
constraint, distribution 131
constraint, fixed or elastic 45
constraint block 65, 76, 78, 119
  definition 45
constraint code 49
constraint mask 49
constraint reaction, nullity 437
constraints
  depending on displacements 394
  dummy 431
  notional 428, 448
  one-sided 394
  of SCOFEM 437
contact forces 395
  in IRFEM 417
  lumped 441
contact non linearity 147, 394, 476, 483
contact only connectors 316
contact pressures 441
contact surface 395
contact surface *see* bearing surface
converge, slow 374
convergence
  difficulties 394
  lack of 392

convergence (*cont'd*)
  no-convergence 389
  reached 392, 398
convex hull 363
  loop 364
cooking recipes 1
coordinate transformation 164
copy 91
corner, sharp 148
corotational approach 393
correlation factor 332
crack growth 148
cracks 452
crime
  connection 416
  in connection-design 3
  equilibrium 3, 8
  static 92
  variational 3
Crisfield M. A. 398
critical multiplier 133, 134, 137, 143, 378, 388, 389, 393, 394
  first 135
  successive 449
criticalness 134–136, 139, 142
critical point 393
critical shape 135, 389
cross-section
  aligned 491
  classification 380
  compound 119
  definition 490
  net (*see* net cross section)
  symmetry 59
crushing stress 358
CSE (Connection Study Environment) 24, 472
CSG (Constructive Solid Geometry) 74, 84
current stiffness parameter 375, 399
cusp 37
cut 84 *see also* work processes
cycles, of loading 391

d

defects, of welding 150
degree of freedom 45
degree of hyperconnectivity 125, 126
Delaunay triangulation 363, 435
demand 320
De Saint Venant principle 115, 347, 443
directional method 332, 333
disassembly, of single weld seam 183
displacement 176
  control 381, 392
  field 175
  high 128
  indicator of improper design 402
  kinematic organization 184
  linear field assumption 452
  notation 164
  organization 237
  of a weld seam as a function of weld layout 176
dissipative zones 152
distance, from the edge 350
distance, of bolt from free edge, computing 352
distribution, external 145
distribution, of internal forces 145, 151
distribution matrix 122, 123
distribution of force packets 144
distribution of forces 137
disturbed region, at near and far extremity 479
domain limit *see* limit domain
drilling algorithm 86
dual geometry 84
dual model 118, 434
ductility 6, 71
ductility, of weld layouts 328

e

earthquake 156, 348, 380
  Emilia 15, 278
  Northridge 71, 451, 452
EBF (Eccentric Bracing Frames) 61
eccentricities 105
  in BFEM 78
  neglect 8, 415
  as a 3D vector 8
edges, free and stiffened 351
effective area, of penetration welds, AISC 330
effective fasteners 381

effective region 380
eigenvalue 389
eigenvalue analysis 132, 133, 138, 377, 393
eigenvalue problem 133, 389
eigenvector 389
elastic critical buckling load 132
elastic critical load multiplier 132
elastic limits 413
  factored 440
elastic peaks 326
elastic perfectly plastic, model 391
end point, true 32
end release 33, 34, 466
engineer of record 343
envelope 8, 9, 118, 147, 388
  over-safety 10
  safety 9
equilibrium 105
  compound members 120
  conditions 125
  connector 124
  force transferrer or stiffener 124
  member stump 124
  nodal zone 118
  rigid bodies 110
  violation 78, 105, 118, 443
  violation, possible consequences 107
equivalent stiffness, bolts in shear 157
error measures 398
errors, in computer models 386
Euler column 134
Eurocode 3 28, 71
EUSPRIG (European Spreadsheet Risk Interest Group) 2
exact distribution of forces 129
exact load multiplier 129
exact plastic distribution 138
exact solution, meaningless 130
expert judgment *see* professional judgment
extremity(ies)
  connected 115
  of connectors 124
  of element 32
  far 115
  far, clamped 480
  far, loaded in PFEM 478
  near 115, 403
extrusions 74, 77

f

face 74
  active 89
  active (of welds) 81
  chosen 86
  in contact 86
  contact condition 85
  coplanar 85
  coplanar and external 85
  coplanarity 79
  drilled 351
  edges 74
  entry 86
  fusion (of welds) 81
  next 86
  outward normal 75
  planar 74
  stack 87
failure mode 5 *see also* check
  buckling, causes 377
  of constituents 347
  fracture 349
  global 349, 402
  local 349, 402
  local, definition 347
  relevant 350
  yielding 373
failure paths 364
fasteners 66
fatigue 127, 129, 481
  at high number of cycles 152
  low number of cycles 152
faying plane 81, 162, 164, 207, 407
faying surface 334, 444, 455
fictitious beam elements 457, 464
fictitious stiff elements 478
fictitious truss elements 465
fillet weld
  bending secant stiffness 202
  initial stiffness 202
  numerical test, axial behavior 209
  numerical test, longitudinal shear behavior 211

fillet weld (*cont'd*)
  numerical tests 207
  secant stiffness 201
  segment, linearized secant stiffness 202
  torsion 201
  transverse and normal secant stiffness 202
fillet weld layouts 81, 160
  AISC curves 197
  contact and friction 199, 233
  directional method 199
  effect of different angles between active faces 207
  force-displacement curve of a single weld 197
  intermittent 166
  isotropic, initial stiffness 232
  isotropic, secant stiffness 231
  isotropic approach 231
  isotropic model, initial stiffness 214
  isotropic model, secant stiffness 214
  limitations 166
  locked and sliding state 234
  locked state 233
  modification factors 225
  normalized secant stiffness 199
  orthotropic, initial stiffness 233
  orthotropic, modified secant stiffness 233
  orthotropic, secant stiffness 232
  orthotropic model 212
  orthotropic model, initial stiffness 213
  orthotropic model, secant stiffness 213
  and penetration weld layouts, difference of behavior 196
  representation 81
  secant stiffness 199, 200
  stiff bearing 199
  stresses computing 231
  throat thickness 196
  transversally loaded 197
  unlocked state 233
final load multiplier 137
finite element, definition 27
finite element analysis 386
finite elements 17
fin plates, graph 68
fittings 65, 78
  primitives 78
flanges, tapered 87
flexibility, uniaxial 18
flexibility index 159, 188, 401, 405, 407, 417
force packet 94, 124, 129, 143
  definition 109
  primary unknowns 402
  redundant 111
  and submembers 119
  transfer 109
forces
  active, distribution 131
  distribution between connectors 159
  flowing in the connectors 157
  generalized 108
  generalized in submembers 121
  losing 114
forces and moments, in a single weld seam 184
force transferrer 76, 78
  definition 65
force vector, generalized 109
fortran 66
foundation block 345
FR, fully resistant bolt, example 314
FR, fully resistant weld seam 185, 189, 191, 193, 195
fracture 127, 129, 147
  failure modes 481
  in isomorphically loaded constituents 150
  propagation 392
friction coefficient 147, 396
  according to Eurocode 3 343
friction constant 10
friction forces 342
functionality, loss of 381
fusion area 334

g

gap
  closed 395
  current 395, 396
  initial 396
  open 395

Gauss point 388, 390, 392, 399
general approach, to buckling check 133
general method 133, 134, 137, 139, 142, 378, 394, 449, 482, 484
general method, in Eurocode 3 133
geometrical effects 136, 137, 389
geometrical imperfections 378
geometrical non linearity 392–394
global failure modes 402
graphs, of connections 66
Green's formula 303
groove welds 160
gusset plate 12, 14, 151, 360, 469

h

hardening model 392
hard node 84
hard point 431–433, 435, 440
  definition 433
Heyman, Jacques 20, 127, 138
hole 434
  coefficient 342
  factor 344
  modeling 465
  normal oversized and slotted, AISC 356
  normal oversized and slotted, Eurocode 3 355
  oversize 342
  slotted 342, 344
hybrid analysis, steps 432
hybrid approach 385, 446
  definition 385
  a path for 449
hyper-connection *see* renode, hyperconnected
hypo-connection *see* renode, hypoconnected
hysteretic behavior 391

i

ICRM 416
  for bolt layouts 306
  for bolts, utilization factor or index 307
  instantaneous center of rotation method 214
  reduction of available shear in bolts 306
  for welds 175, 201–204, 206, 335
  for welds, convergence 223
  for welds, effect of angle between active faces 225
  for welds, effect of angle between active faces, comparison table 227
  for welds, ICRM 215–216
  for welds, incremental-iterative procedure 217, 221
  for welds, number of weld segments 220
  for welds, secant stiffness matrix 220
  for welds, with 3d loading conditions 216
  for welds, utilization factor 221
  for welds, with in plane eccentricity 216
imperfect geometries 394
imperfection 131–133, 137, 377, 378
  factor 135, 378, 379
  factor, table 135
incremental iterative procedure 398, 399
indeterminacy
  connector 124, 160
  external 124, 126, 129, 139, 151
  internal 124, 126, 129
  renode 124
  structural 124
initial conditions 394
initial gap 395
initial renode finite element model *see* IRFEM
initial stresses 133
in place editing 91
insertion 91
insertion point 75, 78
instantaneous center of rotation method *see* ICRM
integration points 390
interaction between plasticity and buckling 144
interaction of buckling and plasticity 131
interdependent jnodes 417
interface, external 113
interface, internal 113, 114
interface, member-to-node 113
internal forces, distribution 19
internal forces, distribution, lack of general tool 20
invariance condition 353, 355

IRFEM, Initial Renode Finite Element Model 385, 400, 430, 437, 440, 449
  assumptions 401
  vs. BFEM 402
  for central jnodes 406
  for constrained jnodes 406
  construction 402
  and contact forces 417
  for cuspidal and tangent jnodes 407
  definition 400
  displaced configuration construction 411
  examples 408
  force packets in the connectors 412
  for hierarchical jnodes 403
  and hyperconnected renodes 401
  and interdependent jnodes 417
  for isoconnected renodes 401
  and jnode instances 413
  load combinations 413
  member forces notional combinations 414
  member forces point of application 415
  nodal displacements 411
  nodes and constraints 403
  pros and cons 416
  rational path 402
  and renode checks 415
  results 411
  secant stiffness matrix 404
  setting member forces 413
  special problems 405
  stiffeners in 416
  zero-length elements 405
iso-connection *see* renode, isoconnected
isomorphically loaded constituents 373

j

jclass, definition 64
jnode
  average connectivity 58
  average frequency 58
  central 43
  classification 42
  classification, summary 46
  complexity 57
  connectivity 56, 57
  constraint 45
  constraint classification 46
  cuspidal 43
  cuspidal, example 47
  data structure 49
  definition 40
  equal 41, 63
  equality, criteria 51
  formal definition 41
  frequency 57
  hierarchical 42
  hierarchical, example 49
  instance 40, 439
  interacting 61
  mark 76
  multiplicity 56
  vs. node 40number of instances 56
  overall complexity 58
  overall connectivity 58
  overall multiplicity 57
  vs. renodes, equal 41
  search 55
  simple 42
  splitting 60
  tangent 44
  tangent, example 49
  topological classification 46
joints, typical 19
Jönsson Jeppe 367

k

Kemeny's report 14
kinematically admissible deformed shapes 130
kinematically admissible force distribution 129

l

layout 70
  connection rules 69
leg length 163
limit domain 5, 10
  of bolt shaft 338
  of a circular section 339
  convexity 6
  linear for T-stub 422
limit load 130

limit load multiplier 5, 6, 133, 134, 137, 391, 432
  upper bound 6
limit load, plastic 11
limit load search 378
linear analysis 388
linearization, of internal effects 4
linear weld group 336
line search 128
link, seismic 61
LO, longitudinal only weld seam 186, 193–195, 213
load case 390
  notional 9
load concentrations 71
loading conditions, most dangerous, evaluation of 481
load multiplier 131, 323, 373, 390
load path 95, 131, 390 *see also* chain
  interrupted 3
Lobatto integration rule 391
Lobatto points 399
local behavior 159
local effects 115
local failure modes 402, 432, 450, 481
local peaks 148
local plasticization 145, 392
local stress peaks 148
loss of convergence 128
lower bound theorem *see* Static Theorem of Limit Analysis

m
matrix
  definition 489
  orthogonal 490
  transpose 489
MCOFEM
  choice of the constraints 448
  constraints 480
  definition 445
  vs. IRFEM 446
  loads 479
  multiple constituent finite element model 445
  rigid 459
  saturated connectors 445
  stabilizing springs 448
mechanism, plastic *see* plastic mechanism
member 3
  automatic detection 31
  classification 36
  curved 34, 36
  cuspidal 37
  data structure 36
  definition 29
  extremities 77
  vs. finite element 28
  generic stress state 3
  interrupted 37
  master 7, 38, 42, 51
  master, automatic detection 44
  master, combined effects 7
  ordering 51
  passing 29, 36
  released 37
  search 56
  shift 78
  slave 7, 38, 42
  splitting 33
  tapered 34, 36
  typical data 36
  unconnected (to a node) 36
member-alignment classification 38
member alignment, using true end points 32
member at a node, classification of 37
member checks 78
member couple 50
  superimposable 60
member detection, rules 31
member equilibrium 117
member forces 78
  as input 116
  limiting the number of combinations 414
  in local reference system 113
  member forces, setting in IRFEM 413
  member forces, typical 29
  neglect 3
  set by design 116
  and third law 113
member stump 439
  initial length 77

member stump (*cont'd*)
length 115
too short 116
Merchant-Rankine formula 136, 139
mesh 84
and bearing surface 434
and bolt holes 436
feature 434
holes 434
size 436, 438, 455
of solid elements 84
tolerance 436
meshing, basic steps 434
meshing needs 28
meshing tools 431
mid surfaces 395
moment, additional 8
moment, in bolt shaft 107
moment, plastic 12
moment resistance, design 19
MRC (Moment Resisting Connection) 11
MRF (Moment Resisting Frames) 3, 451
multiple constituent finite element model *see* MCOFEM
multiplicity of connection 124
total 403
total of connectors 125
multiplier, critical, load, limit *see* critical multiplier, load multiplier, etc,

n

NDSHA, Neo Deterministic Seismic Hazard Assessment 423
net cross section 12, 369, 429
Newton, Isaac 111
Newton-Raphson method 398
nodal forces, and pressure field 441
nodal region 115, 448, 482,
nodal zone extent 115
node 118
node, of element 32
node-match 84
nodes 1, 27
nodes, hard *see* hard node
non linear analysis 28, 392, 461, 468, 484
slip 468
normal-stress path 21
NS, no-shear bolt, example 307, 308
NS, no-shear weld seam 185, 186, 213
Nuclear Regulatory Commission 14
numerical indices 56
numerical integration 390

o

obstacles 87
offended constituents 92
offending score 92
offer value, and jnode complexity 57
offset rigid *see* rigid offset
orientation, notional (dummy) 405
orientation, of members 36
orientation, superimposable 51
orientation matrix 50, 75, 404, 412, 492
orthotropic approach, for weld segments 460
overlapping 85, 87
overlapping check 91

p

parabola rectangle 263
parameterization 103
parametric real node *see* prenode
P-δ effect 152
P-Δ effect 71, 114, 136, 278, 392, 394
penalty constant 395, 396
penetration weld layouts 160
angle of principal axes 170
asymmetrical 194
center 186
clashes 82
cross-section 170
cross-section properties 182
distribution of forces 185
effective length of welds 172
effects of working mode modifiers, example 188
elastic computation 168
elastic distribution 168
extremities 173
generalized forces 167
intermittent 165
modified sectional data 187
organized displacement 172
and plastic design 167

plastic distribution 168
principal axes 170, 186
representation of 82
second area moment 172
single sided partial 165
stresses evaluation, simple methods 168
torsional second area moment 172
perfect plasticity 127
PFEM (Pure Finite Element Model) 385, 401, 416, 451
bolt bearing, force recovery 467
bolt modeling 463
constraints 480
contact non linearity for fillet welds 462
definition 451
fillet welds, example 461
loads 478
modeling of bolt in bearing 464
penetration weld, example 458
preloaded bolt stiffness 468
and renode checks 415
speed 481
speed, example 472
stiffness evaluation 482
strategy 484
physical model 29
physical model, from analytical 31
pin, nominal 71
plane stress 390
plastic design 140
plastic domain 72
plastic hinge 128
plasticity and buckling, interaction 135
plastic limit 413, 431
plastic limits, factored 440
plastic load multiplier 132
plastic mechanism 12, 127, 128, 137, 138, 147, 373, 392
exact 145
true 142
plastic multipliers 141
plastic neutral axis 303, 304, 438
plastic neutral axis, of bearing surface 260
plastic range 326
plastic redistribution 387
plastic region 387
plastic response 131
plastic rotation 128
plastic stress locus 129
plasticization 157
amount 374
local 392
plate bending, importance 238
plate-shell elements 387
plug weld layouts 160
point
entry 87
exit 87
point of application
and third law 114
of forces 105, 431
position, global 75
position in space, need of 73
post buckling configuration 136
post critical behavior 393
precision 156
appearance of 321
significant digits 341
preloaded bolt
finite element models 270
hollow cone model 270
hollow cylinder model 269
models 269
stiffness, is high 476
prenode 64, 102
definition 41
primary unknowns 124, 129
principal, plastic strain 392
principal axes 60, 119, 490
strong and weak 36
probability, of contemporary loads 3
professional judgment 15, 387, 396, 450
proportions sound, importance of 380
prying forces 16, 20, 147, 396, 416, 417
cause 419
effect 252, 419
evaluation 21
factor, table 426
forces factor 424, 449
and kinematic theorem of limit analysis 424
and limit load multiplier 424
methods to address them 424
Thornton's model 425

prying forces (*cont'd*)
  and T-stub 420
  T-stub failure modes 421
pseudocode
  automatic member detection 34
  finding chains 97
  finding equal jnodes 52
  interacting jnodes 62
PSHA, Probabilistic Seismic Hazard Assessment 423
pure fem 157, 182, 202, 206
pure finite element model *see* PFEM
purlin 45

r
radial elements 466
radial return 400
real multiplier 378, 394
real node *see* renode
Reason, James 2
Redistribution 128, 156
  between connectors 160
  of elastic forces 157
reduction factor
  AISC 379
  χ, 133–135, 138
  Eurocode 3 378, 379
reference system, change of 493
reference system, for beam and truss 490
regular grids 80
regulations 15
relative stiffness 401
  of connectors 470
release, end *see* end release
release, internal 33
renode 73
  definition 41
  hyperconnected 125, 143
  hypoconnected 125
  isoconnected 125
  vs. jnode 73
  logic of 100
  parameterization 64
representation, solid and plated 84
residual stresses 137, 147, 378, 394
resilience 147
response spectrum analysis 71
restraint, lack of torsional 466
restraint, of a concrete slab 3
rigid body 405
  assumption 416
  displacement 452
  equilibrium 108
  hypothesis 158
  motion 431, 437, 438, 448
rigid elements 405
rigid ends 77, 405, 443
rigid offset 32, 51, 77, 118, 403, 455
rigid rotation 50
rivets 237
robust design 3
rotation, notation 164
rotation capacity 19
Ruppert's algorithm 436
rupture stress 372

s
S16-14, 142
safe theorem of limit analysis *see* static theorem of limit analysis
safety factor 2, 323
  AISC 327
  Eurocode 3 327
  for preloaded bolt, Eurocode 3 342
safety margin 455
saturation of connectors 445
  and contact non linearity 447
  definition 446
  external and internal 446
  saturated connectors, modeled as forces 447
  unsaturated connectors 446
scene 22, 75, 78, 108, 115, 351
  compliance criteria 93
  readjust 91
  sub part 87
SCOFEM (Single Constituent Finite Element Model) 430
  clamp of member far extremity 439
  clamps for highly compressed welds 445
  constraints 437
  definition 430
  evaluation of the nodal forces exchanged by the bearing surface 441

far extremity force packets 443
incompatibility 431, 444
member stump length 444
nodal forces due to welds 443
notional constraints 438
number of load cases 439
and plate-shell elements 431
self balanced 437
and stiffeners 432
secant stiffness, of fillet weld segments 460-461
second order, terms 393
segment(s) definition 433
seismic loading 71
seismic regions 1, 72
self balance 402, 426, 431, 448, 480
sequence of drilled objects 87, 89
serviceability loads 423
sharp corners 387
shear and torque, interaction with 72
shear area 174
shear connection 3, 114, 145
shear forces 156
shear keys 406
shear rupture 359
shear-stress path 21
shear yielding 359
shift 91
significant digits 155
signs, reversion 6
simplified approaches 348
simplified method, for welds, Eurocode 3 332
single constituent finite element model *see* SCOFEM
singular points 392
sketch model 434, 436
slender members 380
slenderness 134
limit 134
non dimensional 134, 135, 378
sliding limit 478
slip 127, 129
consequences 342
no at service loads 150
and torque 342
no at ultimate loads 150
slip checks *see* checks, slip
slip resistant bolt layouts
axial force and bending 275
bearing surface model 276
bolt axial stiffness 292
bolt shear stiffness 292
on concrete, stiffness 278
on concrete, stiffness evaluation 280
importance of plate flexibility 275
local stiffness 275
only shaft resisting model 275
plate stiffness 269
preloaded bolt model 268
preloading effect 266
shear and torsional stiffness 274
surrounding stiffness 275
are usually isoconnected 274
slotted weld layouts 160
SO, shear only bolt, example 307, 308, 314,
SO, shear only weld seam 185, 186, 191, 213
softening branch 128
software, as a miracle potion 386
splice connection, graph 67
splice joint 33
bolted, example 469
bolted, resisting mechanism 469
spreadsheets 2, 11, 17, 108
springs no-tension 437
stability curves 133-135
stabilizing springs 448
stack of faces 89
standard connections, pre-qualified 159
statically admissible force distribution 129
statically admissible plastic mechanism 142
static condensation 288, 464
static theorem of limit analysis 20, 31, 128, 129, 148, 159, 185, 348, 368, 377, 400, 431
and MCOFEM 449
shelter of 130
step by step analysis 132, 137, 377, 389
with geometrical effects 394
stiffeners 66, 78, 132, 251, 451
reduction of use of 132

stiffness
  connection 71
  evaluation 159, 382
  evaluation, example 482
  evaluation, using PFEM 482
  load dependent 483
  preloaded bolts 476
  ratio 159
  rotational 19, 71
  uniaxial 18
  uniaxial, elastic 17
stiffness matrix 144
  assembled, modified 183
  of bolt layout 283
  of bolt layout, assembled 284
  of connectors 156
  of connectors, in IRFEM 404
  elastic 133, 388
  of fillet weld segments 460
  geometrical 133, 388, 393
  influenced by working mode flags 186
  modified, for bearing bolt layouts 298
  modified, for bolt layouts 297
  modified, for slip resistant bolt layouts 298
  of penetration weld layout 174, 178
  of a penetration weld segment 457
  secant, for bolt layouts 307
  of a single bolt 237, 286, 464
  of single connector 160
  of single preloaded bolts 468
  of single weld seam 177
  of slip resistant bolt layout 292
  tangent, for bolt layouts 307
  tangent, for fillet welds 461
  tangent, of a bolt 503
  tangent, of a bolt layout 505
  tangent, of a weld layout 500
  tangent, of a weld segment 495
  tangent, of a weld segment using contact 499
  of a Timoshenko beam 174
  useful to avoid by hand computation 156
STM (Strut and Tie Method) 12, 130, 429
strain distribution 129
stress concentration 439, 441, 479
  factor 149
  in welds 454
stress distribution, in welds 459
stresses, at far extremity 439
stresses, in single welds 183
stresses, notation for welds 164
stress intensity factors 148
stress maps 481
stress peaks 127, 375, 387, 436
stress perturbation 439
stress raisers 387
stretching 84
structures, nonsymmetrical 3
subconnector 451
subconstituents 69
  acting alone 70
  organization 70
submembers 119
subplates 84
superposition principle 390

t

target surface 395
tetrahedrization 84
theory of elasticity 131
theory of plasticity 130, 145
theory of plates and shells 84
thickness, constant and variable 87
thick plate elements 387
third law *see* action reaction principle
Thornton William, A. 21, 425
3D object, stress analysis 83
Three Mile Island 14
throat plane 164, 333, 334
throat size 83, 162, 166, 332
Timoshenko beam 173, 175, 196, 199, 286
tolerance 51, 63, 436
toponode 49
  equal 51
torque, in bolt shaft 339
traditional approaches 2, 413, 451, 454
  simplifications 3
transfer matrix, in partitioned form 177
true systems 147
truss element 27
  always interrupted 49

T-stub 15, 252
approach 394
definition 16
equivalent in compression 255
in Eurocode 3 16
limit load 11
pulled, failure modes 421
typical use and misuse 17
ubiquity 11
utilization factor and utilization index 423
yield patterns, elementary 12

u

ultimate strains 147
ultimate stress 128
uncertainty 155
undisturbed central region 479
utilization, if more loads are applied 321
utilization factor 324, 423
and seismic loads 423
as wishful thinking 423
utilization index 307, 320
and safety measure 322
significant digits 320
vs. utilization factor 326, 350
utilization ratio 5, 6, 452

v

vector
definition 489
dot product 489
vector product 489
Von Mises envelope maps 482
Von Mises stress 349, 376, 443, 464
map 387, 484

w

washers 79
weight, minimum 3
weld
angle between active faces 82, 89
differences between fillet and penetration 161
limits, AISC 166
limits, Eurocode 3 166
penetration, checks 328
weld group *see* weld layout
weld layout
active faces 89
automatic detection of connected objects 90
classification 160
concentrically loaded 336
connection rules 81
extremity 89
faying plane 89
faying surface 155
fillet (*see* fillet weld layout)
full penetration, as a workaround 329
highly compressed 445
insertion 80
insertion point 90, 162
minimum geometrical information 80
mixed penetration and fillet 235
planar representation 83
principal axes 162, 163
reference system 162
rule for connection 160
using contact 445
weld segments 235
weld layout, penetration *see* penetration weld layouts
weld legs 81, 82
weld seam
effective area 162
high displacements 186
low stiffness 186
stresses 164
working mode and hyperconnectivity 186
working mode modifiers 186
weld segment 337, 455
broken 461
number of 455
tributary areas 456
weld thickness 82
Whitmore section 12, 429
width-to-thickness ratio 13
working mode, no-bending 476

working mode of connector 449
working mode factors 405, 457
  for bolts 464
  misapplication 381
work processes 78, 82, 83, 92, 434

y

yield lines 6
  patterns 11
  typical 12
Young's modulus 19

z

Zavarise G. 396